Series in Display Science and Technology

The *Series in Display Science and Technology* provides a forum for research monographs and professional titles in the displays area, covering subjects including the following:

- optics, vision, color science and human factors relevant to display performance
- electronic imaging, image storage and manipulation
- display driving and power systems
- display materials and processing (substrates, TFTs, transparent conductors)
- flexible, bendable, foldable and rollable displays
- LCDs (fundamentals, materials, devices, fabrication)
- emissive displays including OLEDs
- low power and reflective displays (e-paper)
- 3D display technologies
- mobile displays, projection displays and headworn technologies
- display metrology, standards, characterisation
- display interaction, touchscreens and haptics
- energy usage, recycling and green issues

More information about this series at http://www.springer.com/series/15379

Kyoji Matsushima

Introduction to Computer Holography

Creating Computer-Generated Holograms as the Ultimate 3D Image

Springer

Kyoji Matsushima
Faculty of System Engineering
Kansai University
Osaka, Japan

ISSN 2509-5900 ISSN 2509-5919 (electronic)
Series in Display Science and Technology
ISBN 978-3-030-38434-0 ISBN 978-3-030-38435-7 (eBook)
https://doi.org/10.1007/978-3-030-38435-7

This Springer imprint is published by the registered company Springer Nature Switzerland AG
The registered company address is: Gewerbestrasse 11, 6330 Cham, Switzerland

Preface

It is an undoubted fact that the evolution of digital computers impacts our lifestyles as well as science and technology. It is also reasonable to assume that people will be surprised to see three-dimensional (3D) images produced by a well-made optical hologram, and that they believe that the 3D image is perfect and accurate in its representation. Unfortunately, hologram galleries that exhibit holograms for art and decoration have been disappearing recently, and holograms we see in our daily lives are limited to those on bills and credit cards. Such holograms are interesting because the appearance of the image changes when the observer changes his or her viewing angle, but they do not look like 3D images. This decline of holography for 3D imaging is attributed to the fact that holography has not yet evolved into a digital technology. It is not possible to store the digital data of an optical hologram in a digital medium and it is not possible to transmit it through digital networks.

The idea of creating and handling 3D holographic images using computers has a long history. In fact, the origin of the idea goes back to the days right after the actualization of 3D imaging by optical holography. However, computer-generated holograms (CGH) comparable to optical holograms were not developed until recently. This is entirely due to the tremendous data sizes and computational efforts required to produce CGHs capable of reconstructing perfect 3D images. The creation of outstanding large-scale CGHs is, in a sense, a fight against the availability of computer resources at a time. The required computational capabilities often exceed those of state-of-the-art computers. Algorithms and techniques available in the literature are often ineffective because they are too time and/or resource consuming.

This book does not intend to provide all the various techniques proposed in computer holography extensively. The techniques discussed herein are instead ones that have been confirmed and proven to be useful and practical for producing actual large-scale CGHs whose 3D images are comparable to those of optical holography. Some of these techniques are of an atypical nature as well. It is both unfortunate, and a little delightful for researchers like me that such techniques are still far from completion. Thus, I intended this book to provide just a snapshot of these developing technologies.

When we produced the first large-scale CGH, “The Venus” in 2009, approximately 48 h of computation time was needed. An expensive computer mounted on a server rack was used. The computer was too noisy to be kept in a room where one studies, owing to the cooling mechanism. Now, the same CGH can be calculated in approximately 20 min using a quiet desktop computer. This is mainly due to the continuous development of computer hardware and partially due to the development of our technique with time.

This book is for any researcher, graduate or undergraduate student, who wants to create holographic 3D images using computers. Some of the chapters (e.g., Chaps. 3 and 4) deal with basic topics and researchers already familiar with them can skip those. Some techniques described in this book (e.g., Chaps. 6, 9, and 13) are useful not only for computer holography but also for all fields of research that require techniques for handling wavefields of light. In addition, because of the shortness of wavelength, spatial data of light tends to be of gigantic sizes, exceeding memory sizes of computers. Some techniques introduced in this book (e.g., Chap. 12) give hints on how to handle such large-scale wavefields on computers.

I am sincerely grateful to Prof. Nakahara, who was a colleague at Kansai University, co-author of many papers, and someone with the same enthusiasm for creating CGHs as me. Many CGHs introduced in this book were fabricated by him. We visited many places together to present our work on CGHs. I would also like to thank Dr. Claus E. Ascheron for his strong recommendation to write a book on computer holography. He was an executive editor of Springer-Verlag when I met him for the first time and is now retired. This book would never have been written if he had not patiently persuaded me for over 3 years. I regret that I could not complete this book before his retirement. I am also grateful to Dr. Petr Lobaz who reviewed my manuscript carefully and gave me many suggestions based on his profound knowledge and enthusiasm for holography. I am also thankful to my former and current students: Dr. Nishi, Mr. Nakamoto and Mr. Tsuji, who did reviews of the text and formulae. I am grateful to them, and any mistakes that may have crept in are entirely mine.

Finally, I would like to thank my colleagues at the Department of Electrical and Electronic Engineering at Kansai University. I am particularly thankful to all my Ph.D., masters, and undergraduate students, an assistant, and everyone who belonged to my laboratory over the past 20 years at Kansai University. Their ideas, efforts, and passion has led to our success in computer holography.

Osaka, Japan
October 2019

Kyoji Matsushima

Contents

Chapter 1
Introduction

Abstract Computer holography is a technique to create computer-generated holograms (CGH), for reconstructing 3D images using computers. In this chapter, we discuss the differences between CGHs for 3D displays and optical devices, advantages of the display-CGHs, and difficulties in creating holographic displays as an ultimate 3D display. Early and latest large-scale static CGHs that reconstruct brilliant deep 3D images are demonstrated.

1.1 Computer Holography

Holography is most likely the only technology to reconstruct deep 3D scenes from a plate medium. Holography, invented by Dennis Gabor in 1948 [16], makes it possible to reconstruct light from a subject, recorded by interference with a reference wave. A person recognizes three dimensional (3D) spaces and perceives distances from objects by visual sensation through many clues. The most important depth perceptions for 3D imaging are summarized as follows:

Binocular disparity: Difference between images of an object, seen by the left and right eyes,
Vergence: Movement of eyes in opposite directions to maintain binocular vision,
Accommodation: Process to focus on an object to obtain clear images,
Motion parallax: Change in the image of an object when the viewpoint changes.

The first two perceptions are based on binocular cues, while the last two are owing to monocular cues.

Technologies of 3D imaging which have already been put to practical use, stimulate a sensation of depth by showing different images to the left and right eyes. Therefore, depth sensation is based on the binocular cues, i.e., the first two cues mentioned above. In contrast, the last two cues are not given by current 3D images. More accurately, the depth perception given by monocular cues do not agree with those given by binocular cues in current 3D technologies. In particular, it is said that inconsistency of depth perceptions by vergence and accommodation causes a noticeable problem, called *vergence-accommodation conflict*. The conflict

K. Matsushima, *Introduction to Computer Holography*,
Series in Display Science and Technology,
https://doi.org/10.1007/978-3-030-38435-7_1

sometimes causes severe health problems such as headaches, eyestrain, dizziness, and nausea. Therefore, current 3D displays based on binocular cues cannot produce deep 3D scenes because deeper 3D scenes produce more conflict.

In contrast to 3D displays based on binocular vision, the 3D image reconstructed by a hologram does not cause any health concerns because the hologram physically reconstructs the recorded light itself.[1] As a result, all depth cues are properly reconstructed in holography without any conflict. Accordingly, holograms have the ability to display deep 3D scenes unlike current 3D technologies.

Unfortunately, traditional optical holography is not a modern digital technology but is a type of analog photography. Traditional holography always needs a physical subject to create the 3D image. For example, if we want to create the 3D image of a car by holography, we need a physical entity i.e., the car. Thus, prior to recording the hologram, we have to prepare an actual 3D model of the car, which is commonly a scaled down model of an actual car. In addition, because information of the light emitted by a subject is recorded on a photo-sensitive material in optical holography, the information is not digitized and hence, cannot be transmitted through modern digital networks and stored in digital media.[2] Besides, the recorded image of the subject cannot be edited after recording in general. This property of analog holograms is useful for security purposes but inconvenient for 3D imaging.

As explained in the following chapters, information of the light emitted by an object is recorded in the form of fringe patterns in holography. In a type of hologram called a *thin hologram* (see Sect. 7.2), the fringe pattern is a simple two dimensional (2D) image, which looks like a random pattern. Therefore, we can generate the pattern by some technique using a computer. The created holograms are called *computer-generated holograms* or *CGH* in the abbreviated form. We call the technique to create CGHs *computer holography* in this book.

The idea of CGHs was first proposed by A. W. Lohmann in 1967 [55].[3] Articles on CGHs has been increasing every year after Lohmann's first proposal [27, 121]. However, the CGHs found in the literature of early days are not the ones that we deal with in this book. The word "CGH" is commonly used in two different contexts: an optical device and a 3D display. In the former context, the CGH is a device that makes it possible to filter the field of light or to produce a complex field with amplitude-only or phase-only spatial modulation. In other words, the CGH in the former context is a wavefront conversion device, as shown in Fig. 1.1. In this case, we know the forms of both the input and output fields of the CGH and investigate the technique for conversion with high efficiency and accuracy. This conversion corresponds to the *coding* of the fringe patterns in the latter CGHs for 3D displays (for coding of fringe patterns, see Sects. 2.5, 8.5 and thereafter).

In CGHs for 3D displays, the output field of the CGH is an *object field* that forms the 3D image. We must calculate the object field from the 3D model of the object before coding of the fringe pattern. The calculation of this object field is the most

---

[1] Here, we ignore horizontal-parallax-only (HPO) holograms (see Sect. 8.4).

[2] There is a technique to record the information digitally, as described in Chap. 14.

[3] Although [7] is published before [55], the detailed technique was not provided therein.

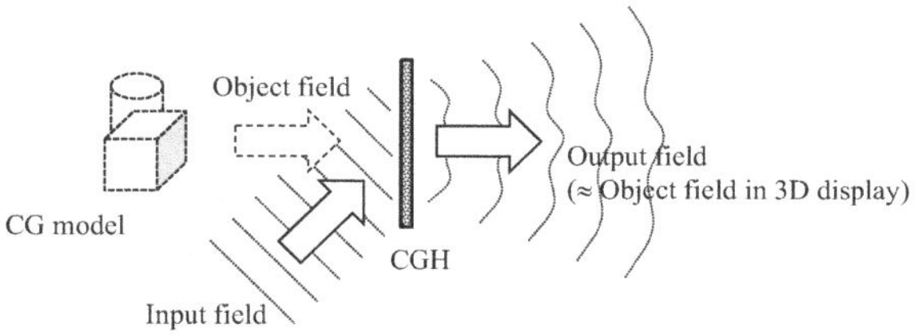

Fig. 1.1 Schematic illustration of the difference between CGHs for an optical device and a 3D display. The target output field is already known in optical devices, while in 3D displays the target field is the object field calculated from a 3D model

important and difficult process in creating CGHs for 3D displays. Therefore, we put little importance on the coding of the fringe pattern in comparison to the calculation of the object field. This is the biggest difference between CGHs for a device and a 3D display.

It should be noted again that the author uses the word "computer holography" in this book to describe the techniques for producing 3D images according to the principle of holography using computers. The 3D image usually comes from 3D models like computer graphics (CG).[4] Computer holography makes it possible to reconstruct holographic motion pictures as well as still pictures. The technique to reconstruct display-CGHs using electronic equipment is sometimes called *electro-holography*. The equipment is called *Holographic 3D display* or simply *Holographic display*. The final goal of the holographic display is undoubtedly to produce an ultimate 3D motion picture in which any conflict of visual sensation does not occur.

1.2 Difficulty in Creating Holographic Display

Computer holography has been suffering from two problems for a long time: calculation and display. The viewing angle of a CGH increases approximately in inverse proportion to the pixel pitch of the fringe pattern (see Sect. 8.2). On the other hand, the size of a CGH is given by the product of the pixel pitch and the number of pixels. Therefore, to combine the size and viewing angle of a CGH, the number of pixels required inevitably takes a gigantic value. This is called the *space-bandwidth product* (SBP) problem (for details, see Sect. 8.3).

For example, supposing that the number of pixels is $2{,}000 \times 2{,}000$ in a CGH and the pixel pitch is 0.1 mm, the CGH size (i.e., the image size) is 20 cm $\times$ 20 cm, but the viewing angle is only 0.36°.[5] If the pixel pitch is reduced to 1 μm, the viewing angle is increased to 36° but the CGH size is only 2×2 mm^2. If we want to create a CGH whose viewing angle is 45°, the pixel pitch must be less than 0.8 μm at a

[4] As mentioned in Chap. 14, the concept of computer holography is extended to include physical objects.

[5] We assume a wavelength of 633 nm (red) in this discussion.

wavelength of 633 nm (red color). Besides, if we want a CGH whose size is more than 5 cm × 5 cm, the total number of pixels required is more than four billion!

Because the object field commonly has exactly the same scale as the fringe pattern, the large scale of the fringe pattern makes it very difficult to calculate the object field and to generate the fringe pattern in a short time. Furthermore, it is difficult to display several billion pixels with a pixel pitch less than or equal to 1 μm. These are the difficulties in displaying CGHs. Although the idea and the study of CGHs has a long history, it had been difficult to create brilliant large-scale CGHs until recently owing to these difficulties.

Coming back to the history of computer holography, the early works by Benton et al. in the late 1980s most likely inspired subsequent research studies on holographic displays [44, 115, 116]. Benton et al. used an acousto-optic modulator (AOM) to produce one dimensional fringe patterns and achieved a horizontal viewing angle of 12° at 633 nm, i.e., an effective pixel pitch of approximately 3.0 μm using demagnification optics. Unfortunately, the system could produce only horizontal parallax. Although the challenge in calculation is eased to a great extent in horizontal-parallax-only (HPO) holograms (see Sect. 8.4), the problem of vergence-accommodation conflict is not resolved in HPO displays.

After the advent of early Benton systems, many display systems have been reported for realizing the ultimate 3D display, i.e., holographic movies/TVs/PC monitors. The reported systems commonly make use of existing display devices such as liquid crystal display (LCD) and digital mirror device (DMD). However, resolu-

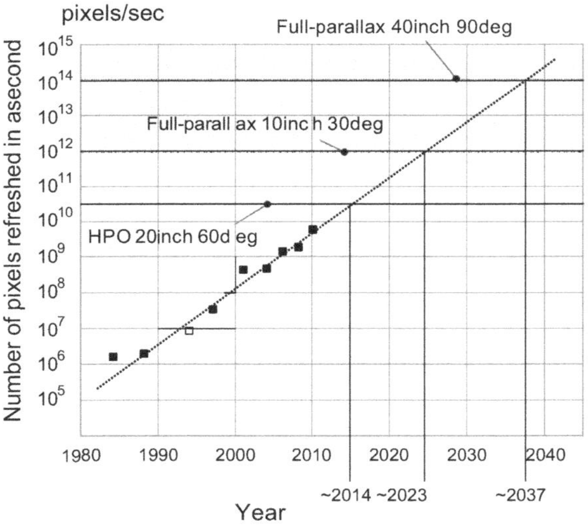

Fig. 1.2 The trend of the spatio-temporal resolution (pixels/second) of display devices from late 1980s to 2010. The square dots represent examples of commercialized or prototyped display devices [132] (From SPIE Proc. 8043, 40 (2011). By courtesy of Prof. Yamaguchi, Tokyo Institute of Technology.)

tion of current display devices are quite far from the SBP requirements in computer holography. Thus, techniques of time or space division multiplexing are often used for producing effective high-resolutions.

The achievable resolution with a time division multiplex system is determined by the number of pixels displayed in a second. Figure 1.2 shows the trend of the spatio-temporal resolutions (i.e., pixels/second) of display devices from the late 1980s [132]. According to extrapolation of the development curve, it is predicted that full-parallax holographic displays with a given size and viewing angle will be realized in the mid-2020s. However, this expectation is very doubtful because the existing top data is of display devices for 8K ultra high-definition television (UHTV) systems and it is unsure if the development of higher resolution devices will continue after the 8K devices.

1.3 Full-Parallax High-Definition CGH

The author and Prof. Nakahara created and reported a large-scale full-parallax CGH named "The Venus" in 2009 [67]. Figure 1.3 shows photographs of the optical reconstruction of The Venus.[6] Although the Venus is a static CGH, it is composed of $2^{16} \times 2^{16}$ pixels; the total number of pixels is approximately four billion. Here, the number of pixels is sometimes represented using a supplementary unit of K; 1K = 1,024 in this book. Thus, The Venus is composed of 64K $\times$ 64K pixels. Because the pixel pitch of the Venus's fringe pattern is 1 μm in both the horizontal and vertical directions, the viewing angle at a wavelength of 633 nm is approximately 37° in both directions. The size of the CGH is approximately 6.5×6.5 cm^2. We call large-scale CGHs like the Venus, which are composed of more than a billion pixels, a *high-definition CGH* (HD-CGH).

The 3D model of The Venus is represented by a polygon-mesh exactly like in computer graphics (CG). The model is composed of 718 polygons[7], arranged at 15 cm from the hologram plane in the CGH. The object field is calculated in full-parallax using the polygon-based method (see Chap. 10), and the occlusion is processed by the silhouette method (see Chap. 11). The fringe pattern was printed using laser lithography (see Sect. 15.3). The CGH can be reconstructed by reflection illumination using an ordinary red LED as well as a red laser. Because the 3D image reproduces proper occlusion, viewers can verify continuous and natural motion parallax, as shown in the online videos. As a result, the reconstructed 3D image is very impressive and gives strong sensation of depth to the viewers.

Because HD-CGHs are holograms, they reconstruct not only the binocular disparity and vergence but also the accommodation. Figure 1.4 shows photographs of

[6]Note that the left and right photographs are interchanged by mistake in Fig. 1.7 of [67].

[7]The mesh data of the Venus statue is provided by courtesy of INRIA by the AIM@SHAPE Shape Repository.

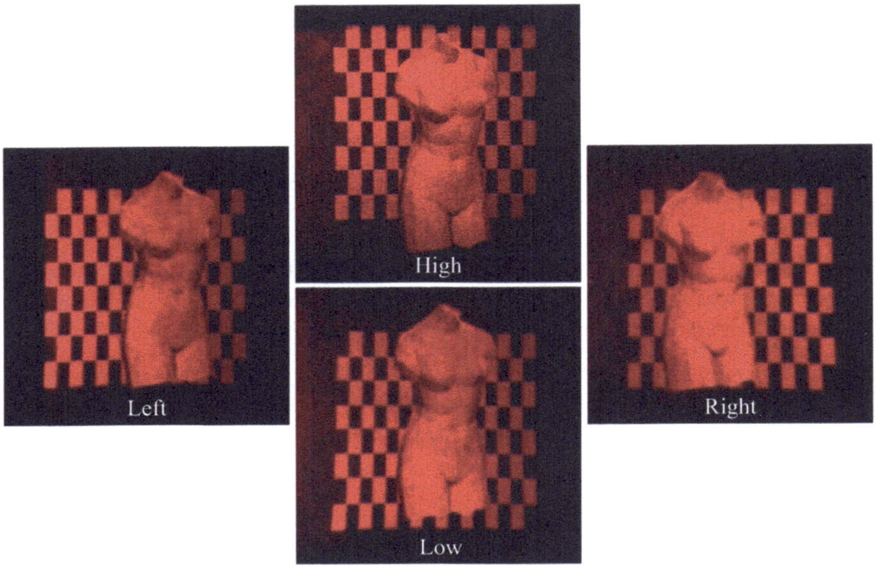

Fig. 1.3 Optical reconstruction of the first high-definition CGH named "The Venus" [67]. A He–Ne laser is used as the illumination light source. The photographs are taken from different viewpoints. The number of pixels and pixel pitches are 4,294,967,296 (=65,536 × 65,536) and 1.0 μm, respectively (see Sect. 10.5 for the detail). Video link, https://doi.org/10.1364/AO.48.000H54.m002 for illumination by a red LED, and https://doi.org/10.1364/AO.48.000H54.m004 for illumination by a He–Ne laser

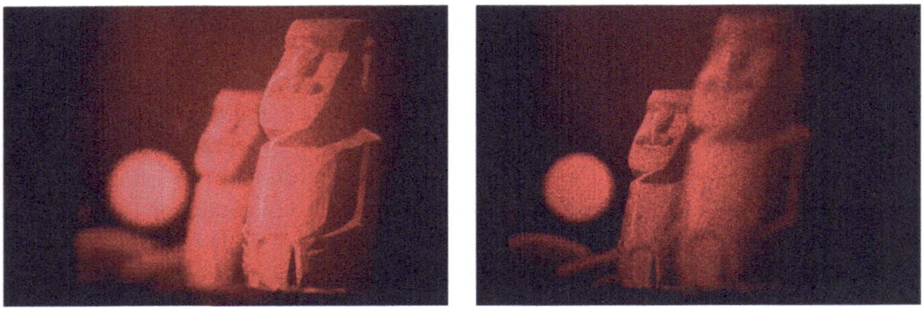

Fig. 1.4 Optical reconstruction of HD-CGH "Moai I" [70]. The camera is focused on **a** the front and **b** rear moai statues. The number of pixels and pixel pitches are 64K × 64K and 1 μm × 1 μm, respectively. Video link, https://youtu.be/DdOveIue3sc

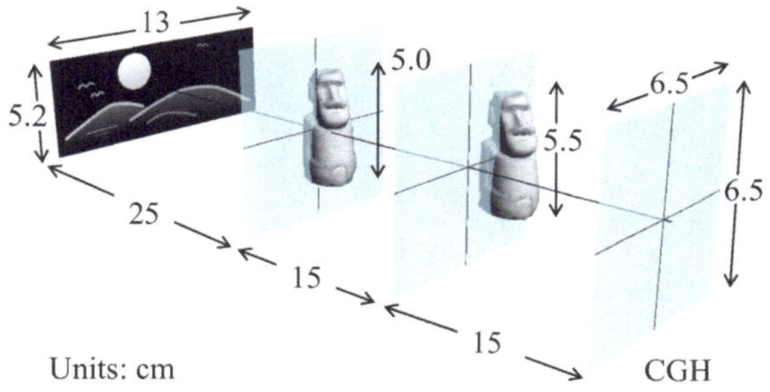

Fig. 1.5 The 3D scene of HD-CGH "Moai I" [70]

the optical reconstruction of another HD-CGH "Moai I." The 3D scene is depicted as a CG image in Fig. 1.5. The background image is a digital image designed with 512 × 512 pixels. Both the moai statues are a polygon-mesh CG model composed of 1220 polygons[8] and arranged at 15 cm from each other. Here, the camera is focused on the front moai in Fig. 1.4a, and on the rear moai in (b). We can clearly verify the change of appearance; the actual moai is out of focus when the front moai is in focus, and vice versa.

As shown in the example above, the 3D image reconstructed by HD-CGHs stimulates all depth perceptions without any inconsistency unlike conventional 3D technologies. As a result, HD-CGHs reconstruct the spatial images as if the viewer were looking at the 3D world spreading beyond the windows formed by the hologram. In practice, we sometimes noticed the viewers of HD-CGHs checking the backside of the HD-CGH in exhibitions to verify that the HD-CGH is a thin plate without depth.

Here, we emphasize that 2D photographs like Figs. 1.3 and 1.4 never convey true impressions of the 3D image reconstructed by the HD-CGHs. One has to look at the actual CGH with their own eyes to learn what is shown by the CGH. Motion pictures usually portray the nature of holographic images better than still pictures. Thus, online movie clips of actual CGHs are referred to as much as possible in this book.

Figure 1.6 shows an early version of "Brothers" that we created for exhibition at Massachusetts Institute of Technology (MIT) Museum in Boston, USA in 2012 [71]. This HD-CGH remained the biggest CGH that we had ever created for a long time. The total number of pixels is 21.5 billion in V1 and 25.8 billion in V2 that is exhibited in the museum. The size is 131 × 105 mm^2 in V1 but 126 × 105 mm^2 in V2.[9] The objects arranged in the 3D scene (see Fig. 11.4, Sect. 11.2.3) are live faces whose 3D shapes were measured by a laser scanner. Photographs taken at the same time as scanning are texture-mapped onto the measured polygon-meshes with the polygon-based method (see Sect. 10.6.4).

[8]The mesh data of the moai statue is provided by courtesy of Yutaka_Ohtake by the AIM@SHAPE Shape Repository.

[9]Brothers V2 is a little smaller than V1 because the horizontal pixel pitch is reduced to 0.64 μm in V2.

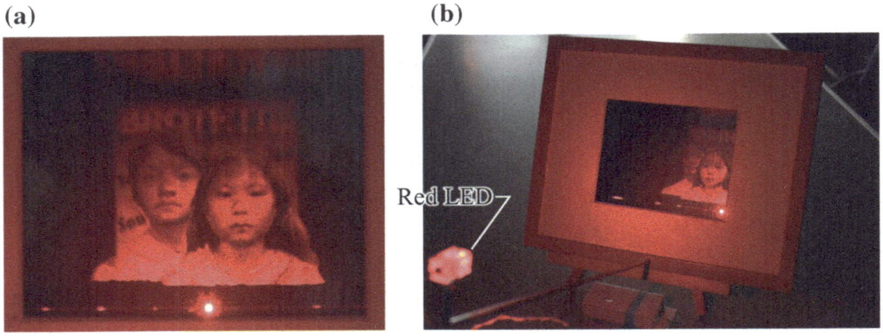

Fig. 1.6 Optical reconstruction of "Brothers" V1. The photographs are taken at **a** a close-up and **b** distant view. The second version of Brothers was on display at MIT Museum from Jul 2012 to Mar 2015. The total number of pixels of V1 is approximately 21.5 billion (=160K × 128K), and the pixel pitch is 0.8 μm in both directions. The CGH size is approximately 131 × 105 mm^2. See Sect. 11.2.3 for the detailed parameters. Video link, https://youtu.be/RCNyPNV7gHM

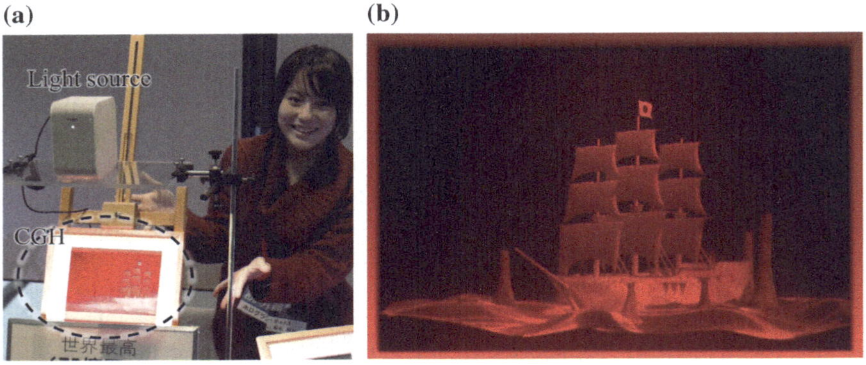

Fig. 1.7 Photographs of "Sailing Warship II" at **a** a distant view and **b** close-up view. The total number of pixels is 67.5 × 10^9 (=225,000 × 300,000). The pixel pitches are 0.8 μm and 0.4 μm in the horizontal and vertical direction, respectively. The CGH size is 18 cm × 12 cm. Video link, https://youtu.be/8USLC6HEPsQ

Fig. 1.8 The 3D scene of Sailing Warship II, depicted by CG. The model is composed of 22,202 polygons

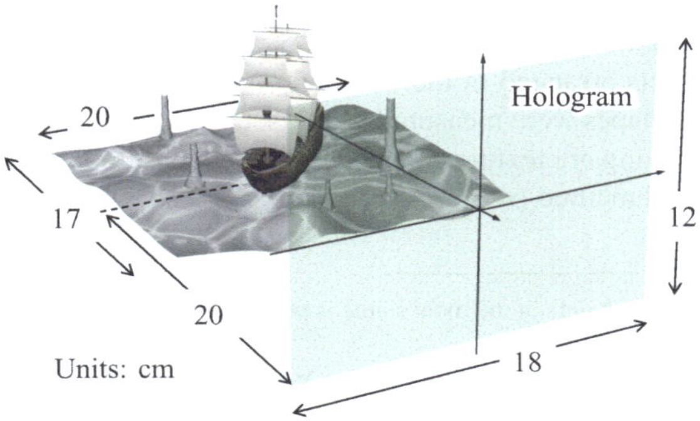

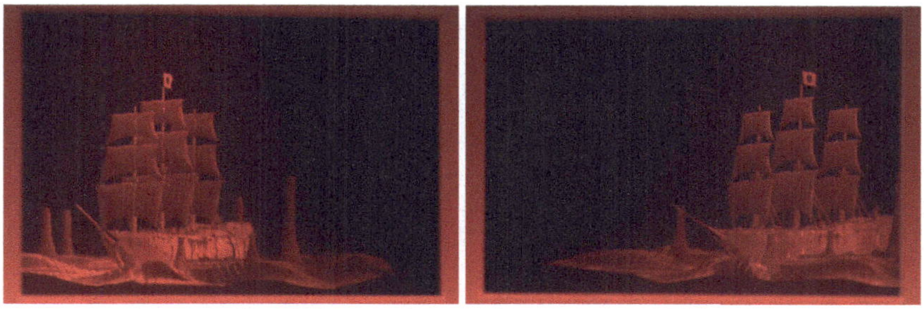

Left Right

Fig. 1.9 Photographs of optical reconstruction of Sailing Warship II, taken from left and right viewpoints. Video link, https://youtu.be/8USLC6HEPsQ

One of the latest HD-CGH is “Sailing Warship II” shown in Fig. 1.7. The total number of pixels is more than 67 billion, and the size is 18 cm $\times$ 12 cm. This HD-CGH was created in 2017. Figure 1.8 is the 3D scene of Sailing Warship II. The 3D model composed of 22,202 polygons has a very complicated shape. It is very difficult to calculate the object field of this kind of intricate model because the model has many self-occlusions (see Sect. 11.1). The silhouette method and the switch-back technique (see Chap. 11) have been developed and are used for the creation of Sailing Warship II. Figure 1.9 shows photographs of the optical reconstruction, taken from left and right viewpoints. It is verified in the photographs that all portions behind the objects are hidden by the front portions. The process of calculation of the object field is called *occlusion processing* or *occlusion culling* in computer holography. This is the counterpart of *hidden surface removal* in CG. Occlusion processing is the most important and difficult technique in computer holography.

An HD-CGH named “Toy Train” was created in 2018 using the same techniques as those of Sailing Warship II. The optical reconstruction and 3D scene are shown in Figs. 1.10 and 1.11, respectively. This CGH is a square whose side length is 18 cm, and the total number of pixels reaches more than 0.1 trillion. Because depth of the 3D scene is more than 35 cm, the CGH gives really strong sensation of depth to the viewers.

Here, note that the vertical pixel pitch is one half of that in the horizontal direction in Toy Train and Sailing Warship II. This is because the illumination light in these two has a larger incident angle than that of other HD-CGHs. We can avoid unnecessary conjugate images and non-diffraction light with a large illumination angle (see Sect. 8.8.3).

Fig. 1.10 Optical reconstruction of "Toy Train," created in 2018. Focus of the camera is changed back and forth. The total number of pixels is 101×10^9 (=225,000 × 450,000). The pixel pitches are 0.8 μm and 0.4 μm in the horizontal and vertical direction, respectively. The CGH size is 18 cm × 18 cm. Video link, https://youtu.be/XeRO7nFvlGc

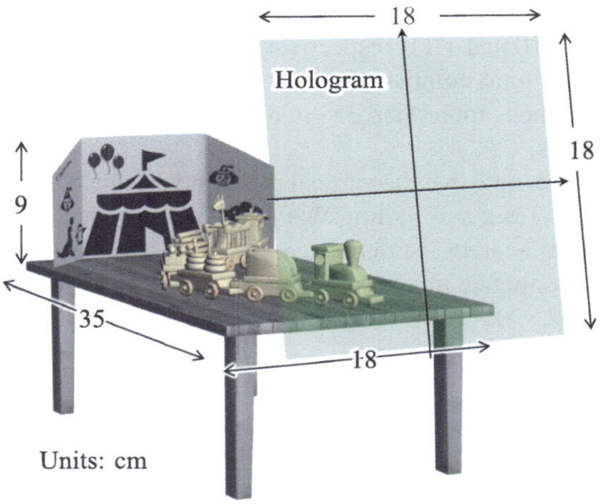

Fig. 1.11 The 3D scene of Toy Train, depicted by CG. The model is composed of 52,661 polygons

Fig. 1.12 Optical reconstruction of a stacked CGVH illuminated by a white LED [46]. The CGH size is approximately 5 × 5 cm. Video link, https://doi.org/10.6084/m9.figshare.8870177.v1

Only monochrome HD-CGHs were created and exhibited for a long time.[10] Recently, it has been possible to produce full-color HD-CGHs by using several methods (see Sects. 15.5–15.7). Figure 1.12 is a photograph of the optical reconstruction of a full-color stacked CGVH (Computer-Generated Volume Hologram). This HD-CGH is created using one of them.

[10] At exhibitions of monochrome HD-CGHs, the author was frequently asked a question, "Is it possible to create a full-color HD-CGH?".

Chapter 2
Overview of Computer Holography

Abstract Techniques in computer holography are summarized and briefly overviewed in this chapter. Several steps are commonly required for creating a large-scale CGH to display vivid 3D images. This chapter deals with the procedure. We also survey other techniques than those explained after this chapter in this book.

2.1 Optical Holography

Holograms are a fringe pattern, and the fringe pattern is recorded on photo-sensitive chemicals in traditional optical holography. Figure 2.1 shows the recording step of an optical hologram. The subject is illuminated by light emitted by a coherent light source, i.e., a laser. The light scattered by the subject reaches to the photo-sensitive material such as a holographic dry plate or film. This is commonly called an *object wave* or *object field*. The output of the light source is also branched into another light path and directly illuminates the same photo-sensitive material. This is commonly called a *reference wave* or *reference field*. The reference field interferes with the object field and generates interference fringes.

The interference fringes are spatial distribution of optical intensity. This distribution has a three-dimensional (3D) structure over the 3D space where the reference and object field are overlapped. If the fringe is recorded on a thin sheet material, only the two-dimensional (2D) cross section of the 3D fringe is recorded on the material. This is called a *hologram* or a *thin hologram* more exactly. An example of the hologram fringe pattern is shown in Fig. 2.2.

In the reconstruction step, we remove the illuminating light for the subject and the subject itself, as shown in Fig. 2.3. When the same light as the reference field illuminates the hologram, the fringe pattern diffracts the illuminating light, and reconstructs the original object field. To be accurate, the field diffracted by the fringes includes the same component as the original object field. As a result, viewers can see the light of the original subject; they look at the subject placed at the original position. This process is called *optical reconstruction* of the hologram.

As mentioned a little more precisely in Sect. 2.3. The illuminating light is diffracted by the 2D fringe pattern and converted into the output light very simi-

K. Matsushima, *Introduction to Computer Holography*,
Series in Display Science and Technology,
https://doi.org/10.1007/978-3-030-38435-7_2

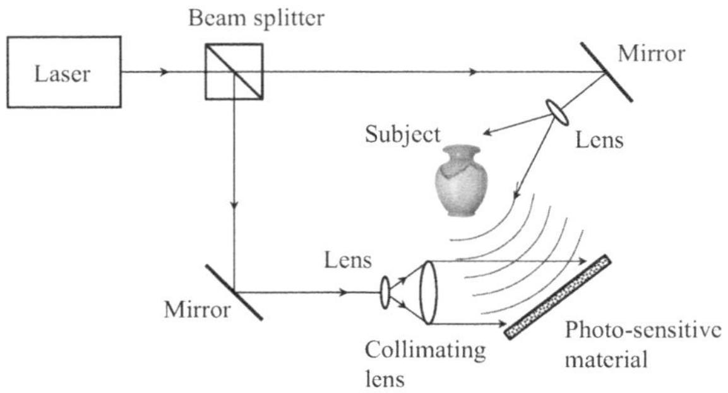

Fig. 2.1 The recording step of a hologram in conventional optical holography

Fig. 2.2 An example of the hologram fringe pattern. Note that this pattern is recorded by using an image sensor in practice

lar to the object field. This means that the whole spatial information of the object field emitted by a subject, which has a 3D structure, is recorded as the 2D image. In other words, information of the original object field is condensed into the sheet material.

Optical reconstruction of a hologram means to playback the original object field itself, which is frozen in the form of the 2D fringe pattern. Therefore, holographic 3D images properly convey the entire depth cues of the subject; the 3D image provides parallax disparity, vergence, accommodation, and motion parallax to the viewers, as mentioned in Sect. 1.1.

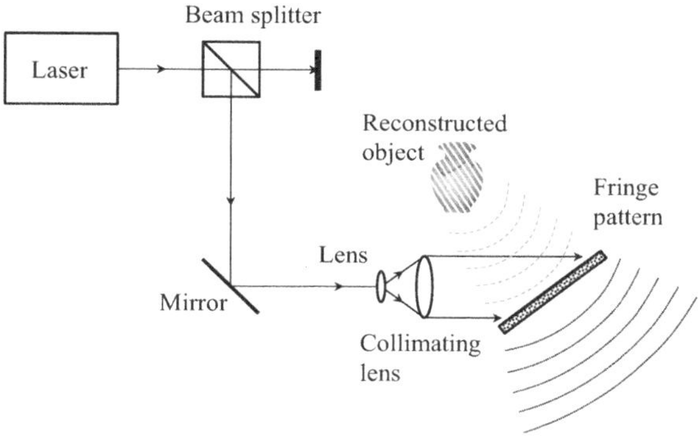

Fig. 2.3 Optical reconstruction of a hologram. The light recorded in the hologram in the form of a fringe pattern is reproduced by diffraction of the illuminating light

2.2 Computer Holography and Computer-Generated Hologram

When the recorded hologram is a thin hologram mentioned above, i.e., the fringe pattern has a simple 2D structure, the fringe pattern is just a monochrome picture whose resolution is unusually high. This type of interference fringes can be recorded digitally using an image sensor as well as a photo-sensitive material if the sensor resolution is high enough for recording. This technique, digital recording of the fringe pattern is commonly called *digital holography* (DH) in limited meaning.[1]

If you own a printer whose resolution is enough high to print the fringe pattern, you can print the hologram recorded by the image sensor. Any special feature is not required for the printer except for the resolution. This means that we can optically reconstruct any object field if we can digitally generate the fringe pattern for the object using a computer. In this technique, any real object is no longer required for producing the 3D image. This type of hologram is commonly called a *computer-generated hologram* or the abbreviation: a *CGH*.

The author also uses a word *computer holography* in this book. Computer holography is the technique to create a CGH that reconstructs a virtual object image from the numerical model in a narrow sense. The author uses "computer holography" in strong consciousness of computer graphics (CG) that also produces 2D images of a virtual object. In a broad sense, computer holography is simply the technique for creating holograms by using digital computers; it is no matter whether a physical object is required for producing the 3D image or not.

We can also use a combination of digital recording and reconstruction of holograms; the fringe pattern of a hologram is digitally recorded by an image sensor, and the 3D image is optically reconstructed from the digital fringe image. This technique is referred to as *digitized holography*, because the technique replaces the whole

[1] The author does not support this limited meaning of the word "digital holography," i.e., not approve of referring to only the recording step as "digital holography." This word should be used more widely.

process in optical holography: recording and reconstruction, with the digital counterparts. The detailed technique of digitized holography is discussed in Chap. 14.

If the photo-sensitive material in optical holography is enough thick and the recorded fringe pattern has a 3D structure, the hologram is called a *thick* or *volume hologram* (for the detail, see Sect. 7.2). In this case, we commonly cannot print the fringes, because ordinary printers only print 2D patterns. However, a special printer, called a *wavefront printer*, has the ability to print 3D fringes. This type of special printer is briefly discussed in Sect. 15.4.

2.3 Steps for Producing CGHs and 3D Images

Figure 2.4 shows a generalized procedure for producing a CGH and reconstructing the 3D image. The first step for producing a CGH is to prepare the object field numerically. Here, suppose that the object field is given by $O(x, y)$ that is two-dimensional distribution of complex amplitudes. The object field is usually produced from a set of model data. This process is referred to as *numerical synthesis of object field* or simply *field rendering* in this book. This is the most important and difficult step

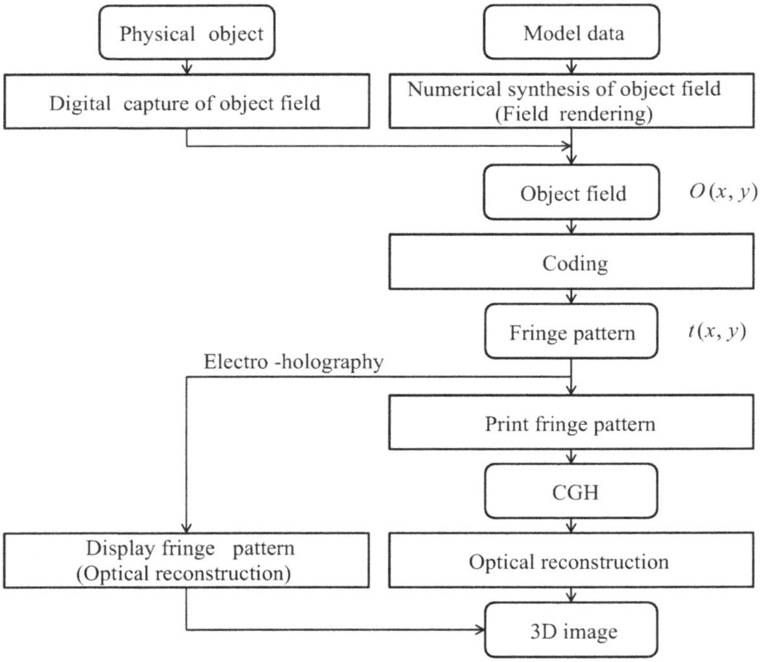

Fig. 2.4 Steps for creating a CGH and reconstructing the 3D images in computer holography

in computer holography. Thus, the overview of the techniques is given in Sect. 2.4, while the detail techniques used in this book are given in Chaps. 10 and 11.

The 3D image that we want to reconstruct is sometimes the image of a physical object. There are several techniques in this case. Photographs of the real object can be mapped onto the polygon mesh that gives the 3D shape of the real object. A 3D scanner or range finder is usually used to measure the real object and produce the polygon mesh. Brothers, shown in Fig. 1.6, is an example of this technique. Otherwise, we can also calculate the object field from multi-viewpoint photographs of the real object. These are regarded as a sort of numerical synthesis of the object field. As mentioned in the preceding section, another technique is to capture the fringe pattern of a physical object using an image sensor. We can extract the object field from the captured fringe pattern and arrange it in a virtual 3D scene that also includes virtual objects given by the numerical model (see Chap. 14).

Object field $O(x, y)$ is a complex function of the position in the hologram plane. The CGH works as a spatial modulator for the illuminating light in its optical reconstruction. However, the real CGH generally modulates either amplitude or phase of the illuminating light. Thus, we need an additional step to convert the complex function $O(x, y)$ into amplitude or phase fringe pattern $t(x, y)$. This step is called *coding* in this book. The detailed coding step is discussed in Sect. 8.5.

In optical holography, the coding process of the object field is provided by optical interference with a reference field. The finalized photo-sensitive material is the hologram. In computer holography, the fringe pattern $t(x, y)$ must be printed using a printer to fabricate the CGH; the 3D image is then reconstructed from the printed fringes. In electro-holography, the fringe pattern is not printed but directly displayed by some electric equipment. The detailed techniques to print the fringe pattern are discussed in Chap. 15.

2.4 Numerical Synthesis of Object Fields

2.4.1 Object Field

The object field is commonly identical to a physical electric field $\mathbf{E}(x, y, z)$. However, we usually ignore the vector nature of the electric field and treat the field value as a dimensionless complex value for simplicity. The object field is also defined as a 2D function in a plane perpendicular to the optical axis.[2] More restricted definition of an object field or a wavefield is given in Sects. 3.1 and 3.4.

Figure 2.5 shows the main coordinate system used in this book. Here, we suppose that the object field commonly travels along z axis in the coordinate system. In many cases, we can numerically calculate object field $g_{\mathrm{obj}}(x, y; z)$ at an arbitrary z position. Thus, we can assume that the hologram is arranged at the position of $z = 0$ without losing generality. In this case, the object field used for coding is simply given as:

[2]Notation $g_{\mathrm{obj}}(x, y; z)$ means that the object field is a function of x and y, while z is a parameter.

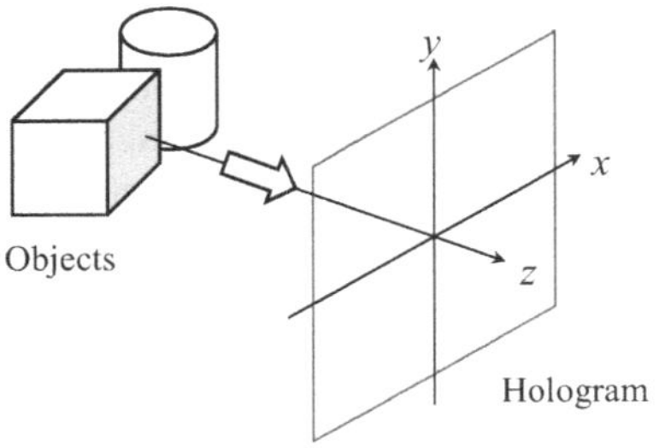

Fig. 2.5 The coordinate system mainly used in this book

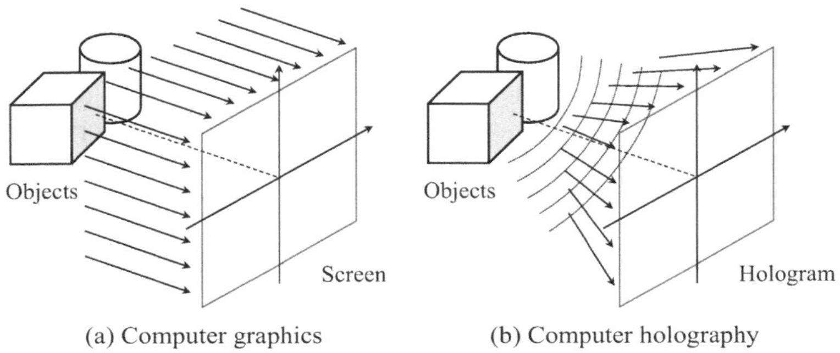

Fig. 2.6 Difference between **a** computer graphics and **b** computer holography

$$O(x, y) = g_{\mathrm{obj}}(x, y; 0). \tag{2.1}$$

The numerical synthesis of object fields is most important and difficult process in the whole process to create a CGH in computer holography, as mentioned repeatedly. This process is, in a sense, very similar to CG that produce 2D images from the model data. There are several types of model, such as a wire frame and surface model. Computer holography as well as computer graphics requires effective algorithms to reconstruct realistic objects, as mentioned in the next section.

There is, however, a big difference between computer holography and computer graphics. CG treats light as a set of rays that only travel perpendicularly to the screen, as shown in Fig. 2.6. In contrast, computer holography treat light as a wavefield that includes not only a component of light perpendicular to the hologram but all components of light traveling to the hologram plane at various angles.

2.4.2 *Field Rendering*

The step of the numerical synthesis of object fields is very similar to CG in a sense. Therefore, the author sometimes call this step *field rendering* as an alias, though the synthesized object field is not the final image unlike CG. We need coding and optical reconstruction steps to see the 3D image after the field rendering, but almost all appearances of the final 3D image is determined by the field rendering.

Several techniques are required for reconstructing realistic 3D images like CG. One of the most important technique is called *occlusion processing* [124] or *occlusion culling* [128]. The effect of occlusion is one of the most important depth cues in computer holography as well as other 3D techniques and CG. An object behind another object is hidden by the front object in the real world unless the front object is see-through.

The process to hide the background objects is called *hidden surface removal* in CG. Occlusion processing in computer graphics is also sometimes called hidden surface removal because of similarity to CG. The considerable difference in computer holography from CG is that viewers can change their viewing point in computer holography, like the real world. The back object must come from or hide behind the front object when the viewpoint is moved in computer holography, even if the hologram is static. This feature of movable viewpoint makes the hidden surface removal very difficult in computer holography. The detailed techniques of occlusion processing used in this book are given in Chap. 11.

Other rendering techniques required in computer holography are, of course, shading and texture-mapping. Because these are very similar to CG, we can use almost the same technique as those in CG. Rendering techniques reflecting surface materials such as specular and transparent properties are also required for realistic rendering. The feature of movable viewpoint, however, also makes these renderings very difficult in computer holography.

2.4.3 Brief Overview of Rendering Techniques

Field rendering coarsely consists of processes of shape forming and occlusion processing. However, these processes are often combined and cannot be separated in many techniques. Several major techniques are briefly introduced in the following.

2.4.3.1 Point-Based Method

A field rendering technique used most widely is definitely the *point-based method*. This is sometimes called a *point cloud*. As shown in Fig. 2.7, it is assumed in this technique that object surfaces are formed by much many point sources of light, and all point saurces emit a spherical wave [117, 129]. Since a spherical wave is given by a simple formula (see Sect. 3.3), we can obtain the object field by superposing the spherical waves in the hologram plane. This technique is very simple in the principle, but very time-consuming because the computation time is proportional to the product of the numbers of point sources and sample points of the object field [81]. Unfortunately, because we need much many point sources to form a surface, and also

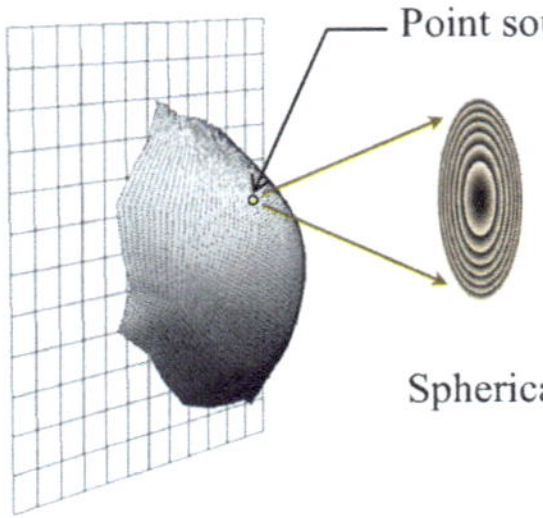

Fig. 2.7 The principle of the point-based method

need much many sample points of the object field to create a large-scale CGH, it is almost impossible to synthesize the object field for full-parallax HD-CGHs using the point-based method.

The simplicity of the point-based method, however, attracts many researcher in computer holography. Many techniques have been proposed to overcome the drawback of slowness: horizontal-parallax-only CGHs [56], look-up tables [41, 56], use of geometric symmetry [36], recurrence formulas [81], and difference formulas [112, 141]. The simplicity in computation also leads to making use of parallel processing and hardware assist [32–34, 101, 104, 114]. Use of graphical processing units (GPU) is absolutely required for computing the object field for creating large-scale CGHs in realistic time [58, 89].

The another problem of the point-based methods is in occlusion processing. In many cases, a variation of the technique called a *visibility-test* is used for occlusion processing in the point-based method [11, 23, 24, 30, 149]. However, simple implements of the visibility-test is too time-consuming as in the point-based method itself. A technique without the visibility test is also proposed in the point-based method [13]. In this technique, the collection of point sources is switched, depending on the viewpoint. This is a useful and practical technique, but the motion parallax is not continuous.

2.4.3.2 Polygon-Based Method

The point-based methods are commonly "ray oriented" because they trace the ray from a point source to a sample point of the object field in the hologram plane. There are also "wave-oriented" methods to calculate object fields. For example, fields emitted from an object defined as planar segments [51, 120] or 3-D distribution of field strength [50] can be calculated on the basis of wave optics. The major advantage of wave-oriented methods is that FFT can be used for numerical calculations.[3] Therefore, the computation time is shorter than point-based methods, especially in full parallax holograms.

---

[3]There is a hybrid technique that makes use of the features of both ray-oriented and wave-oriented methods [113].

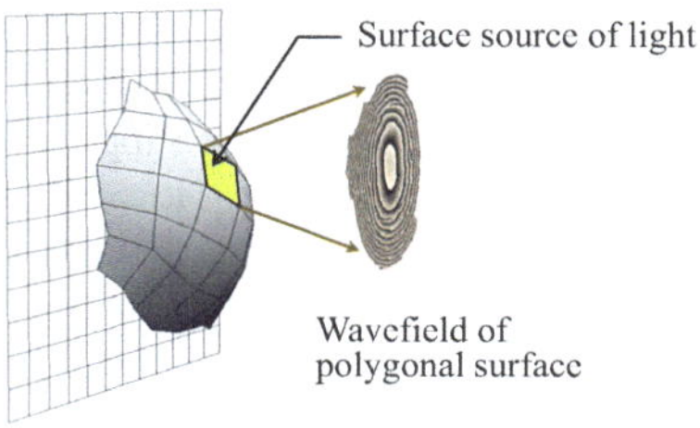

Fig. 2.8 The principle of the polygon-based method

The polygon-based method was proposed for avoiding the problem of long computation time of the point-based methods [59, 67]. In this technique, an object is composed of many polygons instead of discrete points, as shown in Fig. 2.8. Since each polygon is treated as a surface source of light, the object field is obtained by superposing the fields of the polygons. The number of polygons forming an object is definitely much smaller than that of points in the point-based methods. Therefore, this technique drastically speedup computation of the object fields, though the computation of individual polygon fields is complicated and time-consuming.

The complexity of computing a polygon field most likely prevents the technique from being popular. Researchers who intend to implement the polygon-based method in computers may need some more knowledge of wave optics than that in the point-based methods. The introduction of wave optics and detailed techniques of the polygon-based field rendering are given in Chaps. 3 and 10, respectively.

There are many techniques proposed for the polygon-based rendering other than those discussed in this book [1, 29, 40, 52, 98, 99, 131, 151]. Unfortunately, most of these techniques are not practical; we cannot produce high-definition CGHs, which can be practically exhibit, using these techniques. Some of these do not have enough high performance to create HD-CGHs composed of billions pixels [29, 131].

In particular, analytical and mathematical techniques, which are usually based on affine transformations of a triangular mesh, are commonly impractical for producing high-quality HD-CGHs [1, 40, 52, 151]. This is due to lack of diffusibility of the synthesized field. The mesh surfaces in reconstructed 3D images must be uniformly diffusive. Otherwise, the viewers cannot see the surface from various viewpoints. Even if a surface is specular in the 3D model, a certain diffusibility is absolutely required in the surface. A certain amount of randomness is needed to produce diffused surfaces as mentioned in Sect. 10.1.1; however it may be difficult to introduce the randomness into the analytical techniques.

2.4.3.3 Silhouette Methods

The polygon-based method is usually associated with a mask-based occlusion processing technique called the *silhouette method* [45, 66, 67, 75]. This technique is very powerful, especially in use with the *switch-back technique*. This technique, in fact, allows us to synthesize the object fields of very complex-shaped objects with proper occlusion processing in realistic time.

The silhouette method is applicable to not only the polygon-based method but also the point-based methods. However, since the silhouette method is wave-oriented and uses field propagation, the technique surely fits the polygon-based methods that also use field propagation. A type of the point-based methods using field propagation may suit the silhouette method as well [113]. The detail of the silhouette method is given in Chap. 11.

2.4.3.4 Multi-Viewpoint Image Based Techniques

One of the most important techniques of the field rendering is multi-viewpoint image (MVI) based techniques [39, 82, 133, 140]. This technique is the counterpart of holographic stereograms in optical holography. In this technique, the object field is generated from two dimensional (2D) digital images viewed from different viewpoints. If the object is virtual and the shape is given by the model data, the MVIs are provided by ordinary CG rendering. Thus, the rendering techniques such as shading for various types of materials are simply given by CG's techniques.

Occlusion processing peculiar to computer holography is also not necessary in this technique, because MVI-based CGHs are rendered with processing of hidden surface removal in CG. For physical objects, MVIs are obtained by some technique such as a fly's array lens or moving camera. Occlusion processing is also not required in this case. The drawback of the technique is that the accommodation depth cue, which is one of the most remarkable features of holography, is lost in the object.

To get over the problem, a hybrid technique between MVI- and wave-based methods, called a *ray-sampling plane*, is also proposed [127, 128]. In this technique, object fields obtained from MVIs are numerically propagated to the hologram plane. Thus, the vergence-accommodation conflict is greatly lightened.

2.4.3.5 Layer-Based Technique

One of the oldest techniques for numerical synthesis of object fields is the technique of *layered holograms* [31, 53]. In this technique, we slice up an object and make many 2D images corresponding to the cross sections of the object. Since the 2D image is parallel to the hologram, we can easily calculate the diffraction of the 2D image by using conventional techniques of field propagation. Here, the random phase may be added to the amplitude distribution obtained from the 2D image. Furthermore, occlusion processing can be included in this technique. The field of a layer is propagated to the next layer closer to the hologram. The field from the previous layer is shielded by a 2D mask in which the value is corresponding to transmittance of the object. The layer-based techniques are still being developed [9, 10, 148, 150].

2.5 Coding and Reconstruction

A hologram spatially modifies the incident illuminating light and reconstructs the recorded object field. This is the optical reconstruction of the hologram. Holograms are usually able to modify only one component, i.e., either amplitude or phase of the illuminating light.

The simplest coding technique is the same as that in optical holography, i.e., interference between object and reference fields. Suppose that the wavefield of a reference field is given by $R(x, y)$, the distribution of fringe intensity is given by:

$$\begin{aligned} I(x, y) &= |O(x, y) + R(x, y)|^2 \\ &= |O(x, y)|^2 + |R(x, y)|^2 + O^*(x, y)R(x, y) + O(x, y)R(x, y)^*, \end{aligned} \tag{2.2}$$

where symbol '*' denotes complex conjugate along the standard notation.

Here, suppose that a hologram has distribution of amplitude transmittance $t(x, y)$, i.e., the hologram spatially modulates the amplitude of incident light. In addition, suppose that $t(x, y) \propto I(x, y)$, as shown in Fig. 2.9. In optical holography, this is realized by recording the fringe intensity $I(x, y)$ on a photo-sensitive material such as silver halide films. When an illuminating light that is very similar to the reference field illuminates the hologram, the transmitted field is written by:

$$\begin{aligned} t(x, y)R(x, y) \cong\ & |O(x, y)|^2 R(x, y) + |R(x, y)|^2 R(x, y) \\ & + R^2(x, y)O(x, y)^* + |R(x, y)|^2 O(x, y). \end{aligned} \tag{2.3}$$

The first term gives the non-diffraction or 0th order light. The light that comes from the third term gives the conjugate image. These are commonly unnecessary light and degrade the optical reconstruction of the hologram. Only the last term gives proper optical reconstruction of the object field if the intensity distribution of the reference wave $|R(x, y)|^2$ is approximately constant.

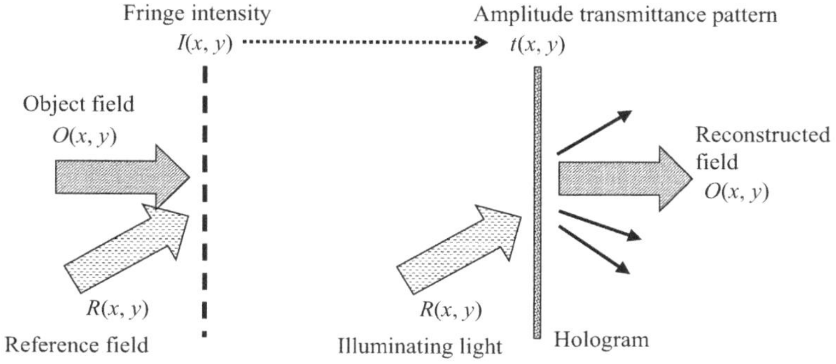

Fig. 2.9 Simple coding of object fields by interference with a reference wave

As a result, (2.2) gives a coding procedure for converting the object field $O(x, y)$ into the fringe pattern $t(x, y)$. In practical process for generating a fringe pattern in computer holography, we usually need several additional process such as quantization. The detailed principle and nature of holography are discussed in Chap. 7. The detailed technique of coding is discussed in Chap. 8.

Chapter 3
Introduction to Wave Optics

Abstract Introduction to wave optics is presented in this chapter. We start from the wave equation that describes the behavior of monochromatic waves and then discuss important wavefields: a plane wave and spherical wave. In particular, we put importance on the properties of the sampled wavefields that often produce aliasing errors and ruin the numerical calculation. An important idea for handling sampled wavefields, i.e., the maximum diffraction angle is also discussed in this chapter.

3.1 Light as Wave

3.1.1 Wave Form and Wave Equation

Many people might image a ripple on a surface of water when they hear the word "wave". Waves generally have nature that it extends or propagates over a certain distance almost without changing its shape. A wave is mathematically represented by the displacement that is a function of the position and time. The displacement is, in general, physical quantity and an instantaneous shift from the static state. For example, the displacement of water-surface waves is instantaneous height of the water surface from the static surface level.

Suppose that the displacement of a wave is given by $f(z)$ at $t = 0$, as shown in Fig. 3.1, the wave that laterally shifts from the origin to $z = \Delta$ is given by $f(z - \Delta)$. Thus, when the shift distance is given by $\Delta = ct$ at time t where c is the velocity of the wave, the displacement of the wave must take the form:

$$u(z, t) = f(z - ct). \tag{3.1}$$

This wave propagates in the z direction if $c > 0$, while it propagates in the opposite direction in cases where $c < 0$. The function $f(\tau)$, where $\tau = z - ct$, is called a *wave form*. Here, note that a wave represented by the form of $f(ct - z)$ also travels in the z direction in $c > 0$, but a wave represented by $f(z + ct)$ travels in the opposite direction.

K. Matsushima, *Introduction to Computer Holography*,
Series in Display Science and Technology,
https://doi.org/10.1007/978-3-030-38435-7_3

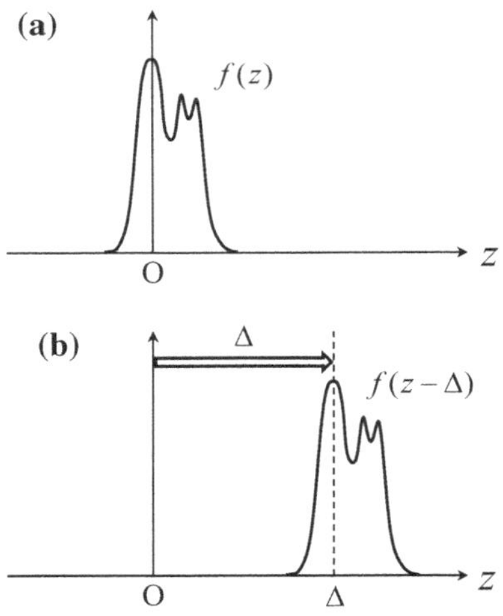

Fig. 3.1 Propagation of a wave

We can derive a differential equation that $u(z, t)$ in (3.1) satisfies. The first derivatives of (3.1) are given by:

$$\frac{\partial u}{\partial z} = \frac{\partial u}{\partial \tau}\frac{\partial \tau}{\partial z} = \frac{\partial u}{\partial \tau}, \tag{3.2}$$

$$\frac{\partial u}{\partial t} = \frac{\partial u}{\partial \tau}\frac{\partial \tau}{\partial t} = -c\frac{\partial u}{\partial \tau}. \tag{3.3}$$

The second derivatives are also given as follows:

$$\frac{\partial^2 u}{\partial z^2} = \frac{\partial}{\partial \tau}\left(\frac{\partial u}{\partial z}\right)\frac{\partial \tau}{\partial z} = \frac{\partial^2 u}{\partial \tau^2}, \tag{3.4}$$

$$\frac{\partial^2 u}{\partial t^2} = \frac{\partial}{\partial \tau}\left(\frac{\partial u}{\partial t}\right)\frac{\partial \tau}{\partial t} = c^2\frac{\partial^2 u}{\partial \tau^2}. \tag{3.5}$$

By eliminating $\partial^2 u/\partial \tau^2$ from (3.4) and (3.5), we get the following partial differential equation:

$$\frac{\partial^2 u}{\partial z^2} = \frac{1}{c^2}\frac{\partial^2 u}{\partial t^2}. \tag{3.6}$$

This is called one-dimensional *wave equation*. Since the solution of the wave equation has the form of (3.1), any physical phenomena, described by the differential equation equivalent to (3.6), is able to propagate as a wave.

We can simply extend the one dimensional wave equation in (3.6) into the three dimensional space as follows:

$$\nabla^2 u = \frac{1}{c^2}\frac{\partial^2 u}{\partial t^2}, \tag{3.7}$$

where the displacement is a function of x, y, z and t, and thus $u = u(x, y, z, t)$. Here, operator ∇^2 is defined as

$$\nabla^2 = \frac{\partial^2}{\partial x^2} + \frac{\partial^2}{\partial y^2} + \frac{\partial^2}{\partial z^2}. \tag{3.8}$$

Three dimensional positions are often represented by position vector $\mathbf{r}$ in this book; that is

$$\mathbf{r} = x\mathbf{e}_x + y\mathbf{e}_y + z\mathbf{e}_z, \tag{3.9}$$

where $\mathbf{e}_x$, $\mathbf{e}_y$ and $\mathbf{e}_z$ are unit vectors in the x, y and z directions, respectively.

Conclusively, the wave equation is represented in the three dimensional space as follows:

$$\nabla^2 u(\mathbf{r}, t) = \frac{1}{c^2}\frac{\partial^2 u(\mathbf{r}, t)}{\partial t^2}. \tag{3.10}$$

3.1.2 Electromagnetic Wave

Light is a wave of electric and magnetic fields governed by Maxwell's equations. In the absence of free charge, Maxwell's equation in standard units are given by:

$$\nabla \times \mathbf{E} = -\mu_0\mu_r \frac{\partial \mathbf{H}}{\partial t} \tag{3.11}$$

$$\nabla \times \mathbf{H} = \varepsilon_0\varepsilon_r \frac{\partial \mathbf{E}}{\partial t} \tag{3.12}$$

$$\nabla \cdot \varepsilon_0\varepsilon_r \mathbf{E} = 0 \tag{3.13}$$

$$\nabla \cdot \mu_0\mu_r \mathbf{H} = 0, \tag{3.14}$$

where vectors $\mathbf{E}(= \mathbf{E}(\mathbf{r}, t))$ and $\mathbf{H}(= \mathbf{H}(\mathbf{r}, t))$ are the electric and magnetic fields, respectively. The symbols $\times$ and $\cdot$ represent a vector cross product and vector dot product, respectively, and

$$\nabla = \frac{\partial}{\partial x}\mathbf{e}_x + \frac{\partial}{\partial y}\mathbf{e}_y + \frac{\partial}{\partial z}\mathbf{e}_z. \tag{3.15}$$

ε_0 and ε_r are the permittivity in vacuum and a relative permittivity, respectively. μ_0 and μ_r are also the permeability in vacuum and a relative permeability. Here, $\mu_r \cong 1$ in optical frequency regions. Hence, we always ignore μ_r in the rest of this book.

Through the whole of this book, we assume that light propagates in the *free space*, i.e., travels in a *uniform* and *isotropic* dielectric medium. In the isotropic media, its properties are independent of the direction of vectors **E** and **H**. In uniform mediums, ε_r is always considered as a constant. Thus, $\nabla \cdot \mathbf{E} \equiv \nabla \cdot \mathbf{H} \equiv 0$ according to (3.13) and (3.14). Applying the $\nabla \times$ operator to the both sides of (3.11) and using the well-known vector formula:

$$\nabla \times \nabla \times \mathbf{E} = \nabla(\nabla \cdot \mathbf{E}) - \nabla^2 \mathbf{E}, \tag{3.16}$$

we get the following differential equation:

$$\nabla^2 \mathbf{E} = \varepsilon_0 \mu_0 \varepsilon_r \frac{\partial^2 \mathbf{E}}{\partial t^2}, \tag{3.17}$$

where we exchanged the order of time and space derivations in the right hand side of (3.11), and then, used (3.12).

The vector of the electric field can be decomposed into three components as follows:

$$\mathbf{E} = E_x \mathbf{e}_x + E_y \mathbf{e}_y + E_z \mathbf{e}_z. \tag{3.18}$$

This means that we actually derived three differential equations:

$$\nabla^2 E_x = \varepsilon_0 \mu_0 \varepsilon_r \frac{\partial^2 E_x}{\partial t^2} \tag{3.19}$$

$$\nabla^2 E_y = \varepsilon_0 \mu_0 \varepsilon_r \frac{\partial^2 E_y}{\partial t^2} \tag{3.20}$$

$$\nabla^2 E_z = \varepsilon_0 \mu_0 \varepsilon_r \frac{\partial^2 E_z}{\partial t^2}. \tag{3.21}$$

By comparing these differential equations with the wave equation in (3.7), we find that electric fields can behave as a wave and the speed is given by

$$c = 1/\sqrt{\varepsilon_0 \mu_0 \varepsilon_r}. \tag{3.22}$$

The speed of light in dielectric media is

$$c = \frac{c_0}{n}, \tag{3.23}$$

where $c_0 = 1/\sqrt{\varepsilon_0 \mu_0}$, and

$$n = \sqrt{\varepsilon_r}. \tag{3.24}$$

This dimensionless value n is called *refractive index*.

In many cases in computer holography, the direction of vector $\mathbf{E}(\mathbf{r}, t)$ is not significant. Thus, we commonly use the scalar wave equation of (3.10) instead of the vector wave equation of (3.17), i.e., we suppose that $u(\mathbf{r}, t)$ represents any one of

$E_x(\mathbf{r}, t)$, $E_y(\mathbf{r}, t)$ or $E_z(\mathbf{r}, t)$. When effect of polarity is essential in a case, multiple components of (3.19)–(3.21) must be handled to resolve the problem.

Applying the same procedure, the following wave equation for magnetic fields is also derived from Maxwell's equations.

$$\nabla^2 \mathbf{H} = \varepsilon_0 \mu_0 \varepsilon_r \frac{\partial^2 \mathbf{H}}{\partial t^2}. \tag{3.25}$$

Let us emphasize that an electric wave is always accompanied with a magnetic wave. However, we usually ignore the magnetic wave in computer holography, because both the direction and magnitude of the magnetic field can be easily obtained from the electric field anytime in free space. Hence, we do not have to care about the magnetic waves.

3.1.3 Complex Representation of Monochromatic Waves

Light is commonly regarded as a temporally coherent wave in computer holography, i.e., we always assume a monochromatic wave. In monochromatic light, the electric field is a sinusoidal wave. Therefore the wave can be represented as

$$u(\mathbf{r}, t) = A'(\mathbf{r}) \cos(\omega t + \phi), \tag{3.26}$$

where ω is an angular frequency and ϕ is an initial phase at $t = 0$.

Since an exponential function is much easier to handle than a trigonometric function, according to Euler's formula we represent the wave function as:

$$u(\mathbf{r}, t) = \mathrm{Re}\left\{A'(\mathbf{r}) \exp(-i\phi) \times \exp(-i\omega t)\right\}, \tag{3.27}$$

where $\mathrm{Re}\{\alpha\}$ means the real part of the complex value α and i is the imaginary unit. Note that we can choose the plus sign as well as minus sign of $i\omega t$ in (3.27). The minus sign is carefully chosen in the above equation so that the several solutions of the wave equation is consistent with the form of $f(z - ct)$. As a result, we suppose that the wave function is given as follows:

$$\begin{aligned} u(\mathbf{r}, t) &= \mathrm{Re}\{g(\mathbf{r}) \exp(-i\omega t)\} \\ &= \frac{1}{2}\left[g(\mathbf{r}) \exp(-i\omega t) + \text{c.c.}\right], \end{aligned} \tag{3.28}$$

where 'c.c.' denotes complex conjugate. In this representation, $g(\mathbf{r})$ is a complex-valued function of a position unlike $u(\mathbf{r}, t)$ that is a real-valued function of a position and time.

The value of $g(\mathbf{r})$ is referred to as a *complex amplitude*. The complex amplitude is usually given as the following amplitude-phase style:

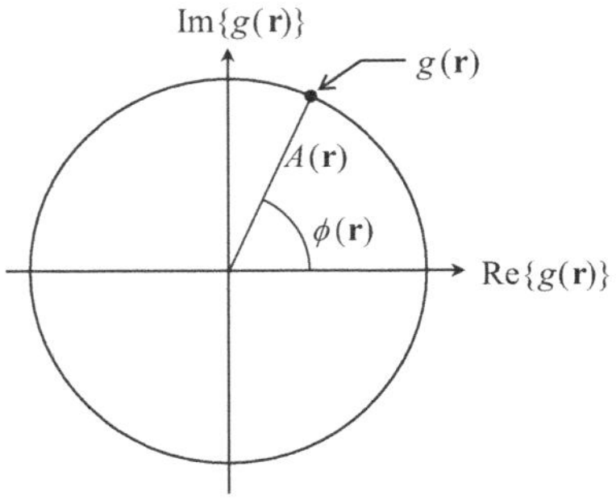

Fig. 3.2 The amplitude and phase of a complex amplitude in the complex plane

$$g(\mathbf{r}) = A(\mathbf{r}) \exp[i\phi(\mathbf{r})], \tag{3.29}$$

where $A(\mathbf{r})$ is the real-valued amplitude (just called "amplitude") and $\phi(\mathbf{r})$ is the phase, as shown in Fig. 3.2.

Substituting (3.28) into (3.10), the wave equation is rewritten by the following Helmholtz equation:

$$\left(\nabla^2 + k^2\right) g(\mathbf{r}) = 0. \tag{3.30}$$

where $k = \omega/c$ is a *wave number*, as mentioned in Sect. 3.2.1. The solution of Helmholtz equation gives spatial distribution of the electric field of monochromatic light.

3.1.4 Wavefield

Two dimensional distribution of complex amplitude in a given plane is called a *wavefield*. In computer holography and wave optics, a wavefield is one of the most important object, because wavefields represent a time-independent form of monochromatic waves.

A wavefield is given in a plane perpendicular to the major optical axis. In many cases of this book, the major optical axis is the z-axis. Therefore, a wavefield is commonly given in the x–y plane. The wavefield placed at $z = z_0$ is, for example, represented by

$$g(x, y; z_0) = A(x, y; z_0) \exp[i\phi(x, y; z_0)]. \tag{3.31}$$

However, in some cases, especially in cases where the rotational transform is concerned, the wavefield is not always given in a plane orthogonal to the z-axis.

3.2 Plane Wave

3.2.1 One-Dimensional Monochromatic Wave

A well-known solution of the one-dimensional wave equation in (3.6) is given as follows:

$$u(z,t) = A\cos\left[2\pi\left(\frac{z}{\lambda} - \frac{t}{T}\right) + \phi_0\right], \tag{3.32}$$

where λ and T are a *wavelength* and period of the wave. The wavelength and period give spatial and temporal length per unit cycle of the monochromatic wave, respectively. A is an amplitude of the wave, and ϕ_0 is the initial phase at $z = t = 0$. These quantities are summarized in Fig. 3.3. Equation (3.32) is also represented as:

$$u(z,t) = A\cos(kz - \omega t + \phi_0), \tag{3.33}$$

$$k = \frac{2\pi}{\lambda}, \tag{3.34}$$

$$\omega = \frac{2\pi}{T}, \tag{3.35}$$

where k and ω are again a wave number and angular frequency of the wave, respectively.

The angular frequency is also given by $\omega = 2\pi f$ using frequency $f = T^{-1}$. The frequency f gives the number of cycles of the wave per unit time, e.g., one second. Hence, the product of the frequency and wavelength gives the distance that the wave travels per unit time; the speed of the wave is

$$c = f\lambda. \tag{3.36}$$

Wave number k is the spatial counterpart of angular frequency ω, and thus, should be represented as $k = 2\pi w$ using a spatial frequency $w = \lambda^{-1}$. But, this is not in general. The quantities mentioned above are summarized in Table 3.1 from the symmetric viewpoint of the space and time.

Equation (3.33) is rewritten by:

$$u(z,t) = A\cos\left[k(z - ct) + \phi_0\right],$$

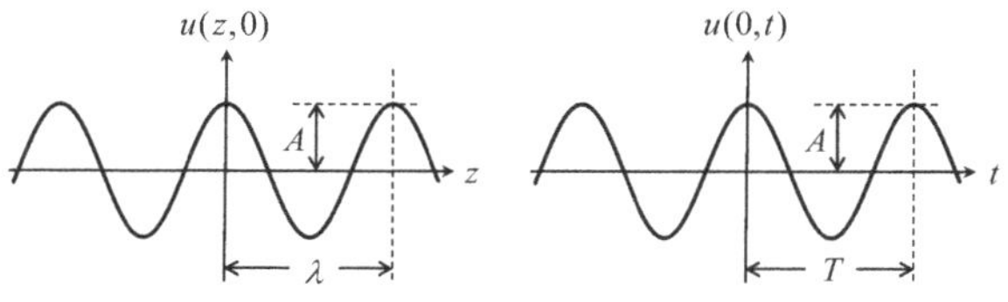

Fig. 3.3 One-dimensional plane wave. $\phi_0 = 0$

Table 3.1 Summary of quantities related to monochromatic waves

Description	Time	Space
Length of a cycle	Period T [s]	Wavelength λ [m]
Number of cycles per unit length	Frequency $f = \dfrac{1}{T}$ [Hz], [1/s]	Spatial frequency $w = \dfrac{1}{\lambda}$ [1/m]
Phase rotation angle per unit length	Angular frequency $\omega = 2\pi f = \dfrac{2\pi}{T}$ [rad/s]	Wave number $k = 2\pi w = \dfrac{2\pi}{\lambda}$ [rad/m]

where $k = \omega/c$ is obtained from (3.34) to (3.36). Therefore, the wave form is given by $f(\tau) = A\cos(k\tau + \phi_0)$ in this wave. Complex representation of this one-dimensional wave is easily obtained from (3.28) and (3.33). The complex amplitude is given by

$$g(z) = A\exp\left[i(kz + \phi_0)\right]. \tag{3.37}$$

3.2.2 *Sampling Problem*

To treat waves in computer, the waves must be sampled with an equidistant sampling in general. According to sampling theorem, sampling rate must be larger than twice the signal frequency. In the time domain, the frequency is f in one-dimensional monochromatic waves. A sufficient sampling rate is therefore $2f$.

In the space domain that is essential in computer holography, the sampling rate must be more than $2w$. Thus, sampling interval Δ in the z direction must satisfy

$$\Delta < \frac{\lambda}{2}. \tag{3.38}$$

Figure 3.4 shows appearance of three waves, sampled at intervals around $\lambda/2$. When a continuous wave is sampled at an interval of $3\lambda/10$ that is slightly smaller than $\lambda/2$, frequency of peaks and valleys in the sampled wave keeps the original frequency, $1/\lambda$ as in (a). In contrast, the spatial frequency of the wave looks lower than the original one in the case of the sampling interval of $7\lambda/10$, i.e., the wavelength looks longer than original, as in (c). This is called an *aliasing error*. The sampling problem imposes limits on the direction of travel of monochromatic waves in the 3D space as shown in Sect. 3.2.4. The detail of the sampling theorem and the aliasing error is discussed in Sect. 4.5.

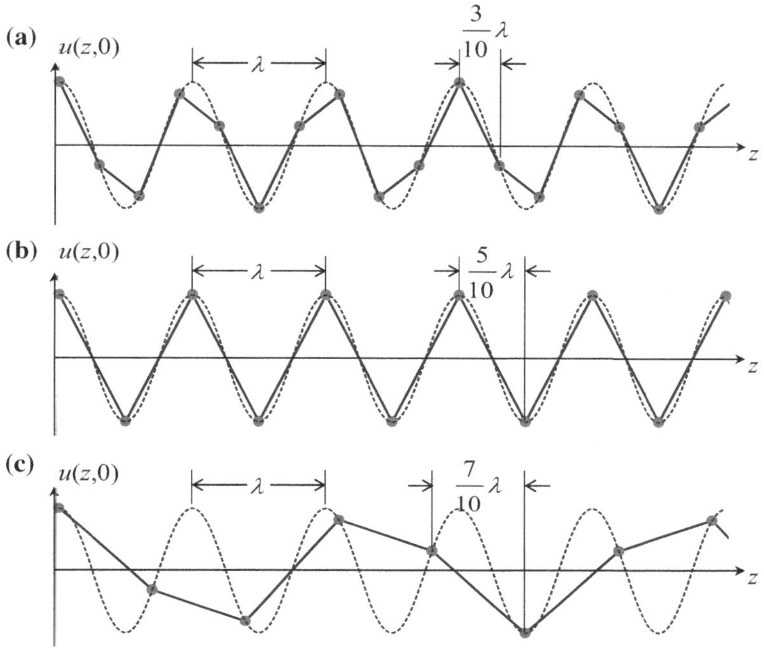

Fig. 3.4 Sampled one-dimensional waves. **a** $\Delta = 3\lambda/10$, **b** $\Delta = \lambda/2$, **c** $\Delta = 7\lambda/10$

3.2.3 Plane Wave in Three Dimensional Space

Suppose that the monochromatic wave represented in (3.33) is in the 3D space, i.e.,

$$u(\mathbf{r}, t) = A\cos(kz - \omega t + \phi_0), \tag{3.39}$$

where vector $\mathbf{r}$ is again a position vector that designates (x, y, z) coordinates in the 3D space. The right hand side of this equation, however, includes only the z-coordinates. This means that the displacement of the wave is independent of both the x- and y-coordinates. Therefore, the wave in a $(x, 0, z)$ plane can be illustrated as in Fig. 3.5, where the white color indicates the peak of the wave (the maximum value of $u(\mathbf{r}, z)$) and the black color indicates the valley. The gray-level[1] is corresponding to the value of $u(\mathbf{r}, z)$.

The dashed lines show the equiphase surfaces in which the phase of the wave satisfies $kz - \omega t + \phi_0 \equiv 0$. This kind of equiphase surfaces are called a *wavefront*. In this case, the wavefront of the monochromatic wave represented by (3.39) definitely forms planes in the 3D space. Therefore, the monochromatic wave is called a *plane wave*.

[1]The image is produced with an encoding gamma of 1/2.2.

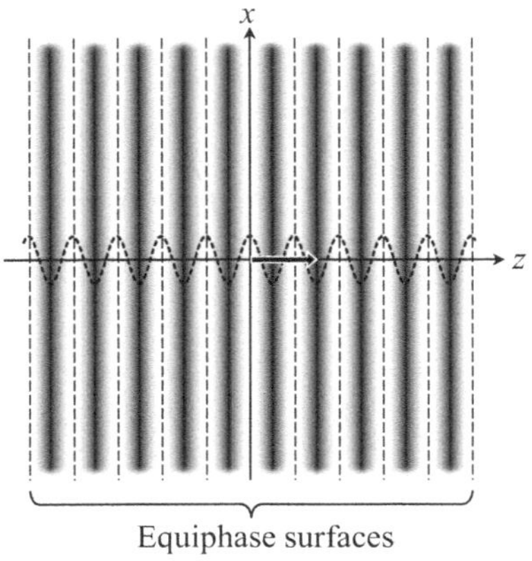

Fig. 3.5 A plane wave traveling along the z-axis in the 3D space

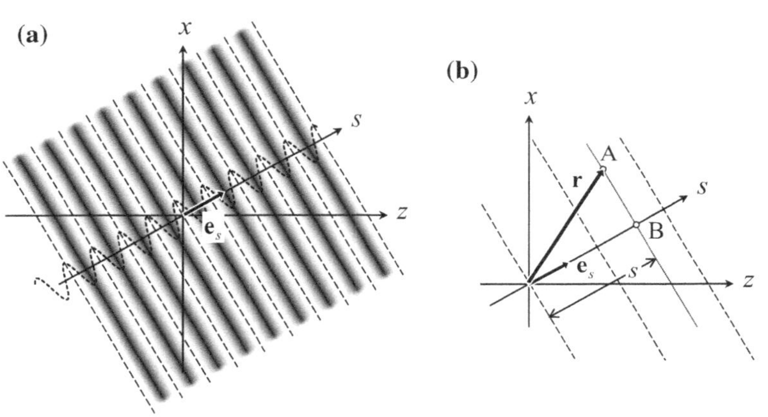

Fig. 3.6 A plane wave traveling along an arbitrary axis

The plane wave shown in Fig. 3.5 travels along the z-axis, while the plane wave in Fig. 3.6a travels along the s-axis. This plane wave is given by

$$u(\mathbf{r}, t) = A\cos(ks - \omega t + \phi_0), \tag{3.40}$$

where s is a distance from the origin. Suppose that $\mathbf{e}_s$ is a unit vector along the s-axis and $\mathbf{r}$ is the position vector of a given point A as in (b). If another point B is in the s-axis and included in the same wavefront as the point A, the distance s is given by

$$s = \mathbf{r} \cdot \mathbf{e}_s, \tag{3.41}$$

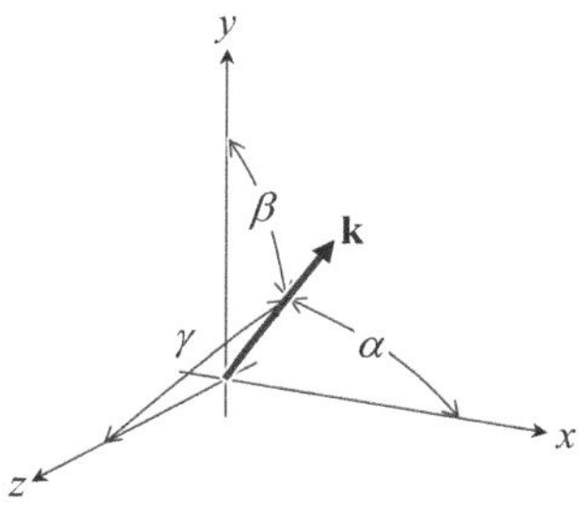

Fig. 3.7 The direction angles in the Cartesian coordinate system

where symbol '·' is again the dot product of two vectors. Therefore, by substituting (3.41) into (3.40), the plane wave is rewritten as

$$u(\mathbf{r}, t) = A\cos(\mathbf{k}\cdot\mathbf{r} - \omega t + \phi_0) \tag{3.42}$$

$$\mathbf{k} = k\mathbf{e}_s. \tag{3.43}$$

Here, **k** is called a *wave vector*. The direction of a wave vector agrees with the traveling direction of the plane wave, and thus, the wave vector always orthogonal to the wavefront. The length of the wave vector agrees with the wave number k.

By substituting (3.9) into (3.42), the plane wave is rewritten as

$$u(x, y, z, t) = A\exp\left[i(k_x x + k_y y + k_z z - \omega t + \phi_0)\right], \tag{3.44}$$

where k_x, k_y and k_z are components of **k** and given by direction cosines:

$$\begin{aligned} k_x &= k\cos\alpha, \\ k_y &= k\cos\beta, \\ k_z &= k\cos\gamma. \end{aligned} \tag{3.45}$$

As shown in Fig. 3.7, the symbols α, β and γ are direction angles; they are angles formed by the wave vector and the x-, y-, and z-axes, respectively. Here, note that these components are not independent of each other. Because $|\mathbf{k}| = k$, the components satisfy

$$k_x^2 + k_y^2 + k_z^2 = k^2 = \left(\frac{2\pi}{\lambda}\right)^2. \tag{3.46}$$

This makes the rotational transform of wavefields very complicated (see Chap. 9).

The complex amplitude of a plane wave is obtained from (3.28) and given as follows:

$$g(\mathbf{r}) = A\exp\left[i(\mathbf{k}\cdot\mathbf{r} + \phi_0)\right]. \tag{3.47}$$

In the amplitude-phase style representation given by (3.29), the plane wave is:

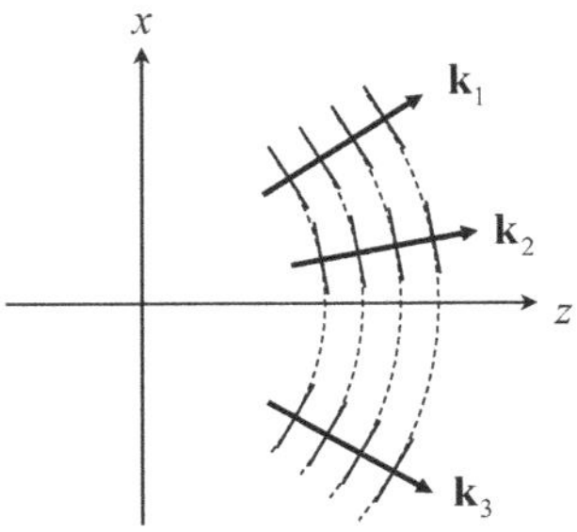

Fig. 3.8 Partial plane waves

$$A(\mathbf{r}) \equiv A, \tag{3.48}$$

$$\phi(\mathbf{r}) = \mathbf{k} \cdot \mathbf{r} + \phi_0. \tag{3.49}$$

Since the amplitude distribution is a constant of A, only the phase distribution provides significant information in this style.

It should be emphasized that plane waves are not physical waves, i.e., any plane wave cannot exist in the real world, because a plane wave spatially extends to infinity. In this point, plane waves are just mathematical solutions of the wave equation in (3.10). However, since any complex wavefront can be regarded as a locally plane wave, as shown in Fig. 3.8, the idea of a plane wave is very useful in computer holography and numerical simulation in wave optics.

3.2.4 Sampled Plane Wave

Supposing that the wavefield of a plane wave at $z = z_0$ is represented by $g(x, y; z_0)$, the wavefield is written by

$$g(x, y; z_0) = A \exp\left[i(k_x x + k_y y + k_z z_0 + \phi_0)\right]. \tag{3.50}$$

To handle a wavefield in computers, the wavefield is uniformly sampled in common and represented by two dimensional array of complex amplitudes. Thus, the sampled wavefield of a plane wave in (3.50) is given by

$$g(x_m, y_n; z_0) = A \exp\left[i(k_z z_0 + \phi_0)\right] \exp\left[i(k_x m\Delta x + k_y n\Delta y)\right], \tag{3.51}$$

where Δx and Δy are sampling intervals in the x and y directions. The sampling position is given by using integer numbers of m and n as follows:

$$\begin{aligned} x_m &= m\Delta x, \\ y_n &= n\Delta y. \end{aligned} \tag{3.52}$$

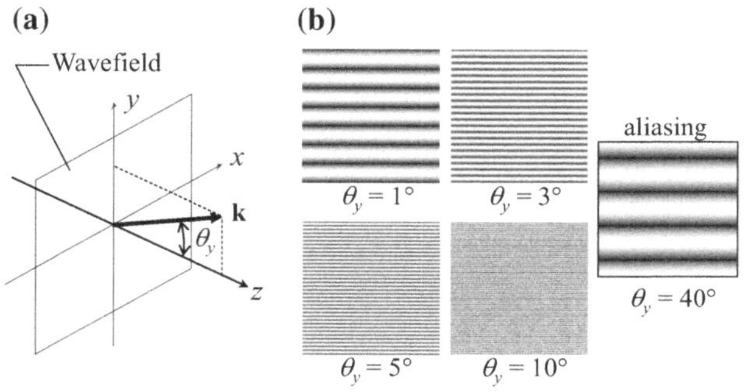

Fig. 3.9 Examples of sampled wavefields of plane waves at different angles of θ_y. $\lambda = 633$ nm, $\Delta x = \Delta y = 1\,\mu$m

Suppose that $k_x = 0$ and the angle formed by **k** and the z-axis is θ_y in the $(0, y, z)$ plane, as shown in Fig. 3.9a. Phase patterns of sampled plane wave $g(x_m, y_n; 0)$ are shown in (b) at different angles of θ_y. Here, it is assumed that the wavelength and sampling intervals are 633 nm and 1 μm, respectively. It is found that the spatial frequency in the y direction increases with increasing θ_y, but it decreases in $\theta_y = 40°$. Aliasing errors mentioned in Sect. 3.2.2 are caused in this case. Therefore, the plane wave traveling at an angle of $\theta_y = 40°$ cannot be handled by the sampled wavefield with a sampling interval of $\Delta y = 1$ [μm]. This fact suggests that there is a limit angle depending on the sampling interval of the sampled wavefield.

3.2.5 Maximum Diffraction Angle

Figure 3.10 shows the wavefront of a plane wave in the $(x, 0, z)$ plane, whose wave vector is $(k_x, 0, k_z)$. Since the wavelength along the x-axis is given by $\lambda / \sin\theta_x$, according to (3.38), the sampling interval must satisfy the following to avoid aliasing errors:

$$\Delta x < \frac{\lambda}{2\sin\theta_x}, \tag{3.53}$$

where Δx is again a sampling interval along the x-axis.

In computer holography, the sampling intervals of wavefields are usually determined before starting the calculation. Therefore, the sampling intervals place a constraint on the direction of plane waves in calculation. When a wavefield is sampled at intervals of Δx and Δy in the x and y directions, respectively, the constraints are given by:

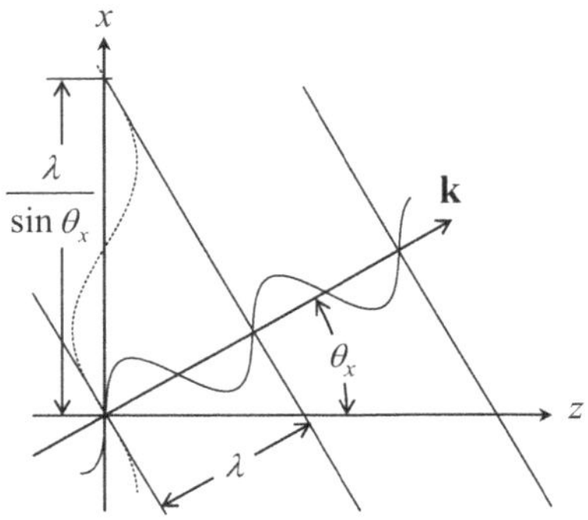

Fig. 3.10 The wavefront of a plane wave traveling at an angle of θ_x in the $(x, 0, z)$ plane

$$\theta_{\max,x} = \sin^{-1}\left(\frac{\lambda}{2\Delta x}\right),$$
$$\theta_{\max,y} = \sin^{-1}\left(\frac{\lambda}{2\Delta y}\right), \tag{3.54}$$

where $\theta_{\max,x}$ and $\theta_{\max,y}$ are maximum angles so that the plane wave can be represented as the sampled wavefield without any aliasing errors. These are called *maximum diffraction angles*.

The maximum diffraction angle is very important idea in computer holography and simulation of wave optics, because any light wave is locally regarded as a plane wave, as shown in Fig. 3.8. If any part of the light wave exceeds the maximum diffraction angle imposed by the sampling interval of the wavefield, aliasing errors may occur in the computation and ruin the whole output or at least a part of it.

3.2.6 More Rigorous Discussion on Maximum Diffraction Angle

The maximum diffraction angles (3.54) are defined in cases where either $k_x = 0$ or $k_y = 0$. If both components of the wave vector have a non-zero value, the maximum angles given in (3.54) are not accurate. We need more rigorous formulation in this case.

Angles θ_x and θ_y are defined in $(x, 0, z)$ and $(0, y, z)$ planes, respectively, as shown in Fig. 3.11. Direction angles α, β and γ in Fig. 3.7 are also depicted in Fig. 3.11. We can derive the following relations among these angles.

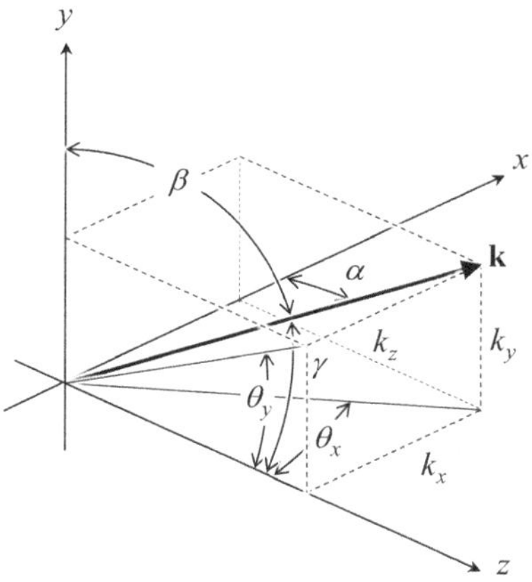

Fig. 3.11 Definition of propagation angles θ_x and θ_y, and their relation to the direction angles

$$\sin\theta_x = \frac{k_x}{\sqrt{k_x^2+k_z^2}} = \frac{\cos\alpha}{\sqrt{\cos^2\alpha+\cos^2\gamma}} = \frac{\cos\alpha}{|\sin\beta|},$$
$$\sin\theta_y = \frac{\cos\beta}{|\sin\alpha|}, \tag{3.55}$$

where formula $\cos^2\alpha+\cos^2\beta+\cos^2\gamma=1$ is used.

Spatial frequencies along the x and y directions are given by $k_x/(2\pi)$ and $k_y/(2\pi)$, respectively. Therefore, substituting (3.45) into these, it is found that the sampling intervals of the wavefield must satisfy the following relation to avoid aliasing errors:

$$\Delta x < \frac{\lambda}{2\cos\alpha} \quad \text{and} \quad \Delta y < \frac{\lambda}{2\cos\beta}. \tag{3.56}$$

Furthermore, substituting $\cos\alpha$ and $\cos\beta$ of (3.55) into above, relations in (3.56) are rewritten as

$$\Delta x < \frac{\lambda}{2\sin\theta_x}\frac{1}{|\sin\beta|} \quad \text{and} \quad \Delta y < \frac{\lambda}{2\sin\theta_y}\frac{1}{|\sin\alpha|}. \tag{3.57}$$

These are more rigorous version of (3.53), which allows $k_x \neq 0$ and/or $k_y \neq 0$. The constraint of the sampling intervals, in this case, is relaxed in comparison with (3.53), because $|\sin\alpha|^{-1}$ and $|\sin\beta|^{-1}$ are always not less than unity; that is, allowable sampling intervals are larger than that in (3.53).

The maximum diffraction angles obtained from (3.57) are

$$\theta_{\mathrm{rig},x} = \sin^{-1}\left(\frac{\lambda}{2\Delta x}\frac{1}{|\sin\beta|}\right) \quad \text{and} \quad \theta_{\mathrm{rig},y} = \sin^{-1}\left(\frac{\lambda}{2\Delta y}\frac{1}{|\sin\alpha|}\right). \tag{3.58}$$

Since $|\sin\beta|^{-1} \geq 1$ and $|\sin\alpha|^{-1} \geq 1$, the following relations are always satisfied:

$$\theta_{\text{max},x} \leq \theta_{\text{rig},x} \quad \text{and} \quad \theta_{\text{max},y} \leq \theta_{\text{rig},y}. \tag{3.59}$$

This means that the maximum diffraction angles given by (3.54) do not provide absolute limit angles, but gives absolutely safe angles. In other words, it is allowed for a diffraction angle to exceed $\theta_{\text{max},x}$ and $\theta_{\text{max},y}$ in many cases. If the diffraction angle of a field keeps under $\theta_{\text{max},x}$ and $\theta_{\text{max},y}$, the field will never cause aliasing errors.

In conclusion, the maximum diffraction angles $\theta_{\text{max},x}$ and $\theta_{\text{max},y}$, derived in the preceding section, give criteria of diffraction angles, which allows that a given sampled wavefield can handle the light in absolute safety.

3.3 Spherical Wave

A point source of light is a wave source where the wave is emitted from the point and spreads out like ripples in water surfaces. The wave is called a *spherical wave* or *spherical field*.

3.3.1 Wave Equation and Solution

When we use a spherical coordinate system (r, θ, ϕ) for the three-dimensional space, the radial distance r is written by:

$$r = |\mathbf{r}| = \sqrt{x^2 + y^2 + z^2}. \tag{3.60}$$

Thus, the first and second derivatives with respect to x are given by:

$$\frac{\partial}{\partial x} = \frac{\partial r}{\partial x}\frac{\partial}{\partial r} = \frac{x}{r}\frac{\partial}{\partial r}$$
$$\frac{\partial^2}{\partial x^2} = \left(\frac{1}{r} - \frac{x^2}{r^3}\right)\frac{\partial}{\partial r} + \left(\frac{x}{r}\right)^2\frac{\partial^2}{\partial r^2}.$$

Thus, the wave equation in (3.10) is rewritten by:

$$\frac{1}{r}\frac{\partial^2}{\partial r^2}ru(\mathbf{r}, t) = \frac{1}{c^2}\frac{\partial^2 u(\mathbf{r}, t)}{\partial t^2}. \tag{3.61}$$

A solution of this differential equation has the form of:

$$u(\mathbf{r}, t) = \frac{1}{r}f(r \pm ct),$$

where c is again speed of the wave and the minus and plus signs give an expanding and contracting wave, respectively. A well-known monochromatic solution is

$$u(\mathbf{r}, t) = \frac{A}{r} \cos(kr \pm \omega t + \phi_0), \tag{3.62}$$

where A is a constant and k and ϕ_0 are again the wave number and initial phase. Supposing that the phase factor is constant ϕ', i.e., $kr \pm \omega t + \phi_0 \equiv \phi'$, and substituting (3.60), we get

$$x^2 + y^2 + z^2 = \frac{\phi' - \phi_0 \mp \omega t}{k}. \tag{3.63}$$

The right-hand side is constant with respect to spatial coordinates. This means that the equiphase surface forms a sphere. Therefore, this solution is called a *spherical wave*. The radius increases in time if the double sign in (3.62) takes a minus. Thus, the spherical wave expands in the minus sign of (3.62) (i.e., plus sign in (3.63)), while the field contracts in the plus sign. The spherical wave plays an important role in field rendering based on the point-based method mentioned in Sect. 2.4.3.1. Helmholtz equation in (3.30) is also rewritten in a spherical coordinates system as follows:

$$\frac{\partial^2}{\partial r^2} r g(\mathbf{r}) + k^2 r g(\mathbf{r}) = 0. \tag{3.64}$$

The solution is given by:

$$g(\mathbf{r}) = \frac{A}{r} \exp\left[\pm i(kr + \phi_0)\right]. \tag{3.65}$$

The meanings of the double sign is a little different from that in (3.62), because $g(\mathbf{r})$ is time-independent. However, we make it rule to take the plus and minus sign for an expanding and contracting wave, respectively. The reason is given in the following section.

3.3.2 Spherical Wavefield and Approximation

The wavefield of a spherical wave given in (3.65) is represented in a plane (x, y, z_0) orthogonal to the z-axis as follows:

$$g(x, y; z_0) = \frac{A}{r} \exp\left[\pm i\,(kr + \phi_0)\right], \tag{3.66}$$

$$r = \sqrt{x^2 + y^2 + z_0^2}, \tag{3.67}$$

where z_0 is a distance between the plane of the wavefield and the origin of spatial coordinates, and $z_0 > 0$. An examples of the wavefield is shown in Fig. 3.12a. The field forms many concentric circles, because the spherical equiphase surface is a circle in the wavefield plane.

When the distance z_0 satisfies a condition $x^2 + y^2 \ll z_0^2$, the following gives a good approximation.

$$g(x, y; z_0) \simeq \frac{A}{z_0} \exp\left[\pm i\,(kr + \phi_0)\right]. \tag{3.68}$$

Furthermore, if z_0 is sufficiently large, $1/z_0$ can be regarded as a constant. Thus, the spherical wave is sometimes represented as:

$$g(x, y; z_0) \simeq A' \exp\left[\pm i\,(kr + \phi_0)\right]. \tag{3.69}$$

Because $x^2 + y^2 \ll z_0^2$, the distance in the exponent can also be approximated by

$$r \simeq z_0 + \frac{(x^2 + y^2)}{2z_0}. \tag{3.70}$$

Therefore, a spherical wave is approximately given by:

$$g(x, y; z_0) \simeq A'' \exp\left[\pm i \left\{\frac{k}{2z_0}\left(x^2 + y^2\right) + \phi_0\right\}\right]. \tag{3.71}$$

Here, coefficient $A'' = A' \exp(\pm ikz_0)$ includes a double sign, but let us consider the coefficient as a simple constant. In this case, the radius of the equiphase circle $\sqrt{x^2 + y^2}$ in the wavefield plane increases with increasing z_0 if we take the plus sign in (3.71), i.e., the spherical field expands with increasing z_0 in the plus sign. Because we always assume that light travels in the positive z direction, the spherical field expands in the positive sigh, and vice versa.

3.3.3 Sampled Spherical Wavefield and Sampling Problem

In practice, the wavefields in computer holography are always sampled wavefields. A sampled spherical wavefield is written by:

$$g(x_m, y_n; z_0) = \frac{A}{r_{m,n}} \exp\left[\pm i\left(kr_{m,n} + \phi_0\right)\right], \tag{3.72}$$

$$r_{m,n} \equiv \sqrt{x_m^2 + y_n^2 + z_0^2}\,, \tag{3.73}$$

where sampled coordinates x_m and y_m are given by (3.52).

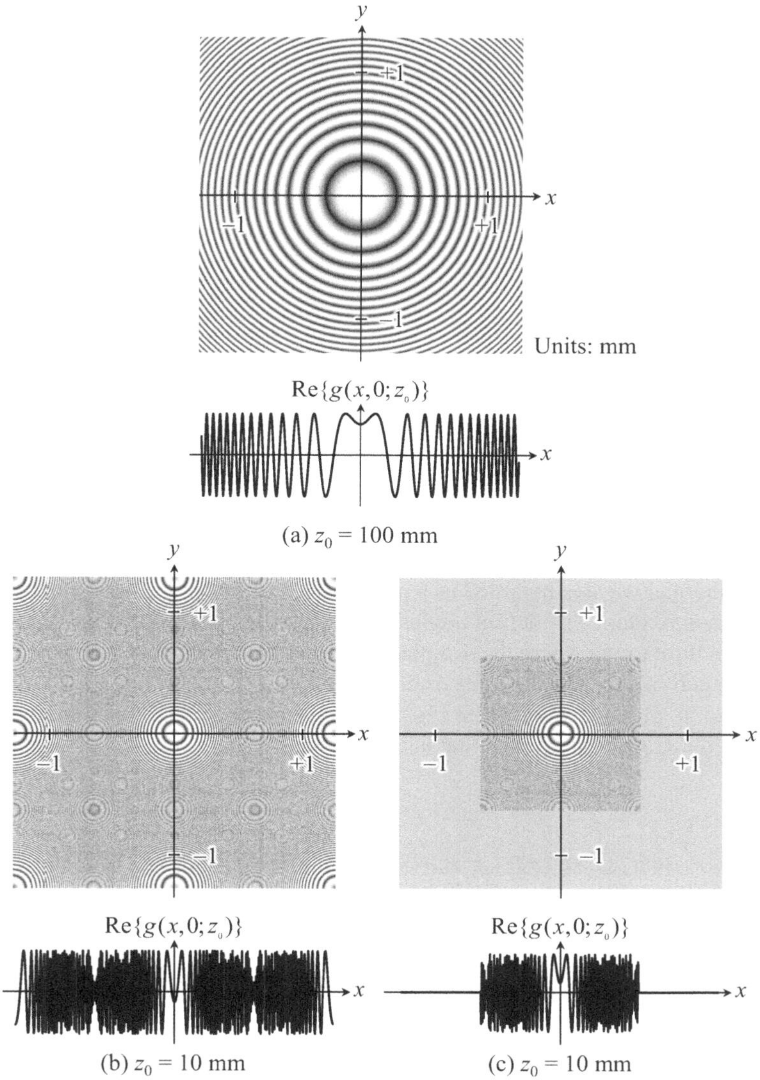

Fig. 3.12 Examples of sampled spherical wavefields. $\lambda = 633$ nm, $\Delta x = \Delta y = 5\,\mu$m. The encoding gamma for grayscale images is 1/2.2

Figure 3.12 shows examples of sampled spherical wavefields. Both sampling intervals in the x and y directions are $5\,\mu$m; $\Delta x = \Delta y = 5\,[\mu\text{m}]$. The curves depicted below the field images show lateral field strength along the x-axis. As in (a), the field takes the lowest spatial frequency in the origin and gradually increases as the position is distant from the origin. It is also verified that smaller z_0 generally gives

Fig. 3.13 A local plane wave as approximation of the spherical wave

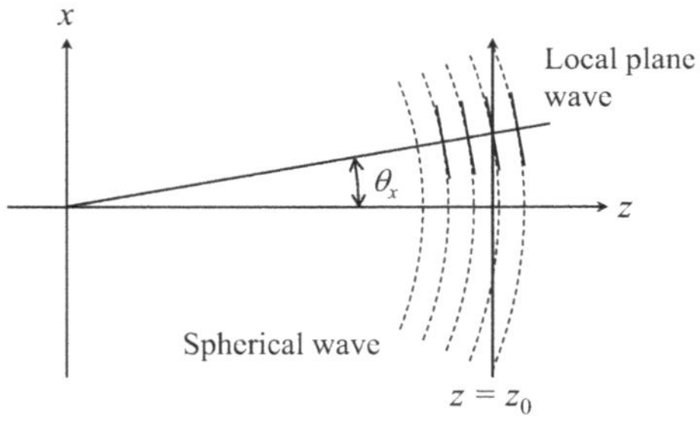

the higher frequency as in (b). In this case, the sampled field actually causes aliasing errors. To avoid the aliasing errors, a spherical wavefield must be limited to a given area as in (c).

3.3.3.1 Rectangular Aliasing-Free Area Based on Maximum Diffraction Angle

A local spherical wave is regarded as a plane wave, as shown in Fig. 3.13. Thus, the idea of the maximum diffraction angles, mentioned in Sect. 3.2.5, can be used for obtaining a limit range where the sampled spherical field does not cause any aliasing errors. We call this an *aliasing-free area* in this book.

In the $(x, 0, z)$ plane, angle θ_x in Fig. 3.13 must not exceed the maximum diffraction angle for x, given in (3.54); that is,

$$\theta_x < \theta_{\max,x} \,. \tag{3.74}$$

Here, suppose that $\Delta x > \lambda/2$, i.e., the sampling interval is not less than a half wavelength. Substituting $\sin\theta_x = x/(x^2 + z_0^2)^{1/2}$ given by the geometry and (3.54) into above, we get the following relation after several steps.

$$|x| < \frac{z_0}{\sqrt{\left(\dfrac{2\Delta x}{\lambda}\right)^2 - 1}}. \tag{3.75}$$

Because the same relation is independently derived in the $(0, x, y)$ plane, a spherical wavefield without aliasing errors is represented by:

$$g(x_m, y_n; z_0) = \frac{A}{r_{m,n}} \exp\left[\pm i \left(k r_{m,n} + \phi_0\right)\right] \times \mathrm{rect}\left(\frac{x_m}{2x_{\max}}\right) \mathrm{rect}\left(\frac{y_n}{2y_{\max}}\right), \tag{3.76}$$

$$x_{\max} = z_0 \left(q_x^2 - 1\right)^{-1/2},$$
$$y_{\max} = z_0 \left(q_y^2 - 1\right)^{-1/2}, \tag{3.77}$$

where rect(ξ) is the rectangle function given in (4.18) and Fig. 4.1 (see Sect. 4.2.3.3). Sampling factors q_x and q_y are defined as follows:

$$q_x = \frac{2\Delta x}{\lambda} \quad \text{and} \quad q_y = \frac{2\Delta y}{\lambda}. \tag{3.78}$$

Note that the sampling factors q_x and q_y are always more than unity, because we assume $\Delta x > \lambda/2$.

The shape of the safety area, where a sampled spherical field does not produce aliasing errors, is a square whose size is $2x_{\text{max}} \times 2y_{\text{max}}$. Figure 3.14 illustrates the rectangular aliasing-free area in the case where $q_x = 2$ and $q_x = 4$. This rectangle is just an approximate safety area but gives a practical criterion enough to be used in computer holography, as described in the next section.

3.3.3.2 Exact Aliasing-Free Area Based on Local Spatial Frequency

The aliasing-free area obtained from the maximum diffraction angles is an approximation of the strict aliasing-free area. Supposing that a spherical field is represented by

$$\begin{aligned} g(x, y; z_0) &= A \exp\left[i\phi(x, y)\right] \\ \phi(x, y) &= \pm kr + \phi_0, \end{aligned} \tag{3.79}$$

local spatial frequencies of the spherical wavefield are given as follows [19]:

$$\begin{aligned} u &= \frac{1}{2\pi}\frac{\partial\phi}{\partial x} = \pm\frac{kx}{2\pi r}, \\ v &= \frac{1}{2\pi}\frac{\partial\phi}{\partial y} = \pm\frac{ky}{2\pi r}, \end{aligned} \tag{3.80}$$

where r is given in (3.67). According to the sampling theorem, both local frequencies must satisfy

$$|u| < \frac{1}{2\Delta x} \quad \text{and} \quad |v| < \frac{1}{2\Delta y}. \tag{3.81}$$

Substituting (3.80) into (3.81), the area is written by

$$(q_x^2 - 1)\left(\frac{x}{z_0}\right)^2 - \left(\frac{y}{z_0}\right)^2 < 1, \tag{3.82}$$

$$-\left(\frac{x}{z_0}\right)^2 + (q_y^2 - 1)\left(\frac{y}{z_0}\right)^2 < 1, \tag{3.83}$$

where relations (3.82) and (3.83) give the limits in x and y coordinates, respectively.

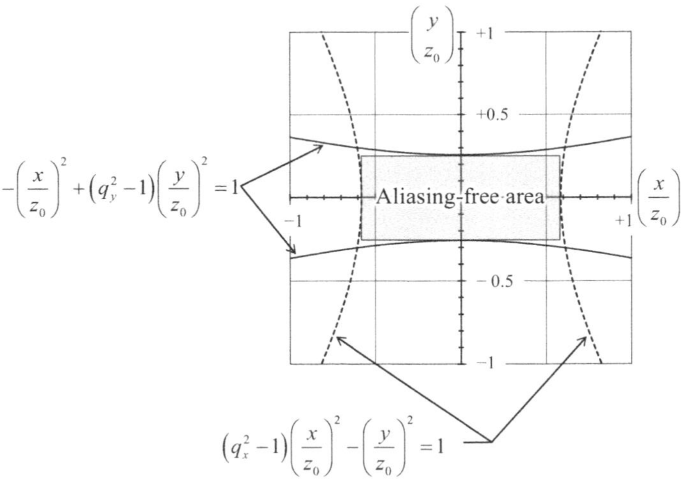

Fig. 3.14 Examples of an approximate rectangular aliasing-free area and the strict safety area in the case where $q_x = 2$ and $q_y = 4$

Figure 3.14 shows an example of the limits in the case where $\Delta x = \lambda$ and $\Delta y = 2\lambda$. Both borders of the limits in x and y coordinates represent a hyperbolic curve. Furthermore, when $y = 0$ in (3.82), the relation is rewritten as

$$|x| < \frac{z_0}{\sqrt{q_x^2 - 1}}. \tag{3.84}$$

This agrees with the relation (3.75) given by the maximum diffraction angle in the $(x, 0, z)$ plane. The relation (3.83) in the case where $x = 0$ also agrees with that by the maximum diffraction angle in the $(0, y, z)$ plane. Therefore, the rectangular aliasing-free area given by the maximum diffraction angles is an inscribed rectangle of the strict safety area, as shown in Fig. 3.14.

Thus, we can conclude that a spherical wavefield given by (3.76) is aliasing-free and the rectangular area gives a good approximation of the strict safety area.

3.4 Optical Intensity of Electromagnetic Wave

The phase of a light wave cannot be observed or detected directly. Only the intensity of light is perceived by human eyes or detected by cameras, sensors and light-sensitive material. Let us consider monochromatic fields given by:

$$\mathbf{E}(\mathbf{r}, t) = \mathrm{Re}\left\{\mathbf{E}(\mathbf{r})\exp(-i\omega t)\right\}, \tag{3.85}$$

$$\mathbf{H}(\mathbf{r}, t) = \mathrm{Re}\left\{\mathbf{H}(\mathbf{r})\exp(-i\omega t)\right\}. \tag{3.86}$$

Substituting these into (3.11) of Maxwell's equation, we get

$$\nabla \times \mathbf{E}(\mathbf{r}) = i\omega\mu_0\mathbf{H}(\mathbf{r}). \tag{3.87}$$

In addition, assume that the electric wave is a plane wave:

$$\mathbf{E}(\mathbf{r}) = \mathbf{E}_0 \exp\left[i(\mathbf{k}\cdot\mathbf{r} + \phi_0)\right]. \tag{3.88}$$

Substituting above into the left-hand side of (3.87), the magnetic wave is

$$\begin{aligned}\mathbf{H}(\mathbf{r}) &= \frac{1}{\omega\mu_0}\mathbf{k}\times\mathbf{E}(\mathbf{r}) \\ &= \mathbf{H}_0 \exp\left[i(\mathbf{k}\cdot\mathbf{r} + \phi_0)\right], \end{aligned} \tag{3.89}$$

$$\mathbf{H}_0 \equiv \frac{1}{\omega\mu_0}\mathbf{k}\times\mathbf{E}_0. \tag{3.90}$$

As a result, it is found that the magnetic wave is also a plane wave and the field is orthogonal to both the electric field and wave vector.

By substituting (3.88) into (3.13) of Maxwell's equation, we find that $i\mathbf{k}\cdot\mathbf{E}_0\exp(i\mathbf{k}\cdot\mathbf{r}+\phi_0) = 0$, because ε_r is constant in free space. Hence, the wave vector is always orthogonal to the electric field, because $\mathbf{k}\cdot\mathbf{E}_0 = 0$. This means that the electric field, magnetic field and the wave vector are orthogonal to each other in free space. Because the wave vector is orthogonal to the electric field, we can easily obtain the strength of the magnetic field from (3.90) as follows:

$$|\mathbf{H}_0| = \frac{k}{\omega\mu_0}|\mathbf{E}_0|. \tag{3.91}$$

As a result, we do not have to take care about magnetic fields, because we can always obtain the strength and direction of the magnetic field in free space from the electric field whenever it is required. Thus, magnetic field is always omitted after this section in this book.

Energy density of an electromagnetic field in free space is, in general, given by:

$$w(\mathbf{r}, t) = \frac{1}{2}\left[\mathbf{E}(\mathbf{r}, t)\cdot\varepsilon_0\varepsilon_r\mathbf{E}(\mathbf{r}, t) + \mathbf{H}(\mathbf{r}, t)\cdot\mu_0\mathbf{H}(\mathbf{r}, t)\right] \tag{3.92}$$

$$= \frac{1}{2}\left[\varepsilon_0\varepsilon_r|\mathbf{E}(\mathbf{r}, t)|^2 + \mu_0|\mathbf{H}(\mathbf{r}, t)|^2\right]. \tag{3.93}$$

Substituting (3.85) and (3.86) into above equation, the time average of the energy density is given as follows:

$$\langle w(\mathbf{r},t)\rangle = \frac{1}{T}\int_0^T w(\mathbf{r},t)dt$$
$$= \frac{1}{4}\left[\varepsilon_0\varepsilon_r|\mathbf{E}(\mathbf{r})|^2 + \mu_0|\mathbf{H}(\mathbf{r})|^2\right]. \tag{3.94}$$

Using (3.91), it is found that $\mu_0|\mathbf{H}(\mathbf{r})|^2 = \varepsilon_0\varepsilon_r|\mathbf{E}(\mathbf{r})|^2$. Hence the average energy density of electromagnetic fields is given by:

$$\langle w(\mathbf{r},t)\rangle = \frac{1}{2}\varepsilon_0\varepsilon_r|\mathbf{E}(\mathbf{r})|^2. \tag{3.95}$$

Optical intensity is defined as energy of the electromagnetic wave that passes through unit area per unit time. Since light travels over a distance of c in unit time, which is the light speed, energy passing through unit area per unit time is given by $c \times \langle w(\mathbf{r},t)\rangle$. Conclusively, optical intensity is given by:

$$I(\mathbf{r}) = \frac{1}{2}c\varepsilon_0\varepsilon_r|\mathbf{E}(\mathbf{r})|^2. \tag{3.96}$$

Units of optical intensity is $\mathrm{W/m^2}$ in standard units according to the definition. However, $\mathrm{W/cm^2}$ is sometimes used as conventional units.

In scalar waves, we use the following simplified version for the optical intensity in this book:

$$I(\mathbf{r}) = |g(\mathbf{r})|^2 = A^2(\mathbf{r}). \tag{3.97}$$

Strictly writing, the scalar field in this case is given by:

$$g(\mathbf{r}) = \sqrt{\frac{c\varepsilon_0\varepsilon_r}{2}}E(\mathbf{r}), \tag{3.98}$$

where $E(\mathbf{r})$ is any one of three components of the electric field in (3.18), or total strength of the electric field. Note that optical intensity and definition of the scalar field are given as follows in cases where the property of medium is significant:

$$I(\mathbf{r}) = \frac{1}{2}nc_0\varepsilon_0|\mathbf{E}(\mathbf{r})|^2, \tag{3.99}$$

$$g(\mathbf{r}) = \sqrt{\frac{nc_0\varepsilon_0}{2}}E(\mathbf{r}), \tag{3.100}$$

where n is again the refractive index of the medium as in (3.24).

In case of a plane wave, since the amplitude is constant as in (3.48), the optical intensity is also constant, i.e.,

$$I(\mathbf{r}) = A^2(\mathbf{r}) \equiv A^2. \tag{3.101}$$

Therefore, let us emphasize that if we observed a plane wave we could not detect any patterns therein.

Chapter 4
The Fourier Transform and Mathematical Preliminaries

Abstract The Fourier transform plays an important role in wave optics and computer holography. Thus, the Fourier transform and related mathematical topics such as convolution, correlation, sampling theorem, symmetry relations, and so forth are introduced in this chapter. These are also preliminaries to discussion on diffraction and field propagation. As with the preceding chapter, we put importance on the discrete form of the Fourier transform, DFT/FFT. The technique to make use of existing FFT libraries is presented for actual computation in wave optics and computer holography.

4.1 Introduction

The Fourier transform plays a great role to handle field propagation in wave optics. This section summarizes its theorems, formulas and mathematical preliminaries related to the Fourier transform briefly. The discrete Fourier transform and fast Fourier transform as well as sampling theory are also mentioned in this chapter. The reader familiar with these topics is recommended to skip this chapter.

4.2 The Fourier Transform of Continuous Function

4.2.1 Definition

We can find many forms of the Fourier transform formula in related references. In this book, one and two dimensional Fourier transforms of a function are defined as

$$\begin{aligned} F(u) &= \mathcal{F}\{f(x)\} \\ &= \int_{-\infty}^{+\infty} f(x)\exp[-i2\pi ux]\mathrm{d}x, \qquad (4.1) \\ F(u,v) &= \mathcal{F}\{f(x,y)\} \end{aligned}$$

K. Matsushima, *Introduction to Computer Holography*,
Series in Display Science and Technology,
https://doi.org/10.1007/978-3-030-38435-7_4

$$= \iint_{-\infty}^{+\infty} f(x, y) \exp[-i2\pi(ux + vy)] \mathrm{d}x\mathrm{d}y, \tag{4.2}$$

where u and v are called Fourier frequencies with respect to x and y, respectively. $F(u)$ and $F(u, v)$ are called a *Fourier spectrum* or simply a *spectrum* of $f(x)$ and $f(x, y)$. The inverse Fourier transform is defined as

$$\begin{aligned} f(x) &= \mathscr{F}^{-1}\{F(u)\} \\ &= \int_{-\infty}^{+\infty} F(u) \exp[i2\pi ux] \mathrm{d}u. \end{aligned} \tag{4.3}$$

$$\begin{aligned} f(x, y) &= \mathscr{F}^{-1}\{F(u, v)\} \\ &= \iint_{-\infty}^{+\infty} F(u, v) \exp[i2\pi(ux + vy)] \mathrm{d}u\mathrm{d}v. \end{aligned} \tag{4.4}$$

Here, these transforms satisfy $f(x) = \mathscr{F}^{-1}\{\mathscr{F}\{f(x)\}\}$ and $f(x, y) = \mathscr{F}^{-1}\{\mathscr{F}\{f(x, y)\}\}$. $f(x)$ and $F(u)(f(x, y)$ and $F(u, v))$ are called a *Fourier transform pair*.

4.2.2 Theorems

There are several useful theorems of the Fourier transform. Here, a and b are constants. Some of these are shown below, and summarized in Table 4.1

Linear theorem

Supposing that

$$F_1(u, v) = \mathscr{F}\{f_1(x, y)\} \text{ and } F_2(u, v) = \mathscr{F}\{f_2(x, y)\}, \tag{4.5}$$

the following is satisfied:

$$\mathscr{F}\{af_1(x, y) + bf_2(x, y)\} = aF_1(u, v) + bF_2(u, v). \tag{4.6}$$

Similarity theorem

A function $f(ax, by)$ commonly represents expansion and contraction of the function $f(x, y)$. The Fourier transform is given by

$$\mathscr{F}\{f(ax, by)\} = \frac{1}{|ab|} F\left(\frac{u}{a}, \frac{v}{b}\right). \tag{4.7}$$

This means that expansion $(a, b < 1)$ and contraction $(a, b > 1)$ of the function $f(x, y)$ cause the opposite effect to the Fourier spectrum.

Table 4.1 Useful Fourier formulas

$f(x) = \int_{-\infty}^{+\infty} F(u) \exp[i2\pi ux]\mathrm{d}u$	$F(u) = \int_{-\infty}^{+\infty} f(x) \exp[-i2\pi ux]\mathrm{d}x$
$f(x, y) = \iint_{-\infty}^{+\infty} F(u, v) \exp[i2\pi(ux + vy)]\mathrm{d}u\mathrm{d}v$	$F(u, v) = \iint_{-\infty}^{+\infty} f(x, y) \exp[-i2\pi(ux + vy)]\mathrm{d}x\mathrm{d}y$
$f(\pm x)$	$F(\pm u)$
$f(\pm x, \pm y)$	$F(\pm u, \pm v)$
$f^*(\pm x)$	$F^*(\mp u)$
$f^*(\pm x, \pm y)$	$F^*(\mp u, \mp v)$
$F(\pm x)$	$f(\mp u)$
$f(ax)$	$\frac{1}{\lvert a\rvert} F\left(\frac{u}{a}\right)$
$f(ax, bx)$	$\frac{1}{\lvert ab\rvert} F\left(\frac{u}{a}, \frac{v}{b}\right)$
$f(x \pm x_0)$	$F(u) \exp[\pm i2\pi x_0 u]$
$f(x \pm x_0, y \pm y_0)$	$F(u, v) \exp[i2\pi(\pm x_0 u \pm y_0 v)]$
$f(x) \exp[=i2\pi u_0 x]$	$F(u \pm u_0)$
$f(x, y) \exp[i2\pi(\mp u_0 x \mp v_0 y)]$	$F(u \pm u_0, v \pm v_0)$

Shift theorem

Shift of function $f(x, y)$ results in the following spectrum:

$$\mathscr{F}\{f(x - a, y - b)\} = F(u, v) \exp[-i2\pi(au + bv)]. \tag{4.8}$$

4.2.3 Several Useful Functions and Their Fourier Transform

4.2.3.1 Dirac Delta Function

The Dirac delta function is a generalized function and have a nature:

$$\delta(x) = \begin{cases} \infty & x = 0 \\ 0 & x \neq 0 \end{cases} \tag{4.9}$$

This function has useful nature in integration as follows

$$\int_{-\infty}^{+\infty} \delta(x) = 1, \tag{4.10}$$

$$\int_{-\infty}^{+\infty} f(x)\delta(x) = f(0), \tag{4.11}$$

$$\int_{-\infty}^{+\infty} f(x)\delta(x - a) = f(a). \tag{4.12}$$

This function is also given by the Fourier transform of unity as follows

$$\delta(u) = \mathscr{F}\{1\} = \int_{-\infty}^{+\infty} \exp[-i2\pi ux]\mathrm{d}x. \tag{4.13}$$

Thus, both the Fourier and inverse Fourier transform of the Dirac delta function are unity.

$$\mathscr{F}\{\delta(x)\} = \mathscr{F}^{-1}\{\delta(u)\} = 1. \tag{4.14}$$

4.2.3.2 Trigonometric Function

The Fourier transform of trigonometric functions is represented by combination of the Dirac delta functions.

$$\mathscr{F}\{\exp[\pm i2\pi u_0 x]\} = \delta(u \mp u_0) \tag{4.15}$$

$$\mathscr{F}\{\sin(2\pi u_0 x)\} = \frac{i}{2}[\delta(u+u_0) - \delta(u-u_0)] \tag{4.16}$$

$$\mathscr{F}\{\cos(2\pi u_0 x)\} = \frac{1}{2}[\delta(u+u_0) + \delta(u-u_0)] \tag{4.17}$$

4.2.3.3 Rectangle Function and Sinc Function

The one dimensional rectangle function is given by

$$\mathrm{rect}(x) = \begin{cases} 1 & \text{if } |x| < \frac{1}{2} \\ \frac{1}{2} & \text{if } |x| = \frac{1}{2} \\ 0 & \text{otherwise} \end{cases} \tag{4.18}$$

Two dimensional rectangle function is a separable function, and thus simply written by

$$\mathrm{rect}(x, y) = \mathrm{rect}(x) \times \mathrm{rect}(y). \tag{4.19}$$

The Fourier transform of the one and two dimensional rectangle functions is given by

$$\mathscr{F}\{\mathrm{rect}(x)\} = \mathrm{sinc}(u), \tag{4.20}$$

$$\mathscr{F}\{\mathrm{rect}(x, y)\} = \mathrm{sinc}(u)\mathrm{sinc}(v), \tag{4.21}$$

where $\mathrm{sinc}(u)$ is called a *sinc* function. This is defined as follows:

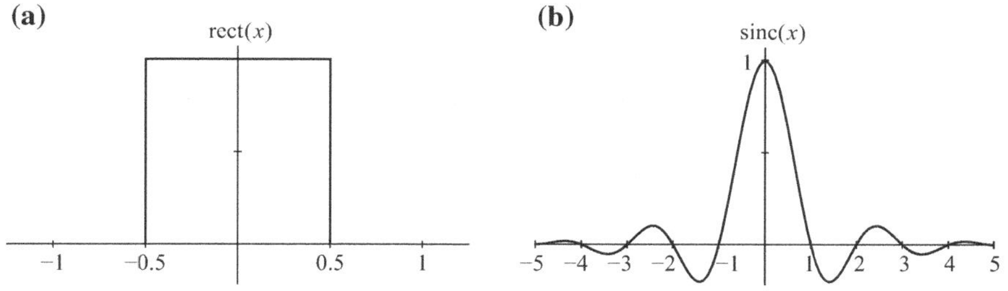

Fig. 4.1 The shapes of **a** the rectangle and **b** sinc function

Table 4.2 Useful Fourier pairs

$f(x) = \int_{-\infty}^{+\infty} F(u)\exp[i2\pi ux]\mathrm{d}u$	$F(u) = \int_{-\infty}^{+\infty} f(x)\exp[-i2\pi ux]\mathrm{d}x$
$f(x, y) = \iint_{-\infty}^{+\infty} F(u, v)\exp[i2\pi(ux + vy)]\mathrm{d}u\mathrm{d}v$	$F(u, v) = \iint_{-\infty}^{+\infty} f(x, y)\exp[-i2\pi(ux + vy)]\mathrm{d}x\mathrm{d}y$
$\delta(x), \delta(x, y)$	1
1	$\delta(u), \delta(u, v)$
$\delta(x \pm x_0)$	$\exp[\pm i2\pi x_0 u]$
$\delta(x \pm x_0, x \pm y_0)$	$\exp[i2\pi(\pm x_0 u \pm y_0 v)]$
$\exp[\pm i2\pi u_0 x]$	$\delta(u \mp u_0)$
$\exp[\pm i2\pi u_0 x \pm i2\pi v_0 y]$	$\delta(u \mp u_0, v \mp v_0)$
$\sin(2\pi u_0 x)$	$\frac{i}{2}[\delta(u + u_0) - \delta(u - u_0)]$
$\cos(2\pi u_0 x)$	$\frac{1}{2}[\delta(u + u_0) + \delta(u - u_0)]$
$\mathrm{rect}(x)$	$\mathrm{sinc}(u)$
$\mathrm{rect}(x, y)$	$\mathrm{sinc}(u)\mathrm{sinc}(v)$
$\exp\left[i\pi x^2\right]$	$\exp\left[i\left(\frac{\pi}{4}\right)\right]\exp\left[-i\pi u^2\right]$
$\exp\left[i\pi(x^2 + y^2)\right]$	$i\exp\left[-i\pi(u^2 + v^2)\right]$

$$\mathrm{sinc}(x) = \frac{\sin(\pi x)}{\pi x}. \tag{4.22}$$

The Fourier transform of a sinc function is again a rect function, i.e.,

$$\mathscr{F}\{\mathrm{sinc}(x)\} = \mathrm{rect}(u), \tag{4.23}$$

$$\mathscr{F}\{\mathrm{sinc}(x)\mathrm{sinc}(y)\} = \mathrm{rect}(u, v). \tag{4.24}$$

Figure 4.1 shows the shapes of the rectangle and sinc functions.

Some useful Fourier pairs including above are summarized in Table 4.2.

4.3 Symmetry Relation of Function

4.3.1 Even Function and Odd Function

The following two functions can be always defined for any real-valued function $f(x)$:

$$f_e(x) \equiv \frac{f(x) + f(-x)}{2}, \tag{4.25}$$

$$f_o(x) \equiv \frac{f(x) - f(-x)}{2}. \tag{4.26}$$

These functions definitely satisfy the following symmetry relations.

$$f_e(-x) = f_e(x), \tag{4.27}$$

$$f_o(-x) = -f_o(x). \tag{4.28}$$

A function having the symmetry represented in (4.27) is called an *even function*, while a function represented in (4.28) is called an *odd function*. Solving (4.25) and (4.26) for $f(x)$, we easily get

$$f(x) = f_e(x) + f_o(x). \tag{4.29}$$

This means that any function can be decomposed into even and odd functions. Infinite integration of these functions has natures:

$$\int_{-\infty}^{+\infty} f_e(x)\mathrm{d}x = 2\int_{0}^{+\infty} f_e(x)\mathrm{d}x, \tag{4.30}$$

$$\int_{-\infty}^{+\infty} f_o(x)\mathrm{d}x = 0. \tag{4.31}$$

These symmetric relations can be extended to a function of two variables. The point symmetry and anti-symmetry of a two-dimensional real-valued function are given by

$$f_e(-x, -y) = f_e(x, y), \tag{4.32}$$

$$f_o(-x, -y) = -f_o(x, y). \tag{4.33}$$

Any two dimensional real-valued function can also be decomposed into these symmetric functions.

$$f(x, y) = f_e(x, y) + f_o(x, y). \tag{4.34}$$

Infinite integration has natures similar to those in the one-dimensional function:

$$\int_{-\infty}^{+\infty}\int_{-\infty}^{+\infty} f_e(x, y)\mathrm{d}x\mathrm{d}y = 2\int_0^{+\infty}\int_0^{+\infty} [f_e(x, y) + f_e(x, -y)]\,\mathrm{d}x\mathrm{d}y, \tag{4.35}$$

$$\int_{-\infty}^{+\infty}\int_{-\infty}^{+\infty} f_o(x, y)\mathrm{d}x\mathrm{d}y = 0. \tag{4.36}$$

4.3.2 Symmetry Relations in the Fourier Transform

The Fourier transform of (4.29) is written by

$$\begin{aligned}\mathscr{F}\{f(x)\} &= \int_{-\infty}^{\infty} [f_e(x) + f_o(x)]\exp(-2\pi ux)\mathrm{d}x \\ &= 2\int_0^{\infty} [f_e(x)\cos(2\pi ux) - i f_o(x)\sin(2\pi ux)]\,\mathrm{d}x \\ &= F_e(u) + i F_o(u),\end{aligned} \tag{4.37}$$

where

$$F_e(u) = 2\int_0^{\infty} f_e(x)\cos(2\pi ux)\mathrm{d}x, \tag{4.38}$$

$$F_o(u) = -2\int_0^{\infty} f_o(x)\sin(2\pi ux)\mathrm{d}x. \tag{4.39}$$

Because it is apparent that $F_e(-u) = F_e(u)$ and $F_o(-u) = -F_o(u)$ in the above, $F_e(u)$ and $F_o(u)$ are even and odd functions, respectively. Accordingly, real and imaginary parts of $\mathscr{F}\{f(x)\}$ are even and odd functions, respectively. A function having this symmetry relation is called a *Hermitian* function. Hermitian function $g_h(x)$ satisfies

$$g_h(-x) = g_h^*(x), \tag{4.40}$$

and thus, the absolute function $|g_h(x)|$ is always an even function.

In two-dimensional real-valued functions, Using (4.35) and (4.36), the Fourier transform is rewritten by

$$\begin{aligned}F(u, v) &= \mathscr{F}\{f(x, y)\} \\ &= \int_{-\infty}^{+\infty}\int_{-\infty}^{+\infty} [f_e(x, y) + f_o(x, y)]\exp[-2\pi(ux + vy)]\,\mathrm{d}x\mathrm{d}y \\ &= 2\int_0^{\infty}\int_0^{\infty} [f_e(x, y)\cos\{2\pi(ux + vy)\} + f_e(x, -y)\cos\{2\pi(ux - vy)\}]\,\mathrm{d}x\mathrm{d}y \\ &\quad -2i\int_0^{\infty}\int_0^{\infty} [f_o(x, y)\sin\{2\pi(ux + vy)\} + f_o(x, -y)\sin\{2\pi(ux - vy)\}]\,\mathrm{d}x\mathrm{d}y \\ &= F_e(u, v) + i F_o(u, v),\end{aligned} \tag{4.41}$$

Table 4.3 Symmetric relations in the Fourier transform

$f(x) = \int_{-\infty}^{+\infty} F(u) \exp[i2\pi ux] \mathrm{d}u$	$F(u) = \int_{-\infty}^{+\infty} f(x) \exp[-i2\pi ux] \mathrm{d}x$
$f(x, y) = \iint_{-\infty}^{+\infty} F(u, v) \exp[i2\pi(ux + vy)] \mathrm{d}u\mathrm{d}v$	$F(u, v) = \iint_{-\infty}^{+\infty} f(x, y) \exp[-i2\pi(ux + vy)] \mathrm{d}x\mathrm{d}y$
Asymmetric complex function	Asymmetric complex function
Asymmetric real function	Hermitian function
Asymmetric imaginary function	Anti-Hermitian function
Hermitian function	Asymmetric real function
Anti-Hermitian function	Asymmetric imaginary function

where the Fourier transformed functions

$$F_e(u, v) = 2 \int_0^{\infty} \int_0^{\infty} [f_e(x, y) \cos\{2\pi(ux + vy)\} + f_e(x, -y) \cos\{2\pi(ux - vy)\}] \,\mathrm{d}x\mathrm{d}y, \quad (4.42)$$

$$F_o(u, v) = -2 \int_0^{\infty} \int_0^{\infty} [f_o(x, y) \sin\{2\pi(ux + vy)\} + f_o(x, -y) \sin\{2\pi(ux - vy)\}] \,\mathrm{d}x\mathrm{d}y, \quad (4.43)$$

have the same symmetries as (4.32) and (4.33), respectively. Therefore, $F(u, v)$ is also an Hermitian function. Here, two-dimensional Hermitian function $g_h(x, y)$ has a symmetry:

$$g_h(-x, -y) = g_h^*(x, y). \quad (4.44)$$

Table 4.3 summarizes several symmetric relations important for computer holography. Here, anti-Hermitian function is a function whose real part is odd but image part is even.

As a result, in the Fourier transform of any two-dimensional real-valued function, the absolute function $|F(u, v)|$ has a point symmetry with respect to the origin. This is very important to understand the conjugate image in holography.

4.4 Convolution and Correlation

Suppose that the Fourier transform of two functions is

$$F_1(u) = \mathscr{F}\{f_1(x)\},$$
$$F_2(u) = \mathscr{F}\{f_2(x)\}.$$

The inverse Fourier transform of the product of these functions is given as follows.

$$\begin{aligned}\mathscr{F}^{-1}\{F_1(u)\cdot F_2(u)\} &= \int\left[\int f_1(x')\exp[-i2\pi ux']\mathrm{d}x'\int f_2(x'')\exp[-i2\pi ux'']\mathrm{d}x''\right]\\ &\quad\times\exp[i2\pi ux]\mathrm{d}u\\ &= \iint f_1(x')f_2(x'')\left[\int\exp\left[i2\pi(x-x'-x'')u\right]du\right]\mathrm{d}x'\mathrm{d}x''\\ &= \iint f_1(x')f_2(x'')\delta(x-x'-x'')\mathrm{d}x'\mathrm{d}x''\\ &= \int f_1(x')f_2(x-x')\mathrm{d}x',\end{aligned} \tag{4.45}$$

where we used properties of the delta function given in (4.12) and (4.14). This integral is referred to as a *convolution integral*.

We represent the convolution integral by

$$f_1\otimes f_2(x)=\int_{-\infty}^{\infty}f_1(x')f_2(x-x')\mathrm{d}x', \tag{4.46}$$

and

$$f_1\otimes f_2(x,y)=\iint_{-\infty}^{\infty}f_1(x',y')f_2(x-x',y-y')\mathrm{d}x'\mathrm{d}y', \tag{4.47}$$

where symbol $\otimes$ denotes a convolution operator of two functions. Using this notation, the Fourier transform of (4.45) is rewritten as follows.

$$\mathscr{F}\{f_1\otimes f_2(x,y)\}=F_1(u,v)\cdot F_2(u,v). \tag{4.48}$$

The inverse relation is also established as follows:

$$\mathscr{F}^{-1}\{F_1\otimes F_2(u,v)\}=f_1(x,y)\cdot f_2(x,y). \tag{4.49}$$

These are commonly called a *convolution theorem*.

In the cases where one of two functions is the delta function, the convolution integral satisfies the followings.

$$f\otimes\delta(x,y)=f(x,y), \tag{4.50}$$

$$f\otimes\delta(x-a,y-b)=f(x-a,y-b), \tag{4.51}$$

where a and b are again constants.

A cross-correlation function has a very similar form to the convolution integral; cross-correlation is represented as follows:

$$f_1\star f_2(x,y)=\iint_{-\infty}^{\infty}f_1(x',y')f_2^*(x'-x,y'-y)\mathrm{d}x'\mathrm{d}y', \tag{4.52}$$

where symbol $\star$ is a correlation operator. The Fourier pairs of a cross-correlation function are given by

$$\mathscr{F}\{f_1 \star f_2(x,y)\} = F_1(u,v) \cdot F_2^*(u,v), \tag{4.53}$$

$$\mathscr{F}^{-1}\{F_1 \star F_2(u,v)\} = f_1(x,y) \cdot f_2^*(x,y). \tag{4.54}$$

4.5 Spectrum of Sampled Function and Sampling Theorem

Suppose that one dimensional function $f(x)$ shown in Fig. 4.2a is sampled at an interval of Δx, and the values are only presented at discrete positions: $x_m = m\Delta x$ as in (3.52). This sampled function can be represented by the following form:

$$f_s(x) = f(x) \times \sum_{m=-\infty}^{+\infty} \delta(x - m\Delta x), \tag{4.55}$$

where $\sum \delta(x - m\Delta x)$ is a periodic delta function whose period is Δx, as shown in left column of Fig. 4.2b. Using the convolution theorem in (4.48), the Fourier transform of the sampled function is

$$\begin{aligned} F'(u) &= \mathscr{F}\{f_s(x)\} \\ &= F(u) \otimes \mathscr{F}\left\{\sum_{n=-\infty}^{+\infty} \delta(x - m\Delta x)\right\}, \end{aligned} \tag{4.56}$$

where $F(u) = \mathscr{F}\{f(x)\}$ as in the right column of (a).

The periodic delta function can be easily expanded to Fourier series, i.e.,

$$\sum_{m=-\infty}^{+\infty} \delta(x - m\Delta x) = \frac{1}{\Delta x} + \frac{2}{\Delta x}\sum_{m=1}^{\infty} \cos\left(\frac{2\pi m x}{\Delta x}\right). \tag{4.57}$$

Hence, the Fourier transform is obtained from Table 4.2 as follows:

$$\begin{aligned} \mathscr{F}\left\{\sum_{m=-\infty}^{+\infty} \delta(x - m\Delta x)\right\} &= \frac{\delta(u)}{\Delta x} + \frac{2}{\Delta x}\sum_{m=1}^{\infty} \frac{1}{2}\left[\delta\left(u - \frac{m}{\Delta x}\right) + \delta\left(u + \frac{m}{\Delta x}\right)\right] \\ &= \frac{1}{\Delta x}\sum_{m=-\infty}^{+\infty} \delta\left(u - \frac{m}{\Delta x}\right). \end{aligned} \tag{4.58}$$

This is also a periodic delta function, but the period is changed to $1/\Delta x$ as shown in the right column of Fig. 4.2b. As a result, using (4.58) and (4.51), (4.56) is rewritten as follows:

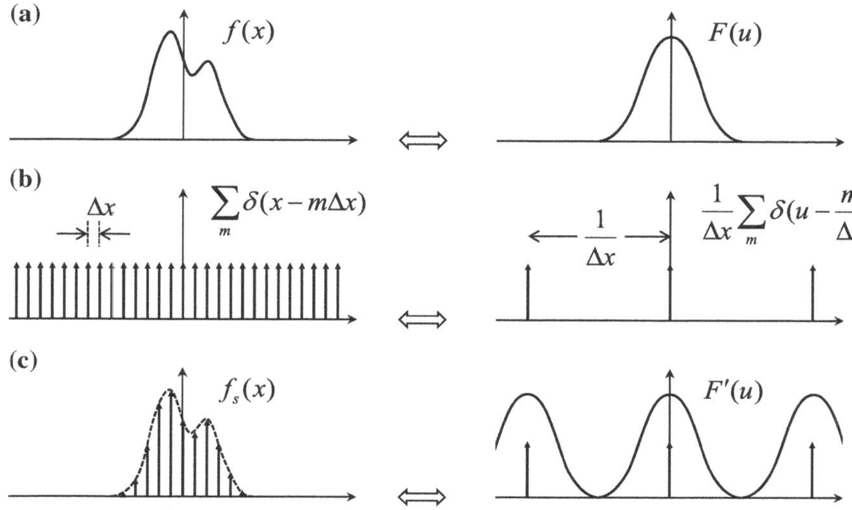

Fig. 4.2 The Fourier transform of a sampled function

$$F'(u) = \frac{1}{\Delta x} \sum_{m=-\infty}^{+\infty} F(u) \otimes \delta\left(u - \frac{m}{\Delta x}\right)$$
$$= \frac{1}{\Delta x} \sum_{m=-\infty}^{+\infty} F\left(u - \frac{m}{\Delta x}\right). \tag{4.59}$$

This means that Fourier spectrum of a sampled function whose sampling interval is Δx has a periodicity of $1/\Delta x$. Figure 4.2c shows the relation schematically.

This nature of sampled functions leads to the sampling theorem represented in (3.38). If the maximum frequency component of $f(x)$ exceeds $1/(2\Delta x)$ as shown in Fig. 4.3b, Fourier spectrum is overlapped with the neighboring period. We cannot retrieve the original signal in this case. On the contrary, when the sampling interval is small enough to satisfy the sampling theorem, i.e., $F(u)$ has a non-zero value only within the region satisfying

$$|u| < \frac{1}{2\Delta x}, \tag{4.60}$$

we can retrieve the original signal by filtering the spectrum as follows

$$F(u) = \text{rect}\left(\frac{u}{\Delta x^{-1}}\right) F'(u), \tag{4.61}$$

where $\text{rect}(\Delta x u)$ is a rectangle function whose width is Δx^{-1}.

The original function is given by the inverse Fourier transform of (4.61).

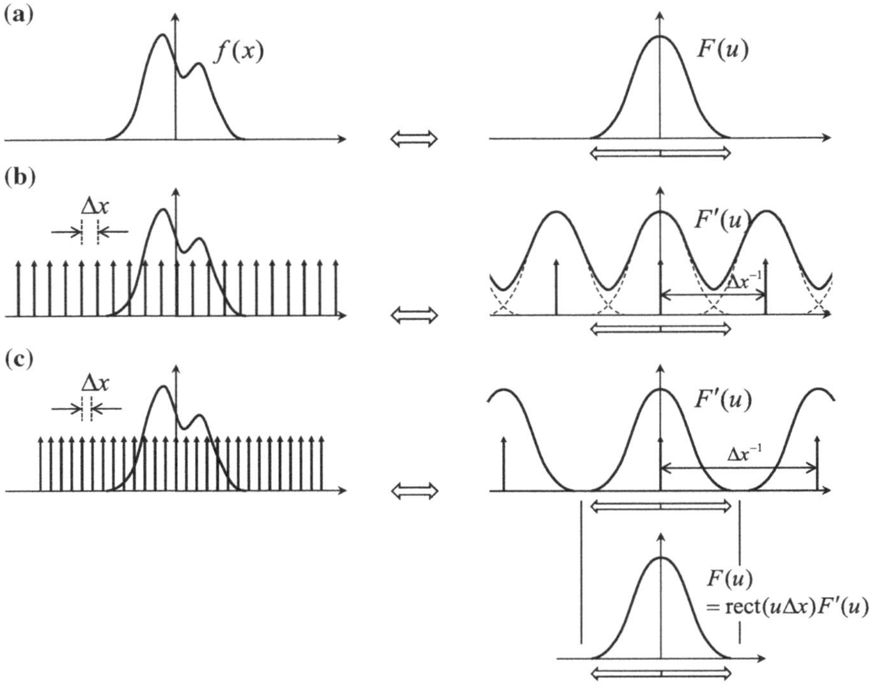

Fig. 4.3 Explanation of the sampling theorem

$$\begin{aligned} f(x) &= \mathscr{F}^{-1}\{F(u)\} \\ &= \frac{1}{\Delta x}\text{sinc}\left(\frac{x}{\Delta x}\right) \otimes f_s(x). \end{aligned} \tag{4.62}$$

Since sampled function $f_s(x)$ in (4.55) includes the Delta function, we can easily perform the convolution and obtain the original function as follows:

$$f(x) = \frac{1}{\Delta x} \sum_{m=-\infty}^{+\infty} f(m\Delta x)\text{sinc}\left(\frac{x - m\Delta x}{\Delta x}\right). \tag{4.63}$$

This result presents that the original function $f(x)$ can be retrieved from the sampled function $f(m\Delta x)$. In other words, we can obtain the value of a continuous function at any point from the series of the sampled values without any loss. This is the *sampling theorem*. Equation (4.63) is sometimes called sinc interpolation.

The relation between a sampled signal and its spectrum can be inverted, as shown in Fig. 4.4. Suppose that a sampled Fourier spectrum is given by

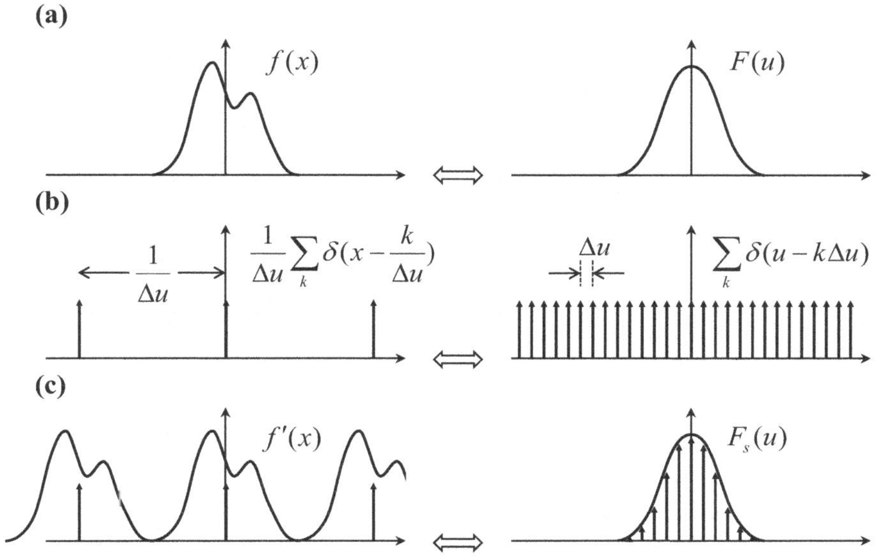

Fig. 4.4 The inverse Fourier transform of a sampled Fourier spectrum

$$F_s(u) = F(u) \times \sum_{k=-\infty}^{+\infty} \delta(u - k\Delta u), \tag{4.64}$$

where Δu is a sampling interval with respect to Fourier frequency u. The inverse Fourier transform is written by

$$\begin{aligned} f'(x) &= \mathscr{F}^{-1}\{F_s(u)\} \\ &= \frac{1}{\Delta u} \sum_{k=-\infty}^{+\infty} f(x) \otimes \delta\left(x - \frac{k}{\Delta u}\right) \\ &= \frac{1}{\Delta u} \sum_{k=-\infty}^{+\infty} f\left(x - \frac{k}{\Delta u}\right). \end{aligned} \tag{4.65}$$

The function $f'(x)$ is also a periodic function whose period is $1/\Delta u$.

4.6 Discrete Fourier Transform (DFT)

On the analogy of the Fourier transform of a continuous function in (4.1), the Fourier transform of a sampled function may be presented by

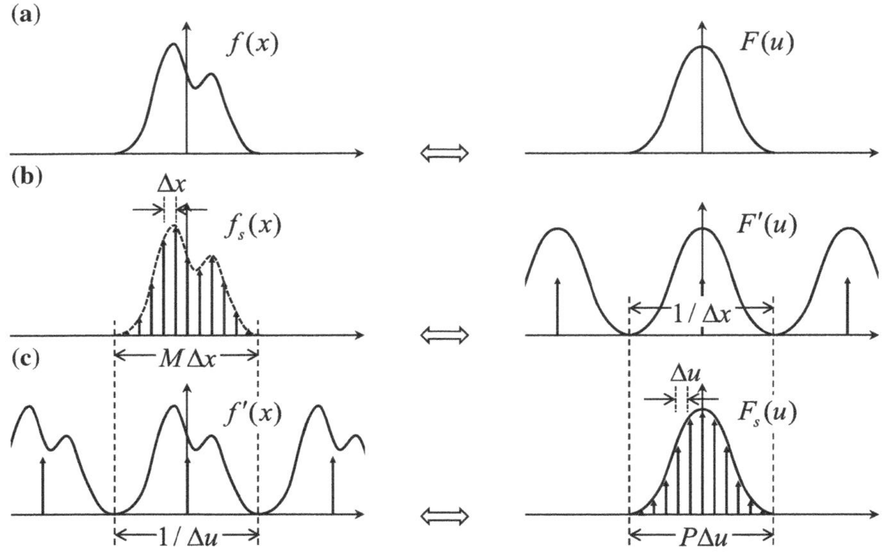

Fig. 4.5 The relation between appropriate sampling areas in the Fourier and real domain

$$F'(u) = \sum_{m=-\infty}^{+\infty} f(x_m) \exp[-i2\pi u x_m]$$
$$= \sum_{m=-\infty}^{+\infty} f(m\Delta x) \exp[-i2\pi u m\Delta x]. \tag{4.66}$$

However, we cannot calculate the infinite sum in practice. Thus, supposing that the number of sample points is even number M, we rewrite above equation as

$$F'(u) = \sum_{m=-M/2}^{M/2-1} f(m\Delta x) \exp[-i2\pi u m\Delta x]. \tag{4.67}$$

Figure 4.5 shows the relation of the sampling manner with the number of samples. Suppose that $F'(u)$ is again sampled spectrum at an interval of Δu and the number of sample points is P, as shown in the right column of (c), the Fourier frequency is represented by

$$u_p = p\Delta u, \quad (p = -P/2, \ \ldots, \ P/2 - 1), \tag{4.68}$$

and the Fourier spectrum is rewritten as

$$F'(u_p) = \sum_{m=-M/2}^{M/2-1} f(m\Delta x) \exp[-i2\pi mp\Delta x \Delta u]. \tag{4.69}$$

Because the period of $F'(u)$ is $1/\Delta x$, sampling $F'(u)$ over a range more than $1/\Delta x$ makes no sense. Extent of sampling in the Fourier domain should agree with the period, i.e., $P\Delta u = 1/\Delta x$ as shown in the right column of Fig. 4.5b, c. There is a similar relation between Δx and M. The extent of sampling in the real space should agree with $1/\Delta u$, i.e., $M\Delta x = 1/\Delta u$, as shown in the left column of (b) and (c), where Δu is again a sampling interval of $F_s(u)$.

As a result, the sampling interval and number of sample points in the Fourier domain are automatically determined as

$$P = M, \tag{4.70}$$

$$\Delta u = \frac{1}{M\Delta x}. \tag{4.71}$$

Therefore, (4.69) is rewritten by

$$F'(p\Delta u) = \sum_{m=-M/2}^{M/2-1} f(m\Delta x) \exp\left(-\frac{i2\pi mp}{M}\right). \tag{4.72}$$

Here, let us adopt the following abbreviated notation for sampled functions.

$$\begin{aligned} F[p] &= F'(p\Delta u), \\ f[m] &= f'(m\Delta x). \end{aligned} \tag{4.73}$$

In this new notation, sampled function is represented using square brackets, and one or two integers in the brackets designates a given sample point.

Using the above notation, the discrete Fourier transform is defined as

$$F[p] = \sum_{m=-M/2}^{M/2-1} f[m] \exp\left(-\frac{i2\pi mp}{M}\right), \quad (p = -M/2, \ldots, M/2-1). \tag{4.74}$$

Inverse transformation of (4.74) is given by

$$f[m] = \frac{1}{M} \sum_{p=-M/2}^{M/2-1} F[p] \exp\left(\frac{i2\pi mp}{M}\right), \quad (m = -M/2, \ldots, M/2-1). \tag{4.75}$$

This is easily proved by substituting (4.75) into the right hand side of (4.74).

$$\sum_{m=-M/2}^{M/2-1}\left[\frac{1}{M}\sum_{l=-M/2}^{M/2-1}F[l]\exp\left(\frac{i2\pi ml}{M}\right)\right]\exp\left(-\frac{i2\pi mp}{M}\right)$$
$$=\frac{1}{M}\sum_{l=-M/2}^{M/2-1}F[l]\left[\sum_{m=-M/2}^{M/2-1}\exp\left(\frac{i2\pi m(l-p)}{M}\right)\right]$$
$$=\frac{1}{M}\sum_{l=-M/2}^{M/2-1}F[l]\times M\delta_{lp}$$
$$=F[p],$$

where δ_{lp} is Kronecker delta that is unity if $l=p$ otherwise zero. Here, we used the following orthogonal relation.

$$\frac{1}{M}\sum_{m=-M/2}^{M/2-1}\exp\left(\frac{i2\pi ml}{M}\right)\exp\left(-\frac{i2\pi mp}{M}\right)=\delta_{lp}.$$

In a 2D function, DFT is also written by

$$F[p,q]=\sum_{m=-M/2}^{M/2-1}\sum_{n=-N/2}^{N/2-1}f[m,n]\exp\left[-i2\pi\left(\frac{mp}{M}+\frac{nq}{N}\right)\right],$$
$$(p=-M/2,\ \ldots,\ M/2-1,\ \text{and}\ q=-N/2,\ \ldots,\ N/2-1),\tag{4.76}$$
$$f[m,n]=\frac{1}{MN}\sum_{p=-M/2}^{M/2-1}\sum_{q=-N/2}^{N/2-1}F[p,q]\exp\left[i2\pi\left(\frac{mp}{M}+\frac{nq}{N}\right)\right],$$
$$(m=-M/2,\ \ldots,\ M/2-1,\ \text{and}\ n=-N/2,\ \ldots,\ N/2-1),\tag{4.77}$$

where N is the number of sample points in the y direction. Here, as in (4.71), the sampling intervals are given by

$$\Delta u=\frac{1}{M\Delta x},$$
$$\Delta v=\frac{1}{N\Delta y}.\tag{4.78}$$

As mentioned above, we introduce notation using square brackets for a sampled function from this chapter on. This is because sampled functions should be distinguished from continuous functions clearly. Let us make a rule that arguments indicated in the square brackets are limited to integers.

4.7 Fast Fourier Transform (FFT)

The discrete Fourier transform is known as very time-consuming operation. When we execute DFT with N sample points, the computational complexity is given by N^2. In HD-CGHs, because N is commonly more than a billion, the computational complexity is mind-boggling huge figures.

Fortunately, algorithm or techniques called fast Fourier transform (FFT) is developed in modern days. When N is given as a power of 2, FFT typically reduces the complexity of DFT to $2N \log_2 N$. The difference may seem to be small but has considerable efficacy in practice. For example, supposing $N = 2^{32}$, the computational complexity reaches to $N^2 \simeq 1.8 \times 10^{19}$ when computing DFT in (4.74) as it is, whereas the complexity is $2N \log_2 N \simeq 2.7 \times 10^{11}$ in FFT. This huge difference of computational complexity allows us to cut down the computation time dramatically.

In this book, we do not go into the detail of FFT algorithm. Nowadays many software libraries of FFT are available for solving scientific and engineering problems. We use one of them and discuss the practical use of FFT in optics. The reader who wants to learn the principle and algorithm are referred to a technical book of FFT.

Since the result of FFT is exactly the same as that of DFT, the nature is also the same as DFT. FFT merely carries out DFT calculation in very short computation time. Therefore, we do not distinguish between DFT and FFT.

4.7.1 Actual FFT with Positive Indexes

In DFT given by (4.74)–(4.77), we represent the DFT by summation with indexes m and p that take positive and negative values. This is because the functions are sampled symmetrically to the origin of the coordinate system. However, actual DFT/FFT is commonly defined in a positive index as follows:

$$\begin{aligned} F[p] &= \sum_{m=0}^{M-1} f[m] \exp\left(-\frac{i2\pi mp}{M}\right) \\ &= \mathbf{RFFT}\{f[m]\}, \quad (p = 0, \ldots, M-1), \end{aligned} \tag{4.79}$$

$$\begin{aligned} f[m] &= \frac{1}{M} \sum_{p=0}^{M-1} F[p] \exp\left(\frac{i2\pi mp}{M}\right) \\ &= \mathbf{RFFT}^{-1}\{F[p]\}, \quad (m = 0, \ldots, M-1). \end{aligned} \tag{4.80}$$

In two dimensional functions, DFT/FFT is also given by

$$F[p, q] = \sum_{m=0}^{M-1} \sum_{n=0}^{N-1} f[m, n] \exp\left[-i2\pi \left(\frac{mp}{M} + \frac{nq}{N}\right)\right]$$

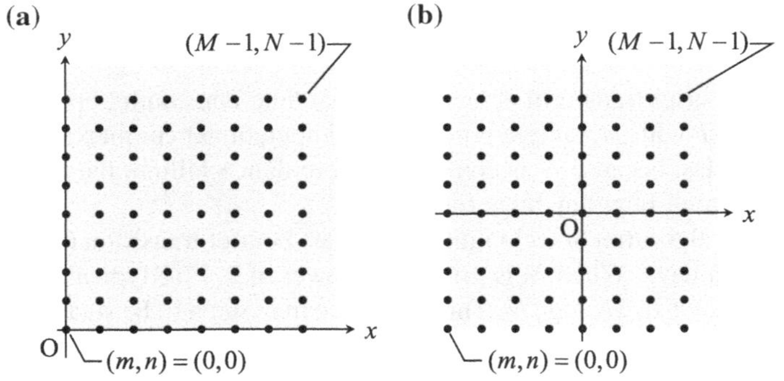

Fig. 4.6 Sampling **a** without and **b** with symmetry. $M = N = 8$

$$= \mathbf{RFFT}\left\{f[m,n]\right\},$$

$$(p = 0,\ \ldots,\ M-1,\ \text{and}\ q = 0,\ \ldots,\ N-1), \qquad (4.81)$$

$$f[m,n] = \frac{1}{MN}\sum_{p=0}^{M-1}\sum_{q=0}^{N-1} F[p,q]\exp\left[i2\pi\left(\frac{mp}{M}+\frac{nq}{N}\right)\right]$$

$$= \mathbf{RFFT}^{-1}\left\{F[p,q]\right\},$$

$$(m = 0,\ \ldots,\ M-1,\ \text{and}\ n = 0,\ \ldots,\ N-1), \qquad (4.82)$$

where M and N are again the numbers of sample points in the x and y direction, respectively. This is the standard definition of DFT/FFT in almost all FFT libraries. We call this type of practical FFT "raw FFT" and denote it by $\mathbf{RFFT}\{\cdot\}$ or $\mathbf{RFFT}^{-1}\{\cdot\}$ in this book.

Because all sampled functions treated in DFT have M- or N-long periodicity such as $f[m+M] = f[m]$ or $F[p+M] = F[p]$, the shift of summation ranges in actual FFT does not cause any severe problem of formulation in general. However, when we represent the sampling positions of a sampled function using (3.52), i.e., $(x_m, y_n) = (m\Delta x, n\Delta y)$, the function is always sampled only in the positive domain, as shown in Fig. 4.6a.

4.7.2 *Use of Raw FFT with Symmetrical Sampling*

It is commonly desired in computer holography that the origin of the coordinate system is placed at the center of the sampling window, as shown in Fig. 4.6b. This sampling manner is referred to as "symmetric sampling" in this book. Therefore, we define the symmetrical FFT operation as

$$F[p] = \mathbf{FFT}\{f[m]\}$$
$$\equiv \sum_{m=-M/2}^{M/2-1} f[m] \exp\left(-\frac{i2\pi mp}{M}\right), \quad (p = -M/2, \ldots, M/2-1), \tag{4.83}$$
$$f[m] = \mathbf{FFT}^{-1}\{F[p]\}$$
$$\equiv \frac{1}{M} \sum_{p=-M/2}^{M/2-1} F[p] \exp\left(\frac{i2\pi mp}{M}\right), \quad (m = -M/2, \ldots, M/2-1). \tag{4.84}$$

In the 2D form, they are

$$F[p,q] = \mathbf{FFT}\{f[m,n]\}$$
$$\equiv \sum_{m=-M/2}^{M/2-1} \sum_{n=-N/2}^{N/2-1} f[m,n] \exp\left[-i2\pi\left(\frac{mp}{M} + \frac{nq}{N}\right)\right],$$
$$(p = -M/2, \ldots, M/2-1, \text{ and } q = -N/2, \ldots, N/2-1), \tag{4.85}$$
$$f[m,n] = \mathbf{FFT}^{-1}\{F[p,q]\}$$
$$\equiv \frac{1}{MN} \sum_{p=-M/2}^{M/2-1} \sum_{q=-N/2}^{N/2-1} F[p,q] \exp\left[i2\pi\left(\frac{mp}{M} + \frac{nq}{N}\right)\right],$$
$$(m = -M/2, \ldots, M/2-1, \text{ and } n = -N/2, \ldots, N/2-1). \tag{4.86}$$

To make use of an FFT library that evaluates DFT of a data array with positive 0-base indexes, we must convert the symmetrical FFT defined in (4.83)–(4.86) into the raw FFT using positive indexes. This is not so difficult work. For example, (4.85) can be rewritten by

$$F(u'_p, v'_q) = \mathbf{FFT}\left\{f(x'_m, y'_n)\right\}$$
$$= \sum_{m=-M/2}^{M/2-1} \sum_{n=-N/2}^{N/2-1} f(x'_m, y'_n) \exp\left[-i2\pi(u'_p x'_m + v'_q y'_n)\right], \tag{4.87}$$

where the sampling positions are given by

$$x'_m = m\Delta x, \quad y'_n = n\Delta y, \quad u'_p = p\Delta u, \quad \text{and} \quad v'_q = q\Delta v. \tag{4.88}$$

In addition,

$$\Delta x \Delta u = \frac{1}{M} \quad \text{and} \quad \Delta y \Delta v = \frac{1}{N}. \tag{4.89}$$

Substituting $m - M/2$ and $n - N/2$ for m and n in (4.87), respectively, we can obtain the following equation that is equivalent to (4.87).

$$F(u_p, v_q) = \mathbf{FFT}\left\{f(x_m, y_n)\right\}$$
$$= \sum_{m=0}^{M-1}\sum_{n=0}^{N-1} f(x_m, y_n)\exp\left[-i2\pi(u_p x_m + v_q y_n)\right]. \tag{4.90}$$

Here, the sampling positions are redefined as follow:

$$x_m = x'_{m-M/2} = \left(m - \frac{M}{2}\right)\Delta x, \quad (m = 0, \ldots, M-1),$$
$$y_n = y'_{n-N/2} = \left(n - \frac{N}{2}\right)\Delta y, \quad (n = 0, \ldots, N-1), \tag{4.91}$$
$$u_p = x'_{p-M/2} = \left(p - \frac{P}{2}\right)\Delta u, \quad (p = 0, \ldots, M-1),$$
$$v_q = x'_{q-N/2} = \left(q - \frac{Q}{2}\right)\Delta v, \quad (q = 0, \ldots, N-1). \tag{4.92}$$

This sampling manner is depicted in 2D coordinates in Fig. 4.6b. Equations (4.91) and (4.92) give the standard method to sample a function in this book. Unless otherwise noted, wavefields are commonly sampled by the same manner from this chapter on.

The symmetrical FFT can be evaluated using the raw FFT. For example, the 1D form of (4.90) is written as follows.

$$F(u_p) = \mathbf{FFT}\left\{f(x_m)\right\}$$
$$= \sum_{m=0}^{M-1} f(x_m)\exp(-i2\pi u_p x_m), \tag{4.93}$$

where x_m and u_p are given in (4.91) and (4.92). Thus, substituting x_m and u_p into above actually, we can rewrite the equation as:

$$F[p] = \mathbf{FFT}\left\{f[m]\right\}$$
$$= \exp(i\pi p)\sum_{m=0}^{M-1} f[m]\exp(i\pi m)\exp\left(-\frac{i2\pi mp}{M}\right)$$
$$= \exp(i\pi p)\mathbf{RFFT}\left\{f[m]\exp(i\pi m)\right\}, \tag{4.94}$$

where we used $\Delta x \Delta u = 1/M$ and assumed that M is a power of 2 and $M \geq 4$. Inverse FFT is also obtained using the same procedure.

$$f[m] = \mathbf{FFT}^{-1}\left\{F[p]\right\}$$
$$= \exp[-i\pi m]\mathbf{RFFT}^{-1}\left\{F[p]\exp[-i\pi p]\right\}. \tag{4.95}$$

Conversion from the raw FFT to the symmetrical FFT needs additional computation, as shown above. Fortunately, there is another technique that gives less

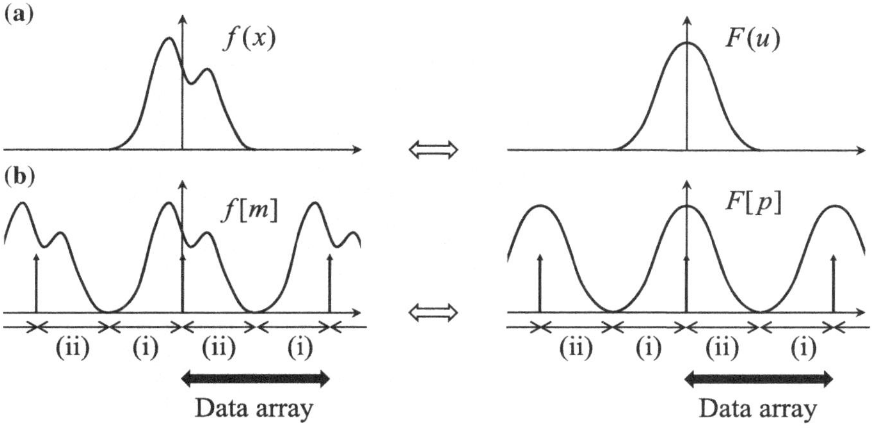

Fig. 4.7 One dimensional data replacement to use raw FFT in symmetric coordinates

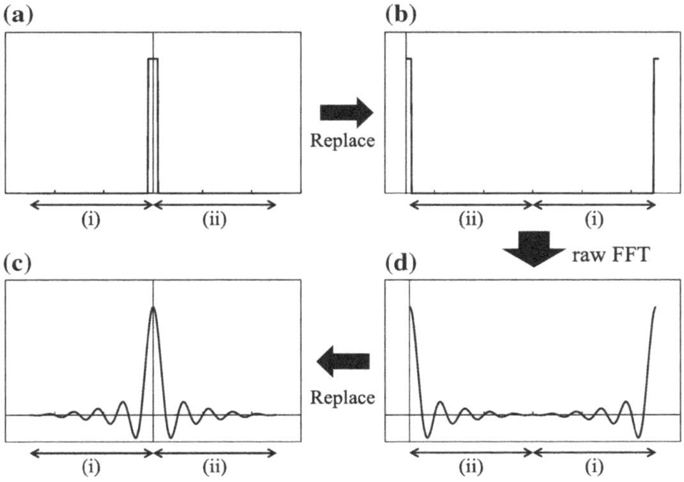

Fig. 4.8 An example of replacement of the 1D data array. $f(x) = \text{rect}(x/2)$, i.e., $f[m] = \text{rect}((m - M/2)\Delta x/2)$, $\Delta x = 0.1$, $M = 512$

computational effort. In this technique, periodicity of DFT is used to execute FFT with symmetric sampling; the data array fed into the FFT package is replaced before executing FFT.

Figure 4.7b shows the arrangement of the data array of a 1D function shown in (a). The data array $f[m]$ is divided into two parts, i.e., negative domain (i): $0 \le m < M/2$, and positive domain (ii): $M/2 \le m < M$. These domains of the input data array are exchanged before performing FFT, as in Fig. 4.7b. The output data array is again replaced to adjust the origin of Fourier spectrum to the center of the output data. Figure 4.8 shows an example of the Fourier transform of a symmetrical 1D function:

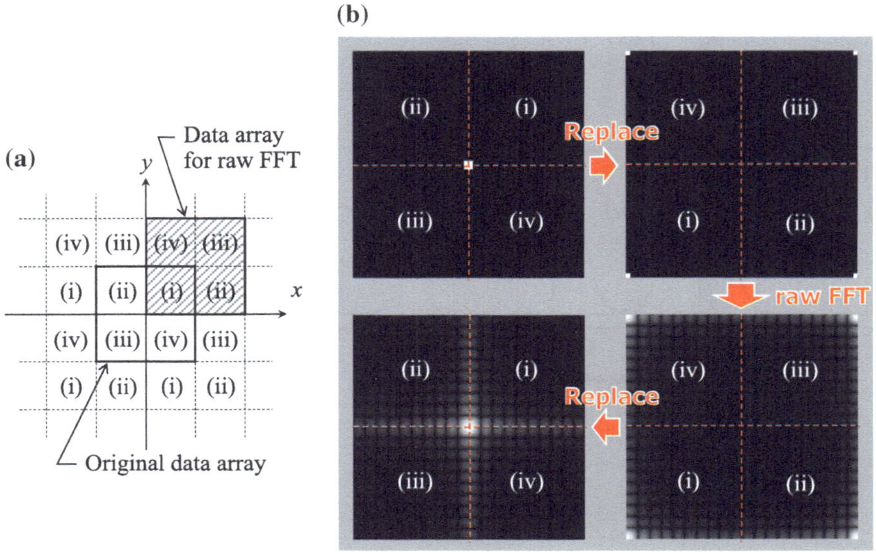

Fig. 4.9 Replacement of a 2D data array. **a** Two dimensional periodicity. **b** The example of replacement of the data array. $f[m, n] = \text{rect}((m - M/2)\Delta x/2)\text{rect}((n - N/2)\Delta y/2)$, $\Delta x = \Delta y = 0.1$, $M = N = 512$. Note that $|F[p, q]|$ is visualized instead of $F[p, q]$ as grayscale images with an encoding gamma of 1/2.2

$f(x) = \text{rect}(x/a)$ using the raw FFT. The domains (i) and (ii) of the data, here $f[m] = \text{rect}(x_m/0.05)$, are switched in Fig. 4.8b, and then FFT is executed in the replaced data array as in (d). Finally, the domains (i) and (ii) are again switched to shift the coordinates origin to the center of the data array.

Data array replacement in a 2D function is shown in Fig. 4.9. Here, Fig. 4.9a shows the periodicity of 2D functions. The data array used in the raw FFT is indicated by hatching. Quadrants (ii)–(iv) must be replaced so that the data array is placed in the positive domain. Figure 4.9b shows an actual example. After the data array of $f[m, n] = \text{rect}(x_m/a)\text{rect}(y_n/b)$ is replaced, the raw FFT is executed in the replaced data array. Then, the result is again replaced to retrieve the original data array.

Let us emphasize that the procedure shown in Fig. 4.10 is always used instead of (4.94) and (4.95) in practice when we evaluate the symmetrical FFT represented by symbol **FFT** $\{\cdot\}$ in this book.

4.7.3 Discrete Convolution Using FFT

The discrete convolution is the convolution integral of sampled functions. The discrete convolution is written by the followings.

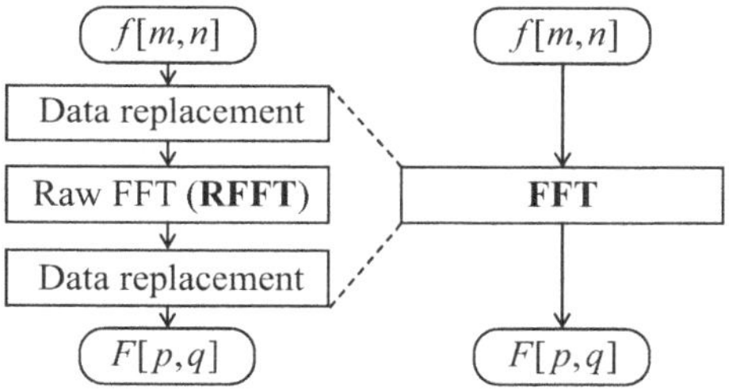

Fig. 4.10 The actual procedure tc evaluate the symmetrical FFT with the symmetrical sampling in Fig. 4.6b

$$f_1 \otimes f_2[m] = \sum_{m'=0}^{M-1} f_1[m']f_2[m-m'], \tag{4.96}$$

$$f_1 \otimes f_2[m,n] = \sum_{m'=0}^{M-1}\sum_{n'=0}^{N-1} f_1[m',n']f_2[m-m',n-n']. \tag{4.97}$$

Because the convolution theorem holds even in sampled functions, discrete convolution can be calculated using FFT.

$$\begin{aligned} f_1 \otimes f_2[m] &= \mathbf{FFT}^{-1}\left\{\mathbf{FFT}\left\{f_1[m]\right\} \times \mathbf{FFT}\left\{f_2[m]\right\}\right\}, \\ f_1 \otimes f_2[m,n] &= \mathbf{FFT}^{-1}\left\{\mathbf{FFT}\left\{f_1[m,n]\right\} \times \mathbf{FFT}\left\{f_2[m,n]\right\}\right\}. \end{aligned} \tag{4.98}$$

This is sometimes called *fast convolution* . Here, note that the sampling manner given in (4.91) and the FFT with data replacement shown in Fig. 4.10 are used for the calculation.

Unfortunately, in some cases, the convolution integral using (4.98) does not give the same result as that by (4.97). For example, supposing $f_1(x) = f_2(x) = \mathrm{rect}(x)$, the result of the analytical convolution integral is well-known:

$$f_1 \otimes f_2(x) = \Lambda(x). \tag{4.99}$$

The function $\Lambda(x)$ is depicted in Fig. 4.11.

Figure 4.12 also shows actual numerical examples calculated by (4.98). In these examples, both the sampled functions have 64 sample points. When the rectangular function has 21-length unity-values sample points, the result agrees with analytical one as in Fig. 4.12a. However, the 41-length case does not reproduce the correct $\Lambda(x)$, as in (b).

The reason is simple. All sampled functions are always periodic when we use the discrete Fourier transform or FFT, as mentioned in Sect. 4.6. Figure 4.13 illustrates the discrete convolution of periodic functions. Product $f_1[m']f_2[m-m']$ in (4.96) has non-zero value in the overlap area of $f_1[m']$ and $f_2[m-m']$. If the unity-values length is less than one half of the total number of sample points, two functions overlap with each other only a region shown in Fig. 4.13a. However, they overlap in two regions in longer unity-value lengths, as in (b). Therefore, the summation in

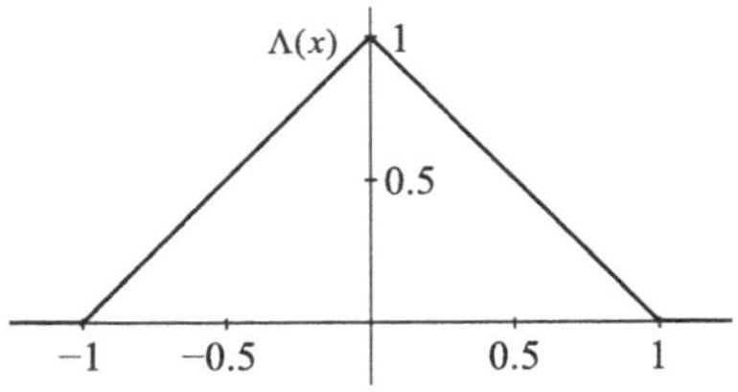

Fig. 4.11 The shape of function $\Lambda(x)$

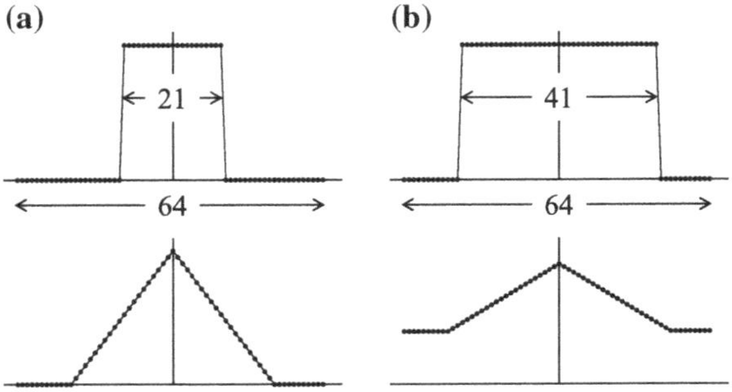

Fig. 4.12 Examples of discrete convolution using FFT. The rectangular function has unity values in **a** 21 and **b** 41 sample points of 64 total points

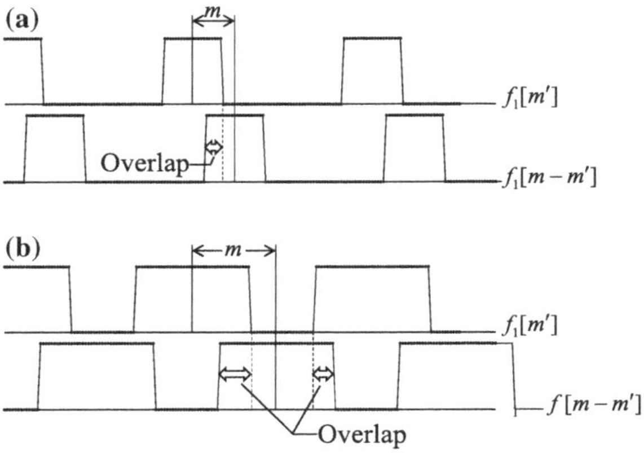

Fig. 4.13 Circular convolution with non-zero-value length **a** less and **b** more than a half of the total number of sample points

(4.96) results in a different value from what we intend. The convolution given in (4.98) is therefore called a different name: *circular convolution*, while that in (4.96) is called *linear convolution*.

Figure 4.13 also gives us a clue to solve the problem; a technique to convert the circular convolution into the linear convolution. Figure 4.13 suggests that either function $f_1[m]$ or $f_2[m]$ must have a longer zero-value length than one half the total number of sample points in order to avoid the problem of the circular convolution. To realize this, we should double the sample points of $f_1[m]$, i.e., take $2M$ sample points, and embed the original $f_1[m]$ at the center of the $2M$ sampling array in cases of symmetric coordinates. Then, sample points outside the original $f_1[m]$ are

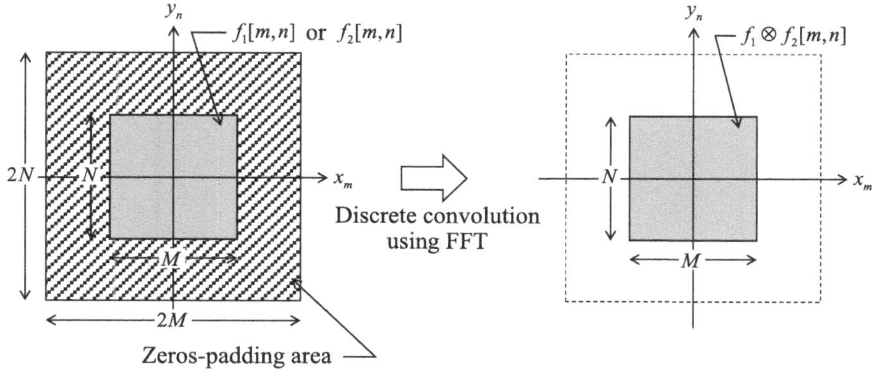

Fig. 4.14 Quadruple extension to convert the circular convolution into the linear convolution

padded with zeros. After calculating the discrete convolution using (4.98), the center M points is extracted from the $2M$ sampling array.

The 2D version of this technique in the symmetric coordinate system is shown in Fig. 4.14. This technique is referred to as *quadruple extension* in this book.

Chapter 5
Diffraction and Field Propagation

Abstract Scalar diffraction theories of monochromatic light are discussed for free space propagation that are very important in computer holography. Important formulas and ideas such as the angular spectrum, Fresnel and Fraunhofer diffraction and the Fourier transform by a lens are introduced as preliminary to the techniques of the numerical field propagation discussed in Chaps. 6, 9, and 12. Thus the equations presented in this chapter are often referred to in the following chapters.

5.1 Introduction

A monochromatic wave in free space is subject to the Helmholtz equation in (3.30), as mentioned in Sect. 3.1.3. The plane and spherical waves discussed in the preceding chapter are special solutions of the Helmholtz equation. We generally need additional information, called a *boundary condition*, to solve the Helmholtz equation and obtain a wavefield that propagates in free space.

5.1.1 Field Propagation

When light passes through a small aperture, the light is diffracted by the aperture, as shown in Fig. 5.1. In this case, the boundary condition is given by the binary function, whose value is unity inside the aperture and zero outside the aperture. Solving the Helmholtz equation using the boundary condition, we can calculate light diffracted by the aperture.

More generally, complex amplitudes in the (x, y, z_0) plane, i.e., a wavefield can be the boundary condition. In numerical simulation, solving the Helmholtz equation using the boundary condition given by the wavefield, we can obtain the complex amplitudes at any place of the given 3D space, as shown in Fig. 5.2. It may look strange from the viewpoint of physics that a diffracted field is obtained in the space where $z < z_0$. Here, remember that any factor with respect to time is lost in the Helmholtz equation. This means that we do not have any information about the

K. Matsushima, *Introduction to Computer Holography*,
Series in Display Science and Technology,
https://doi.org/10.1007/978-3-030-38435-7_5

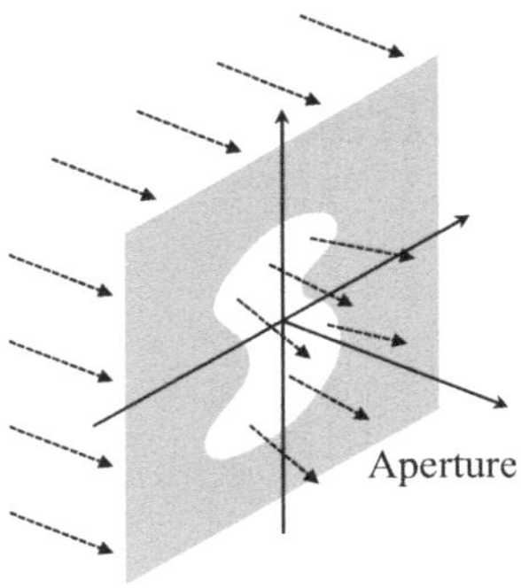

Fig. 5.1 Diffraction by an aperture

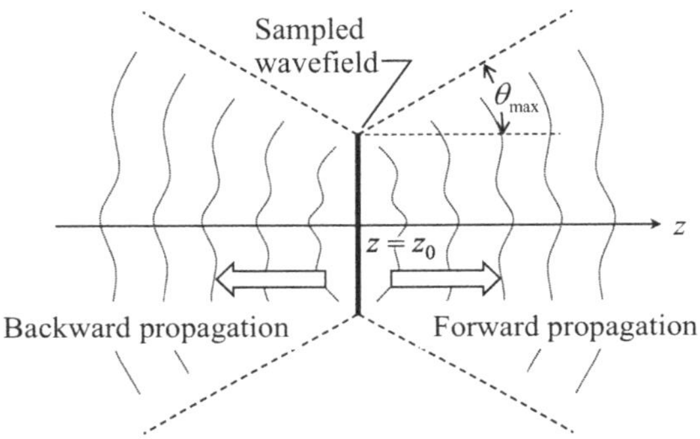

Fig. 5.2 Schematic illustration of numerical field propagation

direction of the light travel. Thus a frozen field is obtained by solving the Helmholtz equation. However, we call the process of obtaining a wavefield from the boundary condition *field propagation*. When the boundary condition is given by a sampled wavefield, and the propagated field is obtained by numerical simulation, the field propagation is particularly referred to as *numerical field propagation* in this book.

When the wavefield after field propagation is placed in a plane at $z > z_0$, we call it *forward propagation*. The opposite is called *backward propagation*. Here, note that a sampled wavefield may not have any information when it travels outside the area limited by the maximum diffraction angle given in (3.54). In other words, we cannot obtain exact wavefields from the given sampled wavefield or boundary condition outside the region designated by the maximum diffraction angles, as shown in Fig. 5.2.

Figure 5.3 shows examples of numerical field propagation. Here, the wavefields after forward and backward propagation are calculated from the binary boundary condition, i.e., an aperture shown in (b), irradiated by a plane wave at a wavelength of 633 nm, whose wave vector makes an angle of 5° with the z-axis. The sampling intervals of the aperture and wavefield are 2 μm in both the x and y directions. The technique of the band-limited angular spectrum, described in Sect. 6.4, is used to calculate this example.

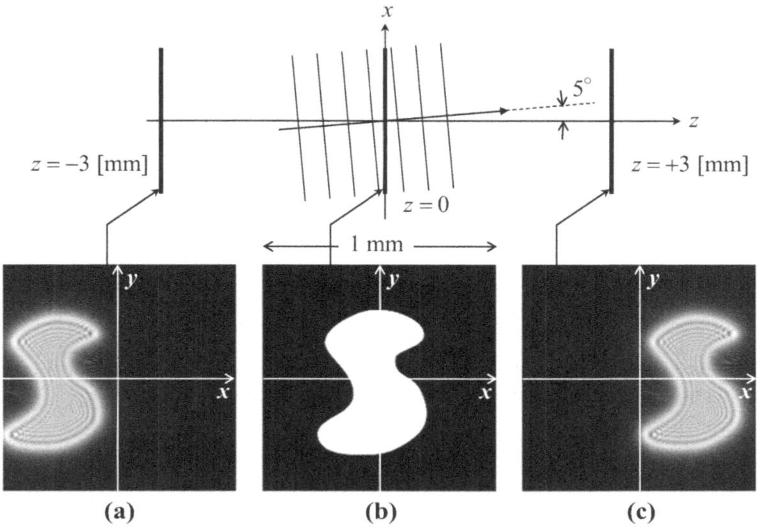

Fig. 5.3 Examples of numerical field propagation: **a** backward propagation, **b** an aperture, and **c** forward propagation. $\Delta x = \Delta y = 2\,[\mu\mathrm{m}]$. $\lambda = 633\,[\mathrm{nm}]$

5.1.2 Classification of Field Propagation

Numerical field propagation is classified into three groups as shown in Fig. 5.4. Here, the wavefield that gives the boundary condition of the Helmholtz equation is referred to as a *source wavefield* or simply *source field*, while the field obtained after numerical propagation is a *destination wavefield* or *destination field*. The first group shown in Fig. 5.4a is the most basic propagation; a source wavefield is propagated to the destination wavefield in a plane parallel to the source wavefield. The lateral position of the sampling window of the destination wavefield is the same as that of the source wavefield in this group. This type of numerical propagation is discussed in Chap. 6.

The second group is called *off-axis propagation* or *shifted propagation*. The destination sampling window is also parallel to the source sampling window but laterally shifts, as shown in Fig. 5.4b. This type of propagation plays an important role to propagate large-scale wavefields, which cannot be kept in the main memory of a computer. This type of numerical propagation is discussed in Chap. 12.

The third group is referred to as *rotational transform* of a wavefield, and the detail is described in Chap. 9. In this propagation, the destination wavefield is not parallel to the source wavefield as in Fig. 5.4c. In the polygon-based method, which is a technique to calculate object fields from the object model, the rotational transform plays an important role to produce surface sources of light because the surface sources are commonly not parallel to the hologram.

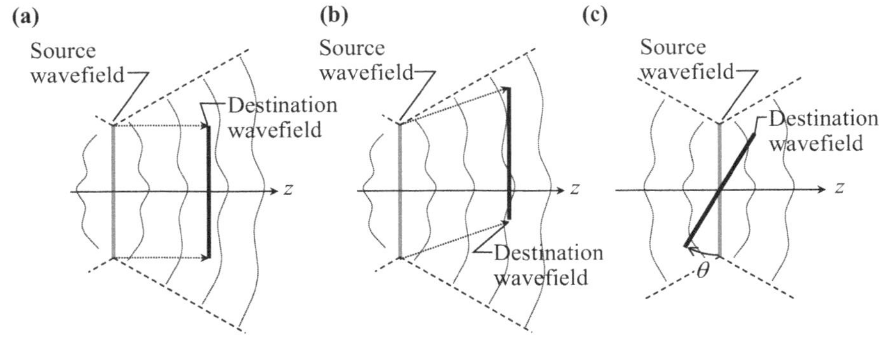

Fig. 5.4 Three types of numerical field propagation: **a** parallel field propagation, **b** off-axis or shifted field propagation, and **c** the rotational transform of a wavefield

In this chapter, we discuss analytical theories of diffraction and field propagation that are the basis of the numerical field propagation.

5.2 Scalar Diffraction Theory

Numerical field propagation is based on the scalar diffraction theory. Figure 5.5 shows the coordinate system used in the formulation. We assume that source wavefield $g(x, y; z_0)$ is given in a plane perpendicular to the z-axis. Here, the plane is called a source plane. Suppose that the source plane is placed at $z = z_0$. The source wavefield is sometimes a simple aperture. In this case, $g(x, y; z_0)$ is a real-valued function. Otherwise, $g(x, y; z_0)$ represents distribution of complex amplitudes, i.e., a wavefield. Since wavefield $g(x, y; z_0)$ provides the boundary condition of Helmholtz equation, our first subject is to solve the Helmholtz equation using the boundary condition and obtain the destination wavefield $g(x, y; z)$ in the destination plane placed at an arbitrary z-position.

We sometimes use source coordinates $(x_s, y_s; z_0)$ instead of $(x, y; z_0)$ if distinction between the source and destination coordinates is desirable for formulation.

5.2.1 Angular Spectrum Method

The angular spectrum method plays an important role in numerical field propagation, because it fits numerical calculation by computer better than traditional techniques such as the Fresnel diffraction and the Fraunhofer diffraction.

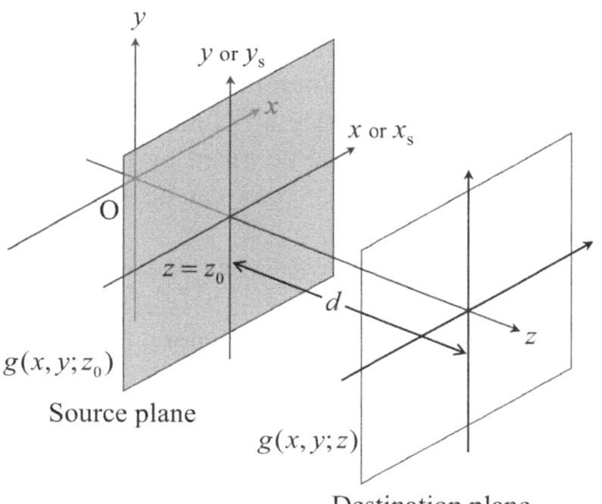

Fig. 5.5 Geometry and the coordinates system used for filed propagation between parallel planes

5.2.1.1 Angular Spectrum of Plane Wave

Supposing that a source wavefield is represented by $g(x, y; z_0)$ in the source plane, the Fourier transform is written as

$$\begin{aligned} G(u, v; z_0) &= \mathscr{F}\{g(x, y; z_0)\} \\ &= \iint_{-\infty}^{+\infty} g(x, y; z_0) \exp[-i2\pi(ux + vy)]\mathrm{d}x\mathrm{d}y. \end{aligned} \tag{5.1}$$

The inverse Fourier transform is also given by

$$\begin{aligned} g(x, y; z_0) &= \mathscr{F}^{-1}\{G(u, v; z_0)\} \\ &= \iint_{-\infty}^{+\infty} G(u, v; z_0) \exp[i2\pi(ux + vy)]\mathrm{d}u\mathrm{d}v. \end{aligned} \tag{5.2}$$

Let us consider the case where the source wavefield is a plane wave given in (3.47) for example:

$$\begin{aligned} G(u, v; z_0) &= \iint_{-\infty}^{+\infty} A \exp[i(k_x x + k_y y + k_z z_0 + \phi_0)] \exp[-i2\pi(ux + vy)]\mathrm{d}x\mathrm{d}y \\ &= A \exp\left[i(k_z z_0 + \phi_0)\right] \delta\left(u - \frac{k_x}{2\pi}\right) \delta\left(v - \frac{k_y}{2\pi}\right). \end{aligned} \tag{5.3}$$

Therefore, the Fourier spectrum of the plane wave has a single peak, which is placed at

$$(u, v) = \left(\frac{k_x}{2\pi}, \frac{k_y}{2\pi}\right). \tag{5.4}$$

This means that the components of a wave vector perform one to one correspondence to the Fourier frequencies.

In practice, a wave number is originally defined as "Phase rotation angle per unit length", while frequency is "Number of cycles per unit length", in Table 3.1. Thus, a wave number should be given by multiplying the spatial frequency by 2π, i.e., components of a wave vector can be written by

$$k_x = 2\pi u, \quad k_y = 2\pi v \quad \text{and} \quad k_z = 2\pi w, \tag{5.5}$$

where w is a spatial frequency along the z-axis.

When $z_0 = 0$, the integrand of (5.2) is rewritten using (5.5) as follows.

$$G(u, v; 0)\exp[i2\pi(ux + vy)] = G\left(\frac{k_x}{2\pi}, \frac{k_y}{2\pi}; 0\right)\exp[i(k_x x + k_y y)].$$

Here, the value of $G(u, v; 0)$ $(=G(k_x/(2\pi), k_y/(2\pi); 0))$ is considered as the amplitude of the plane wave represented by $\exp[i(k_x x + k_y y)]$. Hence, $G(u, v; 0)$ is called an *angular spectrum* of wavefield $g(x, y; 0)$. In this interpretation, the Fourier transform of a wavefield can be considered as disassembling the wavefield to plane waves. Inverse Fourier transform in (5.2) can also be interpreted as assembling the plane waves into the original wavefield.

The frequency w cannot be obtained from the Fourier transform of the wavefield directly. As mentioned in Sect. 3.2.3, however, the components of a wave vector has a restriction represented in (3.46), because $|\mathbf{k}| = k = 2\pi/\lambda$. Therefore, Fourier frequencies of a wavefield always satisfy

$$u^2 + v^2 + w^2 = \left(\frac{1}{\lambda}\right)^2. \tag{5.6}$$

Hence, Fourier frequencies of a wavefield form a sphere whose radius is λ^{-1} in Fourier space (u, v, w), as shown in Fig. 5.6. This is called Ewald's sphere.

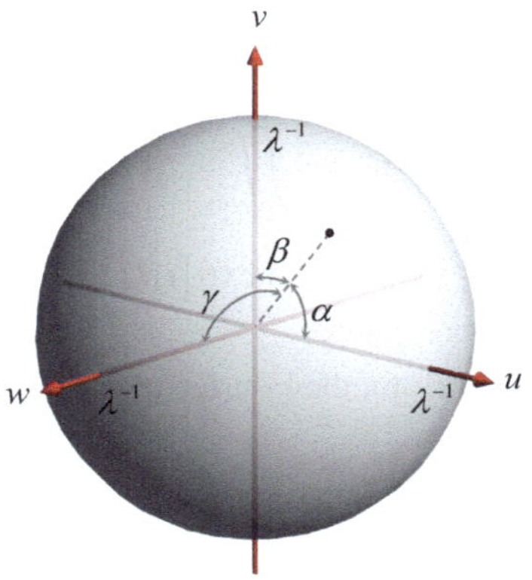

Fig. 5.6 Fourier space (u, v, w) and the point representing the spectrum of a plane wave

As a result, frequency w is not independent of frequencies u and v, but is a function of u and v as follows:

$$\begin{aligned} w &= w(u, v) \\ &= \sqrt{\lambda^{-2} - u^2 - v^2}. \end{aligned} \quad (5.7)$$

Any plane wave is represented by a point on this spherical shell as in Fig. 5.6. Substituting (3.45) into (5.5), the spatial frequencies are rewritten by the direction cosines:

$$u = \frac{\cos\alpha}{\lambda}, \quad v = \frac{\cos\beta}{\lambda} \quad \text{and} \quad w = \frac{\cos\gamma}{\lambda}. \quad (5.8)$$

Therefore, angles α, β and γ indicated in Fig. 5.6 are the same direction cosines as those in (3.45).

5.2.1.2 Formulation Based on Angular Spectrum Propagation

Propagation from the source field to the destination field, shown in Fig. 5.5, can be formulated by propagation of the angular spectrum. In the destination plane, the Fourier transform and inverse Fourier transform are given by

$$\begin{aligned} G(u, v; z) &= \mathscr{F}\{g(x, y; z)\} \\ &= \iint_{-\infty}^{+\infty} g(x, y; z) \exp[-i2\pi(ux + vy)] \mathrm{d}x\mathrm{d}y, \end{aligned} \quad (5.9)$$

$$\begin{aligned} g(x, y; z) &= \mathscr{F}^{-1}\{G(u, v; z)\} \\ &= \iint_{-\infty}^{+\infty} G(u, v; z) \exp[i2\pi(ux + vy)] \mathrm{d}u\mathrm{d}v. \end{aligned} \quad (5.10)$$

Substituting (5.10) into Helmholtz equation in (3.30), we get

$$\mathscr{F}^{-1}\left\{-(2\pi)^2(u^2 + v^2)G(u, v; z) + \frac{\partial^2}{\partial z^2}G(u, v; z)\right\} + k^2\mathscr{F}^{-1}\{G(u, v; z)\} = 0.$$

The Fourier transform of above equation is

$$\left[\frac{\partial^2}{\partial z^2} + (2\pi)^2 w^2(u, v)\right] G(u, v; z) = 0, \quad (5.11)$$

where (5.7) is used. Angular spectrum $G(u, v; z)$ satisfies this second order differential equation. The elementary solution is given by the form of

$$G(u, v; z) = G_0 \exp[i2\pi w(u, v)z]. \quad (5.12)$$

Factor G_0 must be determined by the boundary conditions. In our case, the boundary condition is given by the source wavefield $g(x, y; z_0)$ in the source plane shown in Fig. 5.5. By substituting $z = z_0$ into (5.12), we can easily find

$$G_0 = G(u, v; z_0) \exp\left[-i2\pi w(u, v)z_0\right].$$

Thus, the angular spectrum at an arbitrary z-position is written by

$$G(u, v; z) = G(u, v; z_0) \exp\left[i2\pi w(u, v)d\right], \tag{5.13}$$

where d is a distance between the source and destination planes, i.e.,

$$d = z - z_0. \tag{5.14}$$

Here, note that $w(u, v) = i(u^2 + v^2 - \lambda^{-2})^{1/2}$ when $u^2 + v^2 > \lambda^{-2}$. In this case, (5.13) is rewritten as

$$G(u, v; z) = G(u, v; z_0) \exp\left[-2\pi w(u, v)d\right]. \tag{5.15}$$

This means that the field component gives an evanescent wave and rapidly vanishes as propagating in the z-axis. Because we are not interested in the evanescent components from the viewpoint of numerical simulation, (5.13) should be represented as

$$G(u, v; z) = \begin{cases} G(u, v; z_0) \exp\left[i2\pi w(u, v)d\right] & \text{if } u^2 + v^2 \leq \lambda^{-2} \\ 0 & \text{otherwise} \end{cases}. \tag{5.16}$$

As a result, angular spectrum in the destination plane placed at an arbitrary z-position is represented by

$$G(u, v; z) = G(u, v; z_0) H_{\mathrm{AS}}(u, v; d), \tag{5.17}$$

$$H_{\mathrm{AS}}(u, v; d) \equiv \begin{cases} \exp\left[i2\pi w(u, v)d\right] & \text{if } u^2 + v^2 \leq \lambda^{-2} \\ 0 & \text{otherwise} \end{cases}. \tag{5.18}$$

$H_{\mathrm{AS}}(u, v; d)$ is called a *transfer function*. Using convolution theorem in (4.48), the wavefield in the destination plane is represented by

$$g(x, y; z) = g(x, y; z_0) \otimes h_{\mathrm{AS}}(x, y; d). \tag{5.19}$$

$h_{\mathrm{AS}}(u, v; d)$ is sometimes called a *propagation kernel*. The inverse Fourier transform of the transfer function presents the propagation kernel [47]:

$$
\begin{aligned}
h_{\mathrm{AS}}(x, y; d) &= \mathscr{F}^{-1}\{H_{\mathrm{AS}}(u, v; d)\} \\
&= \frac{\exp(ikr)}{r}\frac{d}{r}\left(\frac{1}{2\pi r} + \frac{1}{i\lambda}\right), \\
r &= \sqrt{x^2 + y^2 + d^2}.
\end{aligned}
\tag{5.20}
$$

This is called the *angular spectrum method* (AS) of field propagation.

5.2.1.3 Formulation Based on Rayleigh-Sommerfeld Formula

The Reyleigh-Sommerfeld formula gives a rigorous solution for field diffraction. The 1st Reyleigh-Sommerfeld solution for the input wavefield $g(x, y; z_0)$ is given by,

$$
\begin{aligned}
g(x, y; z) &= \iint g(x_s, y_s; z_0)\frac{\exp(ikr')}{r'}\frac{d}{r'}\left(\frac{1}{2\pi r'} + \frac{1}{i\lambda}\right)\mathrm{d}x_s\mathrm{d}y_s, \\
r' &= \sqrt{(x - x_s)^2 + (y - y_s)^2 + d^2}.
\end{aligned}
\tag{5.21}
$$

This diffraction integral is also rewritten by a convolution form using propagation kernel $h_{\mathrm{AS}}(u, v; d)$ in (5.20) as follows:

$$
g(x, y; z) = g(x, y; z_0) \otimes h_{\mathrm{AS}}(x, y; d). \tag{5.22}
$$

This is another way to formulate the AS method.

5.2.2 *Fresnel Diffraction*

Formulas of the angular spectrum method are, in general, too difficult to solve the problems of propagation or diffraction analytically. Therefore, the Fresnel approximation of the rigorous formulas is also important. Although the importance is reduced because of development in the computer technology, the technique is still considered as useful from the numerical aspect as well as the historical aspect.

5.2.2.1 Formulation Based on Angular Spectrum

Assume that almost all components of light travel along the z-axis, i.e., the field is paraxial as in Fig. 5.7a. Components of wave vectors satisfy $k_z \gg k_x, k_y$ under this paraxial situation. In the Fourier domain, the angular spectrum localizes around the origin, as in (b). In other words, the angular spectrum has non-zero values around the origin. Thus,

$$
u^2 + v^2 \ll \lambda^{-2}. \tag{5.23}
$$

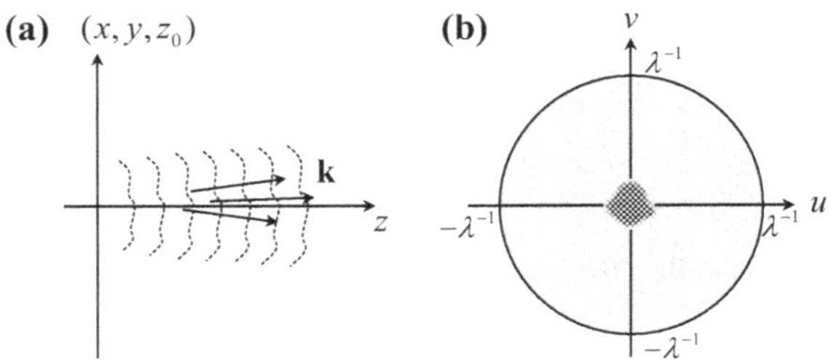

Fig. 5.7 Paraxial light in the **a** real and **b** Fourier space

In this case, (5.7) can be approximated by

$$w(u, v) \approx \frac{1}{\lambda} - \frac{\lambda}{2}\left(u^2 + v^2\right). \tag{5.24}$$

Substituting (5.24), (5.18) and (5.1) into (5.17), the destination angular spectrum is approximated by

$$G(u, v; z) \approx \exp(ikd) \iint g(x_s, y_s; z_0) \exp[-i2\pi\{ux_s + vy_s\}]\mathrm{d}x_s\mathrm{d}y_s \times \exp[-i\pi\lambda(u^2 + v^2)]. \tag{5.25}$$

Here, note that the integration variables x and y are changed to x_s and y_s respectively in order to avoid confusion in the next step. The Fourier transform of (5.25) gives the approximate destination field:

$$g(x, y; z) \approx \exp(ikd) \iint g(x_s, y_s; z_0) \iint \exp[-i\pi\lambda(u^2 + v^2)] \times \exp[i2\pi\{(x - x_s)u + (y - y_s)v\}]\mathrm{d}u\mathrm{d}v\, \mathrm{d}x_s\mathrm{d}y_s. \tag{5.26}$$

The integration of u and v can be performed by using a Fourier pair in Table 4.2. As a result, the destination field is written as

$$\begin{aligned} g_{\mathrm{FR}}(x, y; z) &= A_{\mathrm{FR}}(d) \iint g(x_s, y_s; z_0) \\ &\quad \times \exp\left[i\frac{\pi}{\lambda d}\{(x - x_s)^2 + (y - y_s)^2\}\right] \mathrm{d}x_s\mathrm{d}y_s, \\ A_{\mathrm{FR}}(d) &\equiv \frac{\exp(ikd)}{i\lambda d}. \end{aligned} \tag{5.27}$$

This is well-known formula of the Fresnel diffraction.

Equation (5.27) is also represented by the convolution form as follows:

$$g_{\mathrm{FR}}(x, y; z) = g(x, y; z_0) \otimes h_{\mathrm{FR}}(x, y; d), \tag{5.28}$$

where the propagation kernel and transfer function are given by

$$h_{\mathrm{FR}}(x, y; d) \equiv \frac{1}{i\lambda d} \exp\left[ik\left(d + \frac{x^2+y^2}{2d}\right)\right], \tag{5.29}$$

$$H_{\mathrm{FR}}(u, v; d) \equiv \exp\left[i\pi d\left\{\frac{2}{\lambda} - \lambda(u^2+v^2)\right\}\right]. \tag{5.30}$$

In many textbooks and the following section in this book, above Fresnel approximation is introduced under a condition where the propagation distance is sufficiently large in comparison with the aperture size. However, in formulation based on the angular spectrum method, Fresnel approximation is essentially provided by paraxiality rather than the distance.

5.2.2.2 Formulation Based on Rayleigh-Sommerfeld Formula

In the Rayleigh-Sommerfeld formula represented in (5.21), assume a situation where propagation distance d is long enough to consider r' as $r' \approx d$ in denominators and neglect the term $1/(2\pi r')$. Equation (5.21) is rewritten as

$$g(x, y; z) = \frac{1}{i\lambda d} \iint g(x_s, y_s; z_0) \exp(ikr') \mathrm{d}x_s \mathrm{d}y_s, \tag{5.31}$$

$$r' = \sqrt{(x-x_s)^2 + (y-y_s)^2 + d^2}, \tag{5.32}$$

where $d = z - z_0$ again. Furthermore, because the distances d is much larger than $(x - x_s)$ and $(y - y_s)$, the phase kr' of the exponent can be expanded into Maclaurin series as follows:

$$kr' = kd\sqrt{1 + \frac{(x-x_s)^2 + (y-y_s^2)}{d^2}} \tag{5.33}$$

$$= kd + \frac{k[(x-x_s)^2 + (y-y_s)^2]}{2d} - \frac{k[(x-x_s)^2 + (y-y_s)^2]^2}{8d^3} + \cdots. \tag{5.34}$$

Assuming the third term is much smaller than 2π, i.e.,

$$\left|\frac{k[(x-x_s)^2 + (y-y_s)^2]^2}{8d^3}\right| \ll 2\pi, \tag{5.35}$$

the phase is approximately represented by

$$kr' \approx kd + \frac{k[(x-x_s)^2 + (y-y_s)^2]}{2d}. \tag{5.36}$$

As a result, the destination field is rewritten as

$$g_{\mathrm{FR}}(x, y; z) = \frac{\exp(ikd)}{i\lambda d} \iint g(x_{\mathrm{s}}, y_{\mathrm{s}}; z_0) \times \exp\left[i\frac{\pi}{\lambda d}\{(x_{\mathrm{s}} - x)^2 + (y_{\mathrm{s}} - y)^2\}\right] \mathrm{d}x_{\mathrm{s}}\mathrm{d}y_{\mathrm{s}}. \quad (5.37)$$

In this formulation, the condition necessary for the approximation is given as follows:

$$d^3 \gg \frac{[(x - x_{\mathrm{s}})^2 + (y - y_{\mathrm{s}})^2]^2}{8\lambda}. \quad (5.38)$$

5.2.3 *Fraunhofer Diffraction*

5.2.3.1 Formulation

Equation (5.36) can be rewritten as

$$kr' \approx kd + \frac{k\left[(x^2 + y^2) - 2(xx_{\mathrm{s}} + yy_{\mathrm{s}})\right]}{2d} + \frac{k(x_{\mathrm{s}}^2 + y_{\mathrm{s}}^2)}{2d}. \quad (5.39)$$

Here, note that x_{s} and y_{s} denote coordinates in the source plane, while x and y are coordinates of the destination plane. Assume that the destination plane is even far from the source plane and the area of the source field is not large; that is,

$$\frac{k(x_{\mathrm{s}}^2 + y_{\mathrm{s}}^2)}{2d} \ll 2\pi. \quad (5.40)$$

The third term in the right-hand side of (5.39) can be neglected in this case. Equation (5.37) is rewritten by

$$g_{\mathrm{far}}(x, y; z) = \frac{\exp(ikd)}{i\lambda d} \exp\left[i\frac{k}{2d}(x^2 + y^2)\right] \times \iint g(x_{\mathrm{s}}, y_{\mathrm{s}}; z_0) \exp\left[-i\frac{k}{d}(xx_{\mathrm{s}} + yy_{\mathrm{s}})\right] \mathrm{d}x_{\mathrm{s}}\mathrm{d}y_{\mathrm{s}}. \quad (5.41)$$

Let us introduce new symbols as followings:

$$u_{\mathrm{s}} \equiv \frac{x}{\lambda d}, \quad v_{\mathrm{s}} \equiv \frac{y}{\lambda d}. \quad (5.42)$$

Equation (5.41) is rewritten by

$$g_{\text{far}}(x, y; z) = A_{\text{far}}(x, y; d) \\ \times \iint g(x_s, y_s; z_0) \exp\left[-i2\pi(u_s x_s + v_s y_s)\right] \mathrm{d}x_s \mathrm{d}y_s, \quad (5.43)$$

where

$$A_{\text{far}}(x, y; d) \equiv \frac{\exp(ikd)}{i\lambda d} \exp\left[i\frac{k}{2d}(x^2 + y^2)\right]. \quad (5.44)$$

This is called the *Fraunhofer diffraction*. This type of diffraction is also called the *far-field diffraction*, because this type of diffraction and propagation is only valid in the cases where the destination fields are far from the source field. The terminology of the far-field diffraction is mainly used instead of the Fraunhofer diffraction in this book.

Integral in (5.43) simply agrees with the Fourier transform. We can consider that (5.42) associates spatial coordinates in the far field with the spatial frequencies. Thus, we rewrite (5.43) using the Fourier transform:

$$g_{\text{far}}(x, y; z) = A_{\text{far}}(x, y; d)\mathscr{F}\left\{g(x, y; z_0)\right\}_{u_s = \frac{x}{\lambda d},\, v_s = \frac{y}{\lambda d}} \\ = A_{\text{far}}(x, y; d) G\left(\frac{x}{\lambda d}, \frac{y}{\lambda d}; z_0\right), \quad (5.45)$$

where $G(u_s, v_s; z_0) = \mathscr{F}\{g(x, y; z_0)\}$ and $d = z - z_0$ again. The subscripts in (5.45) denote that Fourier frequencies are replaced by corresponding spatial coordinates after the Fourier transform.

5.2.3.2 Connection of Fourier Frequencies with Direction in Far-Field Propagation

Definitions of (5.42) suggest that there is a closed connection of the Fourier frequencies of a wavefield with the propagation direction. The field having Fourier components u_s reaches an x-position after propagation at a long distance of d, as shown in Fig. 5.8. Therefore, we can represent the propagation angle in $(x, 0, z)$ and $(0, y, z)$ planes as,

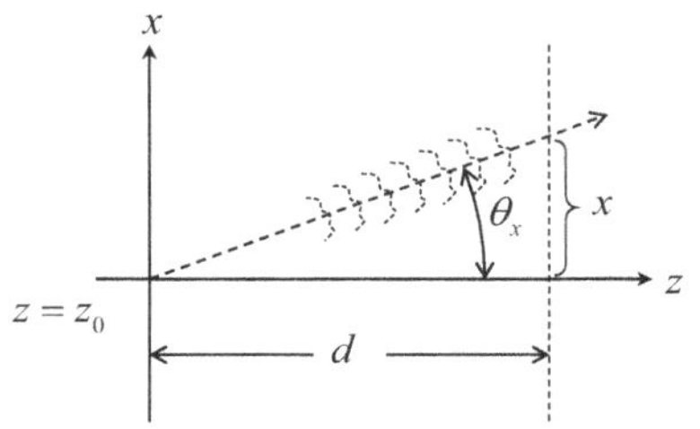

Fig. 5.8 Connection between the field direction and frequencies

$$\sin\theta_x \cong \tan\theta_x = \frac{x}{d}$$
$$\sin\theta_y \cong \tan\theta_y = \frac{y}{d}. \tag{5.46}$$

Therefore,

$$u_s = \frac{1}{\lambda}\tan\theta_x \cong \frac{1}{\lambda}\sin\theta_x,$$
$$v_s = \frac{1}{\lambda}\tan\theta_y \cong \frac{1}{\lambda}\sin\theta_y. \tag{5.47}$$

Here, do not forget that there is a restriction of the propagation distance:

$$x \ll d \quad \text{and} \quad y \ll d. \tag{5.48}$$

The connection between the Fourier frequencies and propagation direction of a wavefield is also suggested by the relation between the spatial frequencies and direction cosines in (5.8). These are very general idea and often used to interpret and understand the behavior of diffracted wavefields.

5.3 Optical Fourier Transform by Thin Lens

In the preceding section, we learned that the far-field propagation is equivalent to the Fourier transform of the source wavefield. This looks a useful technique for some applications, which require the Fourier transform. However, we need a long propagation distance to carry out the actual Fourier transform by the far-field propagation. A lens provides an alternative technique for the optical Fourier transform.

5.3.1 Wave-Optical Property of Thin Lens

If a lens is thin enough that the thickness is negligible, the effect of the lens on light can be represented as three properties: (i) a collimated beam traveling along the optical axis perpendicular to the lens converges on a spot called a *focal point*, (ii) a beam passing through the focal point and entering the lens is converted into a collimated beam, and (c) a beam passing through the center of the lens is not affected by the lens, as shown in Fig. 5.9a. As a result, the field of a point source of light in the object plane at a distance of d_1 from the lens focuses on a point in the image plane at a distance of d_2 from the lens. The distance between the focal point and lens is called *focal length* and represented by f. The relation among d_1, d_2 and f is known as the *thin lens formula*:

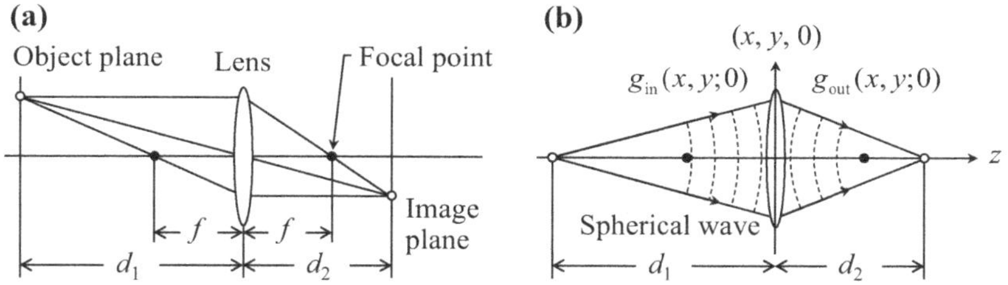

Fig. 5.9 **a** Image formation by a thin lens. **b** Wave-optical model of a thin lens

$$\frac{1}{d_1} + \frac{1}{d_2} = \frac{1}{f}. \tag{5.49}$$

Light emitted from any source point in the object plane converges into a point in the image plane. Thus, suppose a point source is located in the optical axis, as shown in Fig. 5.9b. The divergent spherical field is converted into a convergent spherical field by refraction of the lens. Let $t(x, y)$ represent the complex amplitude of a lens of negligible thickness. The light output from the thin lens is given by

$$g_{out}(x, y; 0) = t(x, y) g_{in}(x, y; 0), \tag{5.50}$$

where $g_{in}(x, y; 0)$ is a spherical wave incident on the lens. Here, we place the coordinates origin at the lens plane. According to (3.71), the incident and output spherical wavefields can be written as follows in the Fresnel approximation:

$$g_{in}(x, y; 0) = A_{in} \exp\left[+i\frac{k}{2d_1}(x^2 + y^2)\right], \tag{5.51}$$

$$g_{out}(x, y; 0) = A_{out} \exp\left[-i\frac{k}{2d_2}(x^2 + y^2)\right]. \tag{5.52}$$

When the lens is transparent and thus the wavefield has the same amplitude after transmission; $A_{in} = A_{out}$, complex transmittance of a lens should be written as

$$\begin{aligned} t_{lens}(x, y) &= t(x, y) p(x, y) \\ &= \exp\left[-i\frac{k}{2}\left(\frac{1}{d_1} + \frac{1}{d_2}\right)(x^2 + y^2)\right] p(x, y), \end{aligned} \tag{5.53}$$

where $p(x, y)$ is called a pupil function. The pupil function is a real binary function corresponding to the shape of the lens, and thus, given by

$$p(x, y) = \begin{cases} 1 & \text{inside lens} \\ 0 & \text{otherwise} \end{cases} \tag{5.54}$$

Using thin lens formula (5.49) the lens transmittance is given by

$$t_{\text{lens}}(x, y) = \exp\left[-i\frac{k}{2f}(x^2 + y^2)\right] p(x, y). \tag{5.55}$$

5.3.2 *Wavefield Refracted by Thin Lens*

Let the incident wavefield into the lens be $g(x, y; 0)$, as shown in Fig. 5.10. Here, we set the coordinates origin at the center of the thin lens again. The wavefield refracted by the lens is given in the space at $z > 0$ using the Fresnel diffraction:

$$\begin{aligned} g(x, y; z) &= [t_{\text{lens}}(x, y) g(x, y; 0)] \otimes h_{\text{FR}}(x, y; z) \\ &= A_{\text{FR}}(z) \iint t_{\text{lens}}(x_{\text{s}}, y_{\text{s}}) g(x_{\text{s}}, y_{\text{s}}; 0) \\ &\quad \times \exp\left[i\frac{\pi}{\lambda z}\{(x_{\text{s}} - x)^2 + (y_{\text{s}} - y)^2\}\right] \mathrm{d}x_{\text{s}}\mathrm{d}y_{\text{s}}. \end{aligned} \tag{5.56}$$

Substituting (5.55) and expanding the exponent in the integrand,

$$\begin{aligned} g(x, y; z) = {} & A_{\text{FR}}(z) \exp\left[i\frac{k}{2z}(x^2 + y^2)\right] \\ & \times \iint g(x_{\text{s}}, y_{\text{s}}; 0) p(x_{\text{s}}, y_{\text{s}}) \exp\left[i\frac{\pi}{\lambda}\left(\frac{1}{z} - \frac{1}{f}\right)(x_{\text{s}}^2 + y_{\text{s}}^2)\right] \\ & \times \exp\left[-i\frac{2\pi}{\lambda z}(x x_{\text{s}} + y y_{\text{s}})\right] \mathrm{d}x_{\text{s}}\mathrm{d}y_{\text{s}}. \end{aligned} \tag{5.57}$$

This can be rewritten using the Fourier transform.

$$\begin{aligned} g(x, y; z) = {} & A_{\text{FR}}(z) \exp\left[i\frac{k}{2z}(x^2 + y^2)\right] \\ & \times \mathscr{F}\left\{g(x, y; 0) p(x, y) \exp\left[i\frac{\pi}{\lambda}\left(\frac{1}{z} - \frac{1}{f}\right)(x^2 + y^2)\right]\right\}, \end{aligned} \tag{5.58}$$

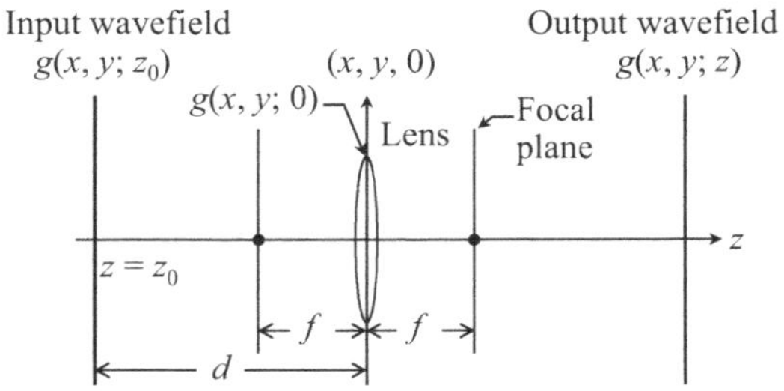

Fig. 5.10 Calculation of the wavefield refracted by a lens whose focal length is f

where the Fourier frequencies are given similarly to (5.42):

$$u_s = \frac{x}{\lambda d} \quad \text{and} \quad v_s = \frac{y}{\lambda d}. \tag{5.59}$$

When obtaining the wavefield at the focal plane that is perpendicular to the optical axis and include the focal point, since $z = f$, the above equation is simplified down to

$$g(x, y; f) = A_{\text{FR}}(f) \exp\left[i\frac{k}{2f}(x^2 + y^2) \right] \times G\left(\frac{x}{\lambda f}, \frac{y}{\lambda f}; 0 \right) \otimes P\left(\frac{x}{\lambda f}, \frac{y}{\lambda f} \right). \tag{5.60}$$

where $P(u, v)$ and $G(u, v; 0)$ are Fourier-transformed spectra of the pupil function and incident field, respectively:

$$P(u, v) = \mathscr{F}\{p(x, y)\}, \tag{5.61}$$

$$G(u, v; 0) = \mathscr{F}\{g(x, y; 0)\}. \tag{5.62}$$

Supposing that $g(x, y; 0)$ is the wavefield that is propagated from the source field $g(x, y; z_0)$ as shown in Fig. 5.10, the spectrum is given by the Fresnel diffraction in (5.28):

$$G(u, v; 0) = G(u, v; z_0) H_{\text{FR}}(u, v; d), \tag{5.63}$$

where d is the propagation distance, and thus, $z_0 = -d$. In this case, the wavefield at the focal plane is

$$g(x, y; f) = A_{\text{FR}}(f) \exp(ikd) \exp\left[i\frac{k}{2f}(x^2 + y^2) \right] \times P\left(\frac{x}{\lambda f}, \frac{y}{\lambda f} \right) \otimes \left\{ G\left(\frac{x}{\lambda f}, \frac{y}{\lambda f}; z_0 \right) \exp\left[-i\pi \frac{d}{\lambda f^2}(x^2 + y^2) \right] \right\}. \tag{5.64}$$

If the lens is large enough that the pupil function is regarded as $p(x, y) \approx 1$, the Fourier transform is

$$\mathscr{F}\{p(x, y)\} = \delta(u, v). \tag{5.65}$$

In this case, the wavefield is simply given by

$$g(x, y; f) = A_{\text{FR}}(f) \exp(ikd) \exp\left[i\left(1 - \frac{d}{f}\right) \frac{k}{2f}(x^2 + y^2) \right] G\left(\frac{x}{\lambda f}, \frac{y}{\lambda f}; z_0 \right).$$

Furthermore, if the source filed is also given at the focal pane; $d = f$, the wavefield is drastically simplified down to

$$\begin{aligned} g(x, y; +f) &= \frac{\exp(2ikf)}{i\lambda f} G\left(\frac{x}{\lambda f}, \frac{y}{\lambda f}; -f\right) \\ &= \frac{\exp(2ikf)}{i\lambda f} \mathscr{F}\{g(x, y; -f)\}_{u=\frac{x}{\lambda f}, v=\frac{y}{\lambda f}} . \end{aligned} \tag{5.66}$$

In conclusion, when both the source and destination wavefields are placed at the focal plane across the lens, the destination field is given by the Fourier transform of the source field. In other words, we can get the same effect by using a lens as that of the far-field propagation. This is a very useful property of a lens and thus used for various optical information processing. In computer holography, this optical Fourier transform using a lens is mainly used to separate non-diffraction light, true and conjugate images (see Sect. 8.10).

5.4 Propagation Operator

Process for obtaining the destination wavefield $g(x, y; z)$ from the source wavefield $g(x, y; z_0)$ can be regarded as a mathematical operation. Thus, we define an operator as follows:

$$\mathscr{P}_d\{g(x, y; z_0)\} = g(x, y; z), \tag{5.67}$$

where $d = z - z_0$ again. We call the symbol $\mathscr{P}_d\{\cdot\}$ a *propagation operator* or *translational propagation operator* more exactly.

5.4.1 Propagation Operator as System

The word *system* is sometimes used to represent mapping of a set of input function into a set of output function. Thus, we can say that the propagation operator presents a kind of system.

It is apparent that the propagation operator presents a liner system:

$$\mathscr{P}_d\{a_1 g_1(x, y; z_0) + a_2 g_2(x, y; z_0)\} = \mathscr{P}_d\{a_1 g_1(x, y; z_0)\} + \mathscr{P}_d\{a_2 g_2(x, y; z_0)\}, \tag{5.68}$$

where a_1 and a_2 are constants. When we consider the field propagation as propagation by the angular spectrum given by (5.16) or the Fresnel diffraction given by (5.27), the propagation operator satisfies the following:

$$\mathscr{P}_d\{g(x - x_0, y - y_0; z_0)\} = g(x - x_0, y - y_0; z), \tag{5.69}$$

where x_0 and y_0 are constants. The system having this property is called a space-invariant system. Accordingly, propagation operator presents a linear space-invariant system. This is the reason that the field propagation can be represented by convolution in the angular spectrum and Fresnel diffraction methods.

When inputting the delta function $\delta(x, y)$ to a system, the output is called impulse response. Therefore,

$$\begin{aligned}\mathcal{P}_d\{\delta(x, y)\} &= h_{\text{AS}}(x, y; d), \quad \text{(Angular spectrum method)} \\ &= h_{\text{FR}}(x, y; d), \quad \text{(Fresnel diffraction)}\end{aligned}$$

give the impulse responses. These are also called *point spread functions* in optics.

Unfortunately, if field propagation is treated as Fourier transform-type propagation such as the Fraunhofer diffraction given in (5.41) and the Fourier transform by a lens given in (5.66), the propagation operator is no longer space invariant. Accordingly, we cannot define the point spread function in these types of propagation.

5.4.2 Backward Propagation

When the z-position of the destination field is larger than that of the source field, the propagation is referred to as "forward" propagation. Supposing that (5.67) represents forward propagation, it is natural that

$$\mathcal{P}_{-d}\{g(x, y; z)\} = g(x, y; z_0), \tag{5.70}$$

represents "backward" propagation. Here, note that $z \geq z_0$, and thus, $d \geq 0$. The propagation operator should satisfy the following:

$$\mathcal{P}_{-d}\{\mathcal{P}_d\{g(x, y; z_0)\}\} = g(x, y; z_0). \tag{5.71}$$

When a space-invariant propagation operator is used, this can be confirmed as follows:

$$\begin{aligned}\mathcal{P}_{-d}\{\mathcal{P}_d\{g(x, y; z_0)\}\} &= [g(x, y; z_0) \otimes h(x, y; d)] \otimes h(x, y; -d) \\ &= g(x, y; z_0) \otimes \mathcal{F}^{-1}\{H(u, v; d)H(u, v; -d)\} \\ &= g(x, y; z_0) \otimes \delta(x, y) \\ &= g(x, y; z_0),\end{aligned} \tag{5.72}$$

where $\mathcal{F}^{-1}\{H(u, v; d)H(u, v; -d)\} = \delta(x, y)$. This can be easily verified both in the angular spectrum method and the Fresnel diffraction method, because

$$H_{\text{AS}}(u, v; d)H_{\text{AS}}(u, v; -d) = H_{\text{FR}}(u, v; d)H_{\text{FR}}(u, v; -d) = 1. \tag{5.73}$$

If propagation is treated as the Fourier-type propagation and the backward propagation is required, an operator specific to backward propagation must be defined to satisfy (5.71) as follows:

$$\begin{aligned}\mathscr{P}_{-d}\{g(x,y;z)\} &= \mathscr{F}^{-1}\{g(u\lambda d, v\lambda d; z)/A_{\text{far}}(u\lambda d, v\lambda d; d)\} \\ &= i\lambda d \exp[-ikd] \\ &\quad \times \mathscr{F}^{-1}\left\{\exp[i\pi\lambda d(u^2+v^2)g(u\lambda d, v\lambda d; z)\right\}. \end{aligned} \tag{5.74}$$

Using the propagation operator, we can move the position of the interested wavefield back and forth along the z-axis. However, it should be noted that numerical errors are more or less produced in practical numerical propagation, as mentioned in the following several chapters. Thus, the field after round-trip propagation does not always agree with the original wavefield exactly.

Chapter 6
Numerical Field Propagation Between Parallel Planes

Abstract In this chapter, we discuss numerical techniques to simulate the most basic field propagation, i.e., the field propagation between parallel planes. There are several techniques in this category, such as single-FFT propagation, and convolution-based techniques. In any technique, restrictions as well as pros and cons are discussed from the aspect of aliasing errors produced in the numerical calculation in many cases. Not only formulas but also plenty of visualized examples of diffracted wavefields are presented in this chapter. The contents are most likely useful for researchers and engineers who deal with wave optics and holography.

6.1 Far-Field Propagation

Numerical field propagation between parallel planes is the most basic numerical propagation and directly derived from the scalar diffraction theory described in Chap. 5. In reverse order of description in Sect. 5.2, let us start from the far field propagation because of simplicity. The far-field propagation is based on the Fraunhofer diffraction described in (5.45). Since the Fraunhofer diffraction is represented by the Fourier transform, the numerical calculation is formulated straightforwardly.

6.1.1 Discrete Formula

To formulate the far-field propagation of sampled wavefields, we use the symmetrical sampling mentioned in Sect. 4.7.2. Thus, sampling positions in the source field are given by

$$\begin{aligned} x_{s,m} &= \Delta x_s \left(m - \frac{M}{2} \right), \quad (m = 0, \ldots, M-1), \\ y_{s,n} &= \Delta y_s \left(n - \frac{N}{2} \right), \quad (n = 0, \ldots, N-1), \end{aligned} \tag{6.1}$$

K. Matsushima, *Introduction to Computer Holography*,
Series in Display Science and Technology,
https://doi.org/10.1007/978-3-030-38435-7_6

and

$$u_{s,p} = \Delta u_s \left(p - \frac{M}{2} \right), \quad (p = 0, \ldots, M-1),$$
$$v_{s,q} = \Delta v_s \left(q - \frac{N}{2} \right), \quad (q = 0, \ldots, N-1), \tag{6.2}$$

where Δx_s, Δy_s, Δu_s and Δv_s are sampling intervals of the source field, and M and N are the numbers of sample points. Here, let both M and N be even numbers.

As in (5.43), the source field, sampled with the symmetrical manner in (6.1) and (6.2), is first Fourier-transformed using FFT in (4.90).

$$G(u_{s,p}, v_{s,q}; z_0) = \sum_{m=0}^{M-1} \sum_{n=0}^{N-1} g(x_{s,m}, y_{s,n}; z_0) \exp\left[-i2\pi(u_{s,p}x_m + v_{s,q}y_n)\right].$$

In the indexed form, this is written by

$$G[p, q; z_0] = \mathbf{FFT}\left\{g[m, n; z_0]\right\}, \tag{6.3}$$

where $g[m, n; z_0] = g(x_{s,m}, y_{s,n}; z_0)$ and $G[p, q; z_0] = G(u_{s,p}, v_{s,q}; z_0)$. Note that FFT is evaluated by the technique shown in Fig. 4.10 of Sect. 4.7.2 in practice. According to (5.45), the sampled destination field is

$$g_{\text{far}}(x_p, y_q; z) = A_{\text{far}}(x_p, y_q; d)G[p, q; z_0],$$
$$(p = 0, \ldots, M-1 \text{ and } q = 0, \ldots, N-1). \tag{6.4}$$

When we use ordinary DFT/FFT, the sampling intervals satisfy (4.78):

$$\Delta u_s = \frac{1}{M\Delta x_s},$$
$$\Delta v_s = \frac{1}{M\Delta y_s}. \tag{6.5}$$

Substituting above into (6.2) and using (5.42), the sampling positions in the destination plane are given as follows:

$$x_p = u_{s,p}\lambda d = \frac{\lambda d}{M\Delta x_s}\left(p - \frac{M}{2} \right),$$
$$y_q = v_{s,q}\lambda d = \frac{\lambda d}{N\Delta y_s}\left(q - \frac{N}{2} \right). \tag{6.6}$$

As a result, the sampling intervals of the destination field are given by

$$\Delta x = x_{p+1} - x_p = \frac{\lambda}{M\Delta x_s}d,$$
$$\Delta y = y_{q+1} - y_q = \frac{\lambda}{N\Delta y_s}d. \tag{6.7}$$

This shows that the sampling interval changes proportionally to the propagation distance d.

6.1.2 Destination Sampling Window

Since the number of sample points is a constant of $M \times N$, the destination sampling window expands as the propagation distance increases, i.e.,

$$W_{\mathrm{far},x} = M\Delta x = \frac{\lambda}{\Delta x_s}d,$$
$$W_{\mathrm{far},y} = N\Delta y = \frac{\lambda}{\Delta y_s}d, \tag{6.8}$$

where $W_{\mathrm{far},x}$ and $W_{\mathrm{far},y}$ are the sizes of the destination sampling window in the far-field propagation in the x and y directions, respectively. Change of the sampling window in the $(x, 0, z)$ plane is schematically shown in Fig. 6.1. Angle $\theta_{\mathrm{far},x}$ indicated in this figure is given by

$$\theta_{\mathrm{far},x} = \tan^{-1}\left(\frac{\lambda}{2\Delta x_s}\right). \tag{6.9}$$

When $\Delta x_s \gg \lambda/2$, this angle is, in practice, the same as the maximum diffraction angle in (3.54);

$$\theta_{\mathrm{far},x} \approx \theta_{\mathrm{max},x} = \sin^{-1}\left(\frac{\lambda}{2\Delta x_s}\right). \tag{6.10}$$

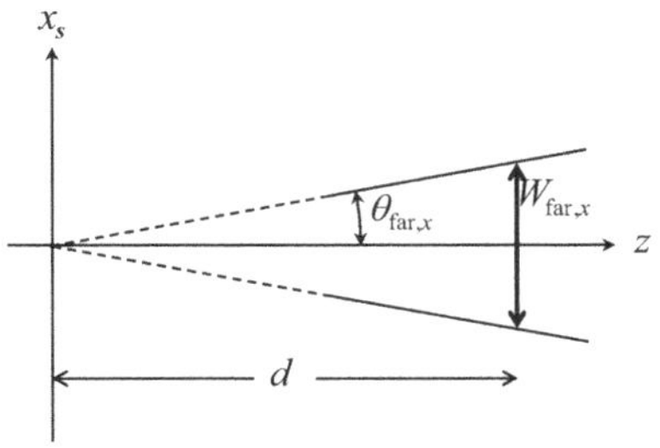

Fig. 6.1 The sampling window of the far-field propagation based on the Fraunhofer diffraction

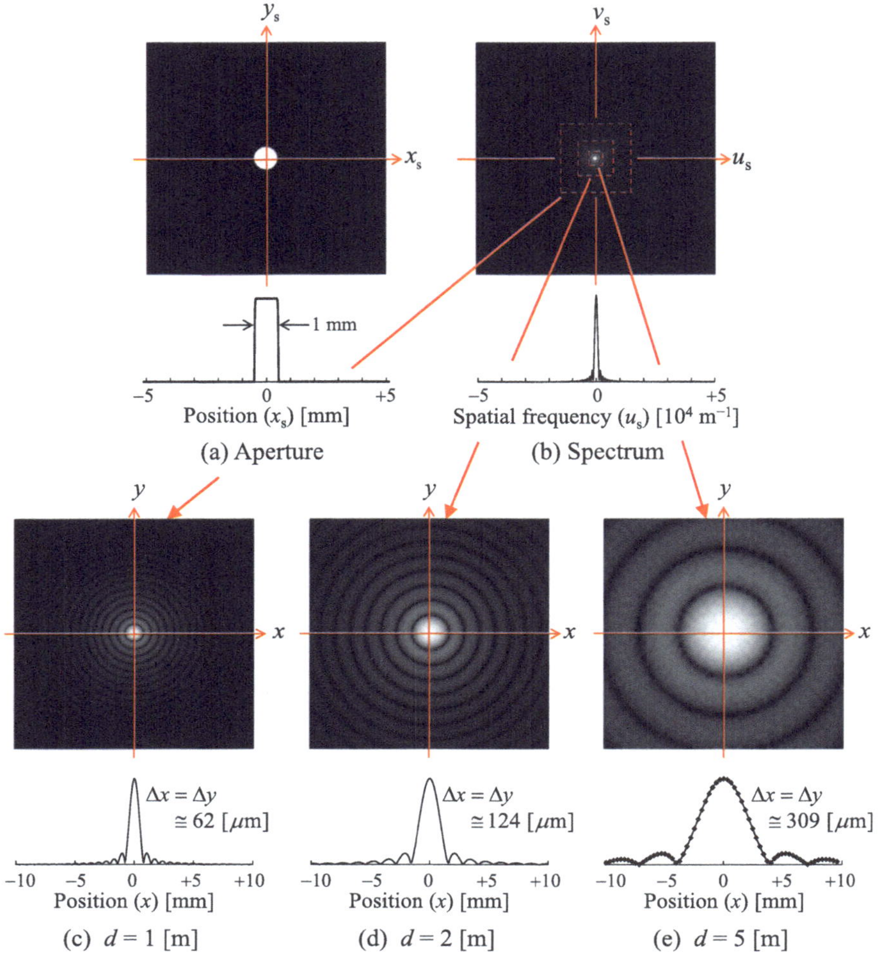

Fig. 6.2 An example of numerical calculation of the far-field diffraction. **a** Source field $|g(x_{s,m}, y_{s,n}; z_0)|$, $M = N = 1024$, $\Delta x_s = \Delta y_s = 10\,[\mu\text{m}]$, $\lambda = 633\,[\text{nm}]$. **b** Spectral amplitude $|G[p, q; z_0]|$. **c**–**d** Amplitude images of the destination field $|g_{\text{far}}(x_p, y_q; z)/(\lambda d)^2|$ for different propagation distances $d = z - z_0$. Note that images of (**c**)–(**d**) are just cut out from (**b**). According to (5.40), the propagation distance must satisfy $d \gg 0.2\,[\text{m}]$ because maximum $|x_s|$ is 0.5 mm. Thus, also note that these examples may not perfectly meet the condition in (5.40)

6.1.3 Numerical Example

Figure 6.2 shows examples of the far field propagation. The source field is shown in (a). In this example, a circular aperture whose diameter is 1 mm is placed at the center of the source plane, besides it is assumed that the aperture is irradiated by a plane wave traveling perpendicularly to the aperture. Therefore, the source wavefield

$g(x_{s,m}, y_{s,n}; z_0)$ is represented by a real-valued function. The source field is shown as a digital image in Fig. 6.2a, whose white and black pixels show 1 and 0, respectively. Here, the sampling interval and number of sample points of the source field are 10 μm and 1024 in both the x and y directions, respectively. Note that the source field is not a simple binary function, but actually given by

$$g(x, y; z_0) = \exp\left[-\left(\frac{\sqrt{x^2+y^2}}{D/2}\right)^{50}\right], \tag{6.11}$$

in order to reduce the effect of jaggies. Here, the aperture has a diameter of $D = 1$ [mm].

The amplitude pattern of the spectrum $G[p, q; z_0]$ of the sampled source field is shown in Fig. 6.2b. The size of the sampling window is $M\Delta u \times N\Delta v = \Delta x^{-1} \times \Delta y^{-1} = 10^5$ [m^{-1}] $\times$ 10^5 [m^{-1}]. Here, the amplitude $|G[p, q; z_0]|$ is normalized so that the peak value is unity. White and black pixels in the image represent 1 and 0 again. The grayscale images are produced using an encoding game of 1/2.2 in practice. The lateral scale is given by (6.2).

Amplitude images of $|g_{\text{far}}(x_p, y_q; z)|$ in the destination plane are shown in Fig. 6.2c–e for different propagation distances. Here, the same manner is used for the visualization, i.e., the amplitude is normalized and gray level is proportional to the normalized amplitude. These all results are actually the same as the spectrum (b) of the source field. Only the lateral scale is changed dependently on the physical coordinates by using (6.6). As the propagation distance d increases, the diffraction patterns seem to extend proportionally. This is because constant square regions of side 20 mm are cut out from the spectrum (b), while the sampling window expands in coordinates $(x, y; z)$ as the propagation distance d increases.

6.1.4 Sampling Problem

As regards amplitude or intensity distribution, there is no sampling problem in this method. However, factor $A_{\text{far}}(x, y; d)$ in (5.44) can cause a problem of the phase pattern.

Redefining $A_{\text{far}}(x, y; d)$ as

$$A_{\text{far}}(x, y; d) = \frac{\exp(ikd)}{i\lambda d}\exp\left[i\phi(x, y)\right],$$
$$\phi(x, y) \equiv \frac{\pi}{\lambda d}(x^2+y^2), \tag{6.12}$$

the local spatial frequency with respect to x is given in the source plane as follows [19]:

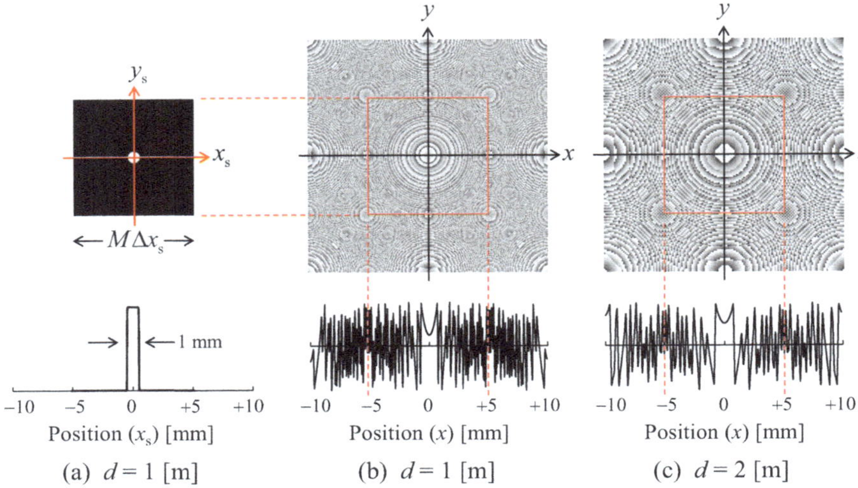

Fig. 6.3 Examples of phase distribution in the far-field diffraction. **a** Source field $|g(x_{s,m}, y_{s,n}; z_0)|$, $M = N = 1024$, $\Delta x_s = \Delta y_s = 10\,[\mu\text{m}]$, $\lambda = 633\,\text{nm}$. **b–c** Phase images of the destination field $g_{\text{far}}(x_p, y_q; z)/(\lambda d)^2$ for different propagation distances. The phase images produce aliasing errors outside the red square

$$f_x = \frac{1}{2\pi}\frac{\partial}{\partial x}\phi(x, y) \\ = \frac{x}{\lambda d}. \tag{6.13}$$

According to the sampling theorem, this local frequency must satisfy $\Delta x^{-1} > 2|f_x|$ to avoid aliasing errors. Therefore, we obtain the following restriction in the sampled field,

$$|x| < \frac{\lambda d}{2\Delta x}. \tag{6.14}$$

Substituting (6.7) into above relation, we can find that the sampling position in the destination plane must be limited to a range:

$$-\frac{M\Delta x_s}{2} < x < \frac{M\Delta x_s}{2}. \tag{6.15}$$

Figure 6.3 shows examples of phase distribution in far-field propagation. Because $M\Delta x_s$ represents the size of the source sampling window, the above relation means that the size of the destination sampling window W_x in (6.8) must not exceed that of the source window; $W_x < M\Delta x_s$ in order to avoid aliasing errors. This relation can be rewritten as

$$M > \frac{\lambda}{\Delta x_s^2}d \text{ and } N > \frac{\lambda}{\Delta y_s^2}d. \tag{6.16}$$

These relations give the minimum numbers of sample points to avoid the sampling problem of phase distribution in the destination plane. Since the minimum numbers of the sample points for each axis are proportional to the propagation distance, the total number of the sampling points MN is proportional to the square of the propagation distance.

This fact seems to reduce usefulness of the technique. However, the technique does not lost the value when only the diffraction image, i.e., intensity distribution, is calculated, because the factor $|A_{\text{far}}(x, y; d)|^2$ is a constant in this case. The sampling problem only arises in cases where the phase pattern in the destination plane is required for some purpose.

6.2 The Fourier Transform by Lens

Many ideas in computer holography are based on the Fourier transform by a lens. Analytical formulas of the optical Fourier transform, given in (5.66), are very similar to the far-field propagation. In fact, replacing d in the far-field propagation by f in the optical Fourier transform, the formulas almost agree with those in the far-field propagation. Discrete formulas are summarized below.

Supposing the sampled input field is $g[m, n] = g(x_m, y_n)$ and the sampling position is given by (6.1), the Fourier transform is

$$G[p, q; -f] = \mathbf{FFT}\{g[m, n; -f]\}, \tag{6.17}$$

and the output field is given by

$$g_{\text{Four}}(x_p, y_q; f) = \frac{\exp(2ikf)}{i\lambda f} G[p, q; -f],$$
$$(p = 0, \ldots, M-1 \text{ and } q = 0, \ldots, N-1), \tag{6.18}$$

where the sampling position of the output field is

$$x_p = \Delta x\left(p - \frac{M}{2}\right) \quad \text{and} \quad y_q = \Delta y\left(q - \frac{N}{2}\right). \tag{6.19}$$

The sampling intervals of the destination field are given by

$$\Delta x = \frac{\lambda}{M\Delta x_s} f \quad \text{and} \quad \Delta y = \frac{\lambda}{N\Delta y_s} f. \tag{6.20}$$

The discrete formulas to simulate the optical Fourier transform are so simple that there is nothing to cause a sampling problem.

6.3 Single-Step Fresnel Propagation

The integral of the Fresnel diffraction in (5.27) can be performed by using FFT only once. This technique is referred to as *single-step Fresnel propagation* in this book.

6.3.1 Formulation

Equation (5.27) can be rewritten by expanding the exponential term as follows:

$$\begin{aligned} g_{\mathrm{FR}}(x, y; z) &= A_{\mathrm{far}}(x, y; d) \iint g(x_{\mathrm{s}}, y_{\mathrm{s}}; z_0) \exp\left[i\frac{\pi}{\lambda d}\left(x_{\mathrm{s}}^2 + y_{\mathrm{s}}^2\right)\right] \\ &\quad \times \exp\left[-i2\pi(u_{\mathrm{s}}x_{\mathrm{s}} + v_{\mathrm{s}}y_{\mathrm{s}})\right] \mathrm{d}x_{\mathrm{s}}\mathrm{d}y_{\mathrm{s}}, \\ A_{\mathrm{far}}(x, y; d) &= \frac{\exp(ikd)}{i\lambda d} \exp\left[i\frac{\pi}{\lambda d}(x^2 + y^2)\right], \\ u_{\mathrm{s}} &= \frac{x}{\lambda d}, \quad \text{and} \quad v_{\mathrm{s}} = \frac{y}{\lambda d}, \end{aligned} \tag{6.21}$$

where $A_{\mathrm{far}}(x, y; d)$, u_{s} and v_{s} are the same as those of the far field propagation in (5.44) and (5.42). Using this representation, we can rewrite the Fresnel integral as follows:

$$\begin{aligned} g_{\mathrm{FR}}(x, y; z) &= A_{\mathrm{far}}(x, y; d) \\ &\quad \times \mathscr{F}\left\{g(x_{\mathrm{s}}, y_{\mathrm{s}}; z_0) \exp\left[i\frac{\pi}{\lambda d}\left(x_{\mathrm{s}}^2 + y_{\mathrm{s}}^2\right)\right]\right\}_{u_{\mathrm{s}}=\frac{x}{\lambda d},\ v_{\mathrm{s}}=\frac{y}{\lambda d}} \end{aligned} \tag{6.22}$$

where the sub-script equations again designate that frequencies u_{s} and v_{s} are replaced using (5.42) and converted to real coordinates x and y after the Fourier transform. Therefore, the following data array must be prepared prior to executing FFT in this technique.

$$g_{\mathrm{s}}[m, n] \equiv g(x_{\mathrm{s},m}, y_{\mathrm{s},n}; z_0) \exp\left[i\frac{\pi}{\lambda d}\left(x_{\mathrm{s},m}^2 + y_{\mathrm{s},n}^2\right)\right] \tag{6.23}$$

Then, the data array is Fourier-transformed;

$$G_{\mathrm{s}}[p, q] = \mathbf{FFT}\left\{g_{\mathrm{s}}[m, n]\right\}. \tag{6.24}$$

The final destination field is given by

$$g_{\mathrm{FR}}(x_p, y_q; z) = A_{\mathrm{far}}(x_p, y_q; d) G_{\mathrm{s}}[p, q], \tag{6.25}$$

where coordinates x_p and y_q are given by the same equations as (6.6) of the far-field propagation;

$$x_p = \frac{\lambda d}{M \Delta x_s}\left(p - \frac{M}{2}\right) \text{ and } y_q = \frac{\lambda d}{N \Delta y_s}\left(q - \frac{N}{2}\right). \tag{6.26}$$

In this technique, only the data array of (6.23) inputted to FFT is different from that of the far-field propagation. Since the process after FFT is exactly the same as the far-field propagation, the numerical properties are very similar to it. Real coordinates are associated with the spatial frequencies by $x = u_s \lambda d$ in (5.42). Because DFT imposes $\Delta u_s = (M \Delta x_s)^{-1}$ on the sampling intervals, as in (4.78), the destination sampling windows and intervals are given exactly like the far-field propagation.

$$W_{\text{far},x} = \frac{\lambda}{\Delta x_s} d \text{ and } W_{\text{far},y} = \frac{\lambda}{\Delta y_s} d, \tag{6.27}$$

$$\Delta x = \frac{\lambda}{M \Delta x_s} d \text{ and } \Delta y = \frac{\lambda}{N \Delta y_s} d. \tag{6.28}$$

As a result, the destination sampling window expands proportionally to the propagation distance d exactly like the far field propagation, as shown in (6.8) and Fig. 6.1.

6.3.2 Numerical Example

Figure 6.4 shows examples of the single-step Fresnel propagation. The source field is again a circular aperture of 1 mm in diameter, which is represented by (6.11). Amplitude images of the destination wavefield are shown in (b)–(e), where a certain square is cut out from the whole destination sampling window, because the sampling window expands with increasing d. Therefore, there are too few sample points to visualize the amplitude pattern smoothly in the case of $d = 2$ [m], as in Fig. 6.4e.

6.3.3 Sampling Problem

The single-step Fresnel propagation has a different sampling problem from the far-field propagation, because the integrand of the Fourier transform includes an exponential function. We can use the same technique as that in the far-field propagation to analyze the sampling problem, i.e., the integrand in (6.21) is redefined as

$$g(x_s, y_s; z_0) \exp\left[i \frac{\pi}{\lambda d}\left(x_s^2 + y_s^2\right)\right] = g(x_s, y_s; z_0) \exp\left[i\phi(x_s, y_s)\right],$$

$$\phi(x_s, y_s) = \frac{\pi}{\lambda d}\left(x_s^2 + y_s^2\right). \tag{6.29}$$

Using (6.13) and sampling theorem, we can find that the local frequency of $\exp\left[i\phi(x_s, y_s)\right]$ imposes restriction on the range of x_s and y_s as follows:

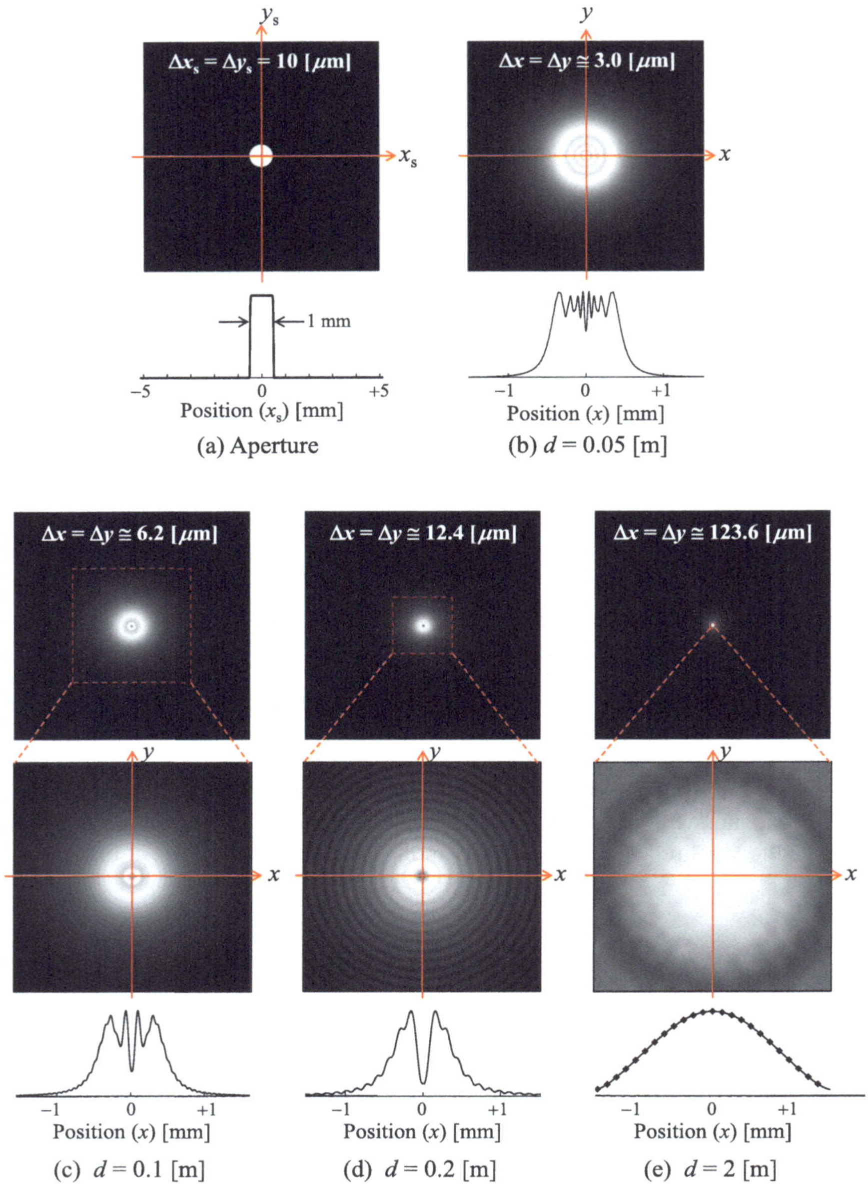

Fig. 6.4 Amplitude images calculated by the single-step Fresnel diffraction. **a** Source field $|g(x_{s,m}, y_{s,n}; z_0)|$, $M = N = 1024$, $\Delta x_s = \Delta y_s = 10$ [μm], $\lambda = 633$ [nm]. **b–d** The destination field $|g_{FR}(x_p, y_q; z)/(\lambda d)^2|$ for different propagation distances $d = z - z_0$

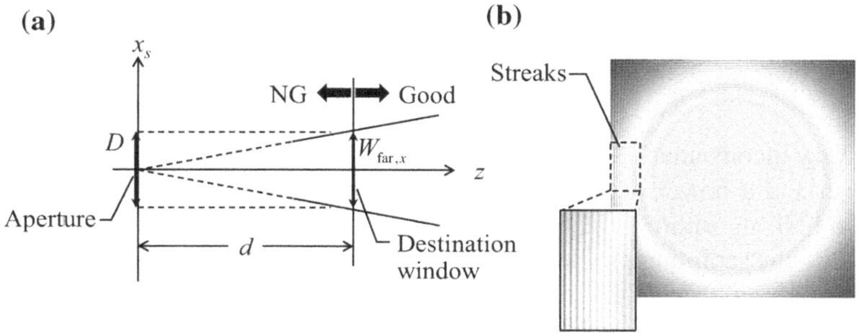

Fig. 6.5 **a** Schematic illustration of aliasing-free condition in the single-step Fresnel propagation. **b** An amplitude image of the destination field diffracted by a circular aperture. $d = 0.015\,[\text{m}]$, $D = 1\,[\text{mm}]$, $M = N = 1024$, $\Delta x = \Delta y = 10\,[\mu\text{m}]$, $\lambda = 633\,[\text{nm}]$

$$|x_s| < \frac{\lambda d}{2\Delta x_s} \quad \text{and} \quad |y_s| < \frac{\lambda d}{2\Delta y_s}. \tag{6.30}$$

The right-hand side of these relations can be rewritten using the destination sampling window in (6.27) as

$$|x_s| < \frac{W_{\text{far},x}}{2} \quad \text{and} \quad |y_s| < \frac{W_{\text{far},y}}{2}. \tag{6.31}$$

When the source field is given by a simple aperture whose diameter is D, the source field $g(x_s, y_s; z_0)$ has non-zero value in $x_s, y_s \leq D/2$. Thus, the sampling problem is caused in the cases where $D \geq W_{\text{far},x}, W_{\text{far},y}$, as shown in Fig. 6.5a. An example of aliasing errors is shown in (b). The size of the destination window is $W_{\text{far},x} = W_{\text{far},y} = 0.95\,[\text{mm}]$ in this case, while the circular aperture has a diameter of $D = 1\,[\text{mm}]$. Horizontal and vertical streaks that appear at the edge of the amplitude image originate in aliasing errors.

The example shown above may be a special case. In many cases, the source field expands in the whole sampling window of the source plane. Thus, the restriction becomes $\Delta x_s M \leq W_{\text{far},x}$ and $\Delta y_s N \leq W_{\text{far},y}$ in this case. As a result, the propagation distance has the following restriction to avoid the aliasing error of amplitude images.

$$d > \frac{\Delta x_s^2 M}{\lambda} \quad \text{and} \quad \frac{\Delta y_s^2 N}{\lambda}. \tag{6.32}$$

In addition, the single-step Fresnel propagation must satisfy restriction of the far-field propagation in (6.16) to avoid aliasing errors of the phase pattern, because the restriction is caused by $A_{\text{far}}(x, y; d)$ after the Fourier transform in common with the single-step Fresnel propagation. Unfortunately, the restriction in (6.16) is exactly opposite to (6.32). Therefore, to avoid both aliasing errors of amplitude and phase distributions, we conclude that the number of sample points is exactly determined by

$$M = \frac{\lambda}{\Delta x_s^2} d \text{ and } N = \frac{\lambda}{\Delta y_s^2} d. \tag{6.33}$$

This is very inconvenient restriction, because we commonly use FFT for the Fourier transform and a power of 2 is the most effective number of the sample points in common FFT algorithm.

Because the sampling problem of (6.33) arises after FFT, a technique to avoid the problem is upsampling of $G_s[p, q]$ in (6.24). The simplest way is to extend the sampling window of $g_s[m, n]$ and pad zeros. By the extension, $G_s[p, q]$ is oversampled after FFT.[1] Here, note that the propagation distance must satisfy (6.32) even in this case.

In the end of this section, it should be noted that above sampling problems of single-FFT type propagation can be avoided by employing the Scaled FFT, as described in Sects. 12.3 and 12.4.

6.4 Convolution-Based Technique: Band-Limited Angular Spectrum Method

Numerical propagation based on the angular spectrum and the Fresnel approximation is represented by the convolution integral, as shown in (5.19) and (5.28). Thus, using the convolution theorem in (4.48) and FFT, we can calculate propagated fields effectively. In this case, the numerical propagation has a desirable property that the sampling interval and window of the destination field is the same as those of the source field. However, this property often causes another problem due to field invasion. This originates in circular convolution of discrete functions.

In addition to the field invasion, a sampling problem is caused in the transfer function. However, this can be avoided by a band-limiting technique. This technique is called a *band-limited angular spectrum method* (BLAS) [79].

6.4.1 Discrete Formula

Using (5.19) and (5.28), Fourier spectrum of the destination field is obtained in the discrete form as follows:

$$G[p, q; z] = G[p, q; z_0] H[p, q; d], \tag{6.34}$$

where $G[p, q; z_0]$ is the Fourier spectrum of the source field given in (6.3):

---

[1] This technique for upsampling is known as the sinc interpolation using FFT.

$$G[p, q; z_0] = \mathbf{FFT}\{g[m, n; z_0]\}. \tag{6.35}$$

The sampling interval is given by (4.78), i.e., in this case,

$$\Delta u = \frac{1}{M \Delta x_s} \quad \text{and} \quad \Delta v = \frac{1}{N \Delta y_s}, \tag{6.36}$$

where Δx_s and Δy_s are again the sampling interval of the source field, and M and N are the numbers of samples.

$H[p, q; d]$ is a sampled transfer function. In the angular spectrum method, since the transfer function is given by (5.18), the sampled transfer function is simply represented by

$$H_{\text{AS}}[p, q; d] = \begin{cases} \exp\left[i2\pi d \left(\lambda^{-2} - u_p^2 - v_q^2\right)^{1/2}\right] & \text{if } u_p^2 + v_q^2 \le \lambda^{-2} \\ 0 & \text{otherwise} \end{cases}. \tag{6.37}$$

Discrete Fourier frequencies u_p and v_q are given by the same manner as (4.92). In the Fresnel diffraction, the transfer function is given by (5.30), and thus, represented by

$$H_{\text{FR}}[p, q; d] = \exp\left[i\pi d \left\{\frac{2}{\lambda} - \lambda(u_p^2 + v_q^2)\right\}\right]. \tag{6.38}$$

The destination field is obtained from the destination spectrum $G[p, q; z]$ using inverse FFT.

$$g[m, n; z] = \mathbf{FFT}^{-1}\{G[p, q; z]\}. \tag{6.39}$$

The sampling intervals are the inverse of $M\Delta u$ and $N\Delta v$, and thus,

$$\Delta x = \Delta x_s \quad \text{and} \quad \Delta y = \Delta y_s. \tag{6.40}$$

As a result, the sampling intervals are independent of the propagation distance and always the same as those in the source field unlike the far-field propagation and single-step Fresnel propagation. These are desirable properties in many cases in computer holography.

6.4.2 Sampling Problem of Transfer Function

Theoretically, the angular spectrum method is valid for any propagation distance. Even in the Fresnel propagation, although the Fresnel approximation places a limit on the propagation distance or the non-paraxial property of the field, the limits are commonly not so strict. However, we often encounter dirty or noisy results in these

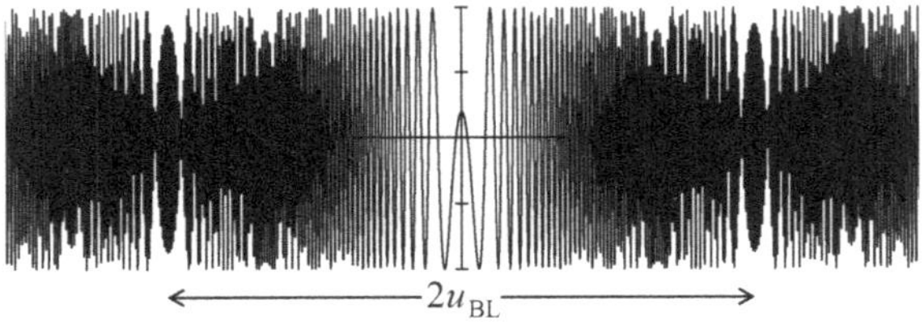

Fig. 6.6 An example of transfer function $\mathrm{Re}\{H_{AS}(u; d)\}$, where $d = 0.01$ m, $N = 2048$, $\Delta x = 1\,[\mu\mathrm{m}]$, and $\lambda = 633\,[\mathrm{nm}]$. Only a range of $5 \times 10^5\,[1/\mathrm{m}]$ is depicted around the origin. $u_{BL} \cong 1.6 \times 10^5\,[1/\mathrm{m}]$

convolution-based techniques. There are usually two origins of the problem. One of them is the sampling problem of the transfer function $H(u, v; d)$.

Supposing that the one-dimensional transfer function of the angular spectrum method in (5.18) is redefined as follows:

$$\begin{aligned} H_{AS}(u; d) &= \exp\left[i\phi_{AS}\right], \\ \phi_{AS} &= 2\pi d\left(\lambda^{-2} - u^2\right)^{1/2}, \end{aligned} \tag{6.41}$$

the local signal frequency of the function $H_{AS}(u; d)$ is given by

$$\begin{aligned} f_u &= \frac{1}{2\pi}\frac{\partial\phi_{AS}}{\partial u} \\ &= \frac{ud}{\left(\lambda^{-2} - u^2\right)^{1/2}}. \end{aligned} \tag{6.42}$$

Here, note that f_u is not a physical spatial frequency because it has the physical dimension of length.

Equation (6.42) shows that the local signal frequency is proportional to the propagation distance d. Therefore, the sampled transfer function inevitably causes aliasing errors over a certain propagation distance. f_u must satisfy the Nyquist condition to avoid aliasing errors as follows:

$$\Delta u^{-1} > 2|f_u|. \tag{6.43}$$

We can easily derive a limit of the propagation distance from this relation. However, the limit most likely decreases the usefulness of the convolution-based techniques. Alternatively, we can limit the frequency range of the transfer function to avoid aliasing errors. This is equivalent to limiting the bandwidth of the source field. The frequency range where the transfer function $H_{AS}(u; d)$ causes no aliasing error is derived from (6.42) and (6.43) as follows.

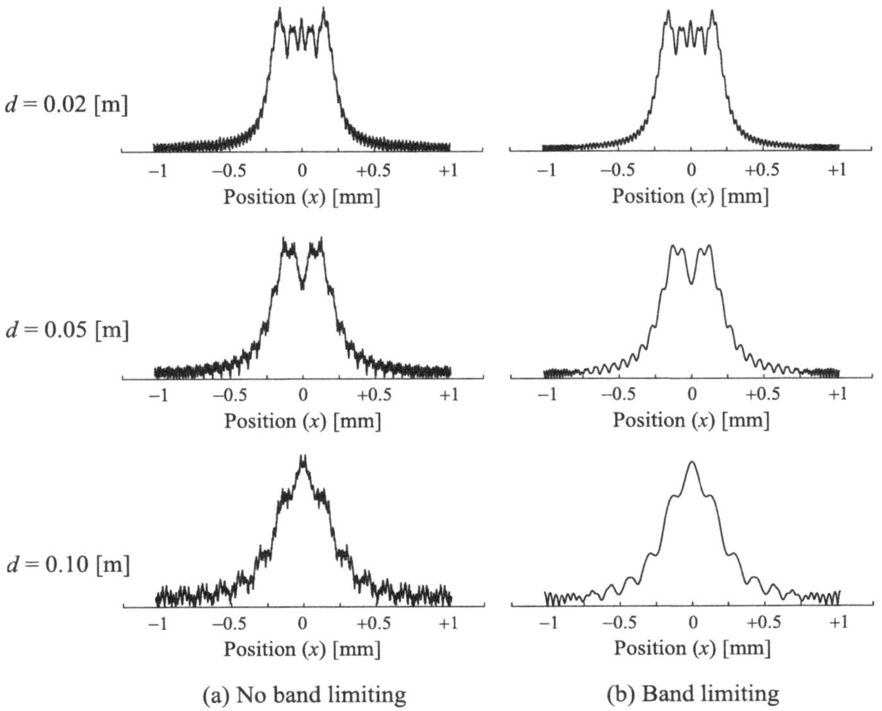

Fig. 6.7 Examples of amplitude distribution $|g(x_m; z)|$ of destination fields calculated by the band-limited angular spectrum method; **a** without and **b** with band limiting. The rectangular aperture has a width of $D = 0.5$ [mm]. $N = 2048$, $\Delta x = 1$ [μm], and $\lambda = 633$ [nm]

$$|u| < u_{\mathrm{BL}},$$
$$u_{\mathrm{BL}} \equiv \frac{1}{\left[(2d\Delta u)^2 + 1\right]^{1/2}\lambda}. \tag{6.44}$$

A numerical example of $H_{\mathrm{AS}}(u; d)$ is shown in Fig. 6.6. Because aliasing errors are caused outside the range of $2u_{\mathrm{BL}}$, the bandwidth should be limited to $2u_{\mathrm{BL}}$. Thus, the sampled transfer function in BLAS is given by

$$H_{\mathrm{BLAS}}[p; d] = H_{\mathrm{AS}}[p; d]\mathrm{rect}\left(\frac{u_p}{2u_{\mathrm{BL}}}\right). \tag{6.45}$$

Numerical examples of one-dimensional diffraction of an aperture, calculated by the angular spectrum method, are shown in Fig. 6.7. Here, the width of the aperture is 0.5 mm. The fields in (a) are calculated using $H_{\mathrm{AS}}(u; d)$ in (6.41), i.e., the bandwidth is unlimited, while the band-limited transfer function $H_{\mathrm{BLAS}}(u; d)$ is used for (b). Definitely, the results using band limit are less noisy than those by unlimited calculation; high-frequency noise produced in (a) is removed in (b) by band limiting with bandwidth $2u_{\mathrm{BL}}$. The cutoff frequency depends on the propagation distance and is reduced with increasing the distance. Physical interpretation of this band limiting and the cutoff frequency are discussed in Sect. 6.4.4.

Band limit in two-dimensional wavefields is also derived from (5.18) using almost the same procedure as the one-dimensional case. Supposing that the transfer function is represented by $H_{\mathrm{AS}}(u, v; d) = \exp[i\phi_{\mathrm{AS}}(u, v)]$, where $\phi_{\mathrm{AS}}(u, v) = 2\pi d \left(\lambda^{-2} - u^2 - v^2\right)^{1/2}$, the local signal frequencies are given by

$$f_u = \frac{1}{2\pi}\frac{\partial \phi_{\mathrm{AS}}(u, v)}{\partial u} = \frac{ud}{\left(\lambda^{-2} - u^2 - v^2\right)^{1/2}}, \tag{6.46}$$

$$f_v = \frac{1}{2\pi}\frac{\partial \phi_{\mathrm{AS}}(u, v)}{\partial v} = \frac{vd}{\left(\lambda^{-2} - u^2 - v^2\right)^{1/2}}. \tag{6.47}$$

Since both Nyquist conditions $\Delta u^{-1} > 2|f_u|$ and $\Delta v^{-1} > 2|f_v|$ must be satisfied simultaneously, the sampled transfer function is limited within the region represented by

$$\frac{u^2}{u_{\mathrm{BL}}^2} + \frac{v^2}{\lambda^{-2}} < 1 \text{ and } \frac{u^2}{\lambda^{-2}} + \frac{v^2}{v_{\mathrm{BL}}^2} < 1, \tag{6.48}$$

where

$$v_{\mathrm{BL}} \equiv \frac{1}{\left[(2d\Delta v)^2 + 1\right]^{1/2}\lambda}. \tag{6.49}$$

Both relations in (6.48) give ellipsoidal regions with a major radius of $a = \lambda^{-1}$ in the (u, v) plane, as shown in Fig. 6.8. The minor radii are given by $b = u_{\mathrm{BL}}$ and v_{BL} in the u and v directions, respectively. The transfer function and the spectrum of the wavefield must be limited within the common region of these ellipsoidal regions.

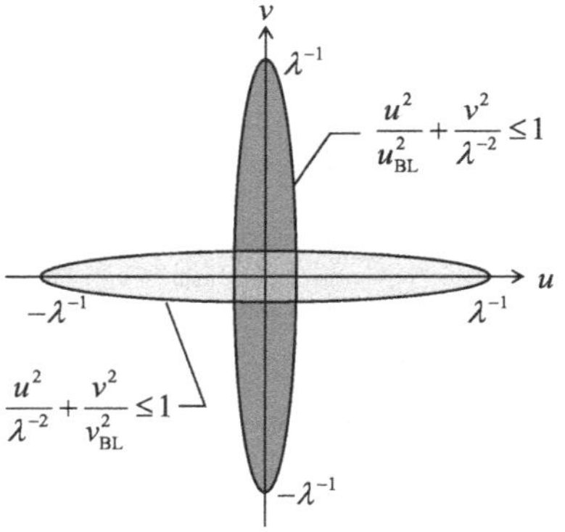

Fig. 6.8 Schematic band limit represented in (6.48) in (u, v) coordinates

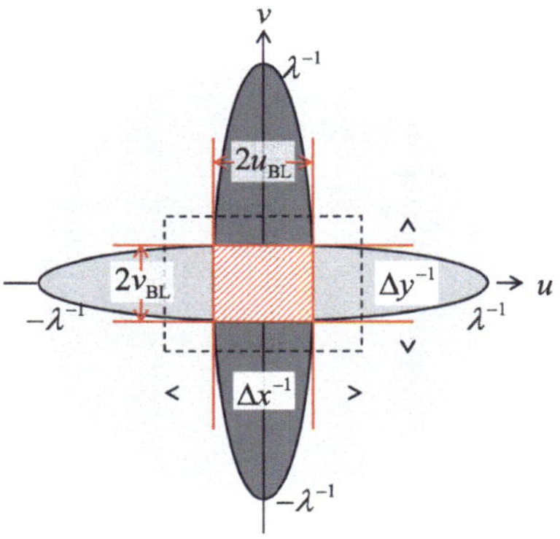

Fig. 6.9 Schematic illustration of rectangular band limiting in (u, v) coordinates

If the ellipses are sufficiently oblate, the ellipsoidal regions in (6.48) can be approximated to a rectangular region, as shown in Fig. 6.9. Oblateness of a ellipse is defined as $f_{\text{oblate}} = 1 - b/a$. When we adopt a value of 1/2 as the criterion of the oblateness, i.e., $f_{\text{oblate}} \gg 1/2$, the approximate regions and the criteria for applying the approximation are given by

$$\begin{aligned} |u| &< u_{\text{BL}} \quad \text{if } |d| \gg \frac{\sqrt{3}}{2} W_x, \\ |v| &< v_{\text{BL}} \quad \text{if } |d| \gg \frac{\sqrt{3}}{2} W_y, \end{aligned} \tag{6.50}$$

where $W_x = 1/\Delta u$ and $W_y = 1/\Delta v$ are again the sizes of the sampling window. Therefore, when $d \gg \sqrt{3}W_x/2$ and $\sqrt{3}W_y/2$ are satisfied, the band-limited transfer function is simply represented by

$$H_{\text{BLAS}}[p, q; d] = H_{\text{AS}}[p, q; d] \text{rect}\left(\frac{u_p}{2u_{\text{BL}}}\right) \text{rect}\left(\frac{v_q}{2v_{\text{BL}}}\right). \tag{6.51}$$

6.4.3 Problem of Field Invasion

In the convolution-based techniques, the convolution is carried out using FFT. This sometimes causes the problem of field invasion. The convolution in (5.19) and (5.28) are actually discrete convolutions defined by

$$g[m; z] = \sum_{m'=0}^{M-1} g[m'; z_0] h[m - m'; d]. \tag{6.52}$$

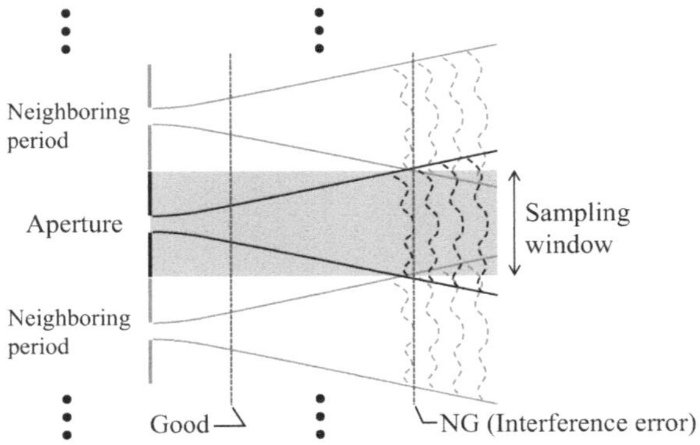

Fig. 6.10 Schematic illustration of the edge effect produced in convolution-based field propagation

Here, we used the one-dimensional formula for simplification. In practice, we do not calculate above equation directly, but use FFT and the convolution theorem to obtain the equivalent result. As mentioned in Sect. 4.6, when we use DFT or FFT, the discrete functions are always considered as a periodic function.

Let's get back to field propagation. When we carry out discrete convolution using FFT for field propagation, both the source and destination fields are considered to be periodic. Figure 6.10 schematically illustrates the periodic aperture and field. There are infinite repetitions of the period defined by the sampling window. If the destination plane is close to the aperture, any problem does not arise because the field diffracted by the aperture does not spread over the whole sampling window, i.e., the field is contained within the window.

However, in a destination plane far from the aperture, the diffracted field expands to the edge of the sampling window and runs over. Fields in neighboring periods also spread and run over exactly like the main window. These fields invade the neighboring region of the sampling window and interact with each other. This is sometimes called the *edge effect* of convolution-based propagation techniques. Since the sampling window is constant in convolution-based methods, errors by the edge effect are inevitably produced in long distance propagation. When the source field has non-zero values over the whole sampling window, the edge effect is always produced even in short propagation.

The edge effect is also interpreted as the result of circular convolution from the viewpoint of fast convolution, as mentioned in Sect. 4.7.3. Therefore, edge effect can be avoided by the quadruple extension of the source field, shown in Fig. 4.14. Examples of destination fields calculated with the quadruple extension are shown in Fig. 6.11b. The parameters are the same as those in Fig. 6.7. Column (a) shows results without the quadruple extension for comparison. Since the bandwidth is limited using (6.45) in both columns, column (a) is actually the same as Fig. 6.7b.

Here, note that the sampling interval in the Fourier domain is not $\Delta u = (M\Delta x)^{-1}$ but $\Delta u = (2M\Delta x)^{-1}$ when using the quadruple extension. Therefore, the limiting bandwidth changes from that without the quadruple extension. In short distance propagation such as $d = 0.02$ [m], there is no remarkable differences in the amplitude

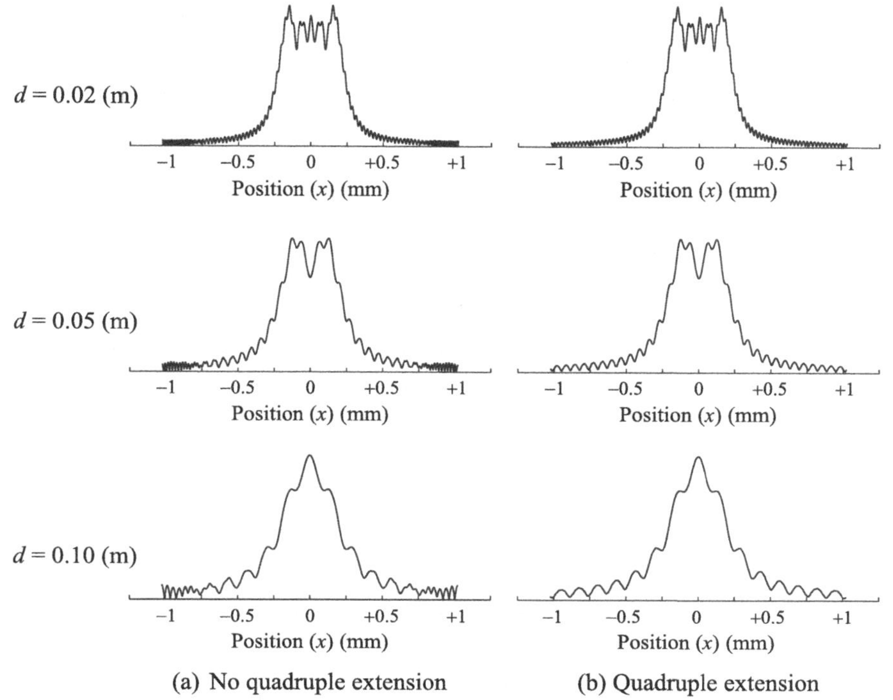

Fig. 6.11 Examples of amplitude distribution $|g(x_m; z)|$ of destination fields calculated by the band-limited angular spectrum method; **a** without and **b** with quadruple extension. The rectangular aperture has a width of $D = 0.5$ [mm]. $N = 2048$, $\Delta x = 1$ [μm], and $\lambda = 633$ [nm]

distribution. However, there are definite differences at the edges of the field in longer propagation of $d = 0.05$ [m] and 0.1 [m].

2D examples of destination fields calculated using the angular spectrum method are also shown in Fig. 6.12. Here, the source field is given by a circular aperture with 0.5 mm in diameter. Column (a) shows results of the original angular spectrum method using the transfer function $H_{\text{AS}}[p, q; d]$ in (6.37). The technique of the quadruple extension is not used in this case. The amplitude images in column (b) is calculated by the band-limited angular spectrum using the transfer function $H_{\text{BLAS}}[p, q; d]$ in (6.51) with the quadruple extension. The amplitude images in (b) are definitely less noisy when compared with those in (a). This example is of diffraction by a simple aperture. Therefore, the source field is localized within the aperture. In many cases of computer holography, however, the source field extends over the whole sampling window. The quadruple extension is always required even in short propagation in that case.

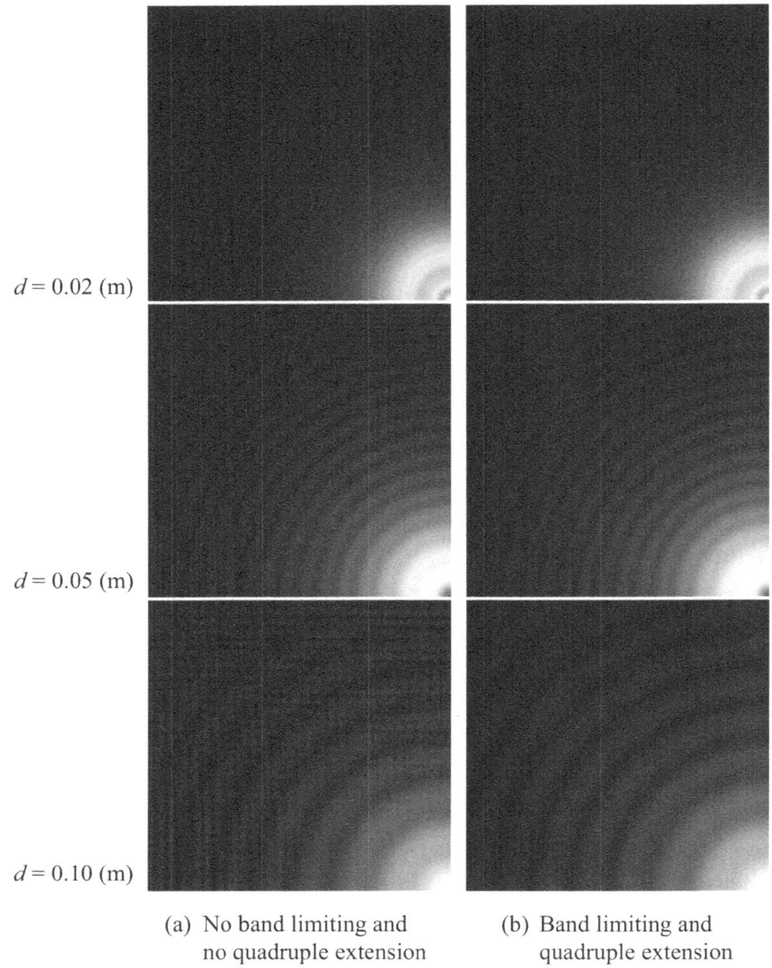

Fig. 6.12 Amplitude images; $|g(x_m, y_n; z)|$ calculated by the angular spectrum method **a** without and **b** with both the band-limiting and quadruple extension. The source field is a circular aperture whose diameter is 0.5 [mm]. $N = 2048$, $\Delta x = 1$ [μm], and $\lambda = 633$ [nm]. Only a quadrant is indicated in order to make the difference clear

6.4.4 Discussion on Band Limiting

We could successfully improve the angular spectrum method by limiting bandwidth of the sampled transfer function. This band limiting is equivalent to limiting bandwidth of the source field. The question arises as to whether the band limiting is physically right. If the band limiting is an adequate technique to calculate the correct

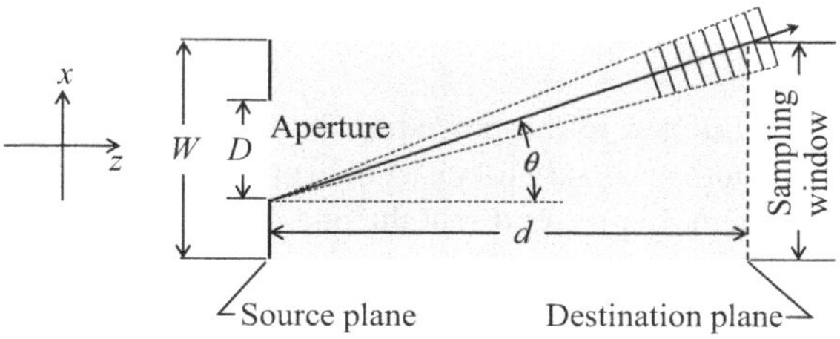

Fig. 6.13 Model for estimating the minimum bandwidth required for exact field propagation

destination field, another question also arises; how much minimum bandwidth is necessary for exact field propagation.

A model for estimating the minimum bandwidth necessary for exact numerical propagation is shown in Fig. 6.13. An aperture with a size of D is placed at the center of a sampling window whose size is W. The highest spatial frequency, observed at the upper end of the destination sampling window, may be given by the field emitted from the point at the lower end of the source aperture. Therefore, the maximum frequency required for field propagation is given by

$$\begin{aligned} u_{\text{req}} &= \frac{\sin\theta}{\lambda} \\ &= \left[\left(\frac{2d}{W+D}\right)^2 + 1\right]^{-1/2} \lambda^{-1}, \end{aligned} \tag{6.53}$$

where θ is the angle shown in Fig. 6.13.

We introduced a band-limit of $2u_{\text{BL}}$ into the source field and the transfer function in order to avoid numerical errors. However, if the cutting frequency is less than the required frequency, i.e., $u_{\text{req}} > u_{\text{BL}}$, we might introduce another physical error into the diffraction field, because the source field loses a part of the bandwidth necessary for exact diffraction.

When the size of the aperture agrees with the sampling window; $D = W$, u_{req} takes the maximum value:

$$u_{\text{req}} = \frac{1}{\left[(d/W)^2 + 1\right]^{1/2} \lambda}. \tag{6.54}$$

On the other hand, $\Delta u = (2W)^{-1}$ because of the quadruple extension. Substituting the sampling interval into (6.44), we find $u_{\text{req}} = u_{\text{BL}}$. Thus, the following relation is always satisfied:

$$u_{\text{req}} \le u_{\text{BL}}. \tag{6.55}$$

As a result, we can conclude that the band-limited angular spectrum method uses only the bandwidth required for exact field propagation. This is also confirmed by measurement of the numerical error [67].

6.4.5 More Accurate Technique

As mentioned in the preceding section, the band limiting in (6.51) preserves the bandwidth necessary for exact field propagation. However, band limiting itself causes a problem. Let us represent the one-dimensional transfer function in (6.45) as

$$H_{\mathrm{BLAS}}(u_p; d) = H_{\mathrm{AS}}(u_p; d)\mathrm{rect}\left(\frac{u_p}{2u_{\mathrm{BL}}}\right). \tag{6.56}$$

Using above transfer function, the spectrum of the destination field is given by

$$G(u_p; z) = G(u_p; z_0)H_{\mathrm{AS}}(u_p; d)\mathrm{rect}\left(\frac{u_p}{2u_{\mathrm{BL}}}\right). \tag{6.57}$$

Thus, the destination field is given by the Fourier transform as follows:

$$g(x_m; z) = 2u_{\mathrm{BL}}\left[g(x_m; z_0) \otimes h_{\mathrm{AS}}(x_m; d)\right] \otimes \mathrm{sinc}(2u_{\mathrm{BL}}x_m), \tag{6.58}$$

where the similarity theorem is used.

In comparison with (5.19), we found that the destination field takes a sinc convolution. Therefore, the destination field does not completely agree with the ideal result of the angular spectrum method. This is caused by neither the sampling problem nor the edge effect, but is produced by the band limiting itself. To avoid the problem, we must produce the transfer function using FFT as follows [111]:

$$H'_{\mathrm{AS}}(u_p; d) = \mathbf{FFT}\left\{h_{\mathrm{AS}}(x_m; d)\mathrm{rect}\left(\frac{x_m}{W_{sx}}\right)\right\}, \tag{6.59}$$

where W_{sx} is again a size of the source sampling window. Therefore, we need three times FFT to perform the improved field propagation.

Chapter 7
Holography

Abstract This chapter mostly deals with the theory of thin holograms. We start the discussion from the difference between thin and volume holograms, and then proceed to the property of thin holograms. We deal with the wave-optical theory of holograms produced with reference plane and spherical waves. Some pages are particularly devoted to analysis of the true and conjugate images reconstructed by holograms. As with the preceding chapter, not only formulas but also plenty of simulated images support the theoretical analysis in this chapter.

7.1 Optical Interference

As mentioned in Sect. 2.1, a hologram is just a fringe pattern generated by optical interference. When two optical fields occupy the same space, the fields interfere with each other, and undulation is produced in the optical intensity.

Suppose that two monochromatic fields are given by $g_1(x, y, z)$ and $g_2(x, y, z)$, the optical intensity is represented by

$$\begin{aligned} I(\mathbf{r}) &= |g_1(\mathbf{r}) + g_2(\mathbf{r})|^2 \\ &= |g_1(\mathbf{r})|^2 + |g_2(\mathbf{r})|^2 + 2\mathrm{Re}\left\{g_1(\mathbf{r})g_2^*(\mathbf{r})\right\}. \end{aligned} \tag{7.1}$$

Here, assume that both fields are plane waves given in (3.47):

$$g_1(\mathbf{r}) = A_1 \exp[i(\mathbf{k}_1 \cdot \mathbf{r} + \phi_1)], \tag{7.2}$$
$$g_2(\mathbf{r}) = A_2 \exp[i(\mathbf{k}_2 \cdot \mathbf{r} + \phi_2)], \tag{7.3}$$

where $\mathbf{k}_1$ and $\mathbf{k}_2$ are wave vectors, and ϕ_1 and ϕ_2 are initial phases. When only a field $g_1(x, y, z)$ or $g_2(x, y, z)$ exists in the space, the intensity is simply a constant A_1^2 or A_2^2. However, if both fields exist in the same space in the same time, the intensity distribution is no longer constant:

$$I(\mathbf{r}) = A_1^2 + A_2^2 + 2A_1A_2\cos(\mathbf{K} \cdot \mathbf{r} + \Delta\phi), \tag{7.4}$$

K. Matsushima, *Introduction to Computer Holography*,
Series in Display Science and Technology,
https://doi.org/10.1007/978-3-030-38435-7_7

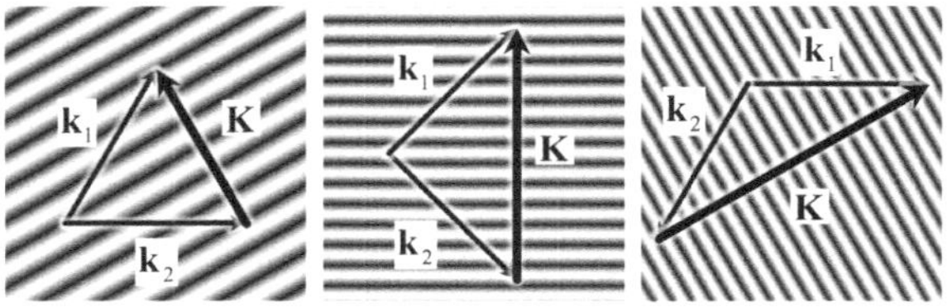

Fig. 7.1 Examples of interference fringe. Intensity distribution $I(\mathbf{r})$ is depicted as a gray-scale image

where

$$\mathbf{K} \equiv \mathbf{k}_1 - \mathbf{k}_2, \tag{7.5}$$

$$\Delta\phi \equiv \phi_1 - \phi_2. \tag{7.6}$$

The third term of (7.4) generates undulation of the intensity distribution. Since $\cos(\mathbf{K} \cdot \mathbf{r} + \Delta\phi) = \mathrm{Re}\,\{\exp[i(\mathbf{K} \cdot \mathbf{r} + \Delta\phi)]\}$, the third term has the same form as a plane wave. Hence, vector $\mathbf{K}$ plays the same role as the wave vector of a plane wave, i.e., the direction of the vector $\mathbf{K}$ gives the direction of the wave of the fringe. The wavelength is also given by

$$\Lambda = \frac{2\pi}{|\mathbf{K}|}, \tag{7.7}$$

as in a plane wave. Thus, $\mathbf{K}$ is sometimes called a fringe wave vector. Figure 7.1 shows examples of interference fringes for different wave vectors.

7.2 Thin Hologram and Volume Hologram

As mentioned in Sect. 2.1, a hologram fringe is recorded by illuminating a photo-sensitive material with an object field and a reference field. Actually, there are two types of recording techniques in optical holography, as shown in Fig. 7.2. These are very similar to each other. However, the natures are considerably different. In (a), the reference field illuminates the photo-sensitive material from the same side as the object field, while the incident side of the reference field is opposite in (b).

The recorded hologram using the setup (a) is called a *thin hologram* in general. Since the recorded object field is reconstructed by irradiating backside of the hologram as in Fig. 2.3 and the illuminating light regenerates the recorded field by transmitting the hologram, the hologram is also called a *transmission hologram*.[1] Holograms recorded by the setup (b) are called a *thick hologram* or a *volume hologram*

[1] A transmission hologram is not reconstructed by reflection illumination in general unless additional optics such as a mirror is used for reconstruction. The exception is CGHs fabricated by laser lithography (see Sect. 15.3).

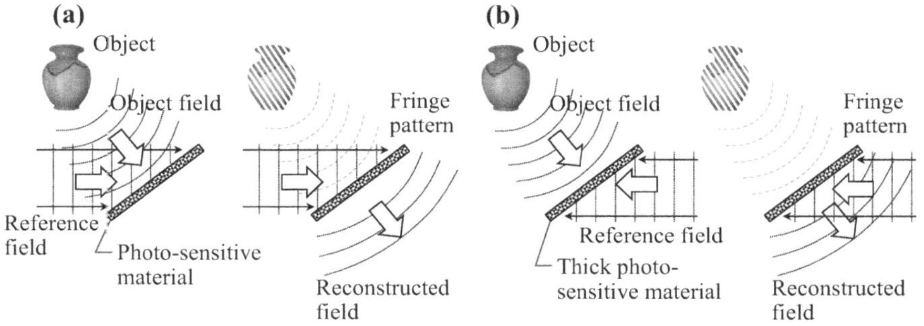

Fig. 7.2 Recording and reconstruction of **a** a thin hologram, and **b** volume hologram

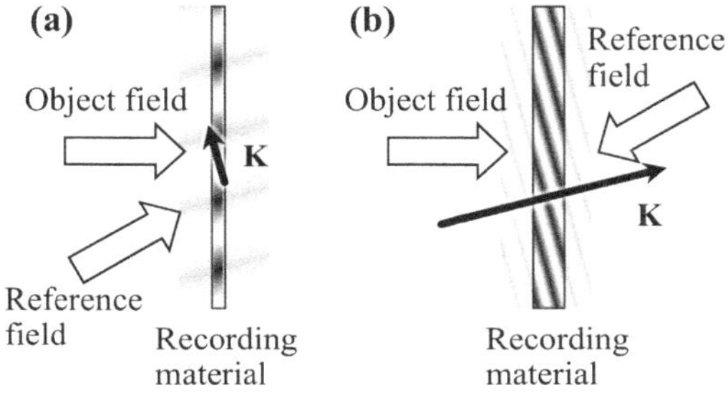

Fig. 7.3 Fringe patterns recorded in **a** a thin hologram, and **b** volume hologram

if the material is thick enough to record the fringe in the depth direction.[2] In this case, the recorded object field is reconstructed by reflection of the illuminating light. Thus, the hologram is sometimes called a *reflection hologram*.

Volume holograms have considerable different properties from thin holograms.[3] This can be understood by the fringe wave vector **K** in (7.5). Figure 7.3 schematically shows the difference between the fringes of two types of hologram. **K** is nearly parallel to the surface of the material in the thin hologram. Since the material records a part of the fringes generated in the 3D space, the fringe pattern can be regarded as a two-dimensional image. In contrast, **K** in the volume hologram is approximately perpendicular to the surface, i.e., the fringes are mostly parallel to the hologram surface. The material therefore must be thick enough to cover several fringes in order to capture their spacing.

The fringe pattern is no longer a two-dimensional image in volume holograms. Diffraction of light illuminating such a structure cannot be described using the scalar theory described in Sect. 5.2. For example, intensity of diffracted light strongly depends on the wavelength. This phenomenon is called *wavelength selectivity*. Such

[2]Typical thicknesses of emulsion and photo-polymer are about 7 μm and 16 μm, respectively.

[3]The border between thin and volume holograms is not very sharp. The volumetric effects are caused even in transmission holograms, although their strength is not so prominent as in reflection holograms.

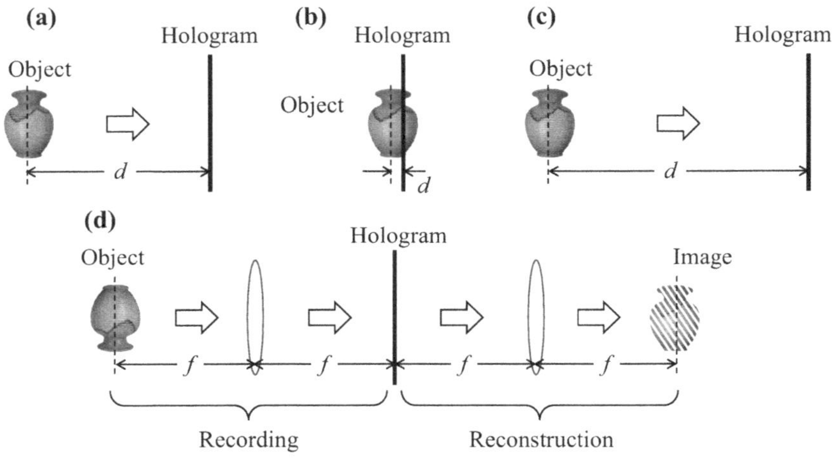

Fig. 7.4 Types of holography. **a** Fresnel, **b** Image, **c** Fraunhofer, and **d** Fourier-transform holography

phenomena, sometimes called "volumetric", cannot be explained by the scalar theory, and more rigorous methods for modeling electromagnetic fields must be employed.[4]

Although volume holograms have several useful natures such as wavelength selectivity, we cannot draw the fringe pattern using ordinary 2D printers. We make use of the nature of volume holograms through contact-copy of the original CGH, as described in Sect. 15.7. Wavefront printers are also being developed to print volume holograms directly, as mentioned in Sect. 15.4.

7.3 Types of Holography

Figure 7.4 shows a classification of holography, based on setup for recording and reconstruction. When the distance between the subject and hologram satisfies the condition of Fresnel diffraction, as in (a), the recorded hologram is commonly called a *Fresnel hologram*. Thus, the distance d is usually several centimeters to several tens centimeters in this case. When d is very small or 0, as in (b), the hologram is called an *image hologram*. In contrast, the hologram recorded with a large d, as in (c), is called a Fraunhofer hologram.

It is not easy in optical holography to record the image hologram in one step, because we cannot make a physical 3D object overlap the hologram. In contrast, creating an image hologram is easy in computer holography, because the difference from Fresnel holograms is simply of the propagation distance. Image holograms have an advantage that severe chromatic aberration does not occur even in thin holograms. In fact, HD-CGHs having an arrangement similar to image holograms can

[4]For example, the coupled-wave theory is well-known for the vector diffraction theory [43, 84–86].

be reconstructed by white-light illumination without any additional technique [137, 138].

It is very difficult in optical holography to record Fraunhofer holograms without any special technique. This is because the optical intensity of the object field is commonly too low to record. As mentioned in Sect. 5.2.3, the Fraunhofer diffraction or far-field diffraction (this term is mainly used in this book) is represented by the Fourier transform. It is definitely better to use a lens to record a Fourier-transform hologram, as shown in Fig. 7.4d. In this case, the object field emitted by the subject is recorded through a lens with $2f$-setup, where f is the focal length of the lens used in the setup. The recorded hologram is called a *Fourier-transform hologram* or simply *Fourier hologram*. The lens is again used in optical reconstruction to Fourier-transform the reconstructed field and reproduce the original image. The reconstructed image is a real image in this case. Fourier holograms feature that the viewing angle is not determined by resolution of the fringe pattern, and thus are often used in computer holography.

Fourier holograms can also be recorded using a reference spherical field without a lens, as mentioned in Sect. 14.2.2. This type of holography, called *lensless-Fourier holography*, is also useful especially in computer holography because it features low fringe frequencies in general.

7.4 Mathematical Explanation of Principle

Since a thin hologram can be considered as a two-dimensional image, we can explain the principle of holography using wavefields that are two-dimensional distribution of optical complex-amplitudes. Suppose that $O(x, y)$ and $R(x, y)$ are object and reference wavefields in a plane placed at $z = 0$. The fringe intensity is given by

$$\begin{aligned} I(x, y) &= |O(x, y) + R(x, y)|^2 \\ &= |O(x, y)|^2 + |R(x, y)|^2 + O(x, y)R^*(x, y) + O^*(x, y)R(x, y). \end{aligned} \tag{7.8}$$

This fringe intensity is recorded on a photo-sensitive material such as silver halide. Transmittance (accurately, power transmittance coefficient) in the surface of the material changes dependently on the fringe intensity after development. Suppose that this transmittance distribution formed in the surface is given by

$$T(x, y) = \eta(I(x, y)), \tag{7.9}$$

where $\eta(I)$ is a sensitivity curve of the material.

Since transmittance T must satisfy $0 \leq T \leq 1$, $\eta(I)$ cannot be a linear function. We assume that the function $\eta(I)$ is a quadratic function in a given range of I and amplitude transmittance $t = \sqrt{T}$ is approximately represented as

$$t(I) = t_0 \pm \beta I \tag{7.10}$$

where constants t_0 and β represent properties of the photo-sensitive material. t_0 is amplitude transmittance of the unexposed material, while β gives sensitivity of the material. The double sign takes '+' in positive-type materials and '−' in negative-type materials.

In the following discussion, let us assume a positive-type material with $t_0 = 0$ and $\beta = 1$ for simplicity. We can make this assumption without loss of generality. Accordingly, distribution of the amplitude transmittance is written as

$$t(x, y) = |O(x, y)|^2 + |R(x, y)|^2 + O(x, y)R^*(x, y) + O^*(x, y)R(x, y). \tag{7.11}$$

When irradiating the hologram with illumination light $P(x, y)$ from behind the hologram, the hologram reconstructs

$$\begin{aligned} g(x, y) &= t(x, y) \times P(x, y) \\ &= |O(x, y)|^2 P(x, y) + |R(x, y)|^2 P(x, y) \\ &\quad + O(x, y)R^*(x, y)P(x, y) + O^*(x, y)R(x, y)P(x, y), \end{aligned} \tag{7.12}$$

right after passing through the hologram. If the illumination light is exactly the same as the reference field, the reconstructed field is

$$\begin{aligned} g(x, y) = {} & |O(x, y)|^2 R(x, y) + |R(x, y)|^2 R(x, y) \\ & + O(x, y)|R(x, y)|^2 + O^*(x, y)R^2(x, y). \end{aligned} \tag{7.13}$$

Furthermore, let us adopt light having a nearly constant amplitude as the reference and illumination field. The candidates are commonly plane waves and spherical waves. The reconstructed field of the hologram, in this case, results in

$$g(x, y) = |O(x, y)|^2 R(x, y) + A_R^2 R(x, y) + A_R^2 O(x, y) + R^2(x, y)O^*(x, y), \tag{7.14}$$

where $A_R^2 = |R(x, y)|^2$ is the constant amplitude of the reference and illumination field.

We find out that the third term of the right-hand side in (7.14) is nothing other than the object field $O(x, y)$. This means that the field diffracted by the fringe pattern $t(x, y)$ reconstructs light of the original object. We can see the figure of the original object when we see the field reconstructed by the hologram. This is the principle of holography.

Other terms of (7.14) are basically unnecessary fields. The second term is exactly the illumination light. The first term is also the illumination light but modified by $|O(x, y)|^2$. These two terms cause a field called *zeroth order light* or *non-diffraction light*. The fourth term includes $O^*(x, y)$, and thus, is called *conjugate light*. The image caused by the conjugate light is called a *conjugate image*. The image caused by the third term is sometimes called a *true image* as opposite to the conjugate image.

Here, it should be emphasized that (7.14) says any type of reference field can be used for recording a hologram if the amplitude can be regarded as a constant. This is true with respect to (7.14). Actual reference fields, however, must have several properties to separate the true image from the conjugate image and non-diffraction light. In practice, a plane wave or spherical wave is commonly used for the reference field because of the simplicity.

7.5 Spatial Spectrum of Amplitude Hologram

Since the amplitude transmittance $t(x, y)$ is a real function, the Fourier transform has the symmetry relation of an Hermitian function, as mentioned in Sect. 4.3.2. Thus, the reconstruction of all amplitude holograms necessarily shows some sort of symmetry, because the direction of field propagation is closely connected with the spatial frequencies as described in Sect. 5.2.3.2. This symmetry relation is the origin of the conjugate image. Therefore, the conjugate image is inevitable attendant of the true image in amplitude holograms unless we use additional optical components to remove it.

For simplicity of the following discussion, suppose that the hologram is placed at $(x, y, 0)$ plane, and the transmittance distribution, called amplitude fringes in this discussion, is simply written as

$$t(x, y) \cong B + O(x, y)R^*(x, y) + O^*(x, y)R(x, y), \tag{7.15}$$

where the terms $|O(x, y)|^2 + |R(x, y)|^2$ in (7.11) are considered as a constant and represented by symbol B. Furthermore, let $G_\mathrm{h}(u, v)$ represent the Fourier transform of the third term that generates the true image:

$$G_\mathrm{h}(u, v) = \mathcal{F}\left\{O(x, y)R^*(x, y)\right\}. \tag{7.16}$$

Accordingly, the Fourier transform of the amplitude fringe is

$$\mathcal{F}\{t(x, y)\} \cong B\delta(u, v) + G_\mathrm{h}(u, v) + G_\mathrm{h}^*(-u, -v), \tag{7.17}$$

where formula $\mathcal{F}\{f^*(x, y)\} = F^*(-u, -v)$ shown in Table 4.1 is used.

When the reference wave is a plane wave traveling along the optical axis, i.e., $R(x, y) \equiv 1$, the spatial spectrum is rewritten as

$$\mathcal{F}\{t(x, y)\} \cong B\delta(u, v) + G_\mathrm{obj}(u, v; 0) + G_\mathrm{obj}^*(-u, -v; 0), \tag{7.18}$$

where $G_\mathrm{obj}(u, v; 0)$ represents the spectrum of the object field in the hologram plane $(x, y, 0)$:

$$G_\mathrm{obj}(u, v; 0) = \mathcal{F}\{O(x, y)\}. \tag{7.19}$$

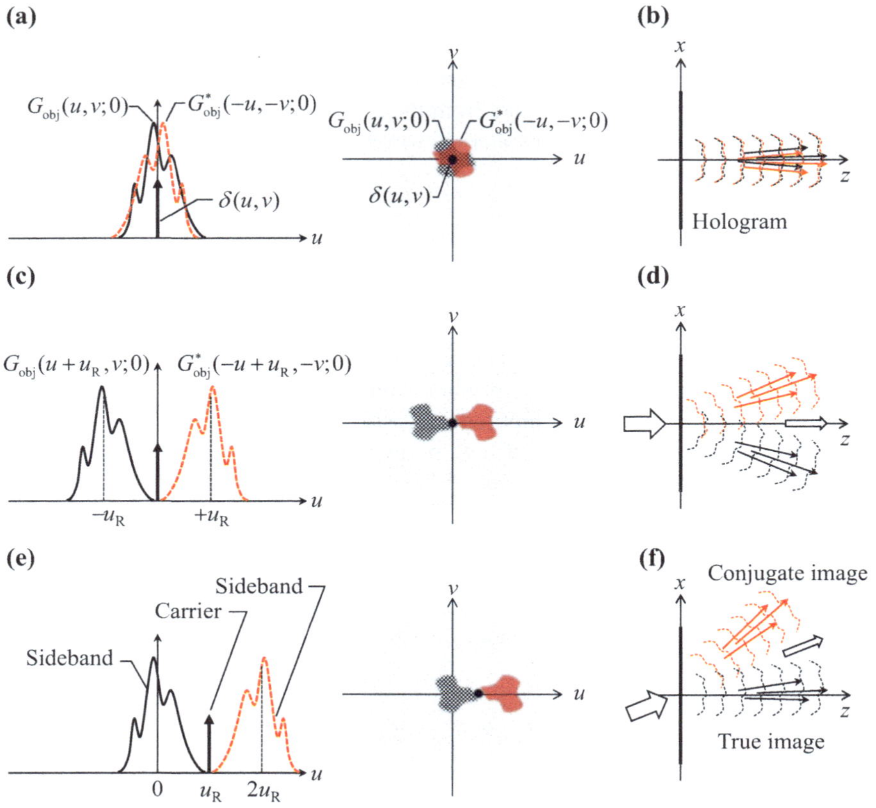

Fig. 7.5 Spatial spectra of hologram fringes and the direction of diffraction

The spectrum of the amplitude hologram is schematically shown in Fig. 7.5a. Since the spectrum of the object field $G_{obj}(u, v; 0)$ is commonly located around the origin of (u, v) coordinates, the components of light for true and conjugate images overlap each other. Therefore, the true image cannot be separated from the conjugate image using any illumination light, as in (b).

This problem can be avoided by using reference plane waves that enter the hologram plane with an angle of θ_R, as shown in Fig. 7.6. Here, represent this reference plane wave as

$$\begin{aligned} R(x, y) &= \exp[ikx \sin\theta_R] \\ &= \exp[i2\pi u_R x], \end{aligned} \tag{7.20}$$

where

$$u_R = \sin\theta_R / \lambda. \tag{7.21}$$

Fig. 7.6 Geometry for recording and reconstruction of a hologram

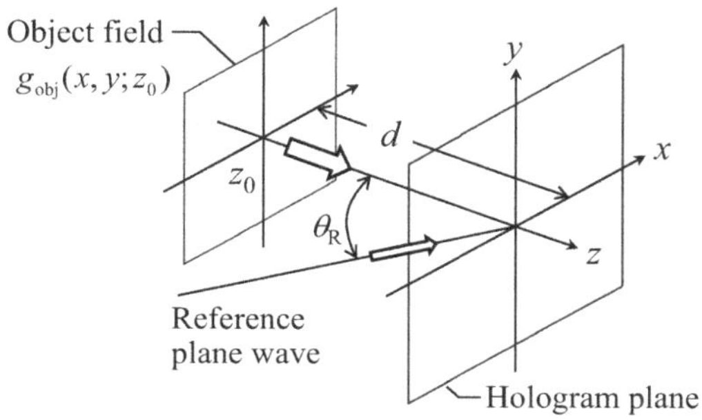

We take unity for the amplitude for simplification. In this case, spatial spectrum of the amplitude transmittance becomes

$$\mathscr{F}\{t(x, y)\} \cong B\delta(u, v) + G_{obj}(u + u_R, v; 0) + G^*_{obj}(-u + u_R, -v; 0). \quad (7.22)$$

Figure 7.5c schematically illustrates the spectrum. Suppose that illumination light $P(x, y)$ is also a plane wave traveling along the optical axis as shown in (d); $P(x, y) = 1$. The reconstructed field is

$$t(x, y)P(x, y) = B + O(x, y)\exp[-i2\pi u_R x] + O^*(x, y)\exp[+i2\pi u_R x]. \quad (7.23)$$

It is found that the lights for true and conjugate images are generated at angles of $-\theta$ and $+\vartheta$, respectively. As a result, we can separate these fields in this case.

In addition, if the illumination light agrees with the reference plane wave; $P(x, y) = \exp[i2\pi u_R x]$, spectrum of the reconstructed field becomes

$$\mathscr{F}\{t(x, y)P(x, y)\} \cong B\delta(u - u_R, v) + G_{obj}(u, v; 0) + G^*_{obj}(-u + 2u_R, -v; 0). \quad (7.24)$$

This spectrum is also shown in Fig. 7.5e. On the analogy of communication engineering, the field corresponding to the first term is called a carrier signal or carrier wave, while the second and third terms are called a *sideband*.

As shown in (7.24) and Fig. 7.5f, the object field is generated in the same direction as the original object field by diffraction of the amplitude fringe, i.e., the hologram. The carrier wave is given by $\mathscr{F}\{\delta(u - u_R, v)\} = \exp[i2\pi u_R x]$. Hence, the carrier wave travels in the same direction of the illumination light, whereas the conjugate light travels with approximately a double angle. The detailed discussion on the position of the reconstructed images are provided by the following section.

7.6 Conjugate Image

Thin amplitude holograms inevitably generate a conjugate image as well as the true image. As mentioned in the preceding section, the direction of conjugate light can be changed by choosing the reference field with a non-zero carrier frequency. However, in computer holography, the reference field with higher frequency generates finer fringe patterns and causes aliasing errors very often. The conjugate light can be removed in thin amplitude holograms by techniques using additional optics, such as the single-sideband method (see Sect. 8.10). However, it is definitely better not to use additional optics in practical exhibition of CGHs. Hence, we have to understand and handle the conjugate image well.

Conjugate images come from the conjugate field $O^*(x, y)$ in the hologram plane. Here, let $g_{\text{obj}}(x, y; z_0)$ represent a wavefield of light at $z = z_0$, which is emitted by the object, and let the hologram lie at $z = 0$, as shown in Fig. 7.7. The object field is represented by

$$\begin{aligned} O(x, y) &= \mathscr{P}_d \left\{ g_{\text{obj}}(x, y; z_0) \right\} \\ &= \mathscr{F}^{-1} \left\{ \mathscr{F} \left\{ g_{\text{obj}}(x, y; z_0) \right\} H(u, v; d) \right\}, \end{aligned} \tag{7.25}$$

where $H(u, v; d)$ is a transfer function and $\mathscr{P}_d\{\cdot\}$ is a propagation operator with distance d, introduced in Sect. 5.4. Here, note that $d = -z_0$, and the propagation is represented by a convolution. Thus, the Fourier transform of the object field is

$$\mathscr{F}\{O(x, y)\} = G_{\text{obj}}(u, v; z_0) H(u, v; d), \tag{7.26}$$

where $G_{\text{obj}}(u, v; z_0) = \mathscr{F}\left\{ g_{\text{obj}}(x, y; z_0) \right\}$. The Fourier transform of the conjugate field is

$$\begin{aligned} \mathscr{F}\left\{ O^*(x, y) \right\} &= G^*_{\text{obj}}(-u, -v; z_0) H^*(-u, -v; d) \\ &= \mathscr{F}\left\{ g^*_{\text{obj}}(x, y; z_0) \right\} H^*(-u, -v; d). \end{aligned} \tag{7.27}$$

where we again use a formula found in Table 4.1. Here, $H^*(-u, -v; d) = H(u, v; -d)$ in both the angular spectrum and Fresnel propagation methods. As a result, the conjugate field is written as

$$\begin{aligned} O^*(x, y) &= \mathscr{F}^{-1} \left\{ \mathscr{F} \left\{ g^*_{\text{obj}}(x, y; z_0) \right\} H(u, v; -d) \right\} \\ &= \mathscr{P}_{-d} \left\{ g^*_{\text{obj}}(x, y; z_0) \right\}. \end{aligned} \tag{7.28}$$

Conclusively, we find out that the conjugate field in the hologram plane is given by propagating $g^*_{\text{obj}}(x, y; z_0)$ with a distance of $-d$, i.e., backward propagation of

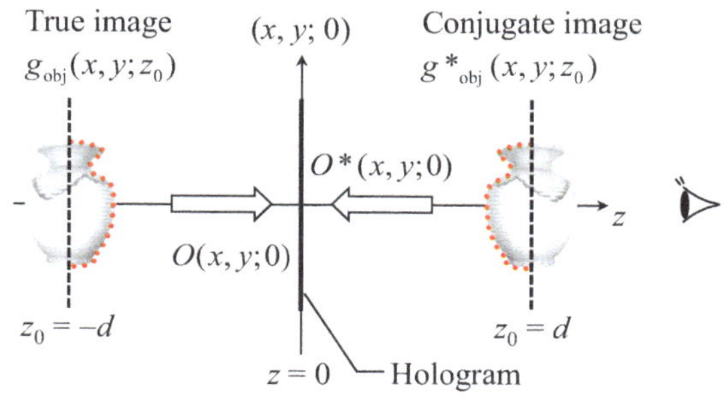

Fig. 7.7 True image and conjugate images reconstructed by thin amplitude holograms

the original conjugate field. This logic can be recursively applied even if the field $g_{obj}(x, y; z_0)$ is given by propagation and summation of other fields. This means that the conjugate image is given by turning the object inside out, as shown in Fig. 7.7. If we see the conjugate image from a viewpoint at $z > d$, an object point closer to the viewer appears at a farther position, i.e., the image is reversed back and forth. This is called a *pseudoscopic image*. The pseudoscopic images are very unnatural images, because we have never seen this kind of images in daily life.

In actual holograms, the position of the conjugate image varies dependently on the reference and illumination field, as mentioned in the following section. If you want, you can easily see the conjugate image by use of a special illumination that agrees with the conjugate field of the reference light; $P(x, y) = R^*(x, y)$. This can be easily realized by illuminating the hologram from the opposite side to that in recording. In this case, the fourth term of (7.14) becomes $A_R^2 O^*(x, y)$ and the hologram reconstructs the conjugate image at the same position as the recorded object.

It would be worthwhile to point out that conjugate images do not appear in phase-modulated holograms unless the hologram has binary phases (see Sect. 8.7). Phase holograms are rather easily realized in computer holography using a surface relief (see Sect. 15.3.6) or phase-only spatial light modulator (SLM) (see Sect. 15.2.2).

7.7 Theory and Examples of Thin Hologram

We discuss theories of several types of thin holograms based on wave-optics in this section. The discussion puts importance on the position of the reconstructed images. The true image is reconstructed at the same position as that of the original object when the illumination field is identical to the reference field. However, the true image shifts in cases where the illumination light does not completely agree with the reference field. In addition, it is also important to expect the position of the conjugate image in order to avoid the overlap between the true and conjugate images in holographic 3D imagings in practice.

7.7.1 Hologram with Plane Wave

In the beginning, we discus the cases where a plane wave is used for the reference field, because plane waves are easy to handle in general. The illumination field is also a plane wave in this case.

7.7.1.1 True Image

Suppose that the amplitude hologram is again placed at the $(x, y, 0)$ plane, as shown in Fig. 7.6. Wave vectors of the reference and illumination plane waves make angles of θ_{R} and θ_{P} with the optical axis in the $(x, 0, z)$ plane, respectively. Accordingly, these wavefields are written by

$$R(x, y) = A_{\mathrm{R}} \exp[iku_{\mathrm{R}}x], \tag{7.29}$$

$$P(x, y) = A_{\mathrm{P}} \exp[iku_{\mathrm{P}}x], \tag{7.30}$$

where

$$u_{\mathrm{R}} = \sin\theta_{\mathrm{R}}/\lambda, \tag{7.31}$$

$$u_{\mathrm{P}} = \sin\theta_{\mathrm{P}}/\lambda. \tag{7.32}$$

Suppose that the object field is represented by (7.25), i.e., $O(x, y)$ is given by propagating $g_{\mathrm{obj}}(x, y; z_0)$ with a distance of d. Assuming the Fresnel diffraction, the third term of (7.12), which presents the reconstructed field that produces the true image, is written as

$$\begin{aligned} &O(x, y)R^*(x, y)P(x, y) \\ &= A_{\mathrm{P}}A_{\mathrm{R}}\mathscr{P}_d\left\{g_{\mathrm{obj}}(x, y; z_0)\right\} \exp[i2\pi(u_{\mathrm{P}} - u_{\mathrm{R}})x] \\ &= A_{\mathrm{P}}A_{\mathrm{R}}A_{\mathrm{FR}}(d) \iint g_{\mathrm{obj}}(x_{\mathrm{s}}, y_{\mathrm{s}}; z_0) \\ &\quad \times \exp\left[i\frac{\pi}{\lambda d}\left\{(x - x_{\mathrm{s}})^2 + (y - y_{\mathrm{s}})^2 + 2\lambda d(u_{\mathrm{P}} - u_{\mathrm{R}})x\right\}\right] \mathrm{d}x_{\mathrm{s}}\mathrm{d}y_{\mathrm{s}}, \end{aligned} \tag{7.33}$$

where the formula of the Fresnel diffraction in (5.27) is used. Here, introducing variable $x_{\mathrm{s}}' = x_{\mathrm{s}} - \lambda d(u_{\mathrm{P}} - u_{\mathrm{R}})$ and replacing x_{s} by x_{s}' in the integral, we can rewrite the above equation as follows:

$$\begin{aligned} &O(x, y)R^*(x, y)P(x, y) \\ &= A_{\mathrm{P}}A_{\mathrm{R}}A_{\mathrm{FR}}(d) \exp[i\pi\lambda d(u_{\mathrm{P}} - u_{\mathrm{R}})^2] \iint g_{\mathrm{obj}}(x_{\mathrm{s}}' + \lambda d(u_{\mathrm{P}} - u_{\mathrm{R}}), y_{\mathrm{s}}; z_0) \\ &\quad \times \exp\left[i2\pi(u_{\mathrm{P}} - u_{\mathrm{R}})x_{\mathrm{s}}'\right] \exp\left[i\frac{\pi}{\lambda d}\left\{(x - x_{\mathrm{s}}')^2 + (y - y_{\mathrm{s}})^2\right\}\right] \mathrm{d}x_{\mathrm{s}}'\mathrm{d}y_{\mathrm{s}}. \end{aligned} \tag{7.34}$$

Since the integral has the form of the Fresnel diffraction, we can rewrite it using the propagation operator again:

$$\begin{aligned} &O(x, y)R^*(x, y)P(x, y) \\ &= A_P A_R \exp[i\pi\lambda d(u_P - u_R)^2] \\ &\quad \times \mathscr{P}_d \left\{ g_{obj}(x + \lambda d(u_P - u_R), y; z_0) \exp\left[i2\pi(u_P - u_R)x\right] \right\}. \end{aligned} \tag{7.35}$$

Using incident angles and propagating backward the reconstructed field at distance d, we can get the wavefield corresponding to the true image:

$$\begin{aligned} &g_{true}(x, y; z_0) \\ &= \mathscr{P}_{-d} \left\{ O(x, y)R^*(x, y)P(x, y) \right\} \\ &= A_P A_R \exp[ikd(\sin\theta_P - \sin\theta_R)^2/2] \\ &\quad \times g_{obj}(x + d(\sin\theta_P - \sin\theta_R), y; z_0) \exp\left[ik(\sin\theta_P - \sin\theta_R)x\right]. \end{aligned} \tag{7.36}$$

As a result, we conclude that the true image basically agrees with the original field and has the same depth as that of the original image. However, the x position shifts with $-d(\sin\theta_P - \sin\theta_R)$. This means that when we illuminate the hologram at a larger incident angle of θ_P than that of the reference field, the true image moves toward $-x$ direction. In addition, plane wave $\exp[ik(\sin\theta_P - \sin\theta_R)x]$ makes the field direction change as if it cancels the position shift. This position shift properly disappears when $\theta_P = \theta_R$. In this case, the reconstructed field is simply

$$g_{true}(x, y; z_0) = A_P A_R g_{obj}(x, y; z_0).$$

7.7.1.2 Conjugate Image

The conjugate light reconstructed by illumination field $P(x, y)$ is also given by

$$\begin{aligned} &O^*(x, y)R(x, y)P(x, y) \\ &= A_P A_R \mathscr{P}_d \left\{ g_{obj}(x, y; z_0) \right\} \exp[i2\pi(u_P + u_R)x] \\ &= A_P A_R A_{FR}^*(d) \iint g_{obj}^*(x_s, y_s; z_0) \\ &\quad \times \exp\left[-i\frac{\pi}{\lambda d}\left\{(x - x_s)^2 + (y - y_s)^2 - 2\lambda d(u_P + u_R)x\right\}\right] dx_s dy_s. \end{aligned} \tag{7.37}$$

Introducing new variable $x_s' = x_s - \lambda d(u_P + u_R)$ and performing the same procedure as that for the true image, we get

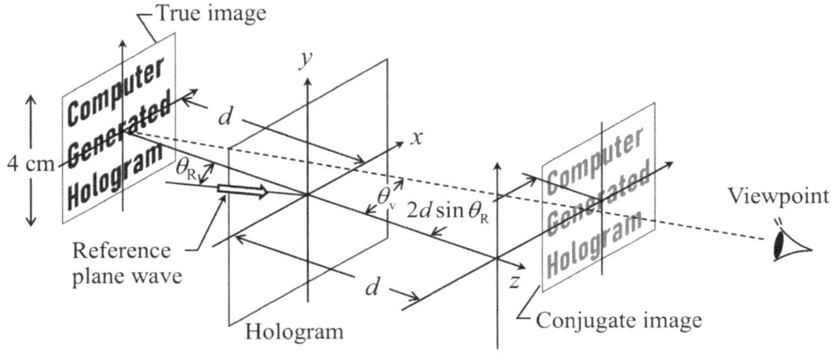

Fig. 7.8 Schematic illustration of the positions of a viewpoint, and the true and conjugate images in the case where $P(x, y) = R(x, y)$

$$
\begin{aligned}
&O^*(x, y)R(x, y)P(x, y) \\
&= A_P A_R \exp[-i\pi\lambda d(u_P + u_R)^2] \\
&\quad \times \mathscr{P}_{-d}\left\{g^*_{\text{obj}}(x - \lambda d(u_P + u_R), y; z_0) \exp\left[i2\pi(u_P + u_R)x\right]\right\}. \quad (7.38)
\end{aligned}
$$

Therefore, the conjugate field is represented by

$$
\begin{aligned}
g_{\text{conj}}(x, y; -z_0) &= \mathscr{P}_d\left\{O^*(x, y)R(x, y)P(x, y)\right\} \\
&= A_P A_R \exp[-i\pi\lambda d(u_P + u_R)^2] \\
&\quad \times g^*_{\text{obj}}(x - \lambda d(u_P + u_R), y; z_0) \exp\left[i2\pi(u_P + u_R)x\right]. \\
&\qquad (7.39)
\end{aligned}
$$

Here, it should be noted that the conjugate image is floating in front of the hologram because $z_0 < 0$, and thus, the conjugate field forms a real image. In addition, the image shifts with $d(\sin\theta_P + \sin\theta_R)$ along the x-axis.

We are usually most interested in the case where $\theta_P = \theta_R$. In this case, the conjugate field is simplified as

$$
\begin{aligned}
g_{\text{conj}}(x, y; -z_0) &= A_P A_R \exp[-i2kd \sin^2\theta_R] \\
&\quad \times g^*_{\text{obj}}(x - 2d\sin\theta_R, y; z_0) \exp\left[i2k\sin\theta_R x\right]. \quad (7.40)
\end{aligned}
$$

Accordingly, we can conclude that the conjugate object field is reconstructed at $z = +d$, and the x position shifts with $2d\sin\theta_R$. The position of the true and conjugate images in this case are schematically illustrated in Fig. 7.8. Here, the object is a square having a pattern of characters on the surface. Thus, the reconstructed image is represented by $|g_{\text{obj}}(x, y; z_0)|^2$.

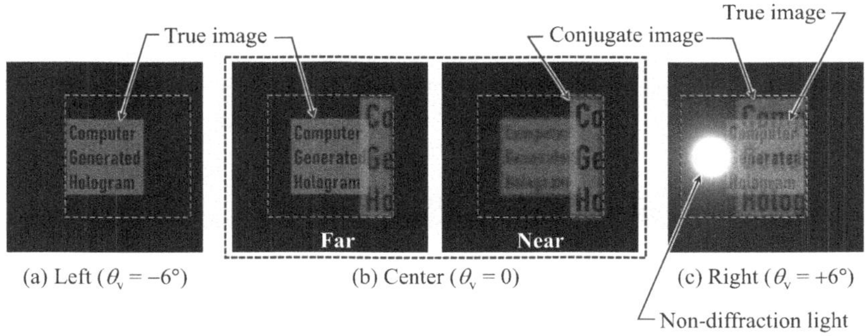

Fig. 7.9 Examples of reconstruction of the hologram created with a reference plane wave. The reconstructed images are calculated at different viewpoints. $d = 10$ [cm] and $\theta_R = 8°$. Red squares indicate the area where the fringe pattern is formed, i.e., the area of the hologram

7.7.1.3 Examples of Reconstruction

Figure 7.9 shows reconstructed images of a thin hologram. The layout of the object, hologram and viewpoint is also shown in Fig. 7.8. Here, note that this hologram is not a real hologram but a CGH, and the reconstructed image is produced by numerical simulation. However, the reconstructed images are exact and realistic enough to understand the nature of thin holograms, because we use the technique of simulated reconstruction based on virtual imaging (see Chap. 13).

The CGH is a square with one side approximately 5 cm. The object in this CGH is also a square with one side 4 cm, where letters are mapped as a binary texture. This square object is placed at a distance of d behind the hologram. The object field is calculated using the polygon-based method (see Chap. 10). The reference field is a plane wave, whose wave vector makes an angle of θ_R with the optical axis.

According to (7.40), the conjugate image appears at a position of $z = +d$ in front of the hologram, but the position shifts in the x direction. Figure 7.9b shows the reconstructed image, which is seen from the center viewpoint. The simulated reconstruction is focused on the true image of the hologram in the 'Far' image, while the conjugate image in the 'Near' image. It is found that the conjugate image floats in front of the true image and comes into the view.

When we move the viewpoint right, the situation is getting worse, as shown in Fig. 7.9c. Non-diffraction light in addition to the conjugate image overlaps the true image, and obstructs the view. Conversely, moving the viewpoint left, the reconstructed image is becoming clearer. We can see only the true image as in (a).

The incident angle of the reference plane wave was $\theta_R = 8°$ in the examples of Fig. 7.9. Reconstructions of other holograms that are created with $\theta_R = 4°$ and $12°$ are shown in Fig. 7.10a and c, respectively. It is obvious that it is better to increase the incident angle θ_R to avoid the overlap of the conjugate image and improve the reconstructed image. However, increasing the incident angle of the reference wave leads to increasing the spatial frequency of the interference fringe as mentioned

(a) $\theta_R = 4°$ (b) $\theta_R = 8°$ (b) $\theta_R = 12°$

Fig. 7.10 Reconstructed images of holograms created at different angles of the reference plane wave. The viewpoint is at the center in common. $d = 10$ [cm]

(a) $d = 5$ [cm] (b) $d = 10$ [cm] (c) $d = 15$ [cm]

Fig. 7.11 Reconstructed images of holograms that have different object distance d. The viewpoint is at the center in common. $\theta_R = 8°$. The apparent size of the true images varies because the distance from the viewpoint changes

in Sect. 8.8.2, and thus, increasing the possibility of aliasing errors in computer holography.

Figure 7.11 also shows reconstruction of similar but different holograms. Here, we vary the distance d between the object and hologram. The problem of overlapping the conjugate image can be improved by increasing the distance d as in (c), but the reconstructed image becomes smaller with increasing the distance. Conversely, by arranging the object closer to the hologram, the reconstructed object appears bigger naturally. However, the overlapped conjugate image disturbs the view and degrades the image, as shown in (a). This problem, caused by placing the object close to the hologram, signifies that it is very difficult to create an image hologram as a thin hologram or a CGH.

7.7.2 Hologram with Spherical Wave

Spherical waves are commonly more useful than plane waves in practical exhibition of holograms, because diffused illumination light is usually easier to generate than collimated illumination light. In addition, if we choose the reference and illumina-

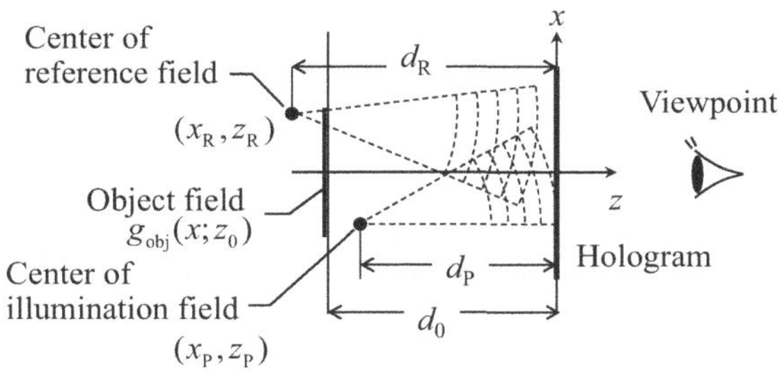

Fig. 7.12 Geometry used for analysis of the reconstructed image of a thin holograms with a reference spherical wave

tion fields carefully, we can considerably reduce the overlap between the true and conjugate images.

7.7.2.1 Formulation

It is very difficult in reference and illumination spherical fields to obtain the positions of reconstructed images. The solution generally has complicate forms. Thus, we formulate only with respect to the x direction for simplifying the formulas. If y components are required, switching the symbol x to y is good enough in many cases.

Assuming that $x^2 \ll z^2$, we treat spherical waves with their Fresnel approximation as in (3.71). Thus, let the reference and illumination fields be written by

$$R(x; d_R) = A_R \exp\left[ik\left\{d_R + \frac{(x - x_R)^2}{2d_R}\right\}\right], \tag{7.41}$$

$$P(x; d_P) = A_P \exp\left[ik\left\{d_P + \frac{(x - x_P)^2}{2d_P}\right\}\right], \tag{7.42}$$

where A_R and A_P are amplitudes of the reference and illumination fields, respectively. The origin of the coordinate system is placed in the hologram plane as in the case of plane waves, as shown in Fig. 7.12. In addition, we suppose the object and light sources are arranged behind the hologram, i.e., in the space where $z < 0$. The centers of the reference and illumination spherical fields are (x_R, z_R) and (x_P, z_P), respectively. Object field $g_{obj}(x; z_0)$ is given at $z = z_0$. Thus, $z_R < 0$, $z_P < 0$ and $z_0 < 0$. Since d_R, d_P and d_0 are distances from the hologram,

$$d_R = -z_R, \; d_P = -z_P, \quad \text{and} \quad d_0 = -z_0. \tag{7.43}$$

Although these should have positive values because they represent distances, let us allow their negative values if necessary for specific arrangements.

Furthermore, we suppose that the object field in the hologram plane is given by the Fresnel diffraction:

$$\begin{aligned} O(x) &= \mathscr{P}_{d_0}\left\{g_{\mathrm{obj}}(x; z_0)\right\} \\ &= g_{\mathrm{obj}}(x; z_0) \otimes h_{\mathrm{FR}}(x; d_0) \\ &= A_{\mathrm{FR}}(d_0) \int g_{\mathrm{obj}}(x_{\mathrm{s}}; z_0) \exp\left[i\frac{\pi}{\lambda d_0}(x_{\mathrm{s}} - x)^2\right] \mathrm{d}x_{\mathrm{s}}. \end{aligned} \tag{7.44}$$

7.7.2.2 True Image

The wavefield reconstructing the true image is

$$\begin{aligned} &O R^* P(x) \\ &= A_{\mathrm{R}} A_{\mathrm{P}} A_{\mathrm{FR}}(d_0) \exp[ik(d_{\mathrm{P}} - d_{\mathrm{R}})] \int g_{\mathrm{obj}}(x_{\mathrm{s}}; z_0) \\ &\quad \times \exp\left[i\frac{\pi}{\lambda d_{\mathrm{t}}}\left\{x^2 - 2\frac{d_{\mathrm{t}}}{d_0}\left(x_{\mathrm{s}} + d\eta^{(-)}\right)x + \frac{d_{\mathrm{t}}}{d_0}\left(x_s^2 + dD^{(-)}\right)\right\}\right] \mathrm{d}x_{\mathrm{s}}, \end{aligned} \tag{7.45}$$

where

$$\frac{1}{d_{\mathrm{t}}} = \frac{1}{d_0} + \frac{1}{d_{\mathrm{P}}} - \frac{1}{d_{\mathrm{R}}}, \tag{7.46}$$

$$\eta^{(\pm)} = \frac{x_{\mathrm{P}}}{d_{\mathrm{P}}} \pm \frac{x_{\mathrm{R}}}{dr}, \tag{7.47}$$

$$D^{(\pm)} = \frac{x_{\mathrm{P}}^2}{d_{\mathrm{P}}} \pm \frac{x_{\mathrm{R}}^2}{dr}. \tag{7.48}$$

Introducing new variable,

$$x_{\mathrm{s}}' = \frac{d_{\mathrm{t}}}{d_0}\left(x_{\mathrm{s}} + d\eta^{(-)}\right), \tag{7.49}$$

the exponent of (7.45) is rearranged to

$$i\frac{\pi}{\lambda d_{\mathrm{t}}}\left[(x - x_{\mathrm{s}}')^2 + \frac{d_0}{d^{(-)}}\left(x_{\mathrm{s}}' - d^{(-)}\eta^{(-)}\right)^2 + d_{\mathrm{t}}\left\{D^{(-)} - d^{(-)}\left(\eta^{(-)}\right)^2\right\}\right].$$

where

$$\frac{1}{d^{(\pm)}} = \frac{1}{d_{\mathrm{P}}} \pm \frac{1}{d_{\mathrm{R}}}, \tag{7.50}$$

and $d_{\mathrm{P}} \neq d_{\mathrm{R}}$ is assumed. Accordingly, (7.45) becomes

$$
\begin{aligned}
&OR^*P(x) \\
&= A_R A_P A_{FR}(d_0)\exp[ik(d_P - d_R)]\exp\left[i\frac{\pi}{\lambda}\left\{D^{(-)} - d^{(-)}\left(\eta^{(-)}\right)^2\right\}\right] \\
&\times \int g_{obj}\left(\frac{x'_s - S_t}{m_t}; z_0\right)\exp\left[i\frac{\pi}{\lambda d_t}\frac{d_0}{d^{(-)}}\left(x'_s - d^{(-)}\eta^{(-)}\right)^2\right] \\
&\times \exp\left[i\frac{\pi}{\lambda d_t}(x - x'_s)^2\right]dx'_s,
\end{aligned} \tag{7.51}
$$

where S_t and m_t represent the shift and magnification of the image, respectively. These are defined by

$$
S_t \equiv d_t \eta^{(-)}, \tag{7.52}
$$

$$
m_t \equiv \frac{d_t}{d_0}. \tag{7.53}
$$

Since (7.51) has the form of the Fresnel diffraction, it can be rewritten as

$$
\begin{aligned}
&OR^*P(x) \\
&= A_R A_P \exp[ik(d_P - d_R)]\exp\left[i\frac{\pi}{\lambda}\left\{D^{(-)} - d^{(-)}\left(\eta^{(-)}\right)^2\right\}\right] \\
&\times \left[g_{obj}\left(\frac{x - S_t}{m_t}; z_0\right)\exp\left[i\frac{\pi}{\lambda d_t}\frac{d_0}{d^{(-)}}\left(x - d^{(-)}\eta^{(-)}\right)^2\right]\right] \\
&\otimes h_{FR}(x; d_t). \\
&= A_R A_P \exp[ik(d_P - d_R)]\exp\left[i\frac{\pi}{\lambda}\left\{D^{(-)} - d^{(-)}\left(\eta^{(-)}\right)^2\right\}\right] \\
&\times \mathscr{P}_{d_t}\left\{g_{obj}\left(\frac{x - S_t}{m_t}; z_0\right)\exp\left[i\frac{\pi}{\lambda d_t}\frac{d_0}{d^{(-)}}\left(x - d^{(-)}\eta^{(-)}\right)^2\right]\right\}.
\end{aligned} \tag{7.54}
$$

As a result, the wavefield corresponding to the true image is presented at a position of $z = -d_t$ as

$$
\begin{aligned}
&g_{true}(x; -d_t) = \mathscr{P}_{-d_t}\left\{OR^*P(x)\right\} \\
&= A_R A_P \exp[ik(d_P - d_R)]\exp\left[i\frac{\pi}{\lambda}\left\{D^{(-)} - d^{(-)}\left(\eta^{(-)}\right)^2\right\}\right] \\
&\times g_{obj}\left(\frac{x - S_t}{m_t}; z_0\right)\exp\left[i\frac{\pi}{\lambda d_t}\frac{d_0}{d^{(-)}}\left(x - d^{(-)}\eta^{(-)}\right)^2\right].
\end{aligned} \tag{7.55}
$$

We found that the original wavefield $g_{obj}(x; z_0)$ is reconstructed at a depth of d_t. The field is shifted with a distance of S_t and magnified by m_t. In addition to these, the field has a curvature represented by the quadratic exponent.

The general formulation (7.55) is too complicated to understand the behavior of the reconstructed true image. Rather, we are more interested in the case $d_P = d_R$. The depth of the reconstructed field agrees with the original field; $d_t = d_0$ in this case. Besides, the field is not magnified; $m_t = 1$, and the position shift is given by

$d_0(x_P - x_R)/d_R$. The wavefield for the true image becomes

$$g_{\text{true}}(x; z_0) = A_R A_P \exp\left[i\frac{k}{2d_R}\left\{\frac{d_0}{d_R}(x_P - x_R)^2 + (x_P^2 - x_R^2)\right\}\right]$$
$$\times g_{\text{obj}}\left(x - \frac{d_0}{d_R}(x_P - x_R); z_0\right)\exp\left[-i\frac{k}{d_R}(x_P - x_R)x\right]. \quad (7.56)$$

Furthermore, when $d_R = d_0$, the position shift is simply $x_P - x_R$. Therefore if we move the illumination light source at a distance of δx along the x-axis, the reconstructed true image also moves at the same distance of δx along the x-axis. It seems as if the true image follow the light source.

According to (7.46), when $d_R = d_0$, the z position of the true image agrees with that of the illumination light source; $d_t = d_P$. This suggests that the true image also follows the illumination in the depth direction. As a result, we can conclude that the true image always follows the illumination light source and seems reconstructed around the bright spot caused by the non-diffraction light of the illumination.

7.7.2.3 Conjugate Image

We can formulate the wavefield reconstructed by the conjugate light of the hologram using very similar procedure to the true image. The wavefield reconstructing the conjugate image is

$$O^*RP(x)$$
$$= A_R A_P A_{FR}^*(d_0)\exp[ik(d_P + d_R)]\int g_{\text{obj}}^*(x_s; z_0)$$
$$\times \exp\left[i\frac{\pi}{\lambda d_c}\left\{x^2 + 2\frac{d_c}{d_0}\left(x_s - d\eta^{(+)}\right)x - \frac{d_c}{d_0}\left(x_s^2 - d_0 D^{(+)}\right)\right\}\right]\mathrm{d}x_s, \quad (7.57)$$

where $\eta^{(+)}$ and $D^{(+)}$ are given in (7.47) and (7.48), respectively, and

$$\frac{1}{d_c} = -\frac{1}{d_0} + \frac{1}{d_P} + \frac{1}{d_R}. \quad (7.58)$$

Changing the variable x_s of integration in (7.57) to

$$x_s' = -\frac{d_c}{d_0}\left(x_s - d\eta^{(+)}\right),$$

the conjugate field becomes

$$
\begin{aligned}
&O^* R P(x) \\
&= A_R A_P A_{FR}^*(d_0) \exp[ik(d_P + d_R)] \exp\left[i\frac{\pi}{\lambda}\left\{D^{(+)} - d^{(+)}\left(\eta^{(+)}\right)^2\right\}\right] \\
&\quad \times \int g_{obj}^*\left(\frac{-x_s' + S_c}{m_c}; z_0\right) \exp\left[-i\frac{\pi}{\lambda d_c}\frac{d_0}{d^{(+)}}\left(x_s' - d^{(+)}\eta^{(+)}\right)^2\right] \\
&\quad \times \exp\left[i\frac{\pi}{\lambda d_c}(x - x_s')^2\right] dx_s',
\end{aligned} \tag{7.59}
$$

where S_c and m_c represent the shift and magnification of the conjugate image, respectively, and are defined by

$$S_c \equiv d_c \eta^{(+)}, \tag{7.60}$$

$$m_c \equiv \frac{d_c}{d_0}. \tag{7.61}$$

Since (7.59) also has the form of the Fresnel diffraction, we can derive the wavefield for the conjugate image as follows:

$$
\begin{aligned}
&g_{conj}(x; -d_c) = \mathscr{P}_{-d_c}\left\{O^* R P(x)\right\} \\
&= -A_R A_P \frac{d_c}{d} \exp[ik(d_P + d_R - d_c - d_0)] \exp\left[i\frac{\pi}{\lambda}\left\{D^{(+)} - d^{(+)}\left(\eta^{(+)}\right)^2\right\}\right] \\
&\quad \times g_{obj}^*\left(\frac{-x + S_c}{m_c}; z_0\right) \exp\left[-i\frac{\pi}{\lambda d_c}\frac{d_0}{d^{(+)}}\left(x - d^{(+)}\eta^{(+)}\right)^2\right].
\end{aligned} \tag{7.62}
$$

This is also too complicated to discuss the property of the conjugate image. Therefore, we consider the most interesting and useful case; that is, the position of the illumination light source agrees with that of the reference field; $x_P = x_R$ and thus $d_P = d_R$.

In addition, we assume that the reference and illumination light sources are arranged in the same plane as that of the object field; $d_0 = d_R = d_P$. This seems to limit the condition too much and abandon generality of the formulation. However, we do not loose so much generality even in this case, because $g_{obj}(x; z_0)$ just represents the object wavefield at $z = z_0$. We can consider $g_{obj}(x; z_0)$ as a field propagated from another field given in another position. For example, it is possible that $g_{obj}(x; z_0) = \mathscr{P}_{z_0 - z_{obj}}\left\{g_{obj}(x; z_{obj})\right\}$. Thus, even if d_0 takes the same value as $d_R (= d_P)$, we can analyze the behavior of the object field and thus conjugate field given at any z position.

Supposing $x_P = x_R$ and $d_0 = d_P = d_R$, the complicated wavefield (7.62) is drastically simplified down to

$$g_{conj}(x; z_R) = -A_R A_P g_{obj}^*(-x + 2x_R; z_0) \exp\left[-i\frac{k}{d_R}(x - x_R)^2\right], \tag{7.63}$$

where remember that $z_R = z_P = z_0 = -d_R$ in this case. We can find two important things in this field promptly. First, the wavefield reconstructing the conjugate image

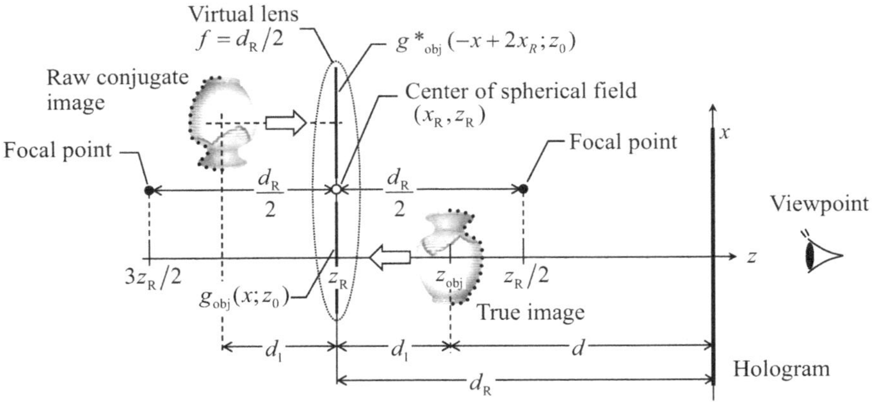

Fig. 7.13 The model for reconstruction of the conjugate image

is given by inverting x coordinates of conjugate object field $g^*_{obj}(x; z_0)$. The field is actually symmetrical to $g^*_{obj}(x; z_0)$ with respect to $x = x_R$. Second, the field is modified by a quadratic phase function. By comparison with (5.55), it is found that this quadratic phase-modulation is actually equivalent to that of a convex lens, whose focal length is $d_R/2$, and the virtual lens is positioned at (x_R, z_R) that is the center of the spherical field.

Figure 7.13 shows a reconstruction model of the conjugate image. Supposing that the recorded object is arranged at

$$z_{obj} = -d, \tag{7.64}$$

and the field is represented by $\hat{g}_{obj}(x)$, the object field $g_{obj}(x; z_0)$ is given by field propagation:

$$g_{obj}(x; z_0) = \mathscr{P}_{-d_1}\left\{\hat{g}_{obj}(x)\right\}, \tag{7.65}$$

where the propagation distance is

$$d_1 = |d_R - d|. \tag{7.66}$$

On the other hand, as mentioned in Sect. 7.6, the conjugate field in the plane at z_0 is given by propagating the field placed in the opposite side. Therefore, the field for the conjugate image is given by

$$g^*_{obj}(-x + 2x_R; z_0) = \mathscr{P}_{+d_1}\left\{\hat{g}^*_{obj}(-x + 2x_R)\right\}. \tag{7.67}$$

We call the source field $\hat{g}^*_{obj}(-x + 2x_R)$ a raw conjugate image in the following discussion. Viewers do not directly see the raw conjugate image, because the conjugate

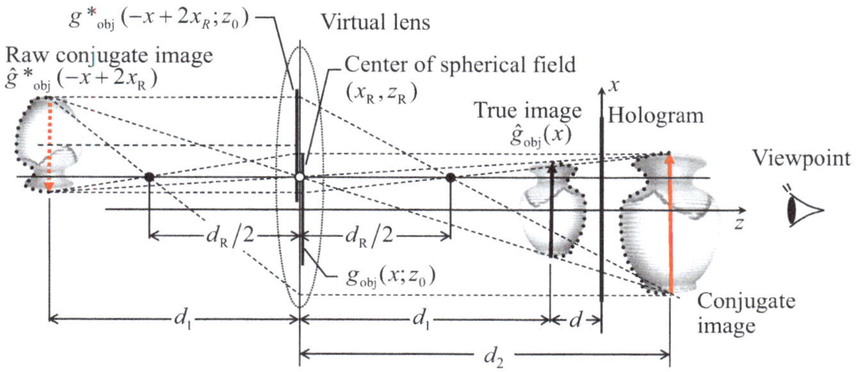

Fig. 7.14 Image formation of the conjugate field by the virtual lens produced by the reference spherical field

field is affected by the virtual convex lens. In practice, viewers see the conjugate image as an image formed by the convex lens whose focal length is $d_R/2$. Therefore, the position and size of the conjugate image is determined by the relation between d_R and d_1.

Figure 7.14 shows how to obtain the position of the conjugate image. The raw conjugate image is first obtained by point-symmetrically projection of the true image with respect to the center of the spherical fields, and then, we can get the position of the conjugate image using ordinary drawing techniques for image-formation by a thin lens.

We can treat the problem by categorizing the object position into several cases, as shown in Fig. 7.15. Here, suppose that $x_R = 0$ for simplification.

Case (i) $\dfrac{z_R}{2} < z_{\rm obj} \quad \left(d < \dfrac{d_R}{2}\right)$

The raw conjugate image is placed beyond the further focal point in this case. Therefore, the conjugate image is formed as a real image by the virtual convex lens. Because the thin lens formula in (5.49) is in this case

$$\frac{1}{d_1} + \frac{1}{d_2} = \frac{1}{d_R/2}, \tag{7.68}$$

we can find

$$d_2 = \frac{d_R d_1}{2d_1 - d_R}. \tag{7.69}$$

Since $d_1 = d_R - d$, the position of the conjugate image is

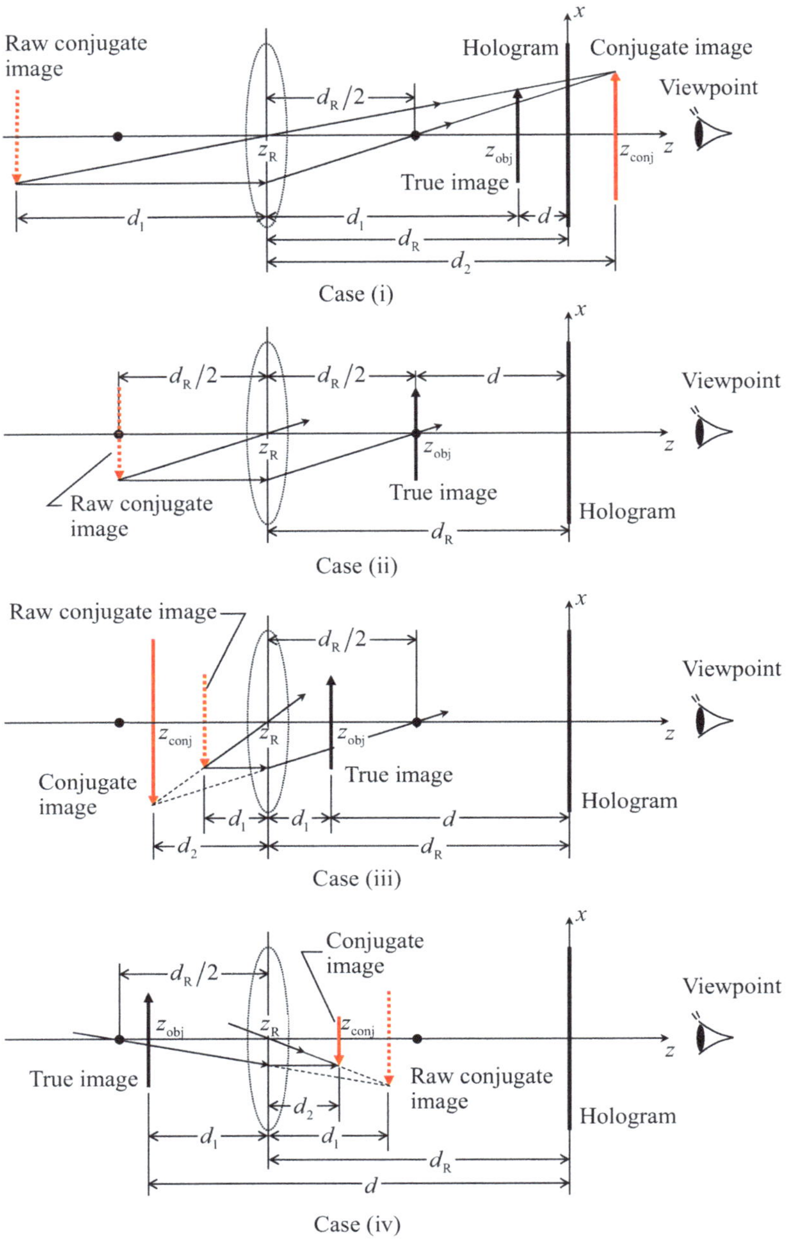

Fig. 7.15 The positions of the conjugate image formed by the virtual lens

$$z_{\text{conj}} = z_{\text{R}} + d_2$$
$$= \frac{d_{\text{R}} d}{d_{\text{R}} - 2d} = \frac{z_{\text{R}} z_{\text{obj}}}{2z_{\text{obj}} - z_{\text{R}}}. \quad (7.70)$$

When $z_{\text{R}}/2 < z_{\text{obj}} \leq 0$; the recorded object is arranged behind the hologram plane, the position of the conjugate image always satisfy $z_{\text{conj}} \geq 0$. Thus, the conjugate image is floating in front of the hologram as a real image. If $z_{\text{obj}} > 0$, the conjugate image exchanges the position with the true image and appears behind the hologram theoretically. However, the hologram is usually recorded as a volume hologram in this case, and thus, may not generate any conjugate image.

Magnification of the conjugate image is given by

$$m_{\text{conj}} = \frac{d_2}{d_1}$$
$$= \frac{d_{\text{R}}}{d_{\text{R}} - 2d} = \frac{z_{\text{R}}}{z_{\text{R}} - 2z_{\text{obj}}}. \quad (7.71)$$

As a result, the magnification increases as the position z_{obj} is closer to $z_{\text{R}}/2$. In that case, as shown in (7.70), the conjugate image appears as a real image far from the hologram and sometimes positioned behind the viewpoint. The viewers no longer recognize any conjugate image in this case, but the conjugate field causes noise.

It should be noted that the conjugate image is always upright, i.e., formed as an erect image in this Case (i). In addition, if the recorded object is arranged behind the hologram ($z_{\text{R}}/2 < z_{\text{obj}} \leq 0$), the conjugate image is magnified more than the true image ($m_{\text{conj}} \geq 1$).

Case (ii) $z_{\text{obj}} = \dfrac{z_{\text{R}}}{2} \quad \left(d = \dfrac{d_{\text{R}}}{2}\right)$

In this case, any conjugate image is not formed by the virtual convex lens. The Fourier transform of the raw conjugate image $g^*_{\text{true}}(x)$ is obtained at the position of the closer focal point, and thus, generates a far-field image.

Case (iii) $z_{\text{R}} \leq z_{\text{obj}} < \dfrac{z_{\text{R}}}{2} \quad \left(d_{\text{R}} \geq d > \dfrac{d_{\text{R}}}{2}\right)$

The conjugate image formed by the virtual lens is a virtual image with respect to the lens in this case. The image is also a virtual image with respect to the hologram and inverted. Therefore, the thin lens formula is

$$\frac{1}{d_1} - \frac{1}{d_2} = \frac{1}{d_{\text{R}}/2},$$

and the position of the conjugate image is given by

Table 7.1 Discriminants of a conjugate image in cases of reference spherical fields

Image formation	$z_{\text{conj}} \geq 0$	Real image
	$z_{\text{conj}} < 0$	Virtual image
Position	$z_{\text{conj}} \geq z_{\text{obj}}$	Before true image
	$z_{\text{conj}} < z_{\text{obj}}$	Behind true image
Direction	$m_{\text{conj}} \geq 0$	Erect image
	$m_{\text{conj}} < 0$	Inverted image
Size	$\lvert m_{\text{conj}} \rvert \geq 1$	Enlarge
	$\lvert m_{\text{conj}} \rvert < 1$	Shrink

$$
\begin{aligned}
z_{\text{conj}} &= -d_{\text{R}} - d_2 \\
&= \frac{d_{\text{R}} d}{d_{\text{R}} - 2d} = \frac{z_{\text{R}} z_{\text{obj}}}{2z_{\text{obj}} - z_{\text{R}}}.
\end{aligned}
$$

This agrees with (7.70). Therefore, we can obtain the position by (7.70) in both cases. However, $z_{\text{conj}} < z_{\text{R}} \leq z_{\text{obj}}$ in this case. Thus, the conjugate image always appears behind the true image unlike Case (i). If we saw a continuous transition from Case (i) to (iii), the conjugate image would appear as if it leaps abruptly from a position before the hologram to another position behind the true image far from the hologram.

Magnification of the conjugate image is given by (7.71) again. However, the value is negative; $m_{\text{conj}} < 0$ in this case. This corresponds to forming the inverted image unlike Case (i), where the conjugate image is erect.

Case (iv) $z_{\text{obj}} < z_{\text{R}}$ $(d > d_{\text{R}})$

The raw conjugate image appears to be generated in front of the virtual lens in this case. However, the raw image is modified by the lens at the position $z = z_{\text{R}}$. Thus thin lens formula is written as

$$
-\frac{1}{d_1} + \frac{1}{d_2} = \frac{1}{d_{\text{R}}/2},
$$

As a result, the conjugate image is positioned at

$$
\begin{aligned}
z_{\text{conj}} &= -d_{\text{R}} + d_2 \\
&= \frac{d_{\text{R}} d}{d_{\text{R}} - 2d} = \frac{z_{\text{R}} z_{\text{obj}}}{2z_{\text{obj}} - z_{\text{R}}},
\end{aligned}
$$

where note that $d_1 = d - d_{\text{R}}$. This also agrees with (7.70).

Magnification is given by (7.71) again, but the value is negative; $m_{\text{conj}} < 0$ like Case (iii). Thus, the formed conjugate image is inverted also in this case.

In conclusion, the following formulas give the z position and magnification of the conjugate image in any case except for Case (ii), where the conjugate image is not formed.

$$z_{\text{conj}} = \frac{d_{\text{R}} d}{d_{\text{R}} - 2d} = \frac{z_{\text{R}} z_{\text{obj}}}{2z_{\text{obj}} - z_{\text{R}}}, \tag{7.72}$$

$$m_{\text{conj}} = \frac{d_{\text{R}}}{d_{\text{R}} - 2d} = \frac{z_{\text{R}}}{z_{\text{R}} - 2z_{\text{obj}}}. \tag{7.73}$$

Here, if $m_{\text{conj}} < 0$, an inverted image is formed, otherwise an erect image is formed as the conjugate image. An object point at $x = x_{\text{obj}}$ is projected into

$$\begin{aligned} x_{\text{conj}} &= (-x_{\text{obj}} + 2x_{\text{R}} - x_{\text{R}}) m_{\text{conj}} + x_{\text{R}} \\ &= -m_{\text{conj}} x_{\text{obj}} + (m_{\text{conj}} + 1) x_{\text{R}}. \end{aligned} \tag{7.74}$$

The conjugate image can be discriminated using the values of z_{conj} and m_{conj}, as summarized in Table 7.1. Figure 7.16a is a diagram that depicts $z_{\text{conj}}/d_{\text{R}}$ and m_{conj} as a function of z_{obj}. Here, let $z_{\text{conj}}/d_{\text{R}}$ and m_{conj} be the left and right vertical axes, respectively. The solid and dashed lines are of z_{conj} and m_{conj}. Variation of the conjugate image is also indicated according to the discriminants in Table 7.1. We can expect the position and property of the conjugate image when moving the recorded object for a fixed reference spherical field. There is a singularity at $z_{\text{obj}} = z_{\text{R}}/2$. As mentioned already, the position of the conjugate image leaps abruptly and the erect and inverted images are switched in the transition.

Figure 7.16b is also a diagram that, however, depicts z_{conj}/d as a function of z_{R} instead of z_{obj}. Here, note that $z_{\text{conj}}/d = d_{\text{R}}/(d_{\text{R}} - 2d) = m_{\text{conj}}$ according to (7.72) and (7.73). Thus, only a single vertical axis is used in this diagram. This diagram is useful to estimate the position and property of the conjugate image in changing the position of the reference spherical field for a fixed object position. The point $z_{\text{R}} = 2z_{\text{obj}}$ gives the singularity, and the direction and position changes abruptly at the position in this case. It should be noted that it is impossible in practice to realize $z_{\text{R}} = 0$, because we cannot arrange the center of the reference field in the same plane as the hologram.

7.7.2.4 Examples of Reconstruction

Examples of reconstruction of holograms recorded with a spherical wave, whose center is located at (−1.5 [cm], 0, −10 [cm]), are shown in Fig. 7.17. The object is a square with a side 3 cm. The position of the object changes correspondingly to Cases (i)–(iv) of the preceding section. These are also simulated reconstruction like the cases of plane waves, but realistic enough to learn the nature of thin holograms with spherical waves, because the simulation is based on virtual imaging (see Chap. 13).

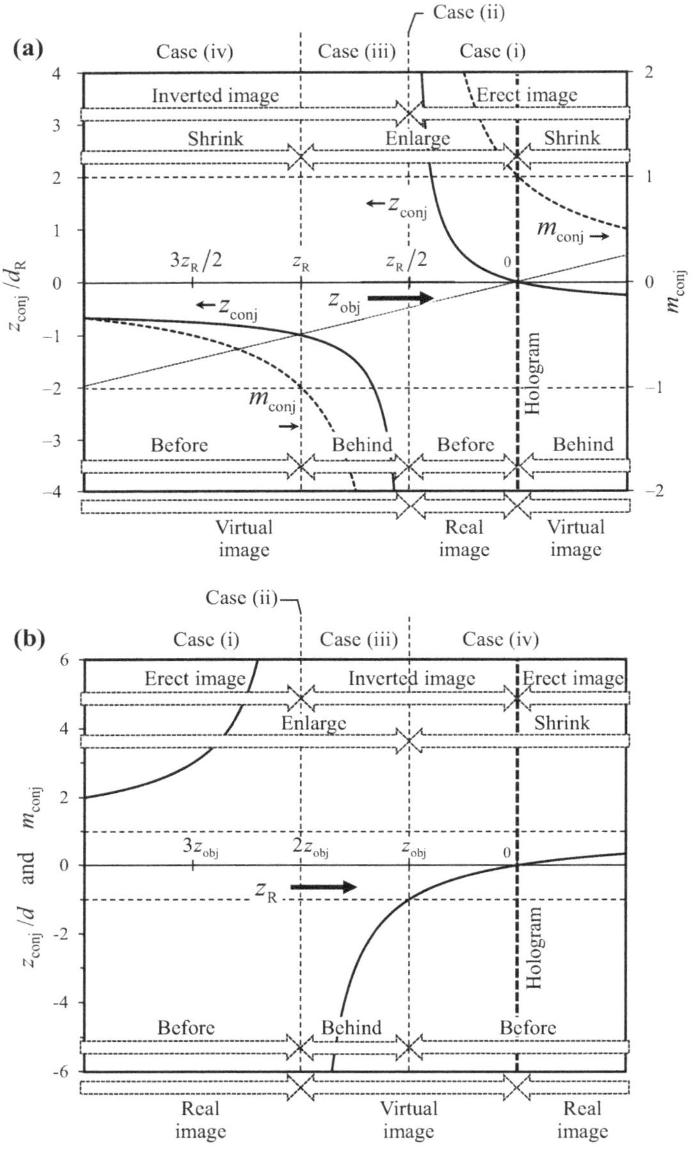

Fig. 7.16 Diagrams depicting the position (z_{conj}/d_R; solid line) and magnification (m_{conj}; dashed line) of the conjugate image; **a** as a function of the position of the object (z_{obj}) in a fixed position of the spherical fields, and **b** as a function of the position of the spherical field (z_R) in a fixed object position

Fig. 7.17 Reconstructed images of the hologram created with a reference spherical wave, whose center is located at $(x, y, z) = (-1.5 \text{ [cm]}, 0, -10 \text{ [cm]})$ and thus $d_R = 10$ [cm]. A square object with a side 3 cm is arranged at $(x, y, z) = (0, 0, -d)$. **a**–**c** and **e** correspond to Cases (i)–(iii) and (iv) of Fig. 7 16a, respectively

Figure 7.17 corresponds to the diagram of Fig. 7.16a, in which the object position varies, while the position of the spherical waves is fixed. The reconstructed images of (a) is corresponding to Case (i). The simulated reconstruction is focused on the true image in the 'Far' image, while the conjugate image in the 'Near' image. $z_{\text{conj}} = +7.5$ [cm] and $m_{\text{conj}} = 2.5$ in this case according to (7.72) and (7.73). Since the conjugate image is floating close to the viewpoint and the center is shifted to $x_{\text{conj}} = 5.25$ [cm], the reconstructed conjugate image is distorted considerably. In Case (ii) shown in (b), the conjugate image is not formed but the far-field is generated. In the case (iii), a magnified conjugate image is generated behind the true image as the inverted image, as in (c). When $d_{\text{obj}} = d_R$, the conjugate image is reconstructed in the same plane of the true image as in (d). The conjugate image shrinks in Case (iv) and reconstructed in front of the true image, as in (e).

Figure 7.18 is corresponding to the diagram of Fig. 7.16b, where the position of the reference spherical wave changes, while the object position is fixed. In (a), $z_{\text{conj}} = +20$ [cm] and $m_{\text{conj}} = 2$ according to (7.72) and (7.73). Thus, the conjugate image considerably blurs in the 'Far' image, in which the simulated reconstruction is focused on the true image. Conversely, in the 'Near' image in which the conjugate image is in focus, the true image is completely out of focus and becomes shapeless blur. The reconstructed image (b) is corresponding to Case (ii) and thus the conjugate image is reconstructed as a far-field. The images (c) and (e) are corresponding to Case (iii) and (iv), respectively. The inverted conjugate image appears behind and in front of the true image in (c) and (e), respectively. The image (d) completely agrees with

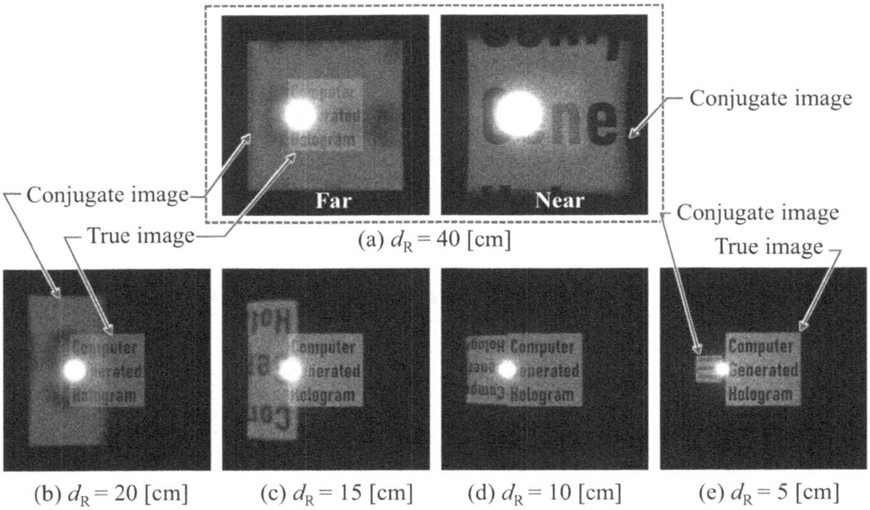

Fig. 7.18 Reconstructed images of the hologram created with a square object and various reference spherical waves, whose center is located at $(x, y, z) = (-1.5\ [\text{cm}],\ 0,\ -d_R)$. A square object with a side 3 cm is arranged with $d = 10$ [cm]. **a–c** and **e** correspond to Cases (i)–(iii) and (iv) of Fig. 7.16b, respectively

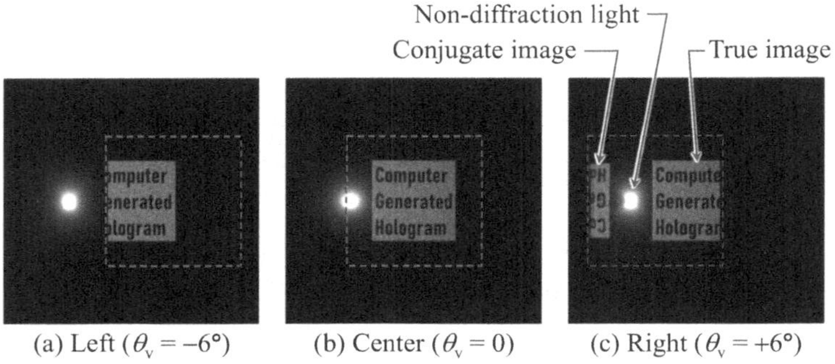

Fig. 7.19 Reconstructed images of the hologram created with a reference spherical wave and square object, whose center is located at $(x, y, z) = (0,\ -4\ [\text{cm}],\ -10\ [\text{cm}])$. $d = 10$ [cm]. The reconstructed images are calculated at different viewpoints. The red square indicates the area of the hologram

(d) in Fig. 7.17. The conjugate image is reconstructed in the same plane as the true image.

Figure 7.19 shows examples in cases where the viewpoint shifts along the x-axis. The side of the square object is 4 cm in length, and the object is arranged with $d = 10$ [cm]. The center of the reference and illumination spherical wave is located at $(x, y, z) = (-3\ [\text{cm}],\ 0,\ -10\ [\text{cm}])$, and thus, $d = d_R$. In this case, as

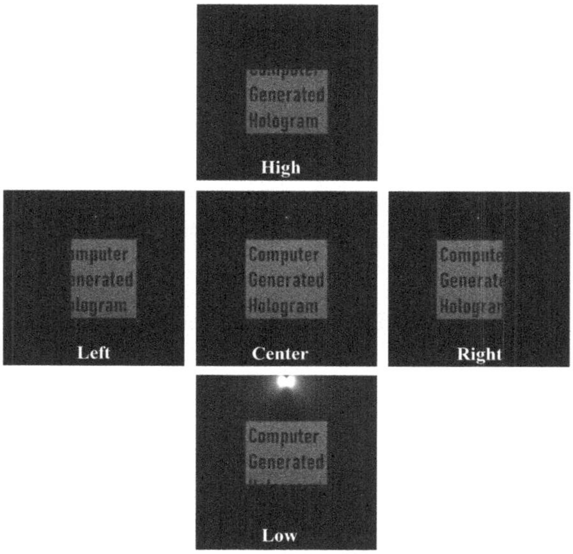

Fig. 7.20 Reconstructed images of the hologram created with a reference spherical wave, whose center is located at $(x, y, z) = (-3\,[\text{cm}],\ 0,\ -10\,[\text{cm}])$. $d = 10$ [cm]. The reconstructed images are calculated at different viewpoints

shown in Fig. 7.18d, the conjugate image appears at the position of point symmetry with respect to the center of the reference spherical wave like Fourier-transform holograms described in the next section. This is, in fact, a lensless-Fourier hologram (see Sect. 14.2.2).

The red squares again indicate the area of the hologram in Fig. 7.19. Since this simulation assumes that the hologram fringe is recorded on a transparent material, non-diffraction light comes into the view even if the apparent center of the spherical wave is placed outside the fringe area, as shown in the image (a) that is reconstruction from a left viewpoint.

Masking the outside of the fringe area and arranging the center of the spherical wave below or above the object, the non-diffraction light and conjugate image can be removed from the reconstructed images at the viewpoints on the left and right. Fig. 7.20 shows the examples. The center of the reference spherical wave is arranged above the object in this example. We do not see the conjugate image when moving the viewpoint left and right in this hologram. Although the non-diffraction light inevitably comes into sight when we watch the reconstructed image from the low viewpoint, arranging a reference spherical wave above or below the object is better than other setups in many cases.

7.7.3 Fourier Transform Hologram

In optical holography, a Fourier-transform hologram is commonly recorded by use of a lens, as shown in Fig. 7.4d. This is because a wavefield placed at the focal plane of the lens is Fourier-transformed at another focal plane, as mentioned in Sect. 5.3. In the reconstruction of the Fourier-transform hologram, we have to use the lens again because the hologram reconstructs Fourier-transformed images unless the lens is used again.

It should be noted that we assume 'on-axis' Fourier holograms in the following discussion; it is assumed that the center of the reference spherical wave is always placed in the optical axis. This is because the on-axis Fourier hologram plays an important role in computer holography in many cases.

7.7.3.1 Size and Direction of Reconstructed Image

Assume that a Fourier-transform hologram is recorded by use of a lens whose focal length is f_1, as shown in Fig. 7.21a. The object field is written by using (5.66) as:

$$\begin{aligned} O(x, y) &\approx \mathscr{F}\left\{g_{\text{obj}}(x, y; -f_1)\right\}_{u=\frac{x}{\lambda f_1}, v=\frac{y}{\lambda f_1}} \\ &= G_{\text{obj}}\left(\frac{x}{\lambda f_1}, \frac{y}{\lambda f_1}; -f_1\right), \end{aligned} \tag{7.75}$$

where the constant coefficient is ignored, and $G_{\text{obj}}(u, v) = \mathscr{F}\left\{g_{\text{obj}}(x, y)\right\}$. Furthermore, supposing that the true image is reconstructed by use of another lens with the focal length of f_2 as in (b), the true image is given again by the Fourier transform in (5.66):

$$\begin{aligned} g_{\text{true}}(x, y; f_2) &\approx \mathscr{F}\left\{O(x, y)\right\}_{u=\frac{x}{\lambda f_2}, v=\frac{y}{\lambda f_2}} \\ &\approx \mathscr{F}\left\{G_{\text{obj}}\left(\frac{x}{\lambda f_1}, \frac{y}{\lambda f_1}; -f_1\right)\right\}_{u=\frac{x}{\lambda f_2}, v=\frac{y}{\lambda f_2}} \\ &\approx g_{\text{obj}}\left(-\frac{f_1}{f_2}x, -\frac{f_1}{f_2}y; -f_1\right), \end{aligned} \tag{7.76}$$

where the constant coefficient is ignored again. As a result, the true image is inverted and the size is changed to f_2/f_1 times as large as the object. When recording an object whose width is W, the image size is $(f_2/f_1)W$ in reconstruction.

Fourier-transform holography is not so often used in optical holography in practice, but it is useful in computer holography because of the property given in (7.76). We can adjust the image size by choosing focal lengths in Fourier-transform holography, and this is more important reason; we can enhance the viewing angle of CGHs in Fourier-transform holography (see Sect. 8.9). One more reason is that the object field in Fourier-transform holograms is easily obtained using the numerical Fourier transform, i.e., FFT in computer holography.

It should be noted that the conjugate image is inevitably reconstructed with the true image by thin amplitude holograms. The conjugate field in Fourier-transform holography is given by

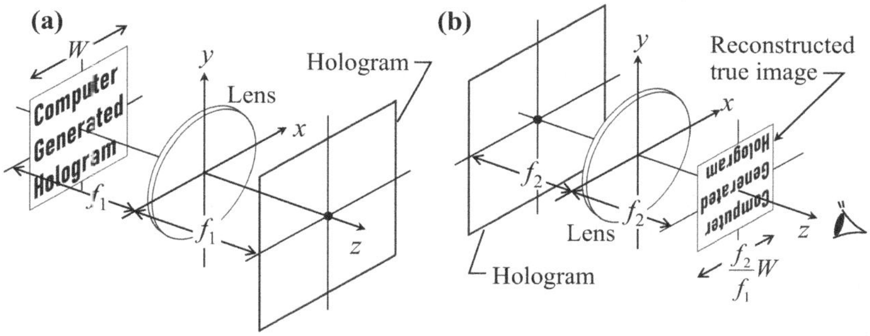

Fig. 7.21 Schematic illustration of **a** recording and **b** reconstruction of a Fourier-transform hologram. Note that only the true image is depicted in (**b**)

$$O^*(x, y) = \mathscr{F}\left\{g^*_{\text{obj}}(-x, -y; -f_1)\right\}_{u=\frac{x}{\lambda f_1}, v=\frac{y}{\lambda f_1}}. \tag{7.77}$$

Therefore, the reconstructed conjugate image is

$$\begin{aligned} g_{\text{conj}}(x, y; f_2) &\approx \mathscr{F}\left\{O^*(x, y)\right\}_{u=\frac{x}{\lambda f_2}, v=\frac{y}{\lambda f_2}} \\ &= \mathscr{F}\left\{G^*_{\text{obj}}\left(\frac{x}{\lambda f_1}, \frac{y}{\lambda f_1}; -f_1\right)\right\}_{u=\frac{x}{\lambda f_2}, v=\frac{y}{\lambda f_2}} \\ &\approx g^*_{\text{obj}}\left(\frac{f_1}{f_2}x, \frac{f_1}{f_2}y; -f_1\right). \end{aligned} \tag{7.78}$$

As a result, although the size is the same as the true image; the size is f_2/f_1 times as large as the object, the conjugate image is an erect image unlike the true image. This is the resultant in a sense from the symmetry relation of the Fourier transform mentioned in Sect. 4.3.2. The combination of the true and conjugate images reconstructed at the same plane is sometimes called a *twin image*.

When the object field is given at a distance of d from the focal point, as shown in Fig. 7.22, the object field in the hologram plane is represented by

$$\begin{aligned} O(x, y) &\approx \mathscr{F}\left\{\mathscr{P}_d\left\{g_{\text{obj}}(x, y; -f_1 - d)\right\}\right\}_{u=\frac{x}{\lambda f_1}, v=\frac{y}{\lambda f_1}} \\ &= G_{\text{obj}}\left(\frac{x}{\lambda f_1}, \frac{y}{\lambda f_1}; -f_1 - d\right) H\left(\frac{x}{\lambda f_1}, \frac{y}{\lambda f_1}; d\right), \end{aligned} \tag{7.79}$$

where $\mathscr{P}_d\left\{\cdot\right\}$ is a propagation operator introduced in Sect. 5.4 and $H(u, v; d)$ is a transfer function for the distance d. The field corresponding to the true image is

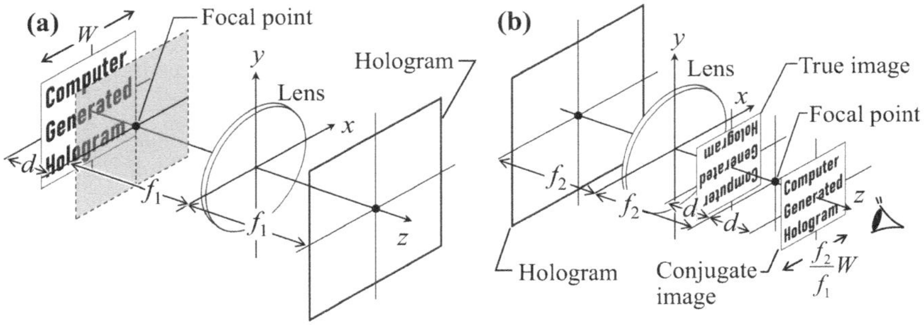

Fig. 7.22 Schematic illustration of **a** recording and **b** reconstruction of a Fourier-transform hologram. Here, the object is placed at a distance of d away from the focal point in (**a**)

$$
\begin{aligned}
g_{\text{true}}(x, y; f_2) &\approx g_{\text{obj}}\left(-\frac{f_1}{f_2}x, -\frac{f_1}{f_2}y; -f_1 - d\right) \otimes h\left(-\frac{f_1}{f_2}x, -\frac{f_1}{f_2}y; d\right), \\
&= \mathscr{P}_d\left\{g_{\text{obj}}\left(-\frac{f_1}{f_2}x, -\frac{f_1}{f_2}y; -f_1 - d\right)\right\},
\end{aligned} \tag{7.80}
$$

and thus, the field for the true image is presented by

$$
g_{\text{true}}(x, y; f_2 - d) = g_{\text{obj}}\left(-\frac{f_1}{f_2}x, -\frac{f_1}{f_2}y; -f_1 - d\right). \tag{7.81}
$$

This means that the object field is reconstructed at the same distance from the focal point as that of recording, as shown in Fig. 7.22b.

On the other hand, the recorded conjugate field is

$$
O^*(x, y) \approx G^*_{\text{obj}}\left(\frac{x}{\lambda f_1}, \frac{y}{\lambda f_1}; -f_1 - d\right) H\left(\frac{x}{\lambda f_1}, \frac{y}{\lambda f_1}; -d\right). \tag{7.82}
$$

Here, note that $H^*(u, v; d) = H(u, v; -d)$. The reconstructed field is

$$
\begin{aligned}
g_{\text{conj}}(x, y; f_2) &\approx g^*_{\text{obj}}\left(\frac{f_1}{f_2}x, \frac{f_1}{f_2}y; -f_1 - d\right) \otimes h\left(\frac{f_1}{f_2}x, \frac{f_1}{f_2}y; -d\right), \\
&= \mathscr{P}_{-d}\left\{g^*_{\text{obj}}\left(\frac{f_1}{f_2}x, \frac{f_1}{f_2}y; -f_1 - d\right)\right\},
\end{aligned} \tag{7.83}
$$

and thus,

$$
g_{\text{conj}}(x, y; f_2 + d) = g^*_{\text{obj}}\left(\frac{f_1}{f_2}x, \frac{f_1}{f_2}y; -f_1 - d\right). \tag{7.84}
$$

Therefore, we conclude that the field corresponding to the conjugate image is reconstructed at a position which is $+d$ away from the focal point, i.e., nearer position to

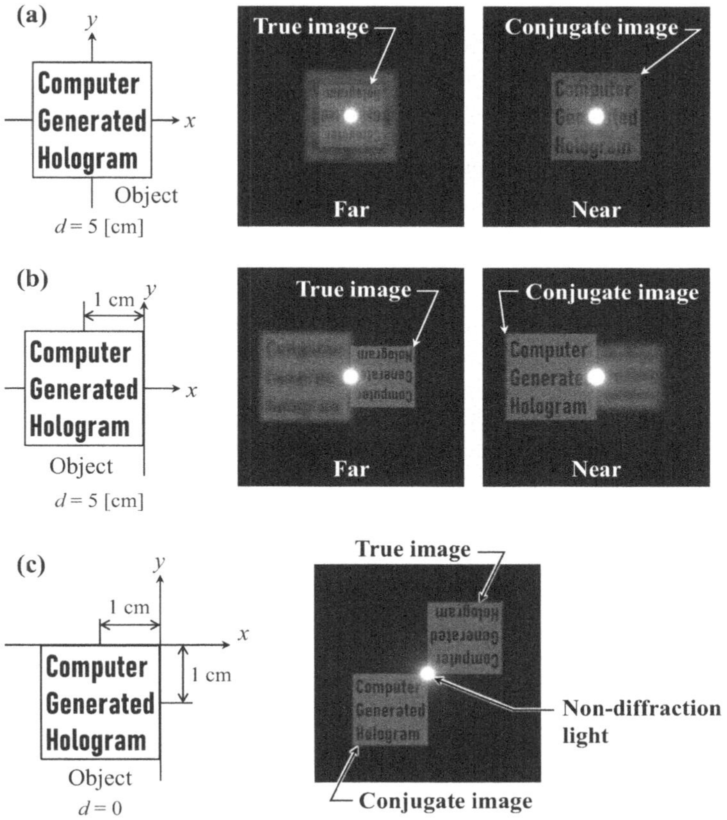

Fig. 7.23 Examples of reconstruction of Fourier-transform holograms. The object is a square with a length 2 cm of one side. **a** The center of the object agrees with the optical axis. $d = 5$ [cm]. **b** The center of the object is shifted -1 cm along the x-axis. $d = 5$ [cm]. **c** The center of the object is shifted -1 cm along both the x- and y-axes. $d = 0$

the viewer, as shown in Fig. 7.22b. This agrees with former discussion on conjugate images in Sect. 7.6.

7.7.3.2 Examples of Fourier Transform Hologram

A square object with 2 cm length on one side is placed at the position shifted from the focal points; $z_{obj} = -f - d$. It is assumed that the focal length in reconstruction is the same as that in recording; $f_1 = f_2 = f$. Hence, the reconstructed image has the same size as the object: 2 cm width. The hologram size is approximately 5 cm. It is supposed that the hologram is recorded with a reference plane wave traveling perpendicularly to the hologram.

Examples of the reconstruction are shown in Fig. 7.23. These are also simulation based on virtual imaging (see Chap. 13) but realistic enough to show the features of Fourier-transform holography. In the example of (a), the center of the square object is located at the optical axis but the z position is shifted from the focal point; $d = 5$ [cm]. The far and near images focus on the true and conjugate images, respectively. The inverted true image overlaps with the erect conjugate images in this case.

To separate the two images, the object position should be shifted from the optical axis, as in (b) and (c). The object is shifted only in the horizontal direction; $x = -1$ [cm] in (b), while the object is positioned at $(x, y, z) = (-1\,[\text{cm}], -1\,[\text{cm}], -f)$ in (c). The depth shifts are $d = 5$ [cm] in (b), while zero in (c). The reconstructed image in (c) is a typical twin image in Fourier transform holography. As shown in the examples of (b) and (c), we can separate the non-diffraction light, and the true and conjugate images of thin holograms in the Fourier plane. This feature of Fourier-transform holography is useful for filtering the true image, and is therefore often used in computer holography. In practice, arranging a proper filter in the focal plane in reconstruction, the unnecessary field components of thin holograms can be removed. This is the principle of the single-sideband method (see Sect. 8.10), and is often the base of holographic displays and wavefront printers (see Sect. 15.4).

Chapter 8
Computer Holography

Abstract This chapter deals with various topics related to computer holography. The topics include the viewing angle, space-bandwidth product problem, parallax, coding fringes, fringe frequencies, and higher-order images in computer holography. In particular, many examples of amplitude and phase coding are presented and discussed to create display CGHs. We also deal with novel and conventional techniques to remove the non-diffraction light and conjugate image in this chapter.

8.1 Introduction

As briefly described in Chap. 2, the process of optical holography is replaced by digital processing in computer holography. The object field is numerically synthesized from a 3D model in many cases, or captured by the technique of digital holography using an image sensor, in some cases. Figure 8.1 shows the process to create the CGH from a 3D model. Here, synthesis of the object field is the most time-consuming and important step in the procedure. Chapters 10 and 11 are devoted to this topic. Creation of HD-CGHs from physical objects is described in Chap. 14.

In this chapter, we assume that the object field has already been synthesized from the object model or captured by an image sensor, and learn how to generate the fringe pattern from the object field.

8.2 Viewing Angle

In 3D imaging, the viewing angle and field of view are very important performance indexes. These are very similar but different ideas, as shown in Fig. 8.2. The field of view is a range that viewers can see the 3D scene at a glance from a fixed viewpoint, while the viewing angle is an angle in which viewers can change their viewpoint.

It is very important to reconstruct a certain wide viewing angle in holography. Since a hologram can reconstruct the object field completely, the viewer perceives motion parallax. This means that the 3D image watched by the viewer changes

K. Matsushima, *Introduction to Computer Holography*,
Series in Display Science and Technology,
https://doi.org/10.1007/978-3-030-38435-7_8

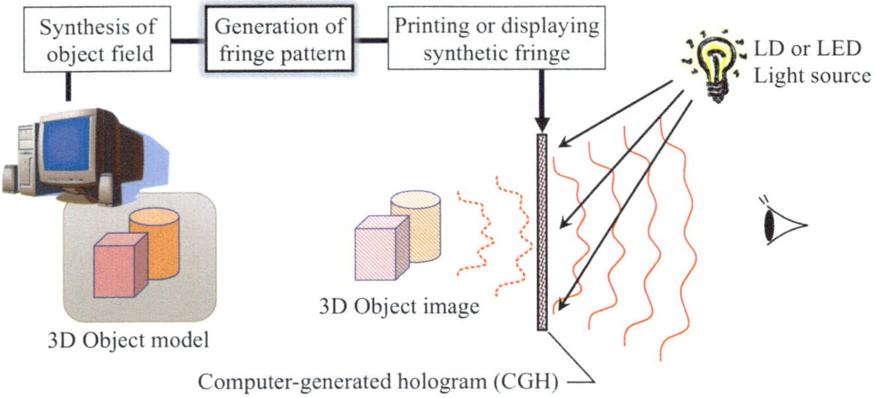

Fig. 8.1 Schematic illustration of process to create a CGH and reconstruct the 3D image from the 3D model

Fig. 8.2 Definitions of viewing angle θ_{view} and field of view θ_{fov}

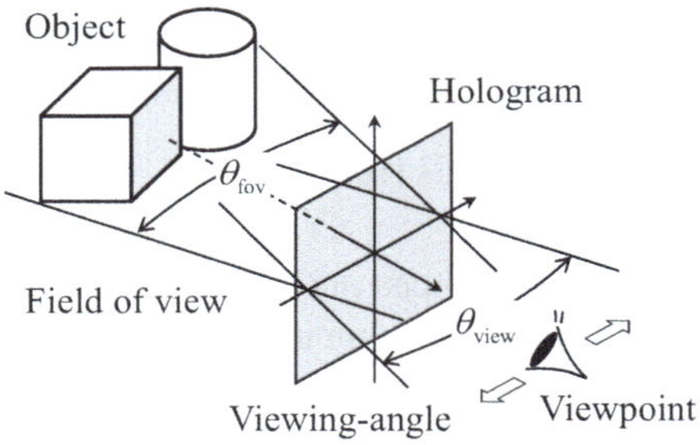

properly as the viewpoint moves; e.g., an object hidden by the front object comes into sight. If the hologram reconstructs full-parallax images, the viewers perceive motion parallax not only in the horizontal direction but also in the vertical direction. The viewing angle should be large enough to cover both eyes of the viewer; otherwise the viewer cannot observe the object by binocular vision and may feel stress.

The viewing angle in optical holography depends on the resolving power of the used material, and is usually very large. In computer holography, the viewing angle is commonly determined by resolution of the digital fringe pattern, i.e., the pixel pitch of the digital fringe image. Assume the simplest case where an amplitude fringe pattern is generated by interference between an object wave and a reference plane wave traveling along the optical axis. In this case, the spatial spectrum of the amplitude fringe is given by (7.18) of Sect. 7.5. Supposing that the illumination field is identical to the reference field; $P(x, y) = R(x, y) = 1$, the spatial spectrum reconstructed by the hologram is again

$$\mathcal{F}\{t(x, y)P(x, y)\} \cong B\delta(u, v) + G_{\text{obj}}(u, v; 0) + G^*_{\text{obj}}(-u, -v; 0). \tag{8.1}$$

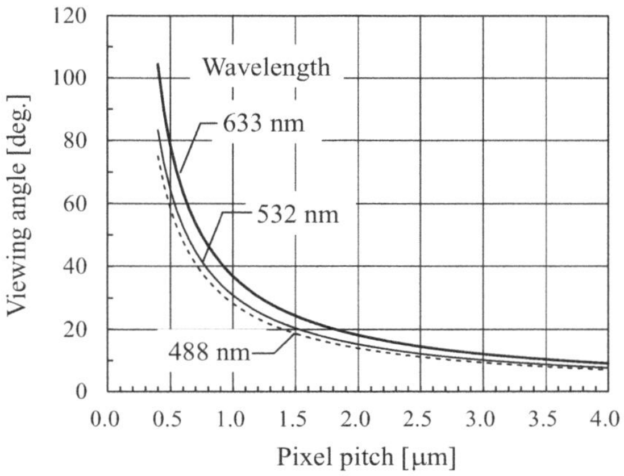

Fig. 8.3 Viewing angles at typical three wavelengths corresponding to red (633 nm), green (532 nm), and blue (488 nm) colors

Thus, the spatial bandwidth is essentially governed by the spectrum of the object field $G_{obj}(u, v; 0)(=\mathcal{F}\{O(x, y)\})$.

Needless to say, the digital fringe pattern is generated from the sampled object field in computer holography, and commonly has the same sampling interval as the object field.[1] Thus, let us define the viewing angle of a CGH as twice the maximum diffraction angle of the object field, given in (3.54):

$$\begin{aligned}\theta_{\mathrm{view},x} &\equiv 2\theta_{\mathrm{max},x} = 2\sin^{-1}\left(\frac{\lambda}{2\Delta x}\right),\\ \theta_{\mathrm{view},y} &\equiv 2\theta_{\mathrm{max},y} = 2\sin^{-1}\left(\frac{\lambda}{2\Delta y}\right),\end{aligned} \tag{8.2}$$

where Δ_x and Δ_y are sampling intervals of the fringe pattern, i.e., pixel pitches of the digital fringe image. Figure 8.3 shows actual viewing angles calculated from (8.2) at typical wavelengths corresponding to red, green and blue colors.

It should be noted that actual viewing angle of a CGH is determined by many factors, such as the position of the object, the angle or form of the reference/illumination waves, and so on. Thus, equation (8.2) gives just a rough estimate.

8.3 Space-Bandwidth Product Problem

Through the discussion in Sect. 4.5, we understand the fact that the spectrum of a spatially-sampled signal is limited within a frequency region:

[1] Note that fringe patterns do not always have the same sampling interval as the object field (see Sect. 8.8.3).

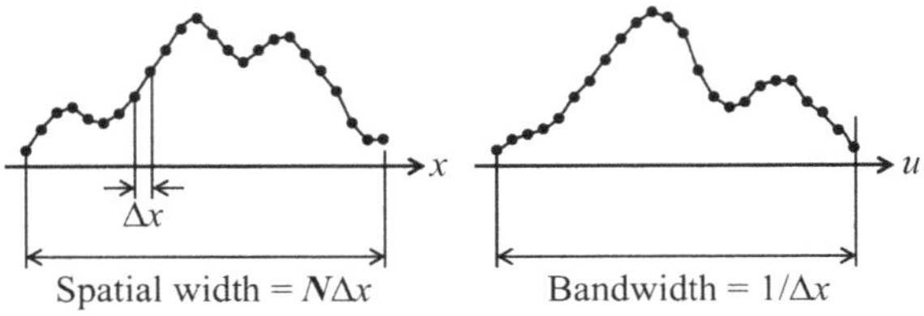

Fig. 8.4 The spatial extent and bandwidth of a spatially-sampled signal

$$-\frac{1}{2\Delta x} < u < \frac{1}{2\Delta x}, \tag{8.3}$$

where Δx is the sampling interval. In other words, the bandwidth of the signal is less than $1/\Delta x$. On the other hand, spatial extent of the sampled signal; the size of the sampling window is given by $N\Delta x$, as shown in Fig. 8.4. Here, N is the number of sample points. As a result, the product of these, called a space-bandwidth product [54], is simply given by N:

$$\begin{aligned}(\text{Spatial width}) \times (\text{Bandwidth}) &= N\Delta x \times \frac{1}{\Delta x} \\ &\equiv N. \end{aligned} \tag{8.4}$$

When the sampled signal is an object field in computer holography, the size of the sampling window commonly gives the size of the CGH in the point-oriented encoding mentioned in the following section. On the other hand, the bandwidth $1/\Delta x$ determines the viewing angle of the CGH according to (8.2). Therefore, we can conclude that there is a trade-off relation between the size and viewing angle in computer holography, i.e., larger viewing angle makes the CGH size smaller in a given number of pixels, and vice versa.

For example, when a CGH requires the viewing angle of 45°, the pixel pitch necessary for the CGH is less than 0.8 μm in red color, according to Fig. 8.3. Therefore, if the CGH is a square having a length of 10 cm of one side, the number of sample points necessary for the object field is $125{,}000^2 = 15{,}625{,}000{,}000$. Accordingly, the fringe image is composed of approximately 16 G pixels. This is roughly 10^3 to 10^4 times larger than state-of-the-art digital images.[2]

The necessity of the huge number of sample points or pixels for reconstructing high-quality 3D images is sometimes called the *space-bandwidth product (SBP) problem* in computer holography. Due to the SBP problem, CGHs for 3D display must be a large-scale image, and therefore it is very difficult to produce holographic displays, as already mentioned in Sect. 1.2.

[2]This book was finalized in 2019.

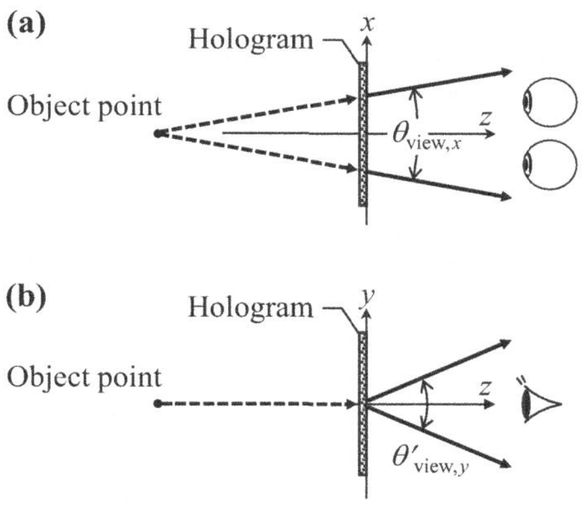

Fig. 8.5 Viewing angles of HPO-CGHs using a vertical diffuser in **a** a top view and **b** side view

8.4 Full-Parallax and Horizontal-Parallax-Only CGH

A technique to get rid of the SBP problem is to restrict the parallax of the reconstructed 3D image to the horizontal direction. This is called a *horizontal-parallax-only* CGH (HPO-CGH).[3] Since an HPO-CGH reconstructs binocular disparity, which is one of the most important depth cue, the CGH is capable of displaying the 3D image properly. On the contrary, CGHs that reconstruct the whole parallax are called *full-parallax* CGHs. Computational complexity of HPO-CGHs is estimated at a square root of that of full-parallax CGHs. Thus, computation time of the object field is drastically decreased in HPO-CGHs.

Holographic displays sometimes reconstruct a HPO-CGH using a vertical diffuser to enhance the vertical viewing angle [116, 119]. In this case, while horizontal viewing angle $\theta_{\text{view},x}$ is given by (8.2), the vertical viewing angle is regarded as zero, i.e., the object field propagates only in the x–z plane [56]. The reconstructed field is commonly spread out in the vertical direction by use of a vertical diffuser placed at the hologram or display plane, as shown in Fig. 8.5. Thus, vertical viewing angle $\theta'_{\text{view},y}$ is determined by the used diffuser.

Figure 8.5 also shows the main drawback of HPO-CGHs using a vertical diffuser. When we see the object reconstructed by a HPO-CGH, a conflict inevitably occurs between the horizontal and vertical accommodation of eyes. Viewer's eyes are focused on the proper object point in the horizontal direction, whereas it is focused on the hologram surface in the vertical direction. This conflict is not so severe if the object is close to the hologram. However, when the object is reconstructed far from the hologram, considerable conflicts occur in the reconstruction. Therefore, a significant advantage of holograms over other 3D techniques, which is ability to reconstruct a deep 3D scene, is lost in HPO-CGHs.

[3] Analog HPO holograms such as a rainbow hologram is beyond the scope of this book. The reader is referred to [4].

Finally, let us emphasize that all actual CGHs introduced in this book are full-parallax holograms, and all algorithms and techniques are developed for creating full-parallax HD-CGHs.

8.5 Coding and Optimization of Fringe Pattern

A CGH is commonly a thin hologram that is represented by the 2D image of the fringe pattern. Generation of the fringe pattern or image that reconstructs a given light field is called *coding* or *encoding* especially in relation to binary CGHs, whose fringe pattern has binary amplitudes or phases. Binary CGHs are easy to fabricate, and thus, very important as optical devices as well as 3D imaging. In practice, CGHs were first proposed for wavefront conversion devices having a binary amplitude pattern [7, 55], as mentioned in Sect. 1.1. It should be emphasized that almost all high-definition CGHs introduced in this book are a binary amplitude CGH. The binary fringe pattern is encoded by a simple technique described in Sect. 8.6.3. Although any optimization technique is not used, the reconstructed 3D images are good enough to be appreciated.

We can find extensive literature on optimization of the fringe pattern of binary CGHs. This is very popular research theme in many applications and, in fact, very important subject in creating CGHs for optical devices. However, as the reader see in the following several sections such as Sects. 8.6.2 and 8.7.2, optimization techniques, in particular iterative optimization techniques are not so important for display HD-CGHs. The reader familiar with device CGHs may think that the 3D images reconstructed by binary HD-CGHs is very 'bad' unless the fringe pattern is optimized, because binary CGHs are not good for an optical device unless it is optimized. However, actual 3D images reconstructed by binary HD-CGHs are not so bad as we anticipate. The author expects that the gigantic space-bandwidth product contributes to unexpected high quality of the optical reconstruction.

To create a display CGH having the size of a large postage stamp and the viewing angle of 45°, we need billion pixels in the fringe image. If the fringe image is a binary image, amount of information is roughly 10^9 bits. In contrast, the number of pixels does not exceed millions in device CGHs in general. If the CGH has binary levels, amount of information is roughly 10^6 bits.

The binary device CGHs having 10^6 bits information may give very bad results such as much noise and low efficiency. We need an optimization technique to improve the output. However, if the fringe image has 8 bits grayscale, the CGH often provides much better outputs without optimization. Even in this case, amount of information of the grayscale CGH is at most 8×10^6 bits and is still much fewer than common HD-CGHs. Therefore, the author expects that plenty of binary pixels in HD-CGHs are effectively transformed into multi-level pixels and makes the binary HD-CGH be equivalent to a grayscale CGH. This is most likely the reason why fringe optimization techniques in computer holography are relatively less significant than those in device CGHs.

In addition, it is clear that iterative optimization is not realistic in creating display CGHs in many cases. The reason is the same as that mentioned above, i.e., due to the large scale of the fringe images. It is commonly too time-consuming to optimize the fringe pattern of display CGHs.[4] Furthermore, evaluation of the 3D image quality is a difficult problem in computer holography because there is no valid evaluation index for assessment. For example, PSNR is used to evaluate quality of a 2D image. However, in HD-CGHs, the 3D image can be observed from various viewpoints because the viewing angle is usually more than 45°. How to evaluate this kind of holographic 3D image? This is still an unresolved problem.

Because of the reasons mentioned above, we do not use any optimization technique to generate fringe patterns in this book. But, optimization of several parameters is discussed to enhance the brightness and reduce noise in the following section.

8.6 Amplitude CGH

8.6.1 *Amplitude Encoding*

In point-oriented coding, the coding technique is theoretically the same as that in optical holography. In Sect. 7.4, we assumed that the amplitude transmittance of the fringe pattern was

$$t(I) = \sqrt{T(I)} = \sqrt{\eta(I)} \approx I, \tag{8.5}$$

where T, I and $\eta(I)$ are transmissivity, fringe intensity and the sensitivity curve of the used material. The purpose of the coding process is to obtain appropriate distribution $T(x, y)$ of the CGH. Thus, let the distribution be represented by

$$T(x, y) = t(x, y)^{\gamma}, \tag{8.6}$$

where $\gamma = 2$ theoretically. Using assumption in (8.5), which is a basis of the theoretical explanation of thin holograms, $T(x, y) \approx I(x, y)^2$. However, photographic density of photo-sensitive materials is generally given by

$$T(x, y) = I(x, y)^{\gamma}, \tag{8.7}$$

Therefore, let us consider γ as a controllable parameter for suitable amplitude encoding in the following discussion.

Suppose that terms $|O(x, y)|^2$ and $|R(x, y)|^2$ in (7.8) are constant and the fringe intensity is represented by

[4] There are exceptions. For example, the technique of error diffusion is usually applicable to optimization even in HD-CGHs [138].

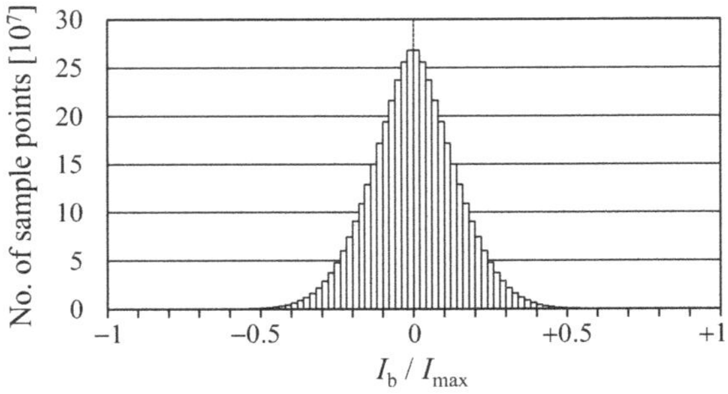

Fig. 8.6 An example of the histogram of normalized bipolar-intensity. The sample CGH is The Venus

$$I(x, y) = \mathrm{Re}\left\{O(x, y)R^*(x, y)\right\} + B, \tag{8.8}$$

where B is a bias to lift up the intensity and make it positive. Introducing bipolar-intensity I_{b} that is an optical intensity but allowed to be negative, the intensity is

$$I(x, y) = I_{\mathrm{b}}(x, y) + B, \tag{8.9}$$

$$I_{\mathrm{b}}(x, y) \equiv \mathrm{Re}\left\{O(x, y)R^*(x, y)\right\}. \tag{8.10}$$

Using the sampled form introduced in Sect. 4.6,

$$I[m, n] = I_{\mathrm{b}}[m, n] + B, \tag{8.11}$$

$$I_{\mathrm{b}}[m, n] \equiv \mathrm{Re}\left\{O[m, n]R^*[m, n]\right\}. \tag{8.12}$$

Normalization is necessary in practical coding. Let the normalized bipolar-intensity be

$$I'_{\mathrm{b}}[m, n] = \frac{I_{\mathrm{b}}[m, n]}{I_{\mathrm{max}}} \tag{8.13}$$

$$I_{\mathrm{max}} \equiv \max |I_{\mathrm{b}}[m, n]|, \tag{8.14}$$

where I_{max} is the maximum value of $|I_{\mathrm{b}}[m, n]|$.

Figure 8.6 is an example of the histogram of the normalized bipolar-intensity in a practical CGH. Here, the CGH used for the example is The Venus[5] introduced in Sect. 1.3. The object field of the CGH has approximately 4.2×10^9 sample points. The reference field is a plane wave, whose incident angle is approximately 9° (see Sect. 10.5 for the detailed parameters of The Venus). We can find that the value of the normalized bipolar-intensity is nearly distributed within ±0.5 in Fig. 8.6. This is because there are a few sample points having a large value of $|I_{\mathrm{b}}[m, n]|$ caused by noise or something. The normalized bipolar intensity is easily lifted up by adding

[5]Note that the object field is not completely the same as that of the original Venus created in 2009, because several new techniques, developed after making the first Venus, are reflected in its computation.

unity: $I'_b[m, n] + 1$. However, the fringe generated from $I'_b[m, n] + 1$ inevitably has a narrow histogram very similar to Fig. 8.6. This means that the fringe image has a very low contrast that most likely leads to low reconstruction efficiency. Therefore, some technique to enhance the contrast is required to increase the reconstruction efficiency in amplitude CGHs.

Although there may be a variety of technique to increase the reconstruction efficiency, we simply let the following represent the fringe intensity.

$$\hat{I}[m, n] = \frac{1}{2}\left(\frac{I_b[m, n]}{I_{\max}} c_{\text{hist}} + 1\right), \tag{8.15}$$

where c_{hist} is an expansion coefficient of the histogram. $\hat{I}[m, n]$ is properly not limited within interval [0, 1] by introducing $c_{\text{hist}}(>1)$. The value outside of the scope is truncated in the quantization process.

In practical CGHs, printing the fringe pattern involves quantization. Therefore, the fringe pattern is represented by

$$T[m, n] = \mathbf{Q}_L\left\{\hat{I}[m, n]^{\gamma}\right\}, \tag{8.16}$$

where $\mathbf{Q}_L\{\xi\}$ is a quantization operator or quantizer that quantizes the value ξ in [0, 1] into L levels. The quantizer used in the following discussion is[6]

$$\mathbf{Q}_L\{\xi\} = \begin{cases} 0 & \xi < 0 \\ 1 & \xi \geq 1 \\ \dfrac{\lfloor \xi L \rfloor}{L - 1} & \text{otherwise} \end{cases}, \tag{8.17}$$

where $\lfloor x \rfloor$ is the floor function of x.

8.6.2 Brightness and Noise

Using "The Venus" CGH, we perform several numerical experiments. Figure 8.7 shows histograms of 8-bit grayscale images of the fringe pattern, i.e., $255 \times T[m, n]$ in $L = 2^8$. The shape of the histogram in (a) is corresponding to the bipolar-intensity in Fig. 8.6. When $\gamma = 1$, the grayscale is distributed into a narrow range around the median value. Thus, the contrast of the fringe image is very low. The center shifts from the median grayscale in $\gamma = 2$ as in (b), but the contrast is still very low. The histogram expanded with $c_{\text{hist}} = 5$ is shown in (c). Since fringe intensities $T[m, n] < 0$ and $T[m, n] > 1$ are rounded to grayscale 0 and 255, respectively, many pixels have these values.

[6]The results in the following sections are not very affected by definition of the quantizer.

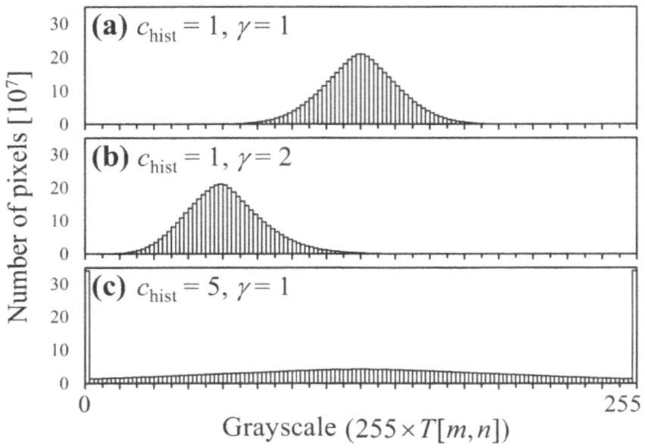

Fig. 8.7 Histograms of 8-bit grayscale fringe images of The Venus CGH

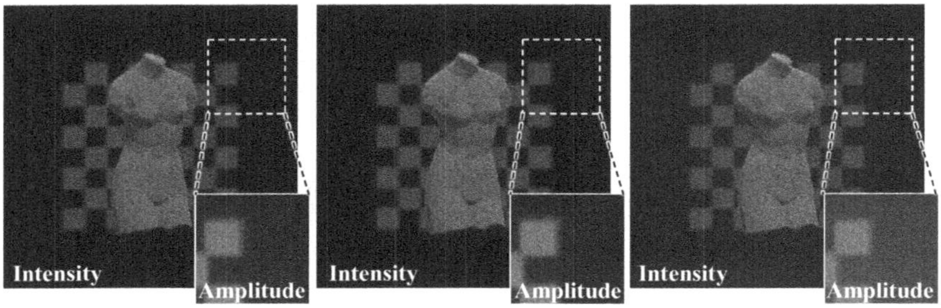

(a) $c_{\text{hist}} = 1,\ \gamma = 1$ (b) $c_{\text{hist}} = 1,\ \gamma = 2$ (c) $c_{\text{hist}} = 5,\ \gamma = 1$

Fig. 8.8 Simulated reconstructions of HD-CGH "The Venus" whose 8-bit fringe images are generated with several different coding parameters. Note that the magnified images are depicted in amplitude to emphasize the background noise. The encoding gamma used for the reconstructed images is 1/2.2

Figure 8.8 shows numerically reconstructed images of the CGHs corresponding to the histograms in Fig. 8.7. Here, the numerical reconstruction is performed using the technique of virtual imaging (see Chap. 13). The viewpoint is positioned at 40 cm right in front of the CGH. It is assumed that the CGH is reconstructed with the same plane wave as that in coding.

No difference is detected in the reconstructed images of Fig. 8.8. However, the background noise in the magnified images increases in (a) to (c). Here, the magnified images are depicted as amplitude pattern $\sqrt{I_{\text{rec}}[p, q]}$ instead of intensity pattern $I_{\text{rec}}[p, q]$ in order to emphasize the background noise.[7] In addition, it must be noted that these simulated reconstructions do not reflect brightness of the images properly,

[7] All images of simulated reconstructions are produced with a standard encoding gamma of 1/2.2 in this book.

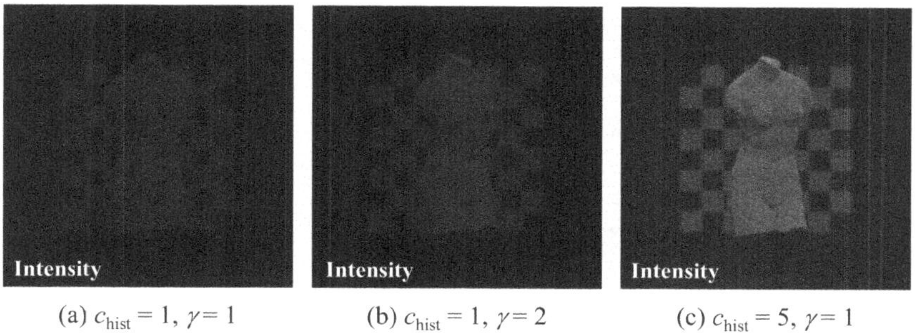

Fig. 8.9 Simulated images that properly reflect the diffraction efficiency of the fringe patterns generated with different coding parameters

Fig. 8.10 Domains where amplitude averages are measured

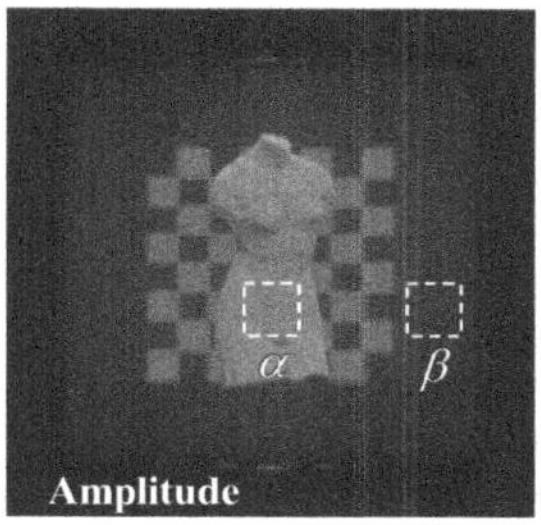

because the maximum intensity in each simulated image is assigned to white in the simulated image. In other word, the brightness is normalized for each image.

Simulated reconstructions that reflect brightness of three CGHs are shown in Fig. 8.9. The reconstructed images are so dark that only a black image may be seen in (a) and (b). This low brightness most likely results from the low contrast of the fringe images, i.e., the low diffraction efficiency of the CGHs. To evaluate brightness and background noise of the reconstructed images, the amplitude averages A_α and A_β of the reconstructed images are measured within the domains α and β in Fig. 8.10, respectively.

Measured brightness and background noise strength of the 3D images reconstructed by 8-bits grayscale fringe images ($L = 256$) are shown in Fig. 8.11a. Here, we evaluate the brightness by subtraction $A_\alpha - A_\beta$, because the background noise is considered to spread over the whole reconstructed image. The brightness increases with increasing c_{hist} and tends to be a constant in $\gamma = 1$, while it has a peak in $\gamma = 2$. The background noise increases with increasing c_{hist} in both $\gamma = 1$ and 2. The brightness in $\gamma = 2$ is higher than that in $\gamma = 1$ when $c_{hist} \lesssim 5$, while the noise is lower in the same range. Thus, the reconstructed images with $\gamma = 2$ are better in this range. However, the relations are inverting in the cases where c_{hist} is more than 5.

Figure 8.11b also shows the measured brightness and background noise strength in different quantization levels. No expansion of the histogram ($c_{hist} = 1$) definitely causes very low brightness except for binary coding ($\log_2 L = 1$). The brightness

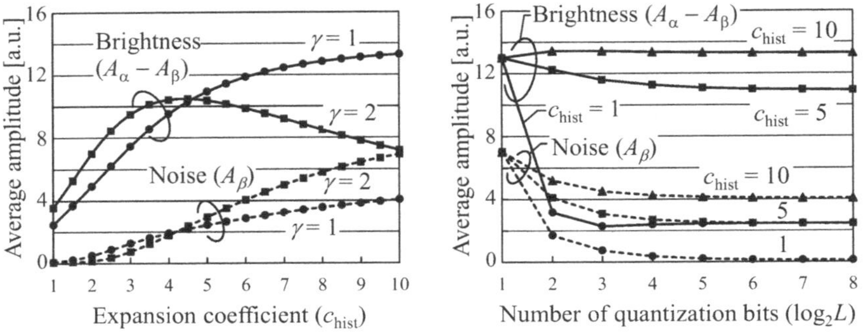

(a) $L = 256$ (b) $\gamma = 1$

Fig. 8.11 Evaluation of brightness and background noise with **a** a constant number of quantization levels: $L = 2^8$ and **b** a constant fringe gamma: $\gamma = 1$

Fig. 8.12 Comparison between the reconstructed images of **a** multi-level and **b** binary amplitude CGHs. Relative brightness of the simulated reconstructions is kept in the images for comparison

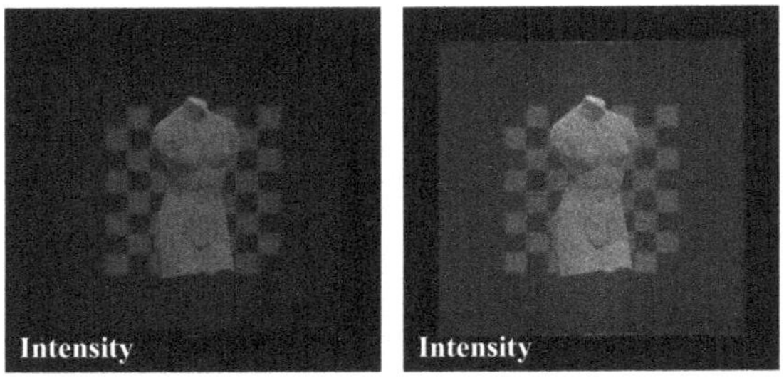

(a) Multi-level ($\gamma = 1$, $L = 256$, $c_{hist} = 5$) (b) Binary level

and noise are nearly constant in quantization approximately more than 4-bits. The fringe with a larger expansion coefficient not only reconstructs a brighter image, but also produces higher noise.

8.6.3 Binary-Amplitude CGH

Binary amplitude coding is the most important in practical fabrication of large-scale CGHs, because the fringe pattern is printed with laser lithography in many cases, which produces a binary transmission pattern or a binary reflectance pattern (see Sect. 15.3). In this case, the fringe pattern can be simply generated by zero threshold of bipolar intensity:

$$T[m, n] = \begin{cases} 0 & \text{for } I_b[m, n] < 0 \\ 1 & \text{otherwise} \end{cases}, \tag{8.18}$$

where $I_b[m, n] = \mathrm{Re}\{O[m, n]R^*[m, n]\}$ again.

As shown in Fig. 8.11b, brightness of the binary amplitude CGH ($\log_2 L = 1$) is not considerably lower than that of multi-level CGHs, but the background noise remarkably increases in the reconstruction. Figure 8.12 shows simulated reconstructions of multi-level and binary level amplitude CGHs for comparison. The number of quantization bits and expansion coefficient are 8 and 5 in the multi-level CGH, respectively. We can perceive the background noise only in the binary CGH.

8.7 Phase CGH

A phase hologram is the hologram that spatially modulates phase of the illumination light. The spatial phase modulation can be accomplished by, e.g., changing local thickness of a thin transparent material or its local index of refraction. Variable thickness hologram (also called surface hologram) is usually made of photoresist, variable refractive index holograms are usually made of photo-polymer, dichromated gelatin or specially treated silver halide emulsions.

In optical holography, the phase modulation is approximately proportional to the fringe intensity; $\phi(x, y) \propto I(x, y)$. Thus, the complex-amplitude transmittance of a hologram is represented by

$$t(x, y) \propto \exp[i\alpha I(x, y)], \tag{8.19}$$

where α is a constant. Such phase holograms are more efficient than amplitude holograms, because $|t(x, y)| = 1$, i.e., no light is absorbed (hologram is fully transparent). On the other hand, phase encoding based on fringe intensity $I(x, y)$ leads to increasing noise, that has its origin in different diffractive properties of phase elements.

In computer holography, phase holograms can be generated more directly using a phase-only SLM or a surface relief fabricated by microfabrication technology (see Sect. 15.3.6). In addition, it is easy to extract phase component $\phi[m, n]$ from calculated object field $O[m, n]$ in computer holography; the complex-amplitude transmittance of a phase CGH is given by[8]

$$t[m, n] = \exp\{i\phi[m, n]\}. \tag{8.20}$$

The CGH is also fully transparent because the amplitude transmittance is unity; $T[m, n] = |t[m, n]|^2 = 1$.

In optical holography, the advantage of phase holograms is often mentioned in the context of the high brightness, because phase holograms do not lose any power of illumination light. However, in computer holography, phase holograms have a greater merit than its brightness; that is no conjugate image. Remember that the origin of

[8]This type of phase hologram is often called the *kinoform*.

generating a conjugate image is the symmetry relation of the Fourier transform; a real function gives an Hermitian function after the Fourier transform (see Sect. 4.3.2). The phase fringe pattern represented by (8.20) is, however, not a real function. Therefore, the Fourier transform can be asymmetry and give no conjugate image.

Actual amplitude CGHs commonly suffer from an overlap of the conjugate image with the true image because it is difficult to give a large incident angle to the reference wave due to limit of the fringe frequency, as mentioned in Sect. 8.8. Phase CGHs can provide a solution of this problem.

8.7.1 Phase Encoding

Let us consider how to determine the phase pattern $\phi[m, n]$. Supposing that $P[m, n]$ represents illumination light again, and a complex fringe pattern reconstructs the object field;

$$t'[m, n]P[m, n] = O[m, n], \tag{8.21}$$

the complex amplitude is

$$\begin{aligned} t'[m, n] &= \frac{O[m, n]}{P[m, n]} \\ &= O[m, n]P^*[m, n], \end{aligned} \tag{8.22}$$

where we assume $|P[m, n]|^2 = 1$. Although there may be many proposed techniques to obtain $\phi[m, n]$ from $t'[m, n]$, we adopt the simplest one[9] here;

$$\phi[m, n] = \arg(t'[m, n]), \tag{8.23}$$

where $\arg(\xi)$ denotes an argument of complex number ξ, i.e., $\xi = |\xi| \exp[i \arg(\xi)]$. An argument is simply calculated by[10]

$$\arg(\xi) = \tan^{-1}\left(\frac{\text{Im}\{\xi\}}{\text{Re}\{\xi\}}\right). \tag{8.24}$$

In addition, the phase value is commonly quantized. As a result, the phase-only fringe pattern is given by

$$\begin{aligned} t[m, n] &= \exp[i Q_{\text{p},L}\{\arg(t'[m, n])\}] \\ &= \exp[i Q_{\text{p},L}\{\arg(O[m, n]P^*[m, n])\}], \end{aligned} \tag{8.25}$$

---

[9]We also assume $|O[x, y]| \simeq$ const.

[10]In practical programming, $\tan^{-1}(x)$ is commonly represented by `atan(x)`. However, the return value of `atan(x)` is limited within interval $[-\pi/2, \pi/2]$ in general. Function `atan2(y, x)` instead of `atan(x)` must be used to obtain a value of $\tan^{-1}(x)$ in $[-\pi, \pi]$.

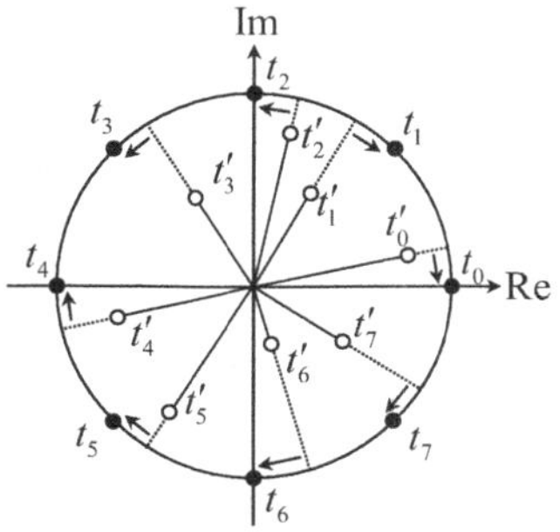

Fig. 8.13 An example of phase encoding in (8.25). Open and solid circles indicate the original and encoded complex-amplitude value, respectively. $L = 8$

where $Q_{p,L}\{\phi\}$ is a quantization operator that quantizes the phase value ϕ given in interval $(-\pi, \pi]$ into L levels.

Figure 8.13 shows an example of the phase encoding by (8.25). Here, the number of quantization levels is eight; $L = 8$. Complex-amplitude values t'_l are rounded to the nearest value of $t_l = \exp[i\pi l/4]$ $(l = 0, \ldots, 7)$ in this case.

8.7.2 Example of Phase CGH

Figure 8.14a and b shows simulated reconstruction of amplitude CGHs of the "Five Rings" model. This CG model has just 5,000 polygons. The 3D scene is shown in Fig. 8.15. The CG model, whose width is 40 mm, is arranged at $z = -90$ [mm]. The parameters of the CGHs are summarized in Table 11.2 of Sect. 11.3.11.

The object field used for generating the fringe patterns is calculated using the polygon-based method (see Chap. 10), and occlusion is processed by the polygon-by-polygon silhouette method using the switch-back technique (see Sect. 11.3). The amplitude CGHs in Fig. 8.14a and b are encoded with different reference spherical waves arranged at $(x_R, y_R, z_R) = (0, -10, -90)$ and $(0, -20, -360)$ [mm], respectively. The fringe patterns are quantized with 8-bits and $c_{\text{hist}} = 5$. The conjugate images overlapping the true images disturb the view in both amplitude CGHs. Here, note that the positions of the reference spherical waves are not practical for the 3D scene. They are chosen for generating distinct conjugate images to make the effect of phase encoding clear. Optical reconstructions of actual "Five Rings" CGH and its true coding parameters are shown in Figs. 11.21 and 11.22, and Table 11.2

Figure 8.14c and d are also simulated reconstructions of CGHs that reconstruct the same 3D scene. The CGHs are however phase CGHs whose phase values are quantized with 8-bit. The phase CGHs have the same parameters as those of the amplitude CGHs other than their encoding. Their reference fields are also spherical waves arranged at the same position as those of the corresponding amplitude CGHs. The images of the simulated reconstruction approximately reflect brightness of the reconstructed image. The phase CGHs definitely reconstruct brighter images than amplitude CGHs. In addition, phase CGHs have a great feature that reconstruct no conjugate image. This advantage allows us to design the 3D scene of CGHs much more freely.

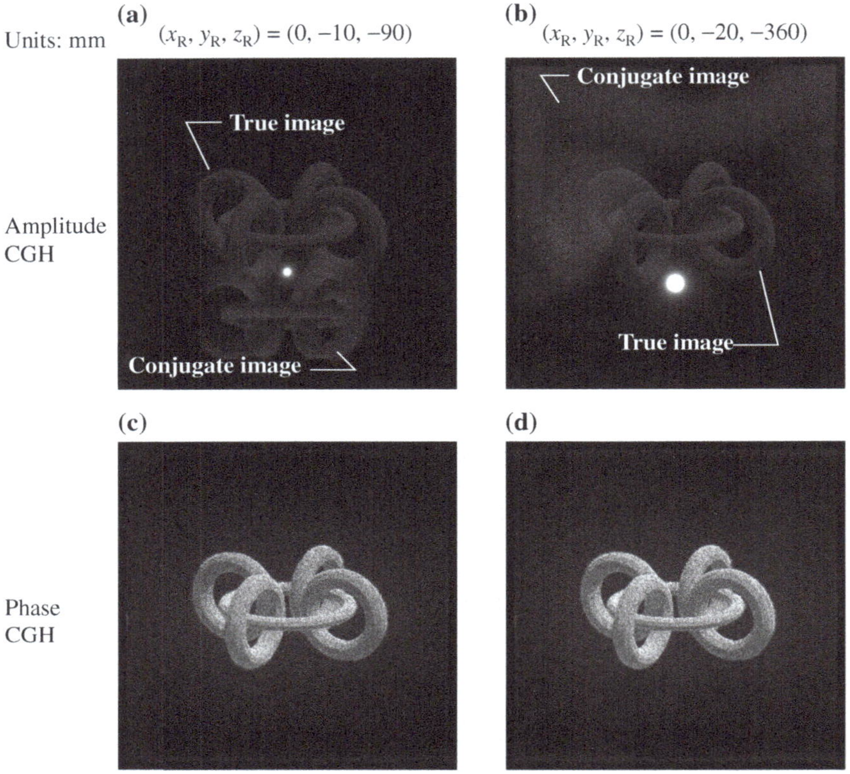

Fig. 8.14 Examples of simulated reconstruction of amplitude and phase CGHs. All CGHs are quantized with $L = 256$ in either encoding

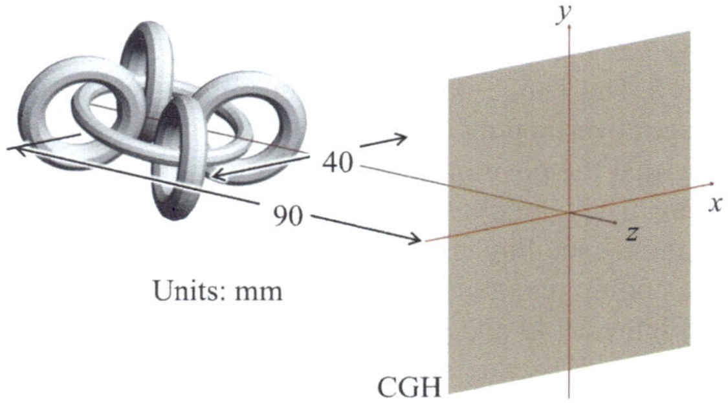

Fig. 8.15 The 3D scene of the "Five Rings" CGH used for the numerical experiment

It should be noted that no bright spot of the non-diffraction light appears in the simulated reconstruction of the phase CGHs. Unfortunately the simulation is not realistic only in this point, because the simulation does not include the effect of

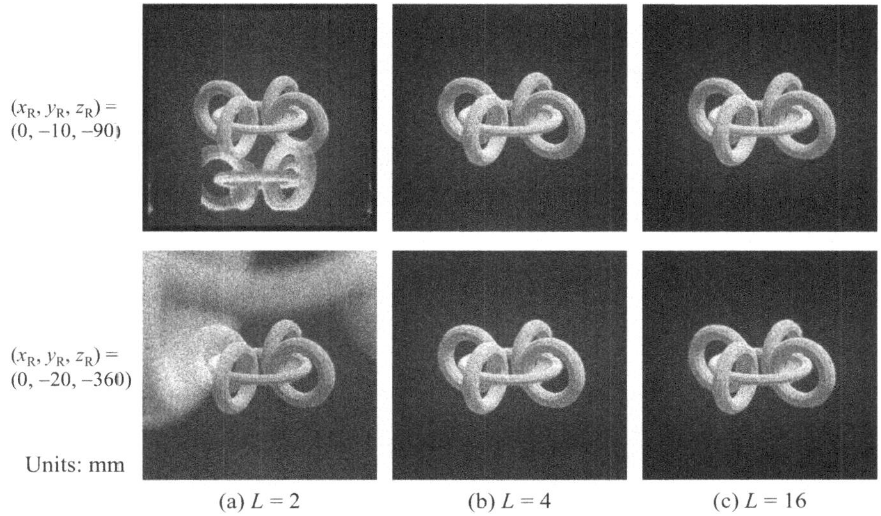

Fig. 8.16 Examples of simulated reconstruction of phase CGHs quantized with several different levels

the surface reflection, phase errors and other actual effects that may produce the non-diffraction light. Non-diffraction light is unavoidably caused by these reasons in practice.

8.7.3 Binary-Phase CGH

As shown in Fig. 8.13, the phase encoding given in (8.25) significantly changes the position of a complex-amplitude in the complex plane. There is a possibility that the phase-encoding causes severe quantization errors. Simulated reconstruction of the CGHs with different numbers of phase quantization levels are shown in Fig. 8.16. When decreasing phase levels, noise appearing in the reconstructed image increases properly. However, the noise is not so severe in the number of quantization levels more than four.[11]

In the binary phase encoding, the situation changes drastically, as in Fig. 8.16a. The reconstructed images are, in fact, the same as those in amplitude CGHs. We can simply explain the reason using Fig. 8.17. An amplitude CGH modulates the

[11]This fact may be doubtful for the reader who studies CGHs as optical devices or digital holography using image sensors. Please remember that the CGH here is composed of more than 4 billion pixels because of the space-bandwidth product problem. The problem of quantization errors may vanish behind the large-scale freedom, as mentioned in Sect. 8.5. Moreover, the object field contains a lot of random values to emulate diffuse surfaces. These random values tend to obfuscate coarse quantization.

Fig. 8.17 Comparison between amplitude modulation and binary-phase modulation

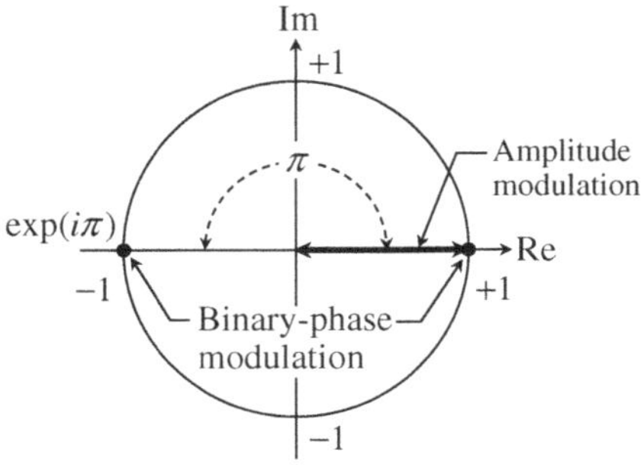

amplitude of the illumination light. Thus, the complex amplitude of the modulated light is located in the real axis of the complex plane. The complex amplitude in binary phase modulation, i.e., $\phi = 0$ or $\phi = \pi$, is also considered to be located in the real axis. Therefore, binary phase CGHs always behave like amplitude CGHs.

8.8 Spatial Frequency of Fringe Pattern

In general, the method to fabricate or display a CGH determines the minimum pixel pitch of the fringe image, and the minimum pixel pitch imposes a limit on the spatial frequency of the displayed fringes. The spatial frequency of the synthesized fringe pattern can not exceed this maximum limit. Therefore, it is important to expect the spatial frequency of the fringe pattern that is generated for reconstructing a given object model and arrangement.

8.8.1 Formulation

Interference fringes formed in the 3D space is determined by the difference between two wave vectors in (7.5), as mentioned in Sect. 7.1. Let $\mathbf{k}_O$ and $\mathbf{k}_R$ represent the wave vectors of an object field and reference field, as shown in Fig. 8.18. The hologram is placed in the $(x, y, 0)$ plane as usual. Here, we consider all vectors to be in the $(x, 0, z)$ plane, but formulation does not lose generality because y dependency can be obtained by switching the symbol x to y anytime. The fringe vector is

$$\mathbf{K} = \mathbf{k}_R - \mathbf{k}_O. \tag{8.26}$$

Supposing that the object field is emitted from an object point whose position vector is $\mathbf{P}_O$, the wave vector at point $\mathbf{X}$ along the x-axis is given as follows.

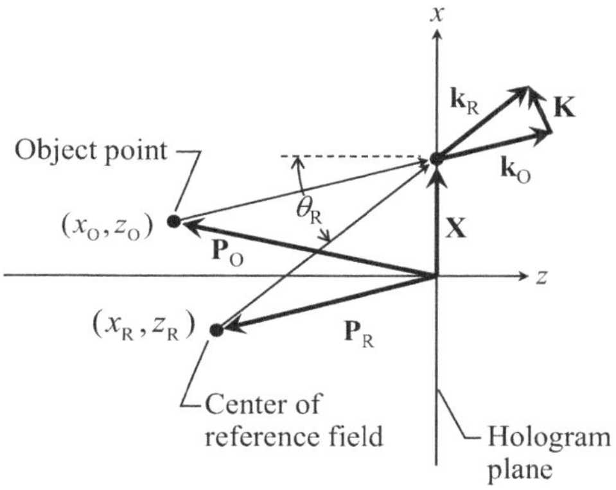

Fig. 8.18 Geometry for formulation of the spatial frequency of fringe patterns

$$\mathbf{k}_O = \frac{\mathbf{X} - \mathbf{P}_O}{|\mathbf{X} - \mathbf{P}_O|} k, \tag{8.27}$$

where k is a wave number again. Suppose that the reference field is emitted from point $\mathbf{P}_R$ in case of a reference spherical wave. The wave vector is in this case

$$\mathbf{k}_R = \frac{\mathbf{X} - \mathbf{P}_R}{|\mathbf{X} - \mathbf{P}_R|} k, \tag{8.28}$$

When the reference field is a plane wave, let θ_R represent the incident angle, as in Fig. 8.18. The wave vector of the reference field is independent of the position in the hologram plane in this case, and written as

$$\mathbf{k}_R = (\sin\theta_R \mathbf{e}_x + \cos\theta_R \mathbf{e}_z)k, \tag{8.29}$$

where $\mathbf{e}_x$ and $\mathbf{e}_z$ are unit vectors with respect to x and z coordinates, respectively.

The spatial frequency in the hologram plane is given by the x-component of $\mathbf{K}$ as follows:

$$u = \frac{1}{2\pi}(\mathbf{K} \cdot \mathbf{e}_x). \tag{8.30}$$

This spatial frequency has units of m^{-1} and the same orders of magnitude as λ^{-1}. Thus, we introduce a dimensionless frequency in units of λ^{-1}:

$$\begin{aligned} \hat{u} &= \frac{u}{\lambda^{-1}} \\ &= \frac{1}{k}(\mathbf{K} \cdot \mathbf{e}_x). \end{aligned} \tag{8.31}$$

Substituting (8.26)–(8.29) into (8.31), the dimensionless spatial frequencies in the hologram plane are written as

Table 8.1 Limit spatial frequency $\hat{u}_{\text{lim}}$ at typical wavelengths corresponding to RGB primary colors

Wavelength [nm]	Pixel pitch [μm]						
	0.4	0.6	0.8	1.0	2.0	4.0	8.0
633	0.791	0.528	0.396	0.317	0.158	0.079	0.040
532	0.665	0.443	0.333	0.266	0.133	0.067	0.033
488	0.610	0.407	0.305	0.244	0.122	0.061	0.031

$$\hat{u} = \begin{cases} \dfrac{1}{r_{\text{R}}}(x - x_{\text{R}}) - \dfrac{1}{r_{\text{O}}}(x - x_{\text{O}}) & \text{(Spherical wave)} \\ \sin\theta_{\text{R}} - \dfrac{1}{r_{\text{O}}}(x - x_{\text{O}}) & \text{(Plane wave)} \end{cases}, \tag{8.32}$$

where

$$r_{\text{R}} = \sqrt{(x - x_{\text{R}})^2 + z_{\text{R}}^2}, \tag{8.33}$$

$$r_{\text{O}} = \sqrt{(x - x_{\text{O}})^2 + z_{\text{O}}^2}. \tag{8.34}$$

When the pixel pitch of a fringe image is Δx_{h}, the sampling frequency is Δx_{h}^{-1} in the fringe pattern. Therefore, the spatial frequency of the fringe pattern must be less than $(2\Delta x_{\text{h}})^{-1}$ to avoid aliasing errors. This limit frequency is represented in units of λ^{-1} as follows.

$$\begin{aligned} \hat{u}_{\text{lim}} &= \hat{u}_{\text{lim}}(\lambda, \Delta x_{\text{h}}) \\ &= \frac{\lambda}{2\Delta x_{\text{h}}}. \end{aligned} \tag{8.35}$$

It should be emphasized that the limit frequency is determined by the wavelength and pixel pitch of the CGH. Table 8.1 summarizes actual values of $\hat{u}_{\text{lim}}$ at typical wavelengths corresponding to three primary colors for typical pixel pitches of fringe patterns. The dimensionless spatial frequency given in (8.32) must satisfy

$$|\hat{u}| < \hat{u}_{\text{lim}}, \tag{8.36}$$

to prevent the fringe pattern from producing aliasing errors.

8.8.2 Example of Fringe Frequency

Figure 8.19 shows the dimensionless frequency $\hat{u}$ in a variety of practical conditions. The horizontal and vertical axes of all graphs are x-coordinates and the dimensionless spatial frequency, respectively. Here, we again assume that the hologram is located in the $(x, y, 0)$ plane.

The horizontal solid red line indicates the limit frequency at 633 nm for a pixel pitch of 0.8 μm. The dashed red line also indicates the limit frequency at 633 nm but for a pixel pitch of 1.0 μm. On the other hand, the horizontal blue lines indicate the limit frequency at 488 nm. The solid and dashed lines are of pixel pitches of 0.8 μm and 1.0 μm, respectively. The left column of Fig. 8.19 show frequencies in cases where plane waves are used for the reference field, while the right column is in cases of using spherical waves for the reference field.

In Fig. 8.19a, the reference plane wave travels exactly along the z-axis. The object point is also in the z-axis. The parameter of the graph (a) is the z-position of the object point, which varies from $z_O = -5$ to -20 [cm]. The situation is schematically illustrated in Fig. 8.20a. In all positions of the object point, the spatial frequency is zero at the center of the hologram. The frequency increases with increasing the distance from the center of the hologram. For example, in $z_O = -5$ [cm], the fringe frequency approximately reaches 0.39 at a position of $x = \pm 2$ [cm]. This value is almost the same as the limit frequency of a CGH with a pixel pitch of 0.8 μm, designed for a red color (633 nm). This means that the size of the CGH should be approximately less than 4 cm. According to the blue horizontal line indicated in the graph, it is found that the maximum size is approximately 3 cm in the blue color CGH. Even if we create a bigger CGH, the fringe pattern of the object point at $z_O = -5$ [cm] cannot be properly generated around the edge of the CGH, and thus, we cannot see the object point through the area of the CGH. This is commonly undesirable in any CGH.

The restriction of the CGH size is relaxed by arranging the object farther from the hologram. If the object point is located at $z_O = -10$ [cm], the maximum size is nearly 9 cm and 5 cm in red and blue, respectively. In conclusion, we should not arrange the object model so close to the hologram especially in a large CGH. The effective CGH size becomes small and the diffraction efficiency most likely decreases, if the object is too close to the hologram.

There is another method to ease restriction on the CGH size. Figure 8.19b shows fringe frequencies in the case of the reference spherical wave whose center is located at a position of $z_R = -20$ [cm] in the optical axis. The fringe frequency is overall smaller than that in the reference plane wave. The CGH size allowed for the object point at $z_O = -5$ [cm] expands to approximately 6.5 cm in red and 5 cm in blue, respectively. The object point arranged at $z_O = -10$ [cm] can be reconstructed and seen through everywhere in the hologram.

Above discussions seem to be not realistic in creation of CGHs, because a bright spot produced by the non-diffraction light is placed at the center of the reconstructed image in both cases where $\theta_R = 0$ and $x_R = 0$. The reference field must be arranged

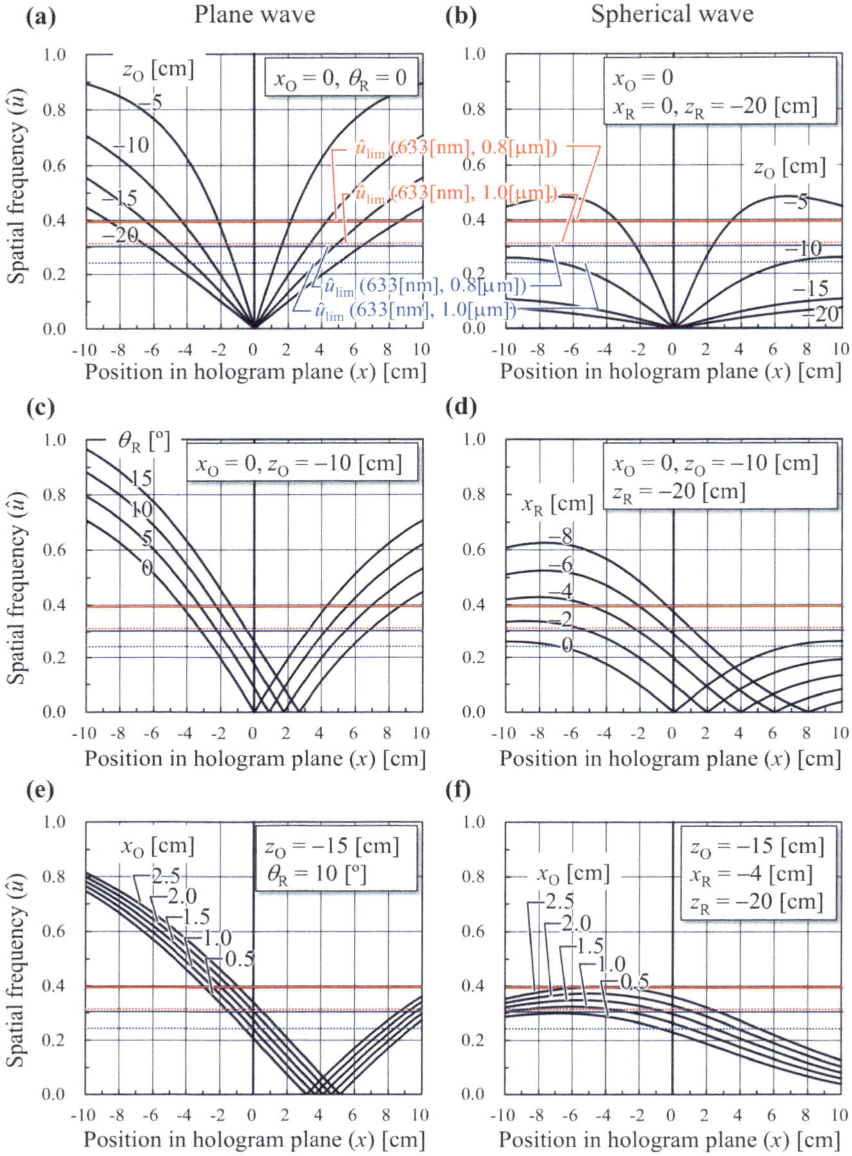

Fig. 8.19 Normalized frequency in the hologram plane under a variety of actual conditions. The variations of the parameter are schematically shown in Fig. 8.20

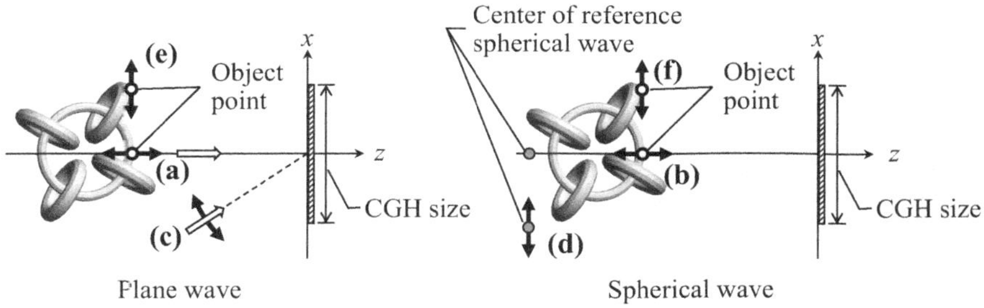

Fig. 8.20 Schematic illustration of variation of parameters in Fig. 8.19. Note that the depicted positions and angle are not realistic in this figure

so that the non-diffraction light does not come into the view of viewers. A non-zero, if possible a large, value of θ_R or x_R is required in the reference field to do this. Therefore, in actual HD-CGHs, we often use a reference field, which has a given large value of θ_R and y_R in the $(0, y, z)$ plane, instead of the $(x, 0, z)$ plane.

Figure 8.19c and d show fringe frequencies under more realistic conditions. In (c), the incident angle of the reference plane wave varies from 0° to 10° as shown in Fig. 8.20c. Here, the object point is arranged at $z_O = -10$ [cm] to reduce the whole fringe frequency. Curves of the fringe frequency are not symmetrical with respect to x coordinates in this case, and the fringe frequency has a higher value in $x < 0$ than that in $x > 0$. The fringe frequency in the negative domain increases with increasing the incident angle. This commonly makes creation of large CGHs more difficult. To avoid the problem of the non-diffraction light, the incident angle should be at least 5°. The allowed CGH size is almost 5 cm in red color in this case. Here, note that we take the origin of the lateral coordinates at the center of the CGH.

In Fig. 8.19d, the x-position of the spherical wave: x_R varies in order to prevent the non-diffraction light from coming into the view. Curves of the fringe frequency are also asymmetrical with respect to x-coordinates like those in (c). But the frequency increase is generally gentler than those in the plane wave.

In above examples, the object point is arranged just in the optical axis. However, actual object models commonly have a given width and height. In this case, the x position of the object point is not zero. Figure 8.19e and f show frequency curves in that case. Here, to relax the restriction on the CGH size even more, the object point is located at $z_O = -15$ [cm], i.e., even further than (a)–(d). However, in the case of reference plane waves of (e), the restriction becomes more severe. If the object has a width of 5 cm; that is corresponding to $x_O = 2.5$ [cm], the fringe frequency exceeds the red solid line at the x-position less than -1 cm. This means that the CGH size must be less than 2 cm to avoid aliasing errors everywhere in the hologram.

In contrast, when we use a spherical wave to generate the fringe pattern, the spatial frequency is always not more than the red solid line. Therefore, a large CGH can be created in this case, where the design wavelength and the pixel pitch are 633 nm and 0.8 μm, respectively.

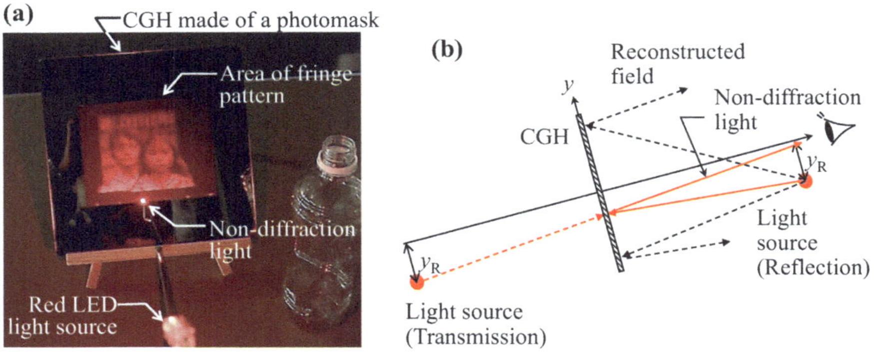

Fig. 8.21 **a** Exhibition of Brothers [123], and **b** geometry of the non-diffraction light

In conclusion, use of a reference spherical wave to generate the fringe pattern generally gives better results than use of a reference plane wave. However, we usually have to pay attention to the conjugate image in amplitude thin holograms. When a spherical wave is used for the reference field, as mentioned in Sect. 7.7.2, the conjugate image appears at an unexpected place and obstructs the view of observers. Therefore, an object model must be arranged very carefully in the 3D scene that we want to reconstruct. In addition, the position of the reference spherical wave is very important to keep the fringe frequency less than the limit value and prevent the bright spot from coming into the view.

8.8.3 *Fringe Oversampling*

Figure 8.21a shows a photograph of actual exhibition of high-definition CGH "Brothers" introduced in Chap. 1. This CGH is made of a photomask in which chromium film coated on the glass substrate forms the fringe pattern (see Sect. 15.3.1). Accordingly, though the CGH is a thin hologram, the CGH can be reconstructed by reflection illumination because of high reflectance of the chromium film. Therefore, the CGH is illuminated by a red LED placed in front of the CGH and below the optical axis ($y_R < 0$).

The bright spot appearing right under the reconstructed image in (a) is caused by the non-diffraction light that is regular reflection of the illumination light as shown in Fig. 8.21b. This non-diffraction light is sometimes so bright that the observer cannot see the 3D image properly. The bright spot is produced by transmission illumination as well as reflection illumination. To prevent the bright spot from disturbing the view of the 3D image, it is important to increase the illumination angle, i.e., to increase $|y_R|$ in the reference spherical wave or θ_R in the reference plane wave. However,

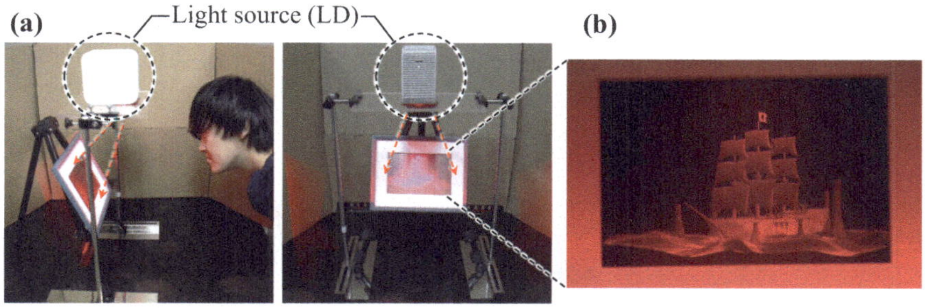

Fig. 8.22 **a** Exhibition of a large-scale CGH: Sailing Warship II, whose size is 18 cm × 15 cm, and **b** the closeup photograph. Here, a laser projector made by SONY™ is used for the monochromatic light source

as shown in Fig. 8.19c and d, increasing the incident angle of the illumination light inevitably increases the fringe frequency.

Even if the printing equipment of the CGH fringes is capable to print very fine fringe pattern required to give a very large illumination angle, we usually cannot adopt so large θ_R or $|y_R|$ in actual CGHs. This is mainly due to the problem of calculation time of the object field. The object field of a CGH commonly have the same sampling intervals as the pixel pitches of the CGH, otherwise we cannot perform arithmetic operations such as (8.12) or (8.25) to generate the fringe pattern. Therefore, reducing pixel pitch to increase the illumination angle leads to decreasing the sampling interval in synthesizing the object field of the object model.

However, it is not easy to decrease the sampling interval of the object field. There are mainly two reasons. First, the size of wavefields emitted from a polygon source or point source becomes large in the calculation because the maximum diffraction angle increases with reducing the sampling interval. Thus, in general, the smaller the sampling interval is, the longer computation time is. Second, the total number of sample points simply increase as the sampling interval reduces. This also leads to increasing the computation time.

The technique to get over the problem is *fringe oversampling*. In this technique, sampling intervals of the object field Δx_O and Δy_O do not agree with that of the hologram; Δx_h and Δy_h. Actually, we choose smaller pixel pitches of the CGH than the sampling intervals of the object field. Figure 8.22a show actual exhibition of "Sailing Warship II" introduced in Chap. 1. The object field of this CGH is calculated with sampling intervals of $\Delta x_O = \Delta y_O = 0.8$ [μm], while the pixel pitches of the CGH are $\Delta x_h = 0.8$ [μm] and $\Delta y_h = 0.4$ [μm]. The object field calculated at the hologram plane is resampled using an interpolation technique such as bicubic interpolation (see Sect. 9.5.4), as shown in Fig. 8.23. In this hologram, one half of the vertical sampling interval of the object field is chosen for the vertical pixel pitch of the fringe image; $\Delta y_h = \Delta y_O/2$ in order to arrange the illumination light source above the CGH and direct the non-diffraction light downward. As a result, any bright spot does not appear in the reconstructed 3D image, as shown in Fig. 8.22b.

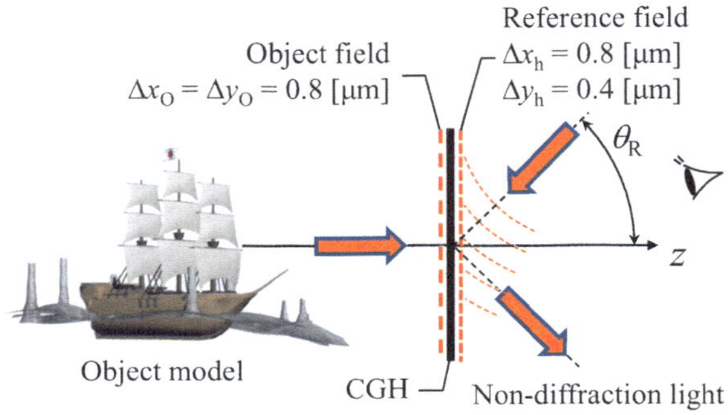

Fig. 8.23 The technique of the fringe oversampling in Sailing Warship II

In addition to the non-diffraction light, the conjugate image nearly disappears from the reconstructed image by use of the fringe oversampling, because the position of the conjugate image is also depending on the illumination angle, as shown in Sects. 7.7.1 and 7.7.2. Thus, this simple technique is a great method if resolution of the printing equipment is sufficiently high. However, it should be noted that a high density fringe pattern tends to cause strong chromatic aberration, and thus blurs the reconstructed image. An illumination light source with a narrow spectrum, such as laser light sources, is commonly required to clear reconstruction in cases where the pixel pitch is approximately less than or equal to the wavelength.

8.9 Fourier-Transform CGH

As shown in Fig. 7.23, Fourier-transform holography has features that the non-diffraction light converges into a point, and the true image is separated from the conjugate image and the non-diffraction light in the focal plane. In addition to these, Fourier-transform holography provides a means to control the viewing angle easily. Since the viewing angle of a CGH is essentially determined by the pixel pitch of the fringe image, as mentioned in Sect. 8.2, this is an advantage of the Fourier-transform holography.

8.9.1 Higher-Order Diffraction Images

In optical holography, the recorded object field $O(x, y)$ is given by optical Fourier transform using a lens, as mentioned in Sect. 7.7.3. We skip this recording process in computer holography because the object field is calculated by numerical Fourier transform; that is FFT:

$$O[m, n] = \mathbf{FFT}^{-1}\left\{g_{\text{obj}}[p, q; +f]\right\}, \tag{8.37}$$

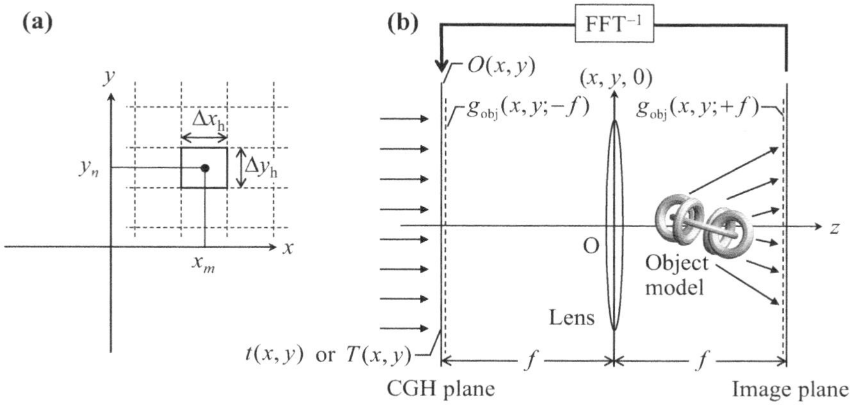

Fig. 8.24 Fourier-transform CGHs: **a** the structure of a fringe pixel and **b** the theoretical model

where $g_{obj}[p, q; +f] = g_{obj}(x_p, y_q; +f)$ and $O[m, n] = O(x_m, y_n)$ are the object field in the image and hologram plane respectively, as shown in Fig. 8.24b. Here, note that inverse FFT is used to avoid inversion of the image in reconstruction.

The Fourier transform using a lens is essentially the same phenomenon as the far-field propagation. The formulas agree with each other by exchanging the propagation distance with the focal length. Thus, according to (6.7), the sampling interval Δx and Δy in the image plane is associated with those in the hologram plane as follows:

$$\Delta x = \frac{\lambda f}{M \Delta x_h} \quad \text{and} \quad \Delta y = \frac{\lambda f}{N \Delta y_h}, \tag{8.38}$$

where Δx_h and Δy_h are pixel pitches of the fringe pattern again, and distance d in (6.7) is replaced by f.

Amplitude fringe $T(x, y)$ or phase fringe $t(x, y)$ is generated from the object field $O(x_m, y_n)$. Suppose that the fringe pattern has the structure shown in Fig. 8.24a and both the reference and illumination fields are a plane wave traveling along the z-axis; $R(x, y) = P(x, y) = 1$. The CGH generates the following diffracted field corresponding to the true image:

$$g_{obj}(x, y; -f) = \sum_{m=0}^{M-1} \sum_{n=0}^{N-1} O[m, n] \text{rect}\left(\frac{x - x_m}{\Delta x_h}\right) \text{rect}\left(\frac{y - y_n}{\Delta y_h}\right). \tag{8.39}$$

The reconstructed field in the image plane is given by the optical Fourier transform represented in (5.66).

$$g_{\mathrm{obj}}(x, y; +f) \cong \sum_m \sum_n O[m, n]\mathcal{F}\left\{\mathrm{rect}\left(\frac{x - x_m}{\Delta x_{\mathrm{h}}}\right)\mathrm{rect}\left(\frac{y - y_n}{\Delta y_{\mathrm{h}}}\right)\right\}$$
$$= \Delta x_{\mathrm{h}} \Delta y_{\mathrm{h}} \mathrm{sinc}(\Delta x_{\mathrm{h}} u)\mathrm{sinc}(\Delta y_{\mathrm{h}} v)$$
$$\times \sum_m \sum_n O(x_m, y_n) \exp[-i2\pi(u x_m + v y_n)], \tag{8.40}$$

$$u = \frac{x}{\lambda f} \quad \text{and} \quad v = \frac{y}{\lambda f}, \tag{8.41}$$

where the constant coefficient is omitted. By comparison between summation in (8.40) and (4.66) in Sect. 4.6, it is found that $g_{\mathrm{obj}}(x, y; +f)$ is represented by the discrete Fourier transform (DFT) of the sampled field $O(x_m, y_n)$, and therefore has periodicity of $1/\Delta x_{\mathrm{h}}$ and $1/\Delta y_{\mathrm{h}}$ with respect to u and v.

According to (8.41), $x = \lambda f u$ and $y = \lambda f v$. Thus, periods of the field reconstructed in the image plane are given by

$$W_{\mathrm{Four},x} = \frac{\lambda f}{\Delta x_{\mathrm{h}}} \quad \text{and} \quad W_{\mathrm{Four},y} = \frac{\lambda f}{\Delta y_{\mathrm{h}}}, \tag{8.42}$$

in the x and y directions, respectively. This means that a Fourier-transform CGH reconstructs not only the object field but also many periodic fields in the image plane. These are called *higher-order* diffraction fields or images, while the primary object field reconstructed around the optical axis is called the *first order* diffraction field or image. $W_{\mathrm{Four},x}$ and $W_{\mathrm{Four},y}$ give the size of the sampling window of the 1st order field in the image plane, and properly agree with $M\Delta x$ and $N\Delta y$ in (8.38).

Although not only Fourier-transform CGHs but also other types of CGH reconstruct higher-order images, it is most remarkable in Fourier-transform CGHs. The amplitude of the 1st and higher-order fields is modulated by sinc function as shown in (8.40):

$$A(x, y) = \mathrm{sinc}\left(\frac{\Delta x_{\mathrm{h}}}{\lambda f} x\right)\mathrm{sinc}\left(\frac{\Delta y_{\mathrm{h}}}{\lambda f} y\right). \tag{8.43}$$

Thus, brightness of the higher-order images is reduced as apart from the origin. However, we commonly need an aperture or filter to eliminate the higher-order field and select only the first order true image in Fourier-transform computer holography. Figure 8.25 shows a schematic example of reconstruction of a Fourier-transform CGH. Here, note that this is a simulation assuming phase encoding and thus there is no conjugate image in this example.

8.9.2 Generation of Fringe Pattern

To create Fourier-transform CGH, object field $g_{\mathrm{obj}}[m, n; +f]$ is first calculated from the object model within the sampling window in the image plane, whose size is

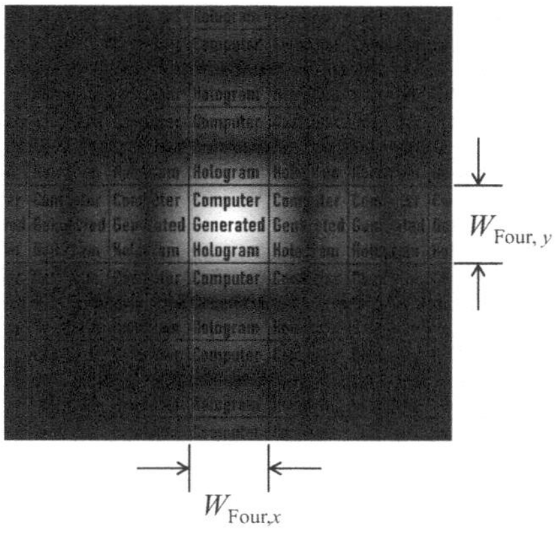

Fig. 8.25 A schematic example of reconstruction of a Fourier-transform CGH. Phase encoding is assumed in this example

$W_{\text{Four},x} \times W_{\text{Four},y}$ in (8.42). The sampling interval of the object field is given by (8.38). This sampling interval determines the viewing angle of the CGH. Thus, the viewing angle is mainly given by the size of the CGH, i.e., $M \Delta x_\text{h} \times N \Delta y_\text{h}$, and the focal distance f of the lens used. The larger the CGH size is and the shorter the focal length is, the larger the viewing angle becomes.

Second, the object field is inversely Fourier-transformed using FFT:

$$\begin{aligned} O(x_p, y_q) &= O[p, q] \\ &= \mathbf{FFT}^{-1}\left\{g_\text{obj}[m, n]\right\}, \end{aligned} \tag{8.44}$$

where the intervals of the sample points represented by x_p and y_q are Δx_h and Δy_h, respectively. The fringe pattern is generated from $O[p, q]$ using the amplitude or phase encoding technique mentioned in the previous sections. When amplitude encoding is used, the conjugate image as well as the true image is reconstructed as in Fig. 7.23. Special technique for amplitude encoding in Fourier-transform CGH is mentioned in the following section.

8.9.3 *Amplitude Fringe Pattern Based on Hermitian Function*

Amplitude fringe patterns are commonly easier to fabricate than phase fringe patterns in computer holography, because the space-bandwidth product problem requires large-scale fringe patterns, and printing amplitude patterns is on the extension of ordinary printing or display technologies. In Fourier-transform holography, there is a special technique for amplitude encoding, which makes use of a symmetric relation of the Fourier transform to encode amplitude fringe patterns.

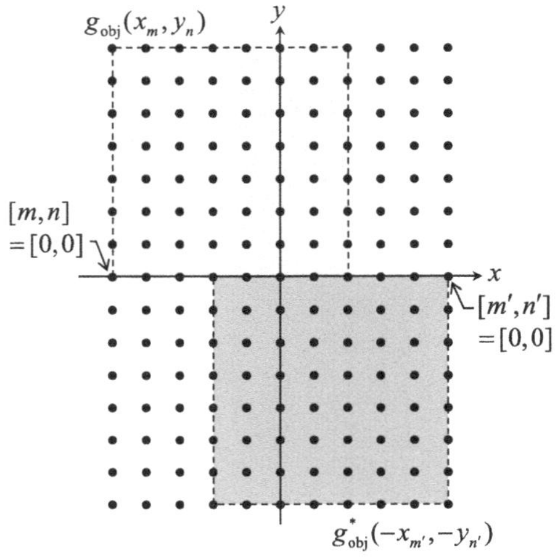

Fig. 8.26 An example to make symmetry of Hermitian functions in sampled object fields

According to Table 4.3, the Fourier transform of asymmetric real functions results in an Hermitian function, which has a symmetry in (4.44); that is

$$g_h(-x, -y) = g_h^*(x, y). \tag{8.45}$$

This is the reason that an amplitude Fourier-transform hologram generates a symmetric image in the Fourier plane. Conversely, the inverse Fourier transform of a Hermitian function gives a real function. Let us make use of this nature. It is easy to confirm that the following function has the symmetry given in (8.45).

$$g'_{\text{obj}}(x, y) = g_{\text{obj}}(x, y) + g^*_{\text{obj}}(-x, -y). \tag{8.46}$$

The inverse Fourier transform of this function; $\mathcal{F}^{-1}\left\{g'_{\text{obj}}(x, y)\right\}$ is a real function, i.e., an amplitude fringe pattern. Therefore, we can generate amplitude CGHs not by interference with reference fields.

In sampled object fields, conversion of (8.46) must be exactly performed as shown in Fig. 8.26. If symmetry is not accurate in the sampled array, the imaginary value does not reduce enough to be ignored in all sample points after FFT.

8.10 Single-Sideband Method in Amplitude CGH

The *single-sideband method* is a technique that makes use of the nature of Fourier-transform holography to remove the conjugate image and non-diffraction light from the reconstruction of amplitude holograms [8]. As shown in Fig. 7.23, non-diffraction

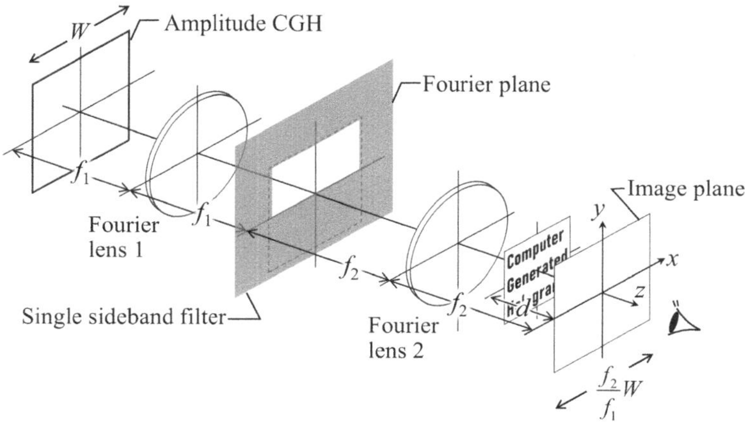

Fig. 8.27 Typical setup for reconstructing an amplitude CGH using the single-sideband method

light, true and conjugate images are isolated in amplitude Fourier holograms. This is a directional visualization of Fourier spectrum of amplitude holograms with a reference plane wave (see Sect. 7.5).

8.10.1 Principle

Figure 8.27 shows a typical setup of the single-sideband method. In this technique, two Fourier lenses are used for removing unnecessary light. Here, note that the origin of the coordinate system is placed in the image plane. For simplicity of the following discussion, let an amplitude fringe pattern be presented by (7.15). In this case, the field reconstructed by the amplitude hologram is simply

$$\begin{aligned} g(x, y; -2f_1 - 2f_2) &= t(x, y)P(x, y) \\ &\approx B + O(x, y) + O^*(x, y), \end{aligned} \tag{8.47}$$

where we assume $R(x, y) = P(x, y) = 1$. The setup in Fig. 8.27 is actually equivalent to recording and reconstruction of a Fourier hologram shown in Figs. 7.21 and 7.22. Therefore, if we do not use any filter in the Fourier plane, the reconstructed field in the image plane is obtained from (7.76) and (7.78):

$$\begin{aligned} g(x, y; 0) &\approx B + g_{\text{true}}(x, y; 0) + g_{\text{conj}}(x, y; 0) \\ &= B + g_{\text{obj}}\left(-\frac{f_1}{f_2}x, -\frac{f_1}{f_2}y; 0\right) + g^*_{\text{obj}}\left(\frac{f_1}{f_2}x, \frac{f_1}{f_2}y; 0\right). \end{aligned} \tag{8.48}$$

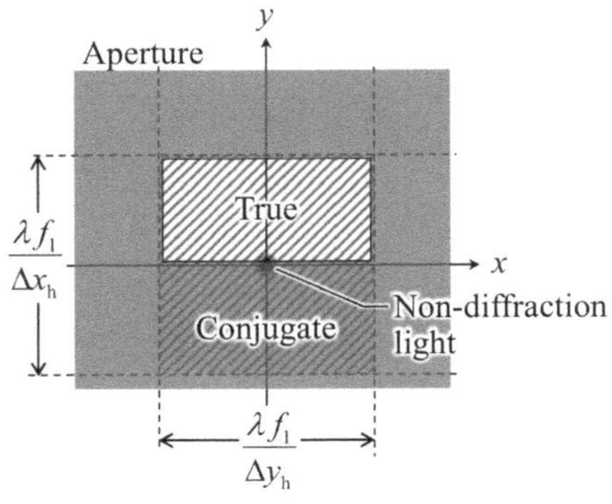

Fig. 8.28 An example of the single-sideband filter placed in the Fourier plane

In this case, non-diffraction light (1st term), true image (2nd term) and conjugate image (3rd term) mix in the image plane. However, as shown in Fig. 7.23c, these are isolated in the Fourier plane if we choose a proper arrangement of the object field. Accordingly, a filter aperture is inserted in the Fourier plane to block the non-diffraction light and conjugate field in the single-sideband method, as shown in Fig. 8.27.

The filter must shield at least one half of the Fourier plane to remove the conjugate image. This is equivalent to cut off a sideband shown in Fig. 7.5c, reconstructed by amplitude holograms. We have freedom to choose which half domain the filter shields in this technique. It depends on the way to generate the fringe pattern. However, shielding lower-half or upper-half domain is usually adopted in many cases. This is because cut-off of the sideband leads to reducing the viewing angle to one half of the original of the CGH. In 3D displays, wider viewing angle in the horizontal direction is more preferred to that in the vertical direction in general. An upper-half or lower-half pass filter does not reduce the viewing angle in the horizontal direction. Here, note that the filter must also shield light converging a small area around the origin of coordinates to block the non-diffraction light.

An example of the single-sideband filter is shown in Fig. 8.28. The aperture size of a sideband-filter must fit the 1st order field in (8.42). Since the filter in Fig. 8.28 is designed to shield the conjugate field in the lower-half of the 1st order diffraction, the vertical size of the aperture is $W_{\mathrm{Four},y}/2$. Here, in practice, the vertical size is a little less than $W_{\mathrm{Four},y}/2$ to block the non-diffraction light simultaneously.

8.10.2 Generation of Fringe Pattern

Figure 8.29 shows the process for generating the fringe pattern of an amplitude CGH that reconstructs only true image using the single-sideband method. The object field $g_{\mathrm{obj}}[m, n]$ is first calculated from the object model using an appropriate method. The sampling interval of the object field must be

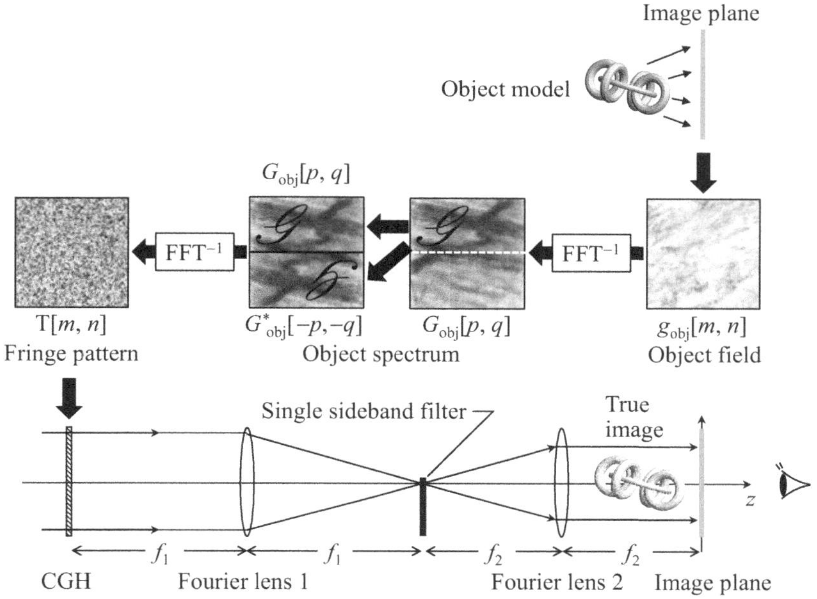

Fig. 8.29 Process for generating the fringe pattern in the single-sideband method

$$\Delta x = \frac{f_2}{f_1}\Delta x_{\mathrm{h}} \quad \text{and} \quad \Delta y = \frac{f_2}{f_1}\Delta y_{\mathrm{h}}, \tag{8.49}$$

where Δx_{h} and Δy_{h} are again the pixel pitches of the fringe image. Object field is inversely Fourier-transformed using FFT:

$$G_{\mathrm{obj}}[p, q] = \mathbf{FFT}^{-1}\left\{g_{\mathrm{obj}}[m, n]\right\}. \tag{8.50}$$

Since the sideband filter cuts off the lower-half domain of the Fourier spectrum in optical reconstruction, we abandon the lower-half of the object spectrum. Supposing that the number of sample points of the object field is $M \times N$, only the data that satisfies $q \geq N/2$ is used for the next step. Using the technique shown in (8.46) and Fig. 8.26, a new Fourier spectrum having the symmetry of Hermitian functions is produced as follows:

$$G'_{\mathrm{obj}}[p, q] = \begin{cases} G^*_{\mathrm{obj}}[M-1-p, N-1-q] & \text{if } q < N/2 \\ G_{\mathrm{obj}}[p, q] & \text{otherwise} \end{cases}, \tag{8.51}$$
$$(p = 0, 1, \ldots, M-1 \quad q = 0, 1, \ldots, N-1),$$

where (4.92) is used for the sampling manner. It is guaranteed that the inverse Fourier transform of this spectrum is a real function. Therefore, the amplitude fringe pattern is given by

$$T[m,n] = \text{Re}\left\{\mathbf{FFT}^{-1}\left\{G'_{\text{obj}}[p,q]\right\}\right\}. \tag{8.52}$$

As a result, amplitude CGHs can reconstruct the 3D image without any conjugate image at the sacrifice of the vertical viewing angle.

Chapter 9
The Rotational Transform of Wavefield

Abstract The rotational transform of wavefields provides numerical propagation between non-parallel planes. This non-traditional technique is based on the angular spectrum method, as with the convolution-based propagation method discussed in Chap. 6. Therefore the rotational transform is performed with a double Fourier transform and resampling the wavefield, and thus practical enough for various applications in wave optics. Not only formulas but also the pseudocode for software implementation and actual examples are presented in this chapter.

9.1 Introduction

The rotational transform of wavefields is very different from the parallel and shifted numerical propagation, as shown in Fig. 5.4. The destination plane is not parallel to the source plane in this technique. The destination plane is sometimes called a *reference plane* in the rotational transform. This is because it is a little difficult to consider the rotational transform to be light propagation from the source plane to destination plane. Rather than light propagation, it is better to consider the rotational transform as a technique that cut out a cross section from the three-dimensionally distributed light field, which satisfies Helmholtz equation with a boundary condition given by the source wavefield, as shown in Fig. 9.1.

We can find various techniques to calculate the field in a tilted reference plane in the literature [12, 17, 25, 51, 100, 103]. The rotational transform of wavefields described in this chapter is based on the angular spectrum method described in Sect. 5.2.1 [61, 78].

9.2 Coordinate Systems and Rotation Matrices

Figure 9.2 shows two coordinate systems used in formulation of the rotational transform. One is the source coordinate system (x_s, y_s, z_s) where the source wavefield is given in the $(x_s, y_s, 0)$ plane. Another is the reference or destination coordinate

K. Matsushima, *Introduction to Computer Holography*,
Series in Display Science and Technology,
https://doi.org/10.1007/978-3-030-38435-7_9

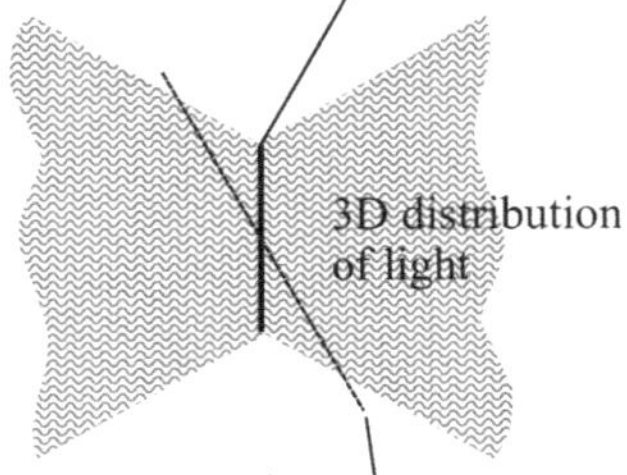

Fig. 9.1 Schematic illustration of the idea of rotational transform

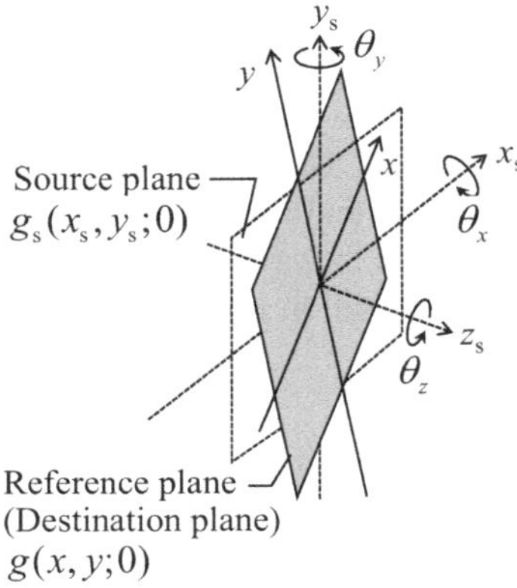

Fig. 9.2 Coordinate systems used in formulation of rotational transform

system (x, y, z) where the destination field is obtained in the $(x, y, 0)$ plane. These coordinate systems share the origin but at least two axes are not parallel to the counterparts in other coordinate system.

A point whose position is given in a coordinate system can be easily represented in other coordinates using rotation matrices. In other words, position vectors $\mathbf{r}_s = (x_s, y_s, z_s)$ and $\mathbf{r} = (x, y, z)$ can be mutually transformed by coordinate rotation using transformation matrix $\mathbf{T}$ as follows:

$$\mathbf{r} = \mathbf{T}\mathbf{r}_s, \tag{9.1}$$

$$\mathbf{r}_s = \mathbf{T}^{-1}\mathbf{r}. \tag{9.2}$$

The matrix $\mathbf{T}$ is, in general, given as a rotation matrix $\mathbf{R}_\xi(\theta_\xi)$ or the product of several rotation matrices as follows:

$$\mathbf{T} = \mathbf{R}_\xi(\theta_\xi)\ldots\mathbf{R}_\eta(\theta_\eta), \tag{9.3}$$

where ξ and η denote axes x, y or z, and θ_ξ and θ_η are the angles of rotation around the axes ξ and η, respectively. Individual rotation matrices are shown in Table 9.1. These commonly possess the following characteristics:

Table 9.1 Coordinates rotation matrices. Note that these are matrices for coordinates rotation, and thus, different from that for position rotation

$$\mathbf{R}_x(\theta_x) = \begin{bmatrix} 1 & 0 & 0 \\ 0 & \cos\theta_x & \sin\theta_x \\ 0 & -\sin\theta_x & \cos\theta_x \end{bmatrix}, \quad \mathbf{R}_y(\theta_y) = \begin{bmatrix} \cos\theta_y & 0 & -\sin\theta_y \\ 0 & 1 & 0 \\ \sin\theta_y & 0 & \cos\theta_y \end{bmatrix}, \quad \mathbf{R}_z(\theta_z) = \begin{bmatrix} \cos\theta_z & \sin\theta_z & 0 \\ -\sin\theta_z & \cos\theta_z & 0 \\ 0 & 0 & 1 \end{bmatrix}$$

$$\mathbf{R}_\xi^{-1}(\theta_\xi) = \mathbf{R}_\xi(-\theta_\xi) = {}^t\mathbf{R}_\xi(\theta_\xi), \tag{9.4}$$

$$\det \mathbf{R}_\xi(\theta_\xi) \equiv 1, \tag{9.5}$$

where $\mathbf{R}_\xi^{-1}(\theta_\xi)$ and ${}^t\mathbf{R}_\xi(\theta_\xi)$ are the inverse and transposed matrix of a rotation matrix, respectively. $\det \mathbf{A}$ is the determinant of a matrix $\mathbf{A}$.

As a result, the inverse matrix of any transformation matrix defined by the product of individual rotation matrices in (9.3) is generally given by

$$\mathbf{T}^{-1} = {}^t\mathbf{T}. \tag{9.6}$$

Thus, $\mathbf{T}$ is an orthogonal matrix and $\det \mathbf{T} \equiv 1$. In general, orthogonal matrices preserve the dot product. For example, supposing $\mathbf{a}$ and $\mathbf{b}$ are vectors in the 3D space, the transformation matrix $\mathbf{T}$ satisfies,

$$\begin{aligned} (\mathbf{Ta}) \cdot (\mathbf{Tb}) &= {}^t(\mathbf{Ta})(\mathbf{Tb}) = {}^t\mathbf{a}{}^t\mathbf{T}\,\mathbf{Tb} = {}^t\mathbf{a}\,\mathbf{b} \\ &= \mathbf{a} \cdot \mathbf{b}. \end{aligned} \tag{9.7}$$

In this chapter, the generalized rotation matrix below is used in formulation as follows.

$$\mathbf{T}^{-1} = \begin{bmatrix} a_1 & a_2 & a_3 \\ a_4 & a_5 & a_6 \\ a_7 & a_8 & a_9 \end{bmatrix}, \tag{9.8}$$

$$\mathbf{T} = \begin{bmatrix} a_1 & a_4 & a_7 \\ a_2 & a_5 & a_8 \\ a_3 & a_6 & a_9 \end{bmatrix}. \tag{9.9}$$

9.3 Principle

Supposing the source wavefield is represented by $g_s(x_s, y_s; 0)$ in the source coordinates, the Fourier spectrum is given by

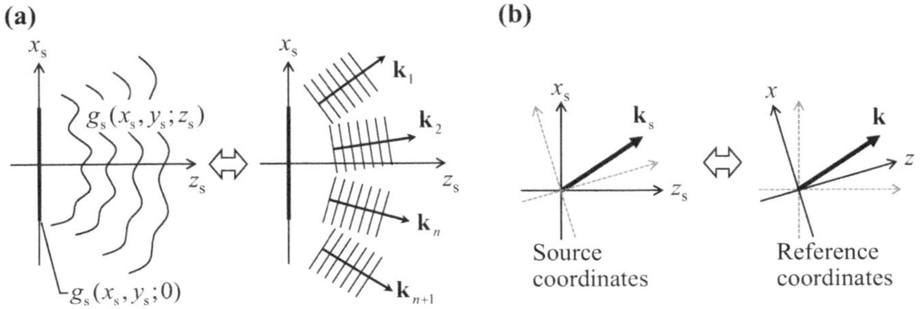

Fig. 9.3 Principle of the rotational transform. **a** Expansion of a wavefield into plane waves, **b** a wave vector represented in the source and reference coordinate systems

$$\begin{aligned} G_s(u_s, v_s; 0) &= \mathscr{F}\{g_s(x_s, y_s; 0)\} \\ &= \iint_{-\infty}^{+\infty} g_s(x_s, y_s; 0) \exp[-i2\pi(u_s x_s + v_s y_s)] dx_s dy_s. \end{aligned} \tag{9.10}$$

The inverse Fourier transform is also given by

$$\begin{aligned} g_s(x_s, y_s; 0) &= \mathscr{F}^{-1}\{G_s(u_s, v_s; 0)\} \\ &= \iint_{-\infty}^{+\infty} G_s(u_s, v_s; 0) \exp[i2\pi(x_s u_s + y_s v_s)] du_s dv_s. \end{aligned} \tag{9.11}$$

As mentioned in Sect. 5.2.1.1, a wave vector is, in general, associated with Fourier frequencies. According to (5.5), we can represent a wave vector in the source coordinates as

$$\mathbf{k}_s = 2\pi(u_s, v_s, w_s). \tag{9.12}$$

Using this representation, since $z_s = 0$, the integrand of (9.11) is rewritten by

$$G_s(u_s, v_s; 0) \exp[i(\mathbf{k}_s \cdot \mathbf{r}_s)].$$

This is nothing but a plane wave. Therefore, we can interpret integration of the inverse Fourier transform in (9.11) as an operation that assembles wavefield $g_s(x_s, y_s; 0)$ from plane waves. The Fourier transform of (9.10) can also be regarded as disassemble of the source wavefield into plane waves whose amplitude is given by $G_s(u_s, v_s; 0)$. This means that a wavefield can be expanded into a series of plane waves, as shown in Fig. 9.3a.

A plane wave is distinguished by its wave vector. Using rotation matrices (9.8) and (9.9), a wave vector represented in the source and reference coordinates can be converted into each other like position vectors, as shown in Fig. 9.3b.

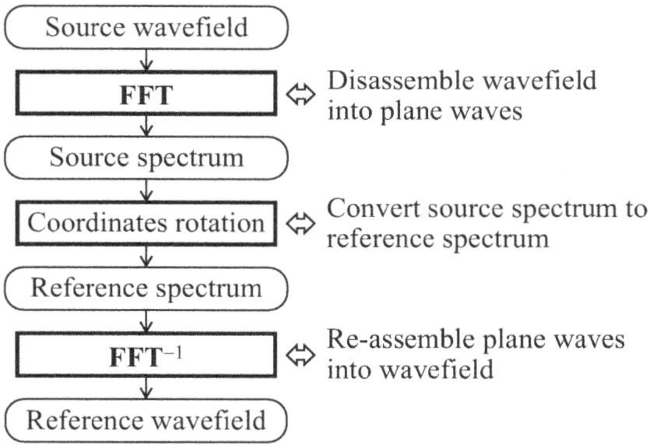

Fig. 9.4 Basic procedure and physical interpretation of the rotational transform

$$\mathbf{k} = \mathbf{T}\mathbf{k}_s, \tag{9.13}$$

$$\mathbf{k}_s = \mathbf{T}^{-1}\mathbf{k}. \tag{9.14}$$

In the reference coordinates, the wave vector is also associated with Fourier frequencies.

$$\mathbf{k} = 2\pi(u, v, w). \tag{9.15}$$

Therefore, substituting the rotation matrix in (9.9) into (9.13), we get

$$\begin{aligned} u &= u(u_s, v_s) = a_1 u_s + a_4 v_s + a_7 w_s, \\ v &= v(u_s, v_s) = a_2 u_s + a_5 v_s + a_8 w_s. \end{aligned} \tag{9.16}$$

Using inverse rotation matrix (9.8), we also get

$$\begin{aligned} u_s &= u_s(u, v) = a_1 u + a_2 v + a_3 w, \\ v_s &= v_s(u, v) = a_4 u + a_5 v + a_6 w. \end{aligned} \tag{9.17}$$

Here, note that both frequencies w and w_s with respect to z and z_s are obtained from (5.7) in each coordinate system as follows:

$$w = w(u, v) = \sqrt{\lambda^{-2} - u^2 - v^2}, \tag{9.18}$$

$$w_s = w_s(u_s, v_s) = \sqrt{\lambda^{-2} - u_s^2 - v_s^2}. \tag{9.19}$$

Equations (9.16) and (9.17) give mapping between Fourier frequencies (u_s, v_s) in the source plane and (u, v) in the reference plane. Therefore, we can obtain Fourier spectrum $G(u, v; 0)$ in the reference plane from $G_s(u_s, v_s; 0)$ as follows:

$$G(u, v; 0) = G_s(a_1 u + a_2 v + a_3 w,\ a_4 u + a_5 v + a_6 w;\ 0). \tag{9.20}$$

In summary, the rotational transform is performed by the following procedure; we first calculate Fourier spectrum of the source wavefield, and then, map the source spectrum onto the reference spectrum. Finally, the wavefield in the reference plane is calculated by the inverse Fourier transform of the reference spectrum. The basic procedure and its physical interpretation are shown in Fig. 9.4.

Here, it should be noted that the mapping between the source and reference spectrum is non-linear. Thus, some technique of interpolation is necessary to perform the rotational transform using FFT in practice. The exact formulation is given in the following section.

9.4 Formulation

9.4.1 General Formulation

In the source coordinate system, the angular spectrum method in Sect. 5.2.1 gives the wavefield at arbitrary z position. This means that we can obtain the complex amplitude of the field at any position. The value of the complex amplitude at a position, which is represented by the source coordinates, is given by the inverse Fourier transform of (5.16) as follows:

$$\begin{aligned} g_s(x_s, y_s; z_s) &= \mathscr{F}^{-1}\{G_s(u_s, v_s; z_s)\} \\ G_s(u_s, v_s; z_s) &= \begin{cases} G_s(u_s, v_s; 0) \exp[i2\pi w_s(u_s, v_s) z_s] & \text{if } u_s^2 + v_s^2 \le \lambda^{-2} \\ 0 & \text{otherwise} \end{cases} \end{aligned} \tag{9.21}$$

When considering the case $u_s^2 + v_s^2 \le \lambda^{-2}$, the Fourier integral is written by

$$g_s(x_s, y_s; z_s) = \iint_{-\infty}^{\infty} G_s(u_s, v_s; 0) \exp[i2\pi\{u_s x_s + v_s y_s + w_s z_s\}] du_s dv_s. \tag{9.22}$$

This can be represented by a vector form;

$$g_s(\mathbf{r}_s) = \iint G_s(u_s, v_s; 0) \exp[i\mathbf{k}_s \cdot \mathbf{r}_s] du_s dv_s. \tag{9.23}$$

By use of coordinates rotation in (9.2), the complex amplitude at a position represented in the reference coordinates is written as

$$\begin{aligned} g(\mathbf{r}) &= g_s(\mathbf{T}^{-1}\mathbf{r}) \\ &= \iint G_s(u_s, v_s; 0) \exp\left[i\mathbf{k}_s \cdot (\mathbf{T}^{-1}\mathbf{r})\right] du_s dv_s. \end{aligned} \tag{9.24}$$

Substituting (9.14) into above,

$$g(\mathbf{r}) = \iint G_s(u_s, v_s; 0) \exp\left[i(\mathbf{T}^{-1}\mathbf{k}) \cdot (\mathbf{T}^{-1}\mathbf{r})\right] du_s dv_s. \tag{9.25}$$

According to (9.7), coordinates rotation preserve the dot product; $(\mathbf{T}^{-1}\mathbf{k}) \cdot (\mathbf{T}^{-1}\mathbf{r}) = \mathbf{k}_s \cdot \mathbf{r}_s = \mathbf{k} \cdot \mathbf{r}$. This is because coordinates rotation does not change the length and relative angles between vectors. Therefore, the complex amplitude in the reference coordinates is rewritten by

$$g(\mathbf{r}) = \iint G_s(u_s, v_s; 0) \exp\left[i\mathbf{k} \cdot \mathbf{r}\right] du_s dv_s. \tag{9.26}$$

Because we are interested in the wavefield in $(x, y, z = 0)$ plane, this should be represented as

$$g(x, y; 0) = \iint G_s(u_s, v_s; 0) \exp[i2\pi(ux + vy)] du_s dv_s. \tag{9.27}$$

Change of variables from u_s and v_s to u and v in the integration can be achieved by substituting (9.17) and $du_s dv_s = \det J(u, v) du dv$ as follows:

$$\begin{aligned} g(x, y; 0) = \iint & G_s(a_1 u + a_2 v + a_3 w, a_4 u + a_5 v + a_6 w; 0) \\ & \times \exp[i2\pi(ux + vy)] \det J(u, v) du dv, \end{aligned} \tag{9.28}$$

where $\det J(u, v)$ is a Jacobian given by

$$\det J(u, v) = \frac{\partial u_s}{\partial u}\frac{\partial v_s}{\partial v} - \frac{\partial u_s}{\partial v}\frac{\partial v_s}{\partial u}. \tag{9.29}$$

As a result, general formulation of the rotational transform is summarized as follows:

$$g(x, y; 0) = \mathscr{F}^{-1}\left\{G(u, v) \det J(u, v)\right\}, \tag{9.30}$$

$$G(u, v) = G_s(a_1 u + a_2 v + a_3 w(u, v), a_4 u + a_5 v + a_6 w(u, v)), \tag{9.31}$$

$$\det J(u, v) = \frac{(a_2 a_6 - a_3 a_5)u}{w(u, v)} + \frac{(a_3 a_4 - a_1 a_6)v}{w(u, v)} + (a_1 a_5 - a_2 a_4), \tag{9.32}$$

$$G_s(u_s, v_s) = \mathscr{F}\left\{g_s(x_s, y_s; 0)\right\}. \tag{9.33}$$

It is worth emphasizing that $g(x, y; 0)$ given by (9.30)–(9.33) is a complete solution of the Helmholtz equation. In addition, only a double Fourier transform is needed to perform this transformation.

9.4.2 Paraxial Approximation

The Jacobian in (9.32) approximates to a constant, and therefore, can be ignored in cases where the field is paraxial in the source or reference coordinates. The paraxial approximation of the rotational transform is briefly mentioned in this section.

Paraxial Field in Reference Coordinates

When fields are paraxial in the reference coordinates, i.e., the field after rotational transform approximately propagates along the z direction of the reference coordinates, u and v are much smaller than $w(u, v)$. As a result, the first and the second term of the Jacobian in (9.32) can be ignored. In that case,

$$\det J(u, v) \simeq (a_1 a_5 - a_2 a_4) \tag{9.34}$$

is a good approximation.

Paraxial Field in Source Coordinates

In contrast, when the waves are paraxial in the source coordinates, the Jacobian remaining after the approximation is a little complex, but all terms that include the Fourier frequencies can be ignored.

When the inverse rotation matrix $\mathbf{T}^{-1}$ is primarily defined by (9.8), suppose that the rotation matrix is given by

$$\mathbf{T} = \frac{1}{\det\left(\mathbf{T}^{-1}\right)} \begin{bmatrix} A_1 & A_2 & A_3 \\ A_4 & A_5 & A_6 \\ A_7 & A_8 & A_9 \end{bmatrix} = \begin{bmatrix} A_1 & A_2 & A_3 \\ A_4 & A_5 & A_6 \\ A_7 & A_8 & A_9 \end{bmatrix}. \tag{9.35}$$

Here, A_i is a cofactor of the matrix $\mathbf{T}^{-1}$, and A_i's in the third column are associated with the matrix $\mathbf{T}^{-1}$ by

$$A_3 = (-1)^{(1+3)} \det \begin{bmatrix} a_2 & a_3 \\ a_5 & a_6 \end{bmatrix} = a_2 a_6 - a_3 a_5, \tag{9.36}$$

$$A_6 = (-1)^{(3+2)} \det \begin{bmatrix} a_1 & a_3 \\ a_4 & a_6 \end{bmatrix} = a_3 a_4 - a_1 a_6, \tag{9.37}$$

$$A_9 = (-1)^{(3+3)} \det \begin{bmatrix} a_1 & a_2 \\ a_4 & a_5 \end{bmatrix} = a_1 a_5 - a_2 a_4. \tag{9.38}$$

Thus, the Jacobian (9.32) is rewritten by using the cofactors:

$$\det J(u, v) = A_3 \frac{u}{w} + A_6 \frac{v}{w} + A_9. \tag{9.39}$$

Moreover, the first term of the Jacobian is rewritten by transforming frequencies (u, v, w) into (u_s, v_s, w_s) through the matrix $\mathbf{T}$:

$$A_3 \frac{u}{w} = A_3 \frac{A_1 u_s + A_2 v_s + A_3 w_s}{A_7 u_s + A_8 v_s + A_9 w_s}. \tag{9.40}$$

If a field is paraxial in the source coordinate, i.e., it propagates in almost the z direction, frequency w_s is much larger than u_s and v_s. Therefore,

$$A_3 \frac{u}{w} = A_3 \frac{A_1 u_s/w_s + A_2 v_s/w_s + A_3}{A_7 u_s/w_s + A_8 v_s/w_s + A_9} \simeq \frac{A_3^2}{A_9} \tag{9.41}$$

is a good approximation. Applying the same procedure, the second term of the Jacobian (9.39) is approximated to

$$A_6 \frac{v}{w} \simeq \frac{A_6^2}{A_9}. \tag{9.42}$$

As a result, when fields are paraxial in the source coordinates, the paraxial approximation of coordinates rotation is given by

$$\det J(u, v) \simeq \frac{A_3^2 + A_6^2 + A_9^2}{A_9}. \tag{9.43}$$

As a result, the Jacobian is also well approximated by a constant in the paraxial source field.

9.5 Numerical Procedure

Implementation of the rotational transform given by (9.30)–(9.33) is not so easy, because coordinates rotation distorts the sampling grid of $G(u, v)$ in (9.31).

9.5.1 Sampling Distortion

Suppose that the input data, i.e., the source field, is represented by $g_s[m, n; 0] = g_s(x_{s,m}, y_{s,n}; 0)$ ($m = 0, \ldots, M-1$ and $n = 0, \ldots, N-1$), where the sampling manner in (6.1) is used as usual. The first step of the rotational transform represented by (9.33) is to calculate the source spectrum by FFT.

$$\begin{aligned} G_s[p, q] &= G_s(u_{s,p}, v_{s,q}) \\ &= \mathbf{FFT}\{g_s[m, n; 0]\}. \end{aligned} \tag{9.44}$$

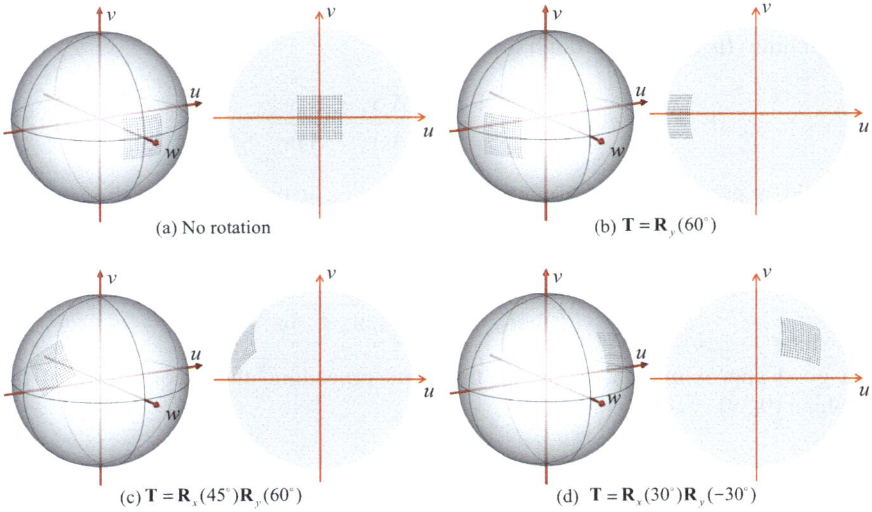

Fig. 9.5 Examples of sampling distortion of $G(u, v)$ in various transform matrices. The right figure shows sample points in (u, v) coordinates, while the left shows sample points indicated in Ewald's sphere of Fig. 5.6

Here, the sampling position $(u_{s,p}, v_{s,q})$ is given by (6.2), and $\Delta u_s = (N\Delta x_s)^{-1}$ and $\Delta v_s = (M\Delta y_s)^{-1}$. Therefore, the source sampling window of the Fourier spectrum has a rectangular shape in the Fourier space. The right figure of Fig. 9.5a schematically shows the sample points of $G_s[p, q]$. More precisely, the sample points are given in the surface of the Ewald's sphere by (5.7), as in the left of (a).

In the reference coordinates (u, v), using (9.16), the sampling position is given by

$$\begin{aligned} u &= u[p, q] = a_1 u_{s,p} + a_4 v_{s,q} + a_7 w_s(u_{s,p}, v_{s,q}), \\ v &= v[p, q] = a_2 u_{s,p} + a_5 v_{s,q} + a_8 w_s(u_{s,p}, v_{s,q}). \end{aligned} \tag{9.45}$$

The sampled spectrum in the reference coordinates is simply represented by $G(u, v) = G[p, q]$ with (9.45). It should be noted that the coordinates rotation rotates the Ewald's sphere, and the sample points in the (u, v) coordinates are regarded as orthogonal projection of the sample points in the rotated spherical shell onto the $(u, v, 0)$ plane. Therefore, the rectangular sampling window gets distorted severely in the reference coordinates (u, v). Figure 9.5b–d schematically shows examples of rotation of the Ewald's sphere and distortion of the reference sampling window.

To obtain the wavefield in the reference plane, it is necessary to Fourier-transform the reference spectrum $G(u, v)$, as in (9.31). However, the sample points are no longer equally spaced in the reference coordinates. Besides, the sampling window is placed far from the origin in many cases. As a result, it is very difficult to carry out FFT of $G[p, q]$ effectively.

9.5.2 Shifted Fourier Coordinates

When the source field travels nearly along the z_s-axis of the source coordinates, as shown in Fig. 9.6a, the angular spectrum is distributed around the origin of the source Fourier coordinates; $(u_s, v_s) = (0, 0)$. Thus, FFT works well for the source field. However, when the same field is observed in the reference coordinates, the field propagates in a direction angled to the reference z-axis. This means that the field has a carrier frequency in the reference coordinates.

As shown in Fig. 9.6b, coordinates rotation projects the origin in the source coordinates to a point far from the origin of the reference coordinates. This reflects the carrier wavefield observed in the reference coordinates. Substituting $u_s = v_s = 0$ into (9.16), we can easily find the carrier frequency.

$$u_c = u(0,0) = \frac{a_7}{\lambda} \quad \text{and} \quad v_c = v(0,0) = \frac{a_8}{\lambda}. \tag{9.46}$$

As a result, the reference field includes high frequency components in many cases of the rotational transform. In this case, the inverse FFT corresponding to (9.30) is very ineffective, because FFT works in the symmetrical sampling. As shown in Fig. 9.6b, a large sampling window required for the inverse FFT may lead to large computational effort in the numerical calculation.

To avoid the problem, the reference coordinates should be shifted so as to cancel the carrier offset and project the origin $(u_s, v_s) = (0, 0)$ to the origin of reference Fourier coordinates again. Thus, we introduce the following shift into the reference coordinates:

$$(\hat{u}, \hat{v}) = (u - u_c, v - v_c). \tag{9.47}$$

We call $(\hat{u}, \hat{v})$ the shifted Fourier coordinates. Figure 9.6c shows the sampling position of the reference spectrum indicated in the shifted Fourier coordinates.

By substituting $u = \hat{u} + u_c$ and $v = \hat{v} + v_c$, (9.30) is rewritten by

$$\begin{aligned} g(x, y; 0) &= \mathscr{F}^{-1}\left\{G(\hat{u}, \hat{v}) \det J(\hat{u}, \hat{v})\right\} \exp\left[i2\pi(u_c x + v_c y)\right] \\ &= \hat{g}(x, y; 0) \exp\left[i2\pi(u_c x + v_c y)\right], \end{aligned} \tag{9.48}$$

where shift theorem of the Fourier transform shown in Table 4.1 is used. The amplitude distribution $|g(x, y; 0)|$ calculated from (9.48) agrees with that by (9.30) of the original formulation. $\hat{g}(x, y; 0)$ is also the reference wavefield, but the carrier offset is removed:

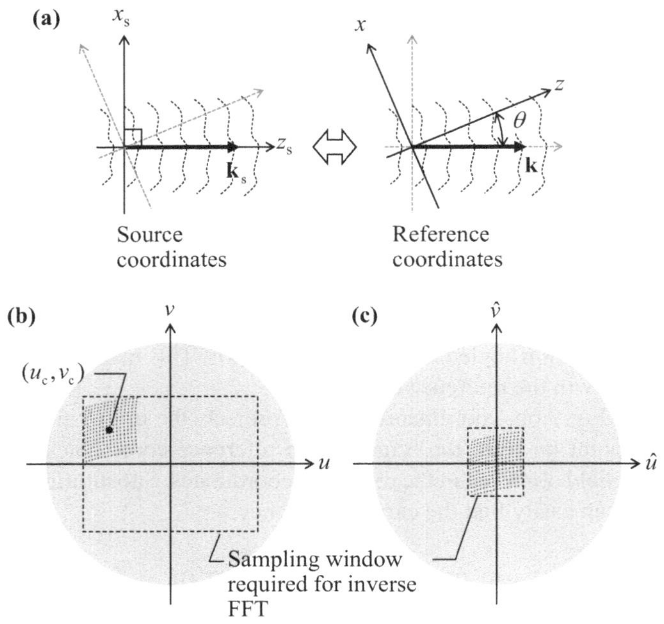

Fig. 9.6 **a** Generation of the carrier wavefield in the reference coordinates. **b** Sampling distortion by transform matrix $\mathbf{T} = \mathbf{R}_x(20°)\mathbf{R}_y(30°)$. **c** Sampling positions in the shifted Fourier coordinates introduced to cancel the carrier offset

$$\begin{aligned}\hat{g}(x, y; 0) &= \mathscr{F}^{-1}\left\{G(\hat{u}, \hat{v}) \det J(\hat{u}, \hat{v})\right\} \\ &= g(x, y; 0) \exp\left[-i2\pi(u_c x + v_c y)\right]. \end{aligned} \tag{9.49}$$

The introduction of shifted Fourier coordinates makes computation much easier in cases where the carrier is not required in the rotational transform for a given purpose. Furthermore, if the carrier field is necessary, we can easily add it by multiplying $\exp[i2\pi(u_c x + v_c y)]$ as in (9.48).

9.5.3 Actual Procedure to Perform the Rotational Transform

The source spectrum $G_s[p', q'] = G_s(u_{s,p'}, v_{s,q'})$ is first calculated using FFT as in (9.44). Here, note that we use indexes p' and q' instead of p and q in the source spectrum in order to distinguish the indexes in the reference spectrum. Figure 9.7a shows an example of sample points and the sampling window of the source spectrum. The spectrum has equally-spaced sample points. However, the sample points projected to the reference coordinates are spaced non-equidistantly even in the shifted Fourier coordinates, as mentioned in the preceding section.

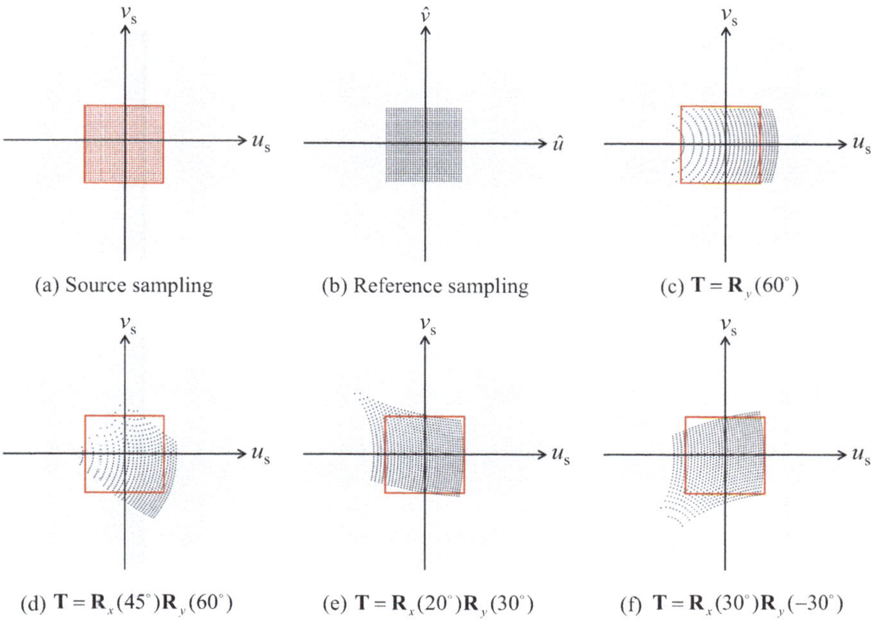

Fig. 9.7 **a** Sample points of the source spectrum, and **b** Sample points of the reference spectrum in the shifted Fourier coordinates. $\Delta x = \Delta y = \Delta x_s = \Delta y_s = 1\,[\mu\text{m}]$, $M = N = 32$, $\lambda = 633\,[\text{nm}]$. **c–f** Resample points in the source coordinates, which are given by projecting the reference sample points onto the source coordinates. The red rectangle is the sampling window of the source spectrum

To implement the rotational transform in software, let us change the point of view and invert the procedure. Our final goal is to calculate wavefield $g(x, y; 0)$ through $\hat{g}(x, y; 0)$ in (9.49), which is uniformly sampled in the reference plane; $\hat{g}[m, n; 0] = \hat{g}(x_m, y_n; 0)$. Thus, (4.91) is used for the sampling manner. To calculate $\hat{g}[m, n; 0]$, the sampled spectrum $G(\hat{u}_p, \hat{v}_q)$ in the shifted Fourier coordinates must also have equally-spaced sample points.

In other word, to calculate equally sampled $g(x_m, y_n; 0)$, it is required that the sampling position $(\hat{u}_p, \hat{v}_q)$ in the shifted Fourier coordinates is given by the same manner as that in (4.92);

$$\hat{u}_p = \left(p - \frac{M}{2}\right)\Delta u, \quad (p = 0, \ldots, M-1)$$
$$\hat{v}_q = \left(q - \frac{N}{2}\right)\Delta v, \quad (q = 0, \ldots, N-1) \tag{9.50}$$

where $\Delta u = (N\Delta x)^{-1}$ and $\Delta v = (M\Delta y)^{-1}$ and Δx and Δy are sampling intervals of the reference field. Figure 9.7b shows sample points in the shifted Fourier coordinates.

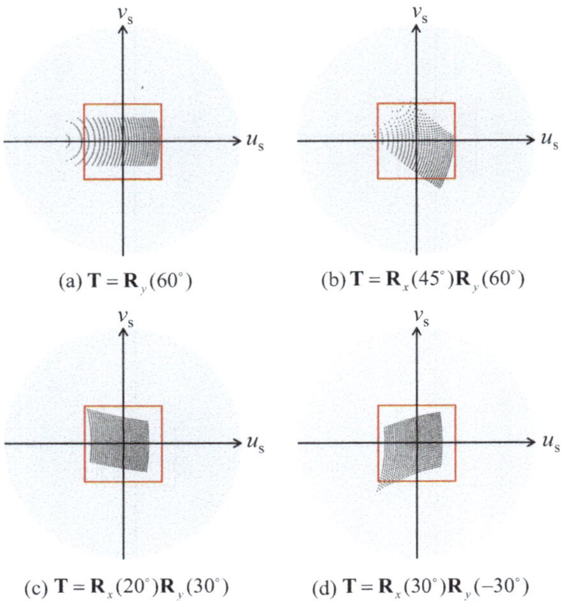

Fig. 9.8 Examples of resample points in the source coordinates for various rotation matrices. $\Delta x = \Delta y = 1.5\,[\mu\text{m}]$, $\Delta x_s = \Delta y_s = 1\,[\mu\text{m}]$, $M = N = 32$, $\lambda = 633\,[\text{nm}]$

When the sampling intervals Δx and Δy are very small, the sampling position $(\hat{u}_p, \hat{v}_q)$ may violate the condition:

$$\hat{u}_p^2 + \hat{v}_q^2 \leq \lambda^{-2}. \tag{9.51}$$

In this case, the sample point is not in the Ewald's spherical shell and the component generates an evanescent wave, as mentioned in Sect. 5.2.1.2. Thus, the sample point that does not satisfy the above condition should not be evaluated.

The sampling position (u_p, v_q) in the reference (non-shifted) Fourier coordinates is obtained by

$$u_p = \hat{u}_p + u_c \quad \text{and} \quad v_q = \hat{v}_q + v_c. \tag{9.52}$$

It is also possible that the sampling position (u_p, v_q) does not satisfy the same condition as that in (9.51). Thus, we should test the condition $u_p^2 + v_q^2 \leq \lambda^{-2}$ again and discard sample points that violate the condition.

Using (9.17), we can find the sampling position in the source coordinates corresponding to $(\hat{u}_p, \hat{v}_q)$ as follows:

$$\begin{aligned} u_s[p, q] &= u_s\left(\hat{u}_p + u_c,\ \hat{v}_q + v_c\right), \\ v_s[p, q] &= v_s\left(\hat{u}_p + u_c,\ \hat{v}_q + v_c\right). \end{aligned} \tag{9.53}$$

Algorithm 9.1 Actual procedure for the rotational transform of wavefields

Require: Functions $u_s(u, v)$ and $v_s(u, v)$ are defined for given rotation matrix **T**
$G_s[p', q'] = \mathbf{FFT}\{g_s[m, n]\}$ // Execute FFT of source wavefield
$u_c \Leftarrow u(0, 0)$ // Calculate carrier frequency
$v_c \Leftarrow v(0, 0)$
for $q = 0$ **to** $N - 1$ **do**
  for $p = 0$ **to** $M - 1$ **do**
    $\hat{u}_p \Leftarrow (p - M/2)\Delta u$
    $\hat{v}_q \Leftarrow (q - N/2)\Delta v$
    if $\hat{u}_p^2 + \hat{v}_q^2 \leq \lambda^{-2}$ **and** $(\hat{u}_p + u_c)^2 + (\hat{v}_q + v_c)^2 \leq \lambda^{-2}$ **then**
      $u_s[p, q] \Leftarrow u_s\left(\hat{u}_p + u_c,\ \hat{v}_q + v_c\right)$
      $v_s[p, q] \Leftarrow v_s\left(\hat{u}_p + u_c,\ \hat{v}_q + v_c\right)$
      get value of $G_s(u_s[p, q], v_s[p, q])$ from source spectrum $G_s[p', q']$ by resampling
      $G[p, q] \Leftarrow G_s(u_s[p, q], v_s[p, q])$
    else
      $G[p, q] \Leftarrow 0$
    end if
  end for
end for
$\hat{g}[m, n] = \mathbf{FFT}^{-1}\{G[p, q]\}$ // Execute inverse FFT of reference spectrum

Examples of sample points in the source coordinates are shown in Fig. 9.7c–f for various rotations. Because of sampling distortion, the sample points given by (9.53) are not spaced equidistantly. The sampling window also does not have a rectangular shape, but the position nearly overlaps the rectangular sampling window of the source spectrum. Therefore, we can obtain the value of $G[p, q] = G_s(u_s[p, q], v_s[p, q])$ from the data array of the uniformly sampled spectrum $G_s[p', q'] = G_s(u_{s,p'}, v_{s,q'})$ by use of some interpolation technique.

Here, we have some freedom to choose sampling intervals of the reference field. Figure 9.8 also shows resample points in the source coordinates for several rotation matrices, but the sampling intervals of the reference field is set to 3/2 times larger than that of the source field. In this case, the sampling interval of the spectrum reduces, and almost all resample point is included inside the source sampling window.

Finally, wavefield $\hat{g}(x_m, y_n; 0)$ in the reference plane is obtained by inverse FFT of $G[p, q]$ as follows.

$$\begin{aligned}\hat{g}(x_m, y_n; 0) &= \hat{g}[m, n] \\ &= \mathbf{FFT}^{-1}\{G[p, q]\}. \end{aligned} \tag{9.54}$$

The procedure for rotational transform described above is summarized in Algorithm 9.1 as a pseudo-code.

9.5.4 Resample of Uniformly Sampled Spectrum

As mentioned in the preceding section, we have to obtain the value of the Fourier spectrum at an arbitrary frequency position from the uniformly sampled source spectrum, i.e., interpolate between sample points of the source spectrum $G_s[p', q']$.

One of the most important interpolation technique is the sinc interpolation given in (4.63) of Sect. 4.5. Let us present the basic formula again for the reader's convenience.

$$f(x) = \frac{1}{\Delta x} \sum_{m=-\infty}^{+\infty} f(m\Delta x) \operatorname{sinc}\left(\frac{x - m\Delta x}{\Delta x}\right). \tag{9.55}$$

$f(m\Delta x)$ is a sampled value and Δx is the sampling interval. Since spatially unlimited interpolation is impossible, the infinite summation is limited to the sampling range in practice. This sinc interpolation is the best way of all according to the sampling theorem, but it is too time-consuming to use in practical computation.

On the analogy of (9.55), interpolation of a 2D function is generally written as,

$$f(x, y) = \sum_{n=n_0}^{n_0+N_i-1} \sum_{m=m_0}^{m_0+N_i-1} f(x_m, y_n) w\left(\frac{x - x_m}{\Delta x}\right) w\left(\frac{y - y_n}{\Delta y}\right), \tag{9.56}$$

where function $w(\xi)$ is called *interpolation kernel* in general or simply *interpolator* when $w(\xi)$ satisfies the following.

$$w(\xi) = \begin{cases} 1 & \xi = 0, \\ 0 & |\xi| = 1, 2, 3, \dots \end{cases} \tag{9.57}$$

This property of an interpolator is required not to modify the value of $f(x, y)$ if it is resampled on the sample point. The sinc function, defined in (4.22), definitely agrees with (9.57). N_i is called a kernel size and denotes the number of points that the kernel supports.

In uniformly sampled 2D functions such as a digital image, the most famous and useful interpolation technique is *bicubic* interpolation. The interpolator is given as follows [49]:

$$w_{\text{bicubic}}(\xi) = \begin{cases} (a+2)|\xi|^3 - (a+3)|\xi|^2 + 1 & \text{for } |\xi| \le 1, \\ a|\xi|^3 - 5a|\xi|^2 + 8a|\xi| - 4a & \text{for } 1 < |\xi| < 2, \\ 0 & \text{otherwise}, \end{cases} \tag{9.58}$$

where a is usually set to -0.5 or -0.75. Since $N_i = 4$ in this interpolation, the value at a position (x, y) is evaluated by 4×4 sample points around the position (x, y), as shown in Fig. 9.9.

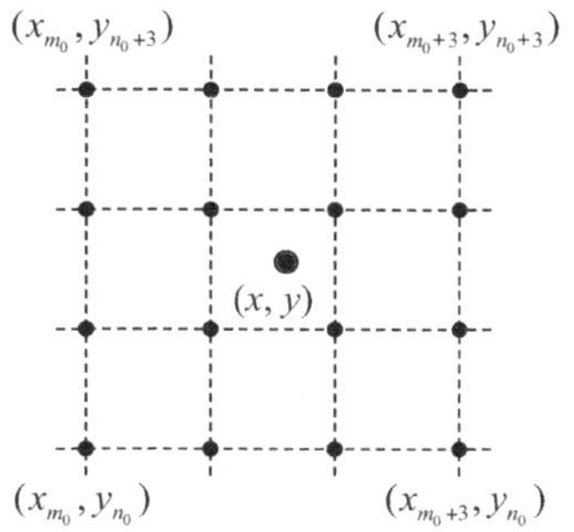

Fig. 9.9 4 × 4 sample points used for bicubic interpolation

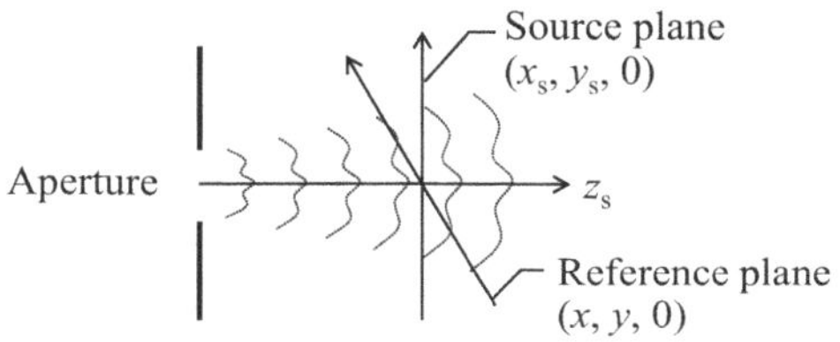

Fig. 9.10 Setup for numerical simulation of rotational transform

When the bicubic interpolation is adopted for the rotational transform, the resample operation in line 11–12 of Algorithm 9.1 is represented by

$$G[p,q] = \sum_{p'=p_0}^{p_0+3} \sum_{q'=q_0}^{q_0+3} G_s[p',q'] w_{\text{bicubic}}\left(\frac{u_s[p,q]-u_{s,p'}}{\Delta u_s}\right) w_{\text{bicubic}}\left(\frac{v_s[p,q]-v_{s,q'}}{\Delta v_s}\right), \tag{9.59}$$

where p_0 and q_0 are chosen so as to surround the position $(u_s[p,q], v_s[p,q])$ with 4 × 4 sample points of $G_s[p',q']$.

Here, note that there are many interpolation techniques in science and engineering. The bicubic interpolation is just one of them. Better techniques are usually more time-consuming, i.e., there is a trade-off between precision and computational effort. We should choose better one for a given purpose.

9.6 Numerical Examples and Errors

9.6.1 Edge Effect and Sampling Overlap

Figure 9.10 shows the setup for numerical experiments of the rotational transform. We calculate wavefields diffracted by an aperture using the band-limited angular spectrum method (BLAS) mentioned in Sect. 6.4. Then, the wavefield in the reference plane, which is tilted at an angle, is obtained from the diffracted field used as the source field.

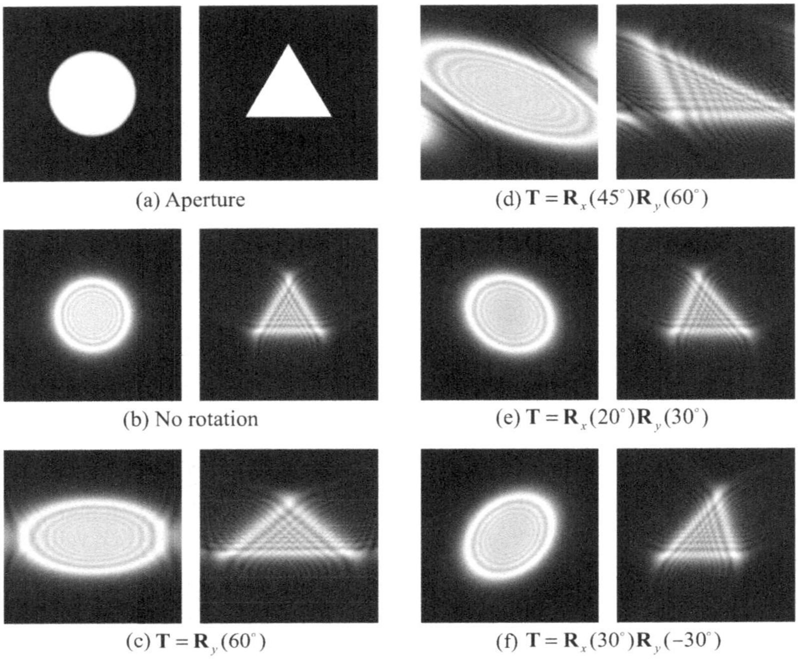

Fig. 9.11 Examples of the rotational transform of wavefields that are translationally propagated 5 mm from the aperture. Diffraction pattern is depicted as intensity image $|g(x, y)|^2$ in the tilted reference plane. $\Delta x = \Delta y = 1\,[\mu\text{m}]$, $\Delta x_s = \Delta y_s = 1\,[\mu\text{m}]$, $M = N = 1024$, $\lambda = 633\,[\text{nm}]$. The grayscale images are depicted with the standard encoding gamma of 1/2.2

Since we use an interpolation in the rotational transform, there is freedom to choose the sampling number and interval of reference fields. The diffraction patterns $|g(x, y; 0)|^2$ calculated in the reference plane are shown in Fig. 9.11. Here, we set the same sampling intervals and numbers as those in the source field and use two apertures of a circle with diameter of 0.5 mm and an equilateral triangle of a side length 0.5 mm.

The results seem to be reasonable in (e) and (f), where the rotation angle is relatively small. However, in large rotation angles as in (c) and (d), strange light appears in the edge of the sampling window. This is most likely an effect of field invasion, i.e., an edge effect, mentioned in Sect. 6.4.3. The field invasion occurs when the field run over the edge of the sampling window, because the field, computed using the discrete Fourier transform or FFT, always has a periodic structure. Figure 9.12 shows the results of rotational transform of Fig. 9.11 again, but two images are tiled without gap. It is verified that the images connect seamlessly. This means that the neighboring fields invade and interfere with each other.

One possible technique to avoid the field invasion is to expand the reference sampling window by changing the sampling intervals. Figure 9.13a–c shows the

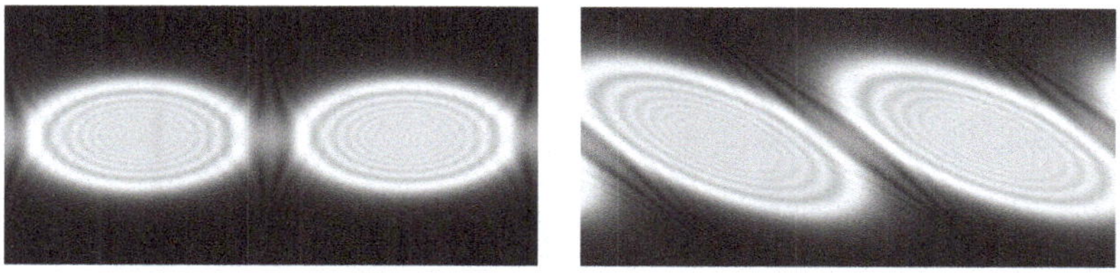

(a) $\mathbf{T} = \mathbf{R}_y(60^\circ)$ (b) $\mathbf{T} = \mathbf{R}_x(45^\circ)\mathbf{R}_y(60^\circ)$

Fig. 9.12 Tiling of the intensity images of the circular aperture in Fig. 9.11

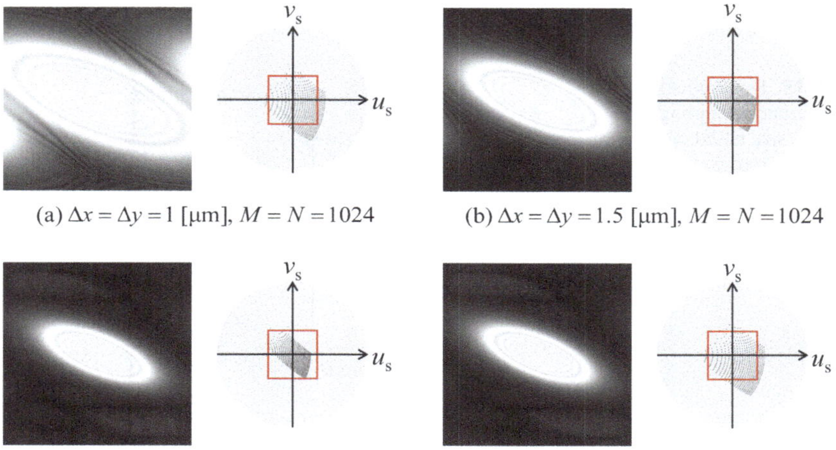

(a) $\Delta x = \Delta y = 1$ [μm], $M = N = 1024$ (b) $\Delta x = \Delta y = 1.5$ [μm], $M = N = 1024$

(c) $\Delta x = \Delta y = 2$ [μm], $M = N = 1024$ (d) $\Delta x = \Delta y = 1$ [μm], $M = N = 2048$

Fig. 9.13 Amplitude images of reference field $|g(x, y; 0)|$ and resample points in the source spectrum in different sampling parameters of the reference field. The red rectangles indicate the source sampling window. $\mathbf{T} = \mathbf{R}_x(45^\circ)\mathbf{R}_y(60^\circ)$

reference field calculated in the same number of sample points but different reference sampling intervals. Here, the amplitude images are depicted instead of intensity to make the errors clear. Larger sampling intervals definitely give better results. This is because the the reference sampling window more extends as increasing the sampling intervals, and thus the reference field less invades the neighboring period. Change of sampling intervals of the reference field also varies sampling overlap between the source and reference spectra. The reference sampling area does not fall within the source area when $\Delta x = \Delta y = 1$ [μm] as in Fig. 9.13a, whereas it almost fits within the source area when $\Delta x = \Delta y = 2$ [μm] as in (c).

Another technique to get over the field invasion is, of course, the sampling extension mentioned in Sect. 4.7.3. The result when using the quadruple extension, i.e., doubling the number of sampling both in x and y coordinates, is shown in Fig. 9.13d. The calculated reference field is almost the same as that in (c), but the computation

(a) $\mathbf{T} = \mathbf{R}_y(60^\circ)$ (b) $\mathbf{T} = \mathbf{R}_x(45^\circ)\mathbf{R}_y(60^\circ)$

Fig. 9.14 Intensity image $|g(x, y; 0)|^2$ calculated in the tilted reference plane using the quadruple extension. The parameters are the same as those in Fig. 9.11. Note that the center $M \times N$ points are clipped out from $2M \times 2N$ points of the reference wavefield after the rotational transform

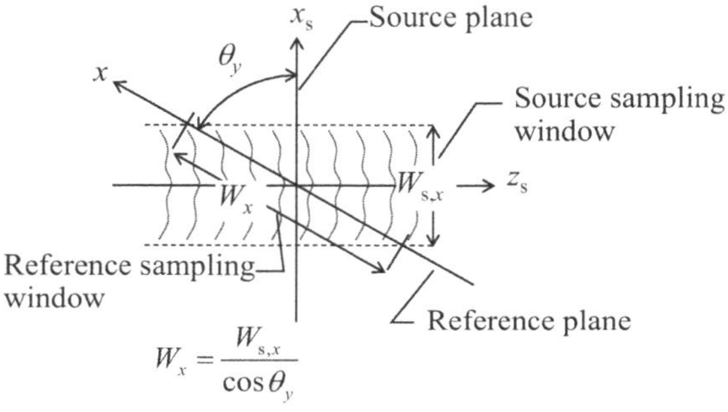

Fig. 9.15 Determination of the reference sampling window size based on simple geometry in paraxial source fields

cost is four times larger. However, we can get the reference field having the same sampling intervals as those of the source field in this case. Here, note that degree of the sampling overlap does not change by the sampling extension. Figure 9.14 shows the intensity images calculated using the quadruple extension in large rotation angles.

In summary, setting an appropriate reference sampling window, which covers the whole field in the reference plane, is most important to obtain good results from the rotational transform. This problem is sometimes resolved by simple geometry especially for paraxial source fields, as shown in Fig. 9.15.

9.6.2 The Rotational Transform with Carrier Offset

Examples of intensity or amplitude images of reference fields are only shown in the preceding section, i.e., examples of $|g(x, y; 0)|^2 = |\mathcal{F}^{-1}\left\{G(\hat{u}, \hat{v})|J(\hat{u}, \hat{v})|\right\}|^2$ are presented so far. However, wavefields accompanied with the phase component are often required in computer holography. Although the carrier offset must be considered when treating phase components in the rotational transform, as discussed in Sect. 9.5.2, we show examples of the phase pattern without the carrier offset at the beginning.

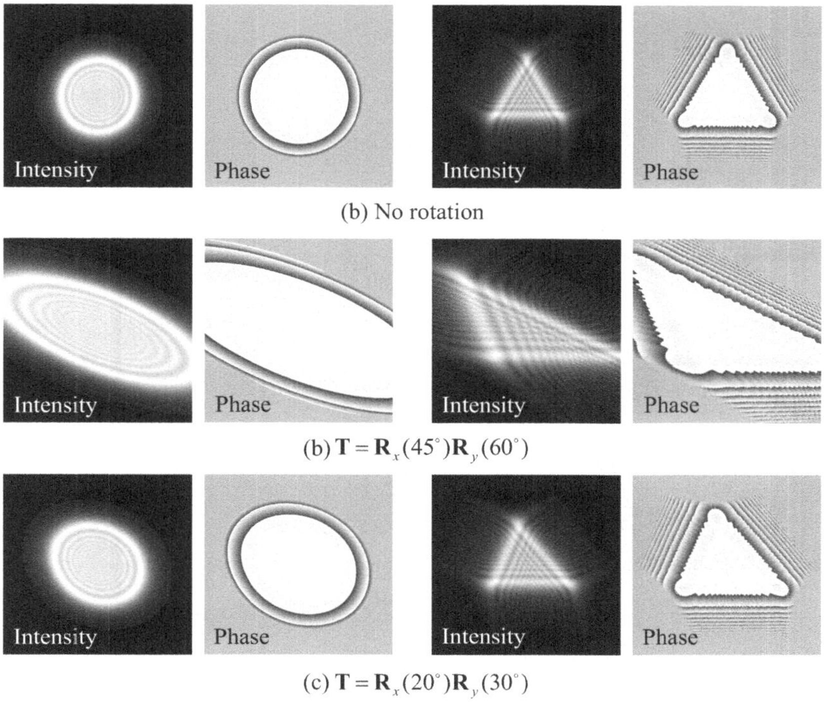

Fig. 9.16 Phase and intensity images of the reference fields without the carrier offset. The parameters are the same as those in Fig. 9.11, but the quadruple extension is used for the rotational transform. The grayscale images are depicted with the standard encoding gamma of 1/2.2

Figure 9.16 shows phase patterns of $\hat{g}(x, y; 0)$, i.e., phase images of the reference field where the carrier phase is canceled. To obtain entire reference wavefields, it is necessary to multiply the carrier phase $\exp[i2\pi(u_c x + v_c y)]$ as in (9.48). This carrier phase has the form of a plane wave. Therefore, we have to examine aliasing errors of the phase distribution carefully before multiplying it. According to the sampling theorem, $2|u_c| < \Delta x^{-1}$ and $2|v_c| < \Delta y^{-1}$ must be satisfied, thus, using (9.46), the aliasing-free condition is given by

$$|a_7| < \frac{\lambda}{2\Delta x} \quad \text{and} \quad |a_8| < \frac{\lambda}{2\Delta y}. \tag{9.60}$$

When calculating the rotational transform only around a single axis such as the x- or y-axis, substituting $\mathbf{R}_x(\theta_x)$ or $\mathbf{R}_y(\theta_y)$ in Table 9.1 into the above, the conditions are written as

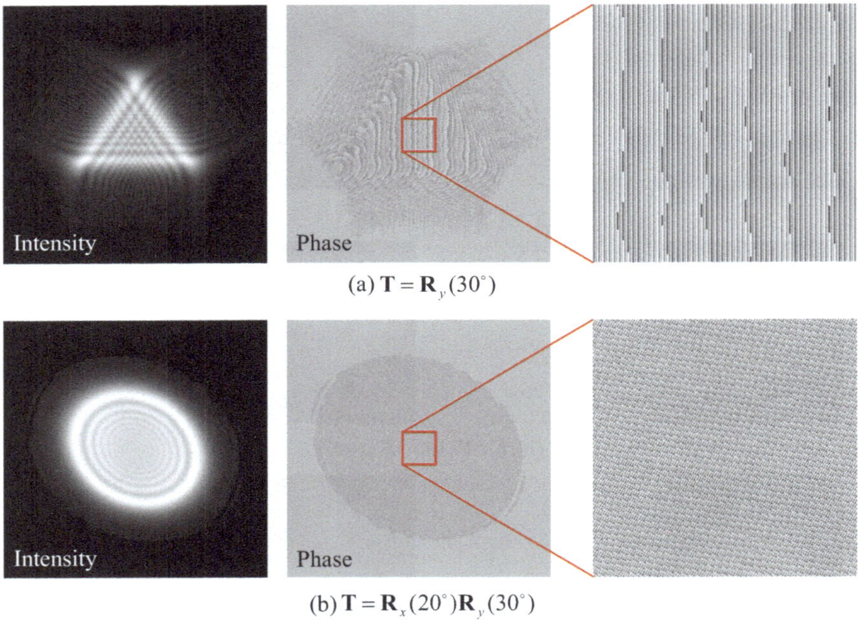

Fig. 9.17 Phase and intensity images of the reference fields with the carrier offset. The parameters are the same as those in Fig. 9.11, but $M = N = 2048$ and $\Delta x = \Delta y = 0.5$ [μm]. The grayscale images are depicted with the standard encoding gamma of 1/2.2

$$|\theta_x| < \sin^{-1}\frac{\lambda}{2\Delta x}, \qquad (x\text{-axis rotation only}) \tag{9.61}$$

$$|\theta_y| < \sin^{-1}\frac{\lambda}{2\Delta y}. \qquad (y\text{-axis rotation only}) \tag{9.62}$$

These exactly agree with the maximum diffraction angles given in (3.54). This is easy to understand by considering the rotational transform of a plane wave traveling along the z-axis.

In case of multiple rotation axes, condition (9.60) must be evaluated. For example, if $\mathbf{T} = \mathbf{R}_x(\theta_x)\mathbf{R}_y(\theta_y)$, the conditions are given by

$$|\sin\theta_y| < \frac{\lambda}{2\Delta x} \quad \text{and} \quad |\sin\theta_x\cos\theta_y| < \frac{\lambda}{2\Delta y}. \tag{9.63}$$

Figure 9.17 shows examples of phase images with the carrier offset. To avoid aliasing errors, sampling intervals of $\Delta x = \Delta y = 0.5$ [μm] are chosen in these examples.

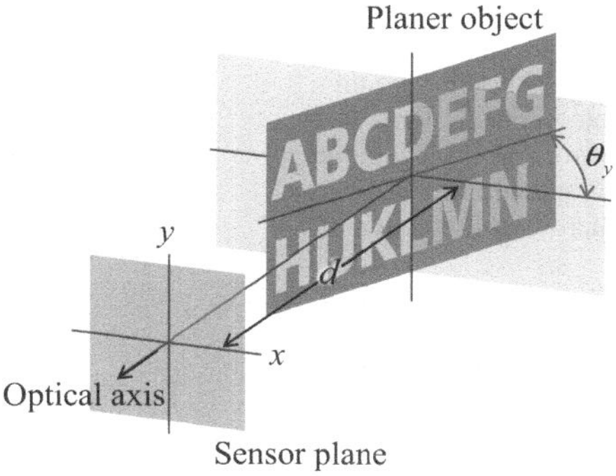

Fig. 9.18 Recording of the wavefield of an inclined planar subject

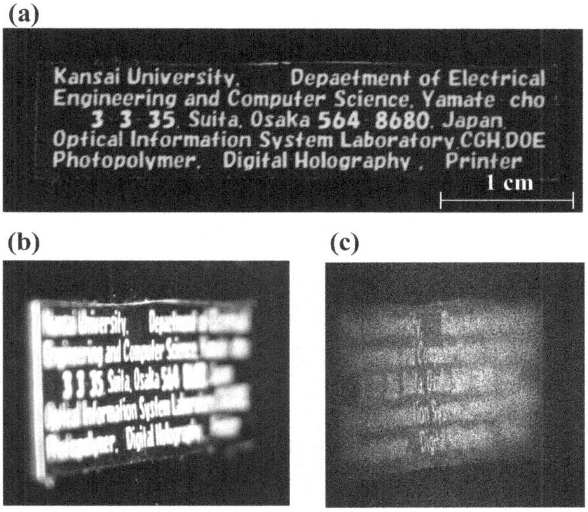

Fig. 9.19 Actual planar object used in the experiment [61]. **a** Photograph of the pattern printed on the planar object. **b** Close-up photograph taken from the direction of the optical axis. **c** Intensity image of translationally propagated wavefield $|g_s(x_s, y_s; 0)|^2$

9.6.3 Examples of the Rotational Transform in Practical Wavefield

The technology of digital holography (DH) allows us to capture physical wavefields digitally (see Sect. 14.2). Actual application of the rotational transform is introduced using DH in this section [61].

Consider a wavefield emitted by a plate object whose surface is not perpendicular to the optical axis, as shown in Fig. 9.18. A photograph of the surface pattern of the actual planar object used in the experiment is shown in Fig. 9.19a. The object was arranged with an angle of $\theta_y \approx 70°$. When we take a close-up photograph of the object using an ordinary lens camera placed in the direction of the optical axis, it is impossible to focus on everywhere of the object surface because of the limited depth of field of the camera. One of the close-up photographs is shown in Fig. 9.19b.

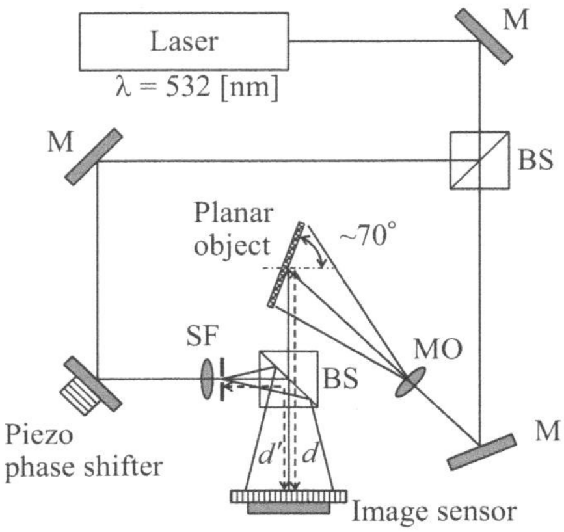

Fig. 9.20 Experimental setup for phase-shifting lensless-Fourier digital holography. L, lens; M, mirror; BS, beam splitter; SF, spatial filter; MO, microscope objective

Fig. 9.21 Amplitude image of the reference field $|g(x, y; 0)|$ after the rotational transform at $\theta_y = 71.8°$

Using DH, we can capture the wavefield of the planar object by an image sensor. Actual experimental setup is shown in Fig. 9.20. In this experiment, the wavefield of the object is captured by using the technique of lensless-Fourier DH. The detailed technique is explained in Sect. 14.2.2. In this technique, the wavefield is obtained at a position of d' distant from the sensor plane after the Fourier transform. Here, d' is the distance between the sensor and pin-hole of the spatial filter, as indicated in Fig. 9.20. Therefore, we get $g_s(x_s, y_s; 0)$ by translational propagation of a distance of $d' - d$.

Figure 9.19c shows the amplitude image $|g_s(x_s, y_s; 0)|$ after translational propagation by employing BLAS. The image is similar to the close-up photograph by the lens camera and not clear because of the limited depth of field. The amplitude image $|g(x, y; 0)|$ after the rotational transform at $\theta_y = 71.8°$ is shown in Fig. 9.21. The whole pattern on the surface of the tilted planar object appears clearly when compared with Fig. 9.19c.

Chapter 10
The Polygon-Based Method

Technique for Shape Forming in Field Rendering

Abstract The polygon-based method is a technique for field rendering. This technique produces virtual surface sources of light in the 3D scene, which has an arbitrary tilt angle and shape. The surface sources are the counterparts of polygons or patches in CG. Thus, using the surface source, we can calculate the field of an object represented by a polygon-mesh as in CG. The detailed technique is discussed in this chapter. We deal with the actual procedure as well as the basic principle to implement the technique in software. Not only generation of simple diffused surfaces but also specular flat and smooth surfaces is presented on the base of the Phong reflection model.

10.1 Surface Source of Light

10.1.1 Generation of Scattered Light

In the real world, we everyday look at surfaces of an object, for example, your palm, tabletop, paper of note, PC monitors and so forth. We can see these surfaces from a variety of viewpoint, i.e., the surfaces commonly keep visible even if we move our viewpoints. This is because the surfaces emit light. To be accurate, the surfaces emit a scattered light; the surfaces may be luminous as shown in Fig. 10.1a, opaque and scatter illumination as in (b), or transparent and diffuse the transmitted light as in (c). There are a few exceptions properly. For example, a mirror does not scatter lights, but reflects them. Many polished materials, such as metals, have similar nature to a mirror. These are discussed later.

The point-based methods emulate the diffusive surface by gathering many point sources of light. In contrast, the polygon-based method use a planar surface source of light, instead of a point source of light, in order to form object surfaces. Thus, in the beginning, let us consider how to create planar surface sources of light computationally. Here, note that we do not need a true physical theory of surface scattering but require some numerical technique that can emulate diffusive surfaces emitting scatter lights. The technique must be simple and have high performance, because it

K. Matsushima, *Introduction to Computer Holography*,
Series in Display Science and Technology,
https://doi.org/10.1007/978-3-030-38435-7_10

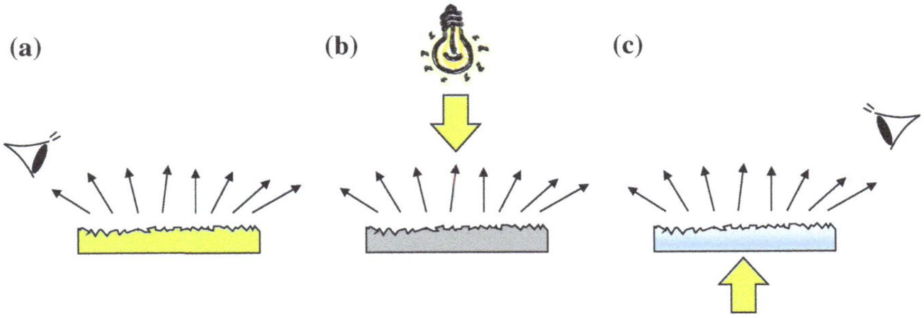

Fig. 10.1 **a** Luminous, **b** opaque, and **c** transparent surfaces that emit scatter light

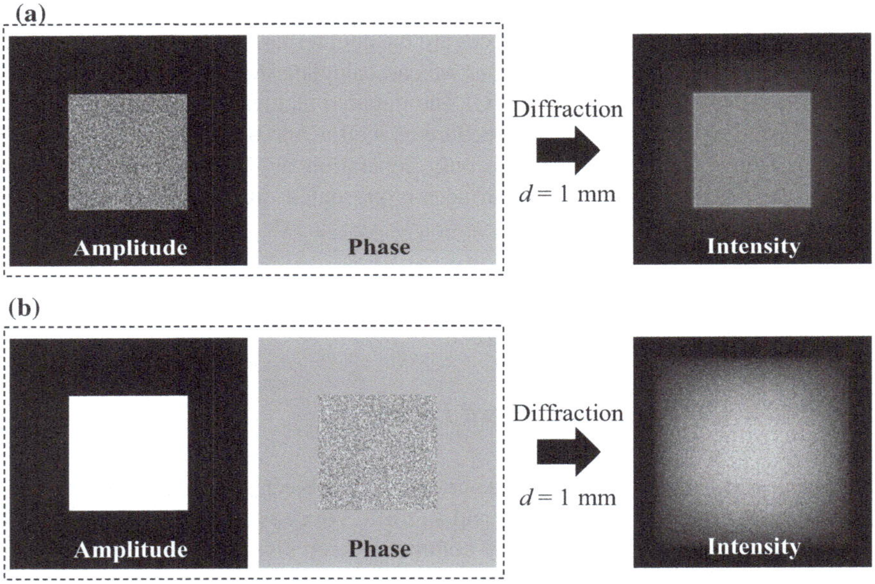

Fig. 10.2 Comparison of diffusiveness between **a** random amplitude and **b** random phase patterns. $\lambda = 633$ [nm], $M = N = 2048$, and $\Delta x = \Delta y = 1$ [μm]. Note that Mersenne twister is used for the random generator

is used to calculate high-definition CGHs, composed of more than billions of pixels. The technique must also have ability to calculate any polygon shape, because an object model is considered to be composed of many polygon shapes. Even if polygons are limited only to a triangular shape, there are many triangles having a variety of shapes.

From aspects of wave optics, it is easy to perceive that scattering may be caused by diffraction, and especially brought by random diffraction. Simple numerical experiments may be valuable in order to confirm this idea. Figure 10.2a shows a random pattern in a square region. Regard the random pattern as the amplitude pattern of a

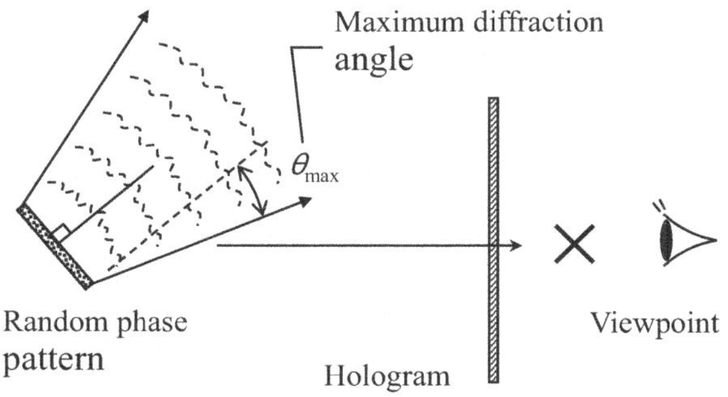

Fig. 10.3 The propagation direction of the field diffracted by a randomized phase pattern

complex-amplitude distribution, i.e., a wavefield. The phase is constant. The wavelength and sampling intervals are 633 nm and 1 μm. The number of sample points is 2048^2. The right column shows the intensity pattern after numerical propagation with a distance of 1 mm. Here, the band-limited angular spectrum method is used for the numerical propagation (see Sect. 6.4). Light is scattered only a little. The diffraction pattern keeps the shape of the original square region.

Diffusiveness of a random phase pattern is shown in Fig. 10.2b. In this case, the phase pattern within the square is randomized, while the amplitude pattern has unity value in the same square. Regard these patterns as a wavefield that has the same parameters as those in (a). The maximum diffraction angle obtained from the parameters is approximately 18.5°. Right column is also the diffraction pattern after propagation at 1 mm. The wavefield spreads over a region whose size approximately agrees with that expected by the maximum diffraction angle.

It is clear that the random phase pattern gives better diffusiveness than the random amplitude pattern. Therefore, we adopt random phase patterns as a diffuser to generate scattered light. This is also convenient to forming the polygon shape. The shape of the amplitude pattern, e.g., the square in Fig. 10.2, gives the shape of the surface source of light.

10.1.2 Theoretical Model of Polygonal Surface Source of Light

There is a problem in the idea of the surface source of light described above. Figure 10.3 explains the problem. The light scattered by the diffuser using a random phase pattern spreads only over the area within the maximum diffraction angle. However, an object is commonly composed of many planar surfaces, i.e., polygons, which have various directions. Thus, light emitted from a random diffuser does not always reach to the hologram plane and viewpoint finally, as shown in Fig. 10.3.

A physical surface in the real world can scatter illumination light very strongly, as shown in Fig. 10.4a. If we emulate this phenomenon faithfully, it is necessary for

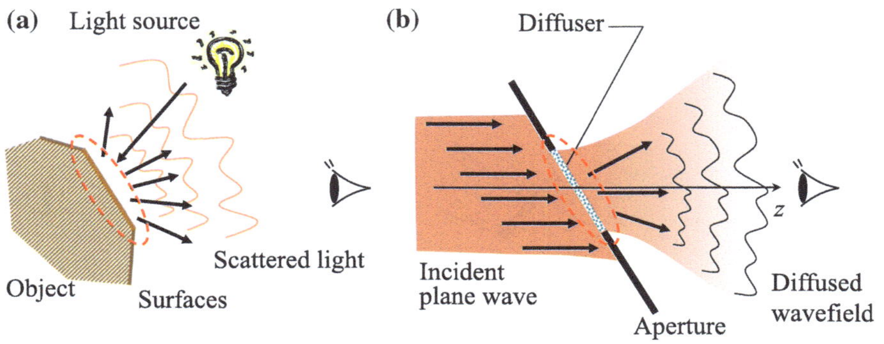

Fig. 10.4 **a** Scattered light from a physical surface, **b** theoretical model of the planar surface source of light

the sampled wavefield with a random diffuser to have a very small sampling interval less than one half wavelength. As a result, the sampled wavefield must have a large number of sample points to produce a given size of the polygon. This requires very large computational effort in synthesis of object fields, and may make it impossible to create large-scale CGHs.

Therefore, the theoretical model shown in Fig. 10.4b is used for the polygonal surface source of light in the polygon-based method in this book. In this model, the polygonal light source is considered as an aperture in which random phase diffuser is installed, and a plane wave traveling along optical axis enters the aperture. Thus, the wavefield diffracted by the diffuser mainly travels along the optical axis, and spreads over a large area. This is the most basic theoretical model of a polygonal surface source of light. The surface created and reconstructed according to this model is a diffusive surface. Specular surfaces, like metal surfaces, are also created by the technique based on this model. However, in that case, the diffusiveness is controlled through the phase pattern. The technique is discussed in later sections.

10.2 Basic Theory for Rendering Diffused Surface

Figure 10.5 shows the process to calculate the object field from an object model. Here, the object model is a triangular prism. Note that the parameters are not realistic in this example. If we use a practical parameter, for example $\Delta x = \Delta y = 1$ [μm], the field is diffused too much to recognize the original shape of the polygons and object model in the amplitude image of the object field.

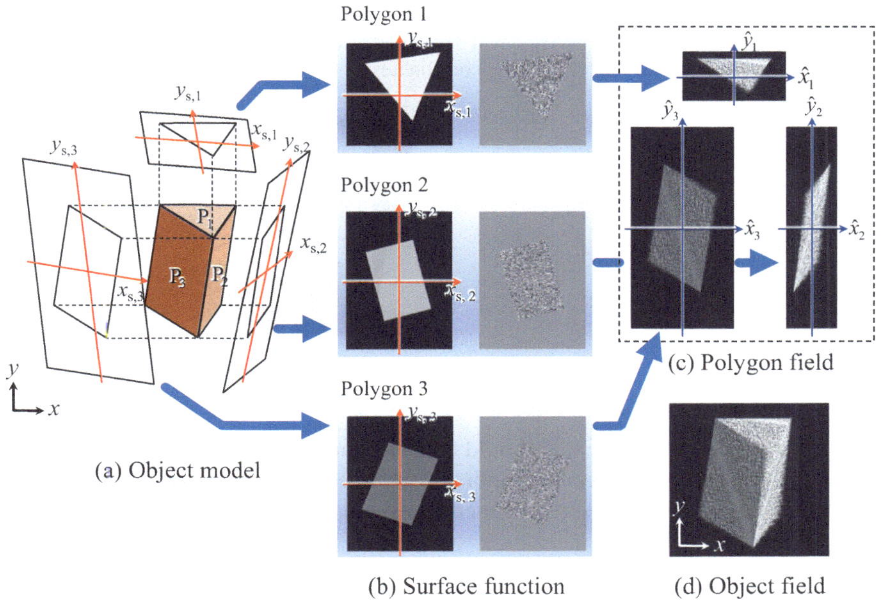

Fig. 10.5 Process to calculate the object field from an object model. The sampling interval and number of samples of the object field are $\Delta x = \Delta y = 4$ [μm] and $M = N = 1024$, respectively

10.2.1 Surface Function

The polygonal surface source of light, modeled by the preceding section, is represented by a surface function:

$$h_s(x_s, y_s; z_s = 0) = a(x_s, y_s) \exp[i\phi(x_s, y_s)], \tag{10.1}$$

where $a(x_s, y_s)$ and $\phi(x_s, y_s)$ are properly amplitude and phase distributions. According to the theoretical model, the amplitude distribution $a(x_s, y_s)$ gives the shape of an aperture, i.e., the shape of a polygon. In contrast, the phase distribution $\phi(x_s, y_s)$ is randomized to simulate a diffuser installed in the aperture like a stained glass, as shown in Fig. 10.5b.

Here, the reason why the coordinates is represented by (x_s, y_s, z_s) is that the surface function is commonly rotationally transformed in order to obtain the wavefield in a plane parallel to the hologram. Thus, the surface function is regarded as a wavefield in the source plane shown in Fig. 9.2. The wavefield after the rotational transform is given in a reference plane (also called a destination plane). The reference plane must be parallel to the hologram, because the field is numerically propagated to the hologram plane to sum all fields. In this sense, the surface function, source coordi-

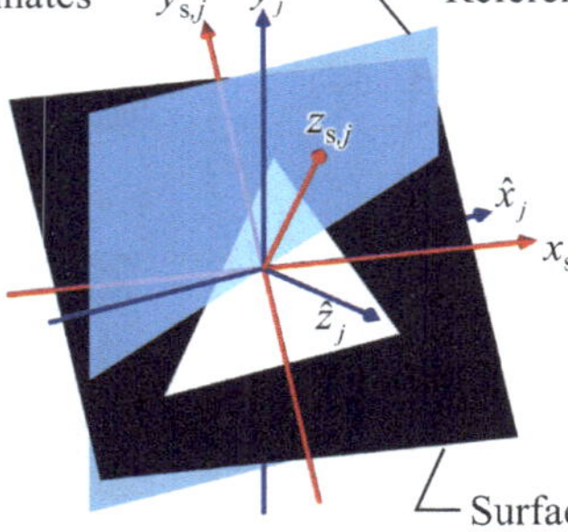

Fig. 10.6 Coordinate systems used in the polygon-based method. The red and blue arrows indicate tilted and parallel local coordinate systems

nates, and reference coordinates should be distinguished among polygons, because polygons comprising an object have individually different shapes, positions, and tilts.

To be more precisely, the surface function of a polygon indexed by j is represented by

$$h_{s,j}(x_{s,j}, y_{s,j}; z_{s,j} = 0) = a_j(x_{s,j}, y_{s,j}) \exp[i\phi_j(x_{s,j}, y_{s,j})]. \tag{10.2}$$

Coordinate system $(x_{s,j}, y_{s,j}, z_{s,j})$ is referred to as a *tilted local coordinate system* for the polygon j, because the plane $(x_{s,j}, y_{s,j}, 0)$, where the polygon is laid, is commonly not parallel to the hologram. Reference coordinates $(\hat{x}_j, \hat{y}_j, \hat{z}_j)$, called a *parallel local coordinate system*, shares the origin with the tilted local coordinates, but the plane $(\hat{x}_j, \hat{y}_j, 0)$, where the destination field is obtained, is parallel to the hologram. These two local coordinate systems are shown in Fig. 10.6.

It should be noted that the phase distribution $\phi_j(x_{s,j}, y_{s,j})$ does not have to be distinguished by index j, because it is simply a random function. This may be true in completely diffusive surfaces. However, when treating specular surfaces in later sections, the phase distributions must be distinguished, because they have different spectra corresponding to the direction of reflection.

10.2.2 Spectrum Remapping by Incident Plane Wave

As shown in Fig. 10.4, the field emitted by a polygonal surface source is forced to direct to the hologram by an incident plane wave. This is equivalent to remap or shift the spatial spectrum of the surface function. This operation is referred to as *spectrum remapping*.

Suppose that a position vector $\mathbf{r}_{s,j} = (x_{s,j}, y_{s,j}, z_{s,j})$ given in the tilted local coordinates is transformed into position vector $\hat{\mathbf{r}}_j = (\hat{x}_j, \hat{y}_j, \hat{z}_j)$ in the parallel local coordinate system by rotation matrix $\mathbf{T}_j$ like (9.1) and (9.2) in Chap. 9:

$$\hat{\mathbf{r}}_j = \mathbf{T}_j \mathbf{r}_{s,j}, \quad \text{and} \quad \mathbf{r}_{s,j} = \mathbf{T}_j^{-1} \hat{\mathbf{r}}_j, \tag{10.3}$$

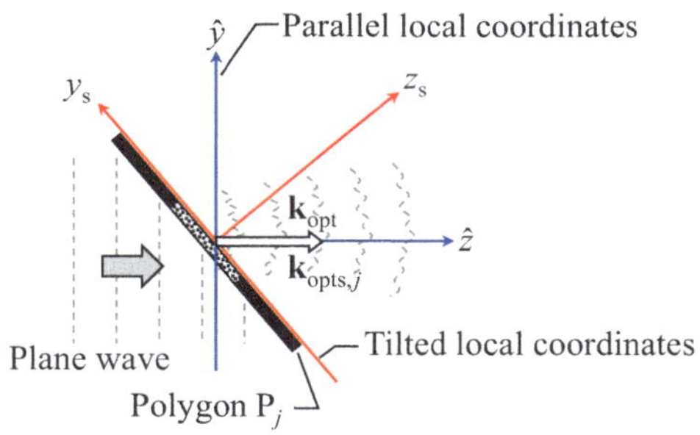

Fig. 10.7 The wave vector of the incident plane wave. The red and blue allow indicate tilted and parallel local coordinates again

where transformation matrix $\mathbf{T}_j$ is defined similarly to (9.9) as

$$\mathbf{T}_j = \begin{bmatrix} a_{1,j} & a_{4,j} & a_{7,j} \\ a_{2,j} & a_{5,j} & a_{8,j} \\ a_{3,j} & a_{6,j} & a_{9,j} \end{bmatrix} \quad \text{and} \quad \mathbf{T}_j^{-1} = \begin{bmatrix} a_{1,j} & a_{2,j} & a_{3,j} \\ a_{4,j} & a_{5,j} & a_{6,j} \\ a_{7,j} & a_{8,j} & a_{9,j} \end{bmatrix}. \tag{10.4}$$

As mentioned above, all position vectors (coordinate systems) $\hat{\mathbf{r}}_j$ and $\mathbf{r}_{s,j}$, and matrices $\mathbf{T}_j$ related to a polygon P_j has a suffix of j, and thus, their components and elements must also have the suffix. However, these suffixes appended to the components and elements make readability decrease considerably. Therefore, in the following discussion, we omit the suffix j from vector components and matrix elements even if they depend on a given polygon. In contrast, the suffix j, appended to vectors and matrices denoted by a boldface, is not omitted in order to avoid confusion.

A plane wave traveling along the optical axis with unity amplitude is written in parallel local coordinates as

$$g(\hat{x}, \hat{y}; 0) = \exp[i\, \mathbf{k}_{opt} \cdot \hat{\mathbf{r}}_j], \tag{10.5}$$

where $\mathbf{k}_{opt}$ is a wave vector along the optical axis, and simply

$$\begin{aligned} \mathbf{k}_{opt} &= k\mathbf{e}_z \\ &= (0,\ 0,\ k). \end{aligned} \tag{10.6}$$

$\mathbf{e}_z$ is the unit vector of the $\hat{z}$-axis of the parallel local coordinate system. Here, note that all parallel coordinate systems share the same unit vector and the vector $\mathbf{e}_z$ also represents the direction of the optical axis.

The same wave vector has different components in tilted coordinates, as shown in Fig. 10.7. The wave vector of the incident plane wave is given in the tilted local coordinates by the rotational transform:

$$\begin{aligned} \mathbf{k}_{opts,j} &= \mathbf{T}_j^{-1}\mathbf{k}_{opt} \\ &= k(a_3,\ a_6,\ a_9). \end{aligned} \tag{10.7}$$

Accordingly, the incident plane wave is

$$\begin{aligned} g_j(x_s, y_s; 0) &= \exp[i\mathbf{k}_{\mathrm{opts},j} \cdot \mathbf{r}_{s,j}] \\ &= \exp[ik(a_3 x_s + a_6 y_s)]. \end{aligned} \tag{10.8}$$

By multiplying wavefield (10.8), the wavefield diffracted by the polygon-shaped diffuser is represented in tilted local coordinates as follows:

$$h'_{s,j}(x_s, y_s; 0) = h_{s,j}(x_s, y_s; 0) \exp[ik(a_3 x_s + a_6 y_s)]. \tag{10.9}$$

The spatial spectrum of the diffracted field is

$$\begin{aligned} H'_{s,j}(u_s, v_s) &= \mathscr{F}\left\{h'_{s,j}(x_s, y_s; 0)\right\} \\ &= H_{s,j}\left(u_s - \frac{a_3}{\lambda}, \ v_s - \frac{a_6}{\lambda}\right), \end{aligned} \tag{10.10}$$

where

$$H_{s,j}(u_s, v_s) = \mathscr{F}\left\{h_{s,j}(x_s, y_s; 0)\right\}. \tag{10.11}$$

As a result, the spectrum of the surface function $h_j(x_s, y_s; 0)$ is remapped in order to direct the field toward the hologram.

10.2.3 Rotation Matrix

Transformation matrix (10.4) converts a position represented in parallel coordinates into that in local coordinates, as in (10.3). There is an arbitrariness in determining the matrix $\mathbf{T}_j$. We use the following Rodrigues' rotation formula:

$$\mathbf{T}_j = \cos\theta_j \mathbf{I} + (1 - \cos\theta_j)^t\mathbf{s}_j \cdot \mathbf{s}_j + \sin\theta_j [\mathbf{s}_j]_\times, \tag{10.12}$$

where symbols $\mathbf{I}$ and '$\cdot$' is the 3×3 identity matrix and the scalar product of vectors. Symbols $^t\mathbf{b}$ and $[\mathbf{b}]_\times$ denote the transposed and cross product matrix for the vector $\mathbf{b}$, respectively. The cross product matrix is the vector product in matrix form; that is,

$$[\mathbf{b}]_\times = \begin{bmatrix} 0 & -b_3 & b_2 \\ b_3 & 0 & -b_1 \\ -b_2 & b_1 & 0 \end{bmatrix}, \tag{10.13}$$

where $\mathbf{b} = (b_1, b_2, b_3)$.

Vector $\mathbf{s}_j$ is the unit vector of the axis of rotation that makes the normal vector $\mathbf{N}_j$ of the polygon coincident with the unit vector $\mathbf{e}_z$ of the optical axis, as shown in Fig. 10.8. Therefore, supposing that the normal vector of the polygon P_j is

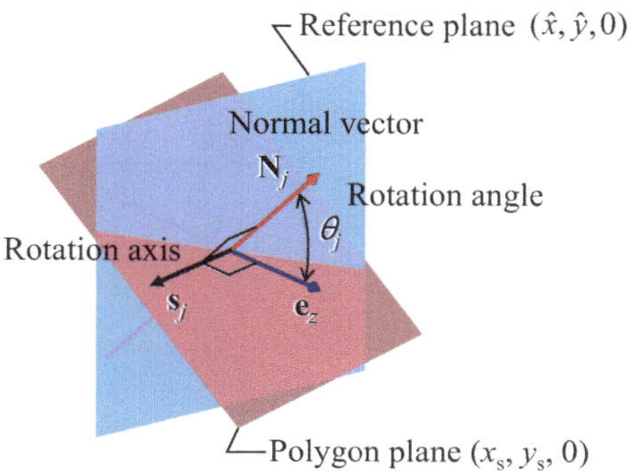

Fig. 10.8 The rotation axis and rotation angle

$$\mathbf{N}_j = (x_\mathrm{N},\ y_\mathrm{N},\ z_\mathrm{N}), \tag{10.14}$$

the vector $\mathbf{s}_j$ is obtained by the vector product of the normal vector and the unit vector $\mathbf{e}_z = (0, 0, 1)$ as follows:

$$\begin{aligned} \mathbf{s}_j &= \frac{\mathbf{e}_z \times \mathbf{N}_j}{|\mathbf{e}_z \times \mathbf{N}_j|} \\ &= (-y_\mathrm{N},\ x_\mathrm{N},\ 0)\frac{1}{\sqrt{x_N^2 + y_N^2}}. \end{aligned} \tag{10.15}$$

Here, note that the components of the vector $\mathbf{N}_j$ is given in the parallel local coordinate system. The trigonometric functions in (10.12) are also given by

$$\begin{aligned} \cos\theta_j &= \mathbf{e}_z \cdot \mathbf{N}_j \\ &= z_\mathrm{N}, \end{aligned} \tag{10.16}$$

$$\begin{aligned} \sin\theta_j &= |\mathbf{e}_z \times \mathbf{N}_j| \\ &= \sqrt{y_\mathrm{N}^2 + x_\mathrm{N}^2}. \end{aligned} \tag{10.17}$$

As a result, the rotation matrix is obtained by the normal vector of the polygon as follows:

$$\mathbf{T}_j = \mathbf{T}_j(\mathbf{N}_j) = \begin{bmatrix} a_1 & a_4 & a_7 \\ a_2 & a_5 & a_8 \\ a_3 & a_6 & a_9 \end{bmatrix}$$

$$
= \begin{bmatrix} z_N - \dfrac{(z_N - 1)y_N^2}{\sqrt{x_N^2 + y_N^2}} & \dfrac{(z_N - 1)x_N y_N}{\sqrt{x_N^2 + y_N^2}} & x_N \\ \dfrac{(z_N - 1)x_N y_N}{\sqrt{x_N^2 + y_N^2}} & z_N - \dfrac{(z_N - 1)x_N^2}{\sqrt{x_N^2 + y_N^2}} & y_N \\ -x_N & -y_N & z_N \end{bmatrix}. \tag{10.18}
$$

10.2.4 Rotational Transform of Remapped Spectrum

The remapped spectrum (10.10) of the surface function is transformed into the spectrum given in the $(\hat{x}, \hat{y}; 0)$ plane that is parallel to the hologram. This transformation is essentially equivalent to the rotational transform described in Chap. 9, but there is a little difference caused by the spectral remapping. In this method, spatial frequencies (u_s, v_s) of the source field or surface function are associated with and converted into frequencies (u, v) represented in the reference plane or parallel plane.

Substituting (9.17) into $H'_{s,j}(u_s, v_s)$ in (10.10), the spectrum of the surface function is given in the parallel plane:

$$
\begin{aligned}
H_j(u, v) &= H_{s,j}\left(u_s(u, v) - \frac{a_3}{\lambda},\ v_s(u, v) - \frac{a_6}{\lambda}\right) \\
&= H_{s,j}\left(a_1 u + a_2 v + a_3\left(w - \frac{1}{\lambda}\right),\ a_4 u + a_5 v + a_6\left(w - \frac{1}{\lambda}\right)\right),
\end{aligned} \tag{10.19}
$$

where the spatial frequency with respect to the z-axis is again $w = \sqrt{\lambda^{-2} - u^2 - v^2}$ as in (9.18). Thus, we can obtain the spectrum in the $(x, y, 0)$ plane by substituting matrix elements $a_1, a_2, \ldots, a_6$ calculated from the normal vector of a given polygon using (10.18). The polygon field created from the surface function is calculated in the parallel plane by the inverse Fourier transform: $h_j(\hat{x}, \hat{y}; \hat{z} = 0) = \mathcal{F}^{-1}\{H_j(u, v)\}$. However, as mentioned in the next section, a transfer function of translational propagation is multiplied before the Fourier transform in practice.

In the rotational transform described in this section, shifted Fourier coordinates introduced in Sect. 9.5.2 is not used, because the spectrum remapping gains a similar effect to use of the shifted Fourier coordinates, where the spectrum is shifted in order to remove the carrier offset in the reference coordinates and reduce the computational load of FFT.

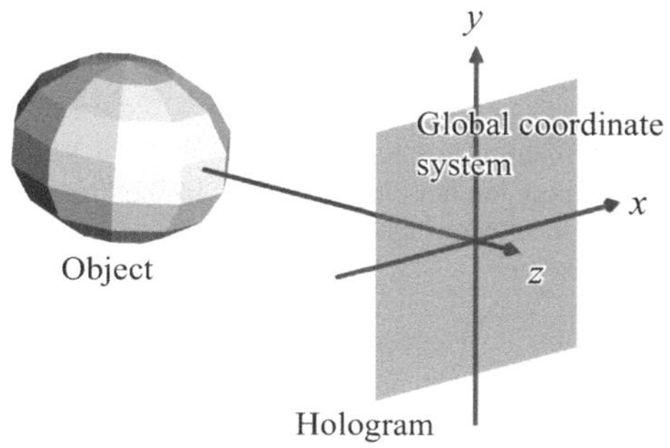

Fig. 10.9 The global coordinate system

10.2.5 Short Propagation to Object Plane

An object is commonly composed of many polygons. Hence, all polygon fields created by the surface functions are superposed in the hologram plane.[1] Here, we introduce a *global coordinate system* (x, y, z). Unlike tilted and parallel local coordinate systems defined for each polygon, the global coordinate system is defined as only one coordinate system in a given CGH. The positions of all objects and polygons arranged in the 3D scene are designated by the global coordinates. The hologram is located in the $(x, y, 0)$ plane of the global coordinate system as usual, and the coordinate origin usually agrees with the center of the CGH, as shown in Fig. 10.9.

The objects are arranged not close to the hologram plane in general. This is because the fringe pattern of objects placed close to the hologram is limited to a small area due to the problem of the fringe frequency mentioned in Sect. 8.8.2. When we arrange objects far from the hologram, a problem arises in superposition of the polygon fields. Figure 10.10 explains the problem. The red sold lines indicate the front-face polygon surfaces, while red broken lines are back-face polygons removed by the back-face culling process. A polygon field calculated in the parallel local coordinates, for example $h_j(\hat{x}, \hat{y}; 0)$, spreads over a large area in the hologram plane. In addition to $h_j(\hat{x}, \hat{y}; 0)$, the field of the neighboring polygon $h_{j+1}(\hat{x}, \hat{y}; 0)$ also spreads over a large area, and overlaps with the field of $h_j(\hat{x}, \hat{y}; 0)$. Therefore, it is very inefficient to calculate the sum total of the polygon fields in the hologram plane.

To solve or reduce the problem, the polygon fields are superposed in a plane close to the object. This plane is called an *object plane*. The object plane is parallel to the hologram and arranged close to the object or to overlap with the object. Since the z positions of the polygons are different from each other, the polygon fields should be translationally propagated back or forth along the z-axis before superposition. We use the band-limited angular spectrum method (BLAS) described in Sect. 6.4 for translational propagation. Supposing that the object plane is placed at z_{obj} and the origin of the local coordinates j is positioned in $(x_j^{(0)}, y_j^{(0)}, z_j^{(0)})$, translational

[1] Here, we assume that the technique of polygon-by-polygon silhouette shielding is not used for occlusion processing (see Sect. 11.3.1).

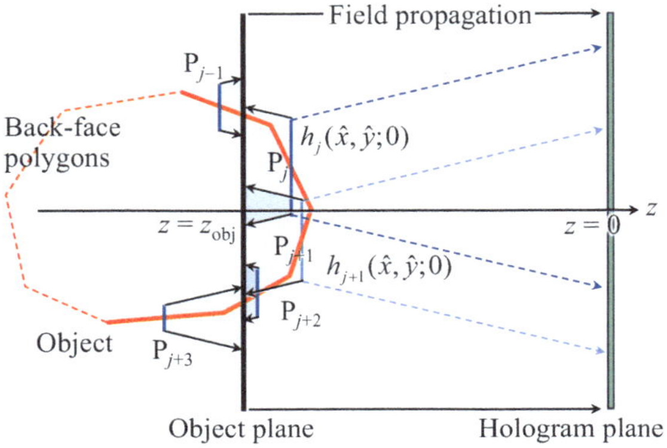

Fig. 10.10 The object plane and its role in calculating the object field. The red and blue solid lines indicate the tilted surface functions (polygons) and the polygon field in the parallel local coordinates, respectively

propagation in the Fourier space is given as:

$$H_j(u, v; z_{obj} - z_j^{(0)}) = H_j(u, v; 0) H_{BLAS}(u, v; z_{obj} - z_j^{(0)})], \tag{10.20}$$

where $H_{BLAS}(u, v; d)$ is the transfer function of the angular spectrum method. Here, note that the transfer function of the BLAS method must be used instead of the simple angular method to reduce noise in actual computation. Choosing the position of the object plane as close to the polygons as possible, the quadruple extension given in Sects. 6.4.3 and 4.7.3 can be avoided in many cases. As a result, computation is accelerated and the used memory is saved in actual numerical calculation of the object field.

The polygon field in the object plane is given by the inverse Fourier transform:

$$h_j(\hat{x}, \hat{y}; z_{obj} - z_j^{(0)}) = \mathscr{F}^{-1}\left\{ H_j(u, v; z_{obj} - z_j^{(0)}) \right\}. \tag{10.21}$$

10.2.6 Superposition of Polygon Fields and Propagation to Hologram Plane

The wavefield of the object is calculated in the object plane by superposition of the polygon field given in the parallel local coordinates as follows:

$$h(x, y; z_{obj}) = \sum_{j=0}^{N_p} h_j(x - x_j^{(0)}, y - y_j^{(0)}; z_{obj} - z_j^{(0)}), \tag{10.22}$$

where N_p is the number of polygons.

Finally, the object field is obtained in the hologram plane by field propagation of the whole field:

Table 10.1 Summary of input data and controllable parameters

Data/Parameter	Symbol	Description
Design wavelength	λ	Wavelength of reconstructed 3D image
Sampling interval of object field	Δx, Δy	Same as pixel pitches of CGH in common
Position of vertexes		Designated by 3D model
Diffraction rate	r_{dif}	$0 < r_{\text{dif}} \leq 1$, Sect. 10.3.5
Culling rate	r_{cull}	$0 < r_{\text{cull}} \leq 1$, Sect. 10.3.8.3

$$h(x, y; 0) = \mathscr{P}_{-z_{\text{obj}}} \left\{ h(x, y; z_{\text{obj}}) \right\}, \tag{10.23}$$

where propagation distance is $d = 0 - z_{\text{obj}}$. This final numerical propagation has much larger scale than that of short propagation to the object plane, and thus, requires a large computational resources, such as memory capacity. A technique of band limiting described in Sect. 10.4 is useful for reducing the memory size required in the final propagation.

10.3 Practical Algorithm for Rendering Diffused Surface

Sampled wavefields and surface functions, and problems related to sampling are not discussed in the preceding section. The actual process for computing object fields by the polygon-based method is described in this section.

10.3.1 Input Data and Controllable Parameters

We suppose that sampling intervals of an object field are Δx and Δy with respect to x and y of the global coordinate system, respectively. Design wavelength λ and positions of vertexes composing polygons, which are given in the global coordinates system, are necessary for computing the object field. Besides, several controllable parameters are required or should be introduced in practical calculation. These parameters determine the quality of the 3D image and computation time. As expected, high quality leads to long computation time in general. These input data and controllable parameters are summarized in Table 10.1. The meaning and role of the controllable parameters are described in the following sections.

10.3.2 Tilted and Parallel Frame Buffers

We calculate the polygon field in the $(\hat{x}_j, \hat{y}_j; 0)$ plane from the surface function in the $(x_j, y_j; 0)$ plane in computing the field of polygon P_j. The complex data array that keeps the sampled wavefield in the parallel plane of $(\hat{x}_j, \hat{y}_j; 0)$ is referred to as a *parallel frame buffer* (PFB), while the data array for the sampled surface function in tilted plane $(x_j, y_j; 0)$ is a *tilted frame buffer* (TFB). These are named after computer memory containing a frame shown on the computer display.

The array size of PFB are represented by M_{PFB} and N_{PFB} with respect to $\hat{x}_j$ and $\hat{y}_j$, respectively, while those of TFB are M_{TFB} and N_{TFB} with respect to x_j and y_j. These array sizes are peculiar to the polygon P_j properly. The ways to determine the array sizes are discussed in Sect. 10.3.5 for PFB and Sect. 10.3.6 for TFB.

10.3.3 The Fourier Transform of Surface Function

A polygon shape is given by the surface function in the polygon-based method. Thus, the shape is depicted in TFB. Suppose the sampled surface function of polygon P_j is

$$\begin{aligned} h_{\text{s},j}[m,n;0] &= h_{\text{s},j}(x_{\text{s},m}, y_{\text{s},n}; z_{\text{s}}=0) \\ &= a(x_{\text{s},m}, y_{\text{s},n}) \exp[i\phi(x_{\text{s},m}, y_{\text{s},n})], \end{aligned} \tag{10.24}$$

where sample positions $x_{\text{s},m}$ and $y_{\text{s},n}$ are given by (6.1) in the tilted local coordinate system. Suffix j should be appended to the coordinates, but is omitted according to the convention adopted in Sect. 10.2.2. The numbers of sample points are M_{TFB} and N_{TFB}, as mentioned above. Accordingly, $m = 0, \ldots, M_{\text{TFB}} - 1$ and $n = 0, \ldots, N_{\text{TFB}} - 1$.

Sampling intervals of the surface function are $\Delta x_{\text{s},j}$ and $\Delta y_{\text{s},j}$ in the tilted local coordinates. These are not always the same as those of the parallel coordinates. Rather, they should be changed in order to reduce the computational load. We also have to determine the proper offset of the local coordinates: $(x_j^{(0)}, y_j^{(0)}, z_j^{(0)})$. As mentioned in the basic theory, $(x_j^{(0)}, y_j^{(0)}, z_j^{(0)})$ represents the origin of the local coordinates in the global coordinate system, and is the center of rotation. These subjects of the surface function and TFB storing it are discussed in Sect. 10.3.6.

The surface function is Fourier-transformed using FFT:

$$\begin{aligned} H_{\text{s},j}[p',q'] &= H_{\text{s},j}(u_{\text{s},p'}, v_{\text{s},q'}) \\ &= \mathbf{FFT}\left\{h_{\text{s},j}[m,n;0]\right\}, \\ p' &= 0, \ldots, M_{\text{TFB}} - 1 \quad \text{and} \quad q' = 0, \ldots, N_{\text{TFB}} - 1, \end{aligned} \tag{10.25}$$

where the sampling positions $u_{\text{s},p'}$ and $v_{\text{s},q'}$ are given by the same manner as (6.2):

$$u_{s,p'} = \Delta u_{s,j}\left(p' - \frac{M_{\mathrm{TFB}}}{2}\right) \quad \text{and} \quad v_{s,q'} = \Delta v_{s,j}\left(q' - \frac{N_{\mathrm{TFB}}}{2}\right). \tag{10.26}$$

Here, note that we use indexes p' and q' instead of p and q in the spectrum of the surface function in order to distinguish the indexes of the spectrum in the parallel coordinates.

The sampling intervals are properly given by the size of sampling window in TFB:

$$\Delta u_{s,j} = \frac{1}{M_{\mathrm{TFB}}\Delta x_{s,j}} \quad \text{and} \quad \Delta v_{s,j} = \frac{1}{N_{\mathrm{TFB}}\Delta y_{s,j}}. \tag{10.27}$$

10.3.4 Basic Procedure for the Rotational Transform and Short Propagation

In this section, basic procedure for numerical processing is introduced along the basic theory described in Sect. 10.2. Advanced techniques and algorithms are discussed in Sect. 10.3.9.

The actual algorithm of the rotational transform of a surface function is shown in Algorithm 10.1. In polygon P_j, normal vector $\mathbf{N}_j$ is obtained from the vertex positions of the polygon. Then, the elements of the transformation matrix: $a_1, \ldots, a_6$ are calculated from vector components x_N, y_N and z_N using (10.18).

The spectrum must be sampled uniformly in the parallel coordinates, and the sampling window must be rectangular to perform the inverse Fourier transform using FFT. Thus, suppose the sampling position of the spectrum is represented in the parallel local coordinates by the same manner as (4.92):

$$\begin{aligned} u_p &= \left(p - \frac{M_{\mathrm{PFB}}}{2}\right)\Delta u_j, \quad (p = 0, \ \ldots, \ M_{\mathrm{PFB}} - 1) \\ v_q &= \left(q - \frac{N_{\mathrm{PFB}}}{2}\right)\Delta v_j. \quad (q = 0, \ \ldots, \ N_{\mathrm{PFB}} - 1) \end{aligned} \tag{10.28}$$

The sampling intervals are properly given by

$$\Delta u_j = \frac{1}{M_{\mathrm{PFB}}\Delta x} \quad \text{and} \quad \Delta v_j = \frac{1}{N_{\mathrm{PFB}}\Delta y}, \tag{10.29}$$

where Δx and Δy are the sampling intervals of the parallel field, which are the same as those of the object field to perform short propagation to the object plane. The array size $M_{\mathrm{PFB}} \times N_{\mathrm{PFB}}$ is discussed in the following section.

The sampling position (u_p, v_q) is converted into that in the (u_s, v_s) coordinates by (10.19):

Algorithm 10.1 The basic procedure for the part of the rotational transform and short propagation

Require: matrix elements $a_1, \ldots, a_6$ are obtained from vertex positions of polygon P_j through normal vector

determination of array sizes $M_{\mathrm{PFB}} \times N_{\mathrm{PFB}}$ and $M_{\mathrm{TFB}} \times N_{\mathrm{TFB}}$
determination of sampling intervals $(\Delta u_j, \Delta v_j)$ and $(\Delta u_{\mathrm{s},j}, \Delta v_{\mathrm{s},j})$
generate surface function $h_{\mathrm{s},j}[m,n;0]$ in TFB
$H_{\mathrm{s},j}[p',q'] \leftarrow$ **FFT** $\{h_{\mathrm{s},j}[m,n;0]\}$ // execute FFT of surface function in TFB
for $q = 0$ **to** $N_{\mathrm{PFB}} - 1$ **do**
  for $p = 0$ **to** $M_{\mathrm{PFB}} - 1$ **do**
    $u_p \leftarrow (p - M_{\mathrm{PFB}}/2)\Delta u_j$
    $v_q \leftarrow (q - N_{\mathrm{PFB}}/2)\Delta v_j$
    if $u_p^2 + v_q^2 \le \lambda^{-2}$ **then**
      $w[p,q] \leftarrow (\lambda^{-2} - u_p^2 - v_q^2)^{1/2}$
      $u_{\mathrm{s}}[p,q] \leftarrow a_1 u_p + a_2 v_q + a_3(w[p,q] - \lambda^{-1})$
      $v_{\mathrm{s}}[p,q] \leftarrow a_4 u_p + a_5 v_q + a_6(w[p,q] - \lambda^{-1})$
      get value of $H_{\mathrm{s},j}(u_{\mathrm{s}}[p,q], v_{\mathrm{s}}[p,q])$ from spectrum $H_{\mathrm{s},j}[p',q']$ by resampling
      $H_j[p,q;0] \leftarrow H_{\mathrm{s},j}(u_{\mathrm{s}}[p,q], v_{\mathrm{s}}[p,q])$ // $H_j[p,q;0]$ is stored in PFB
    else
      $H_j[p,q;0] \leftarrow 0$
    end if
  end for
end for
$H_j[p,q;z_{\mathrm{obj}}] \leftarrow H_{\mathrm{BLAS}}[p,q;z_{\mathrm{obj}} - z_j^{(0)}]H_j[p,q;0]$ // multiply transfer function
$h_j[m,n;z_{\mathrm{obj}}] \leftarrow$ **FFT**$^{-1}$ $\{H_j[p,q;z_{\mathrm{obj}}]\}$ // execute inverse FFT in PFB

$$
\begin{aligned}
u_{\mathrm{s}}[p,q] &= a_1 u_p + a_2 v_q + a_3\left(w[p,q] - \frac{1}{\lambda}\right), \\
v_{\mathrm{s}}[p,q] &= a_4 u_p + a_5 v_q + a_6\left(w[p,q] - \frac{1}{\lambda}\right), \\
w_{\mathrm{s}}[p,q] &= a_7 u_p + a_8 v_q + a_9 w[p,q],
\end{aligned}
\tag{10.30}
$$

where

$$
w[p,q] = \sqrt{\lambda^{-2} - u_p^2 - v_q^2}. \tag{10.31}
$$

Since this conversion is non-linear, we need some interpolation technique to get $H_j(u_p, v_q)$ like the rotational transform of wavefields described in Chap. 9. Here, we can also use the bicubic interpolation in (9.59) to obtain the value of $H_j(u_p, v_q) = H_{\mathrm{s},j}(u_{\mathrm{s}}[p,q], v_{\mathrm{s}}[p,q])$ in PFB from $H_{\mathrm{s},j}[p',q']$ in TFB. Here, note that $w_{\mathrm{s}}[p,q]$ is not used for this process, but used for discrimination of the side of the field component in back-face culling (see Sect. 10.3.8 and Algorithm 10.2).

After the rotational transform, the polygon field is shortly propagated to the position of the object plane before inverse FFT. Here, because the BLAS method is used for the propagation, the spectrum $H_j[p,q;0]$ is multiplied by the transfer function of BLAS in (6.51) in PFB:

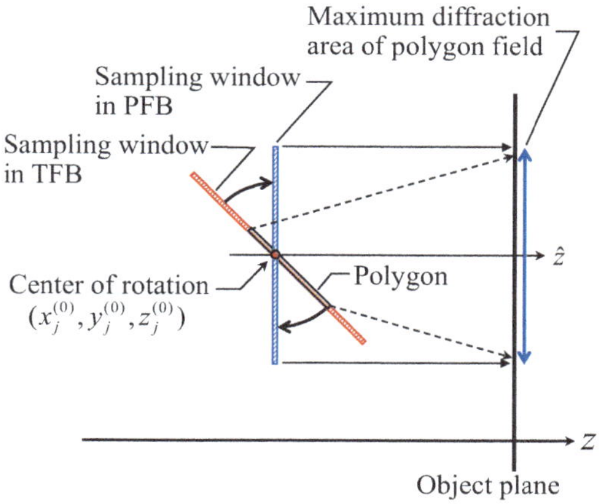

Fig. 10.11 Schematic illustration of the sampling window of the wavefield stored in TFB and PFB, and the center of rotational transform

$$H_j[p, q; z_{\text{obj}} - z_j^{(0)}] = H_{\text{BLAS}}[p, q; z_{\text{obj}} - z_j^{(0)}]H_j[p, q; 0]. \tag{10.32}$$

Then, FFT is performed in PFB to obtain the polygon field in the object plane:

$$h_j[m, n; z_{\text{obj}} - z_j^{(0)}] = \mathbf{FFT}^{-1}\left\{H_j[p, q; z_{\text{obj}} - z_j^{(0)}]\right\}. \tag{10.33}$$

Here, we should choose the size of the field in PFB to avoid the quadruple extension. The technique is described in the following section.

10.3.5 Maximum Diffraction Area of Polygon

The rotational transform in the polygon-based method is a little different from the ordinary rotational transform where we can freely choose the sampling interval of the reference field in many cases. The sampling interval after rotation should agree with that of the object field in the polygon-based method, because the field in PFB is added to the existing object field in the object plane. In contrast, we can freely choose the number of samples of the filed in the parallel coordinates.

It is better to determine the size of PFB so as to avoid the quadruple extension in short field propagation to the object plane (see Sects. 4.7.3 and 6.4.3 for the quadruple extension). The quadruple extension is necessary in cases where a diffracted field expands over the sampling window and invades the neighboring periods of the sampled filed, as shown in Fig. 6.10. Therefore, PFB should have enough size for the sampling window to cover the maximum diffraction area of the polygon field in the object plane, as shown in Fig. 10.11.

It is not easy to estimate the exact diffraction area of a polygon in the object plane, and the exact diffraction area is not absolutely necessary for determination of the

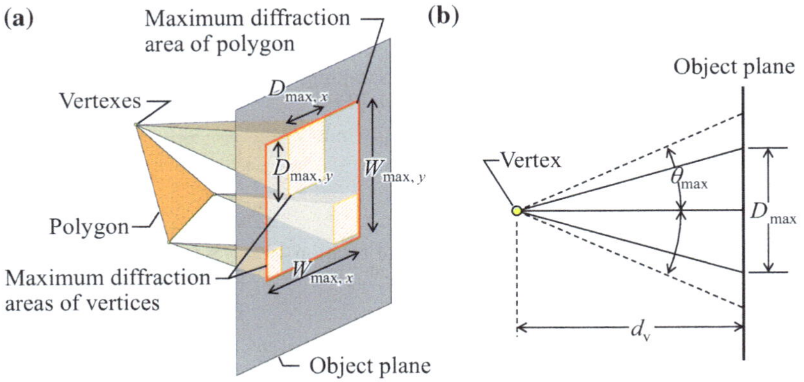

Fig. 10.12 The maximum diffraction area of **a** a polygon and **b** vertices. Symbol θ_{max} indicated in (**b**) is either $\theta_{max,x}$ or $\theta_{max,y}$. Symbol D_{max} is also either $D_{max,x}$ or $D_{max,y}$

buffer size. Since the maximum diffraction area is estimated for all the polygons, the estimate should be made as fast as possible. We use a simple approach to estimate the maximum diffraction area shown in Fig. 10.12. The maximum diffraction area of a vertex of a polygon can be obtained from the maximum diffraction angle given by (3.54) using the values of sampling intervals of the object field: Δx and Δy.

The field emitted from a vertex spreads within a rectangular pyramid whose apex angles are $2\theta_{max,x}$ and $2\theta_{max,y}$, and expands into the area in the shape of a rectangle in the object plane. Suppose the side of the rectangle is

$$D_{max,x} = 2r_{dif}d_v \tan\theta_{max,x} \quad \text{and} \quad D_{max,y} = 2r_{dif}d_v \tan\theta_{max,y}, \tag{10.34}$$

where d_v is the distance between the vertex and object plane, as shown in Fig. 10.12b. r_{dif} is a controllable parameter defined in a range: $0 < r_{dif} \leq 1$. This is called a *diffraction rate*. D_{max} should be simply given $D_{max} = 2rd_v \tan\theta_{max}$. However, the maximum diffraction angle gives a criterion of diffraction angle in which the given sampled wavefield can handle the light in absolutely safety, as mentioned in Sect. 3.2.6. The field does not expand into so large area in actual numerical diffraction in many cases. Thus, the diffraction rate is introduced for limiting the diffraction area and preventing long computation time due to over-quality.

We define the maximum diffraction area of a polygon field as the bounding box of diffraction rectangles of all vertexes in the object plane. This is not the same as the exact diffraction area of the polygon but meets the purpose for determination of the size of the sampling window in PFB.

The polygon field stored in PFB has the same sampling intervals as those of the object field. Therefore, the numbers of sample points are

$$M' = \frac{W_{max,x}}{\Delta x} \quad \text{and} \quad N' = \frac{W_{max,y}}{\Delta y}, \tag{10.35}$$

where $W_{\mathrm{max},x}$ and $W_{\mathrm{max},y}$ are the sizes of the maximum diffraction area shown in Fig. 10.12a. Since the polygon field in PFB is numerically propagated by the BLAS method using FFT, the array size of PFB; $M_{\mathrm{PFB}} \times N_{\mathrm{PFB}}$ is given by rounding up M' and N' to the minimum values of the power of two, which exceed M' and N'.

Unfortunately, the final size of PFB cannot be fixed at this stage, because it is related to the way to determine the center of the rotational transform, as described in Sect. 10.3.7.1.

10.3.6 Determination of Sampling Interval of Surface Function by Probing Sample Points

Complex amplitudes of a surface function is stored in TFB. Thus, the sampling window of the wavefield in TFB has the size corresponding to the polygon; the sampling window covers at least the polygon in the tilted $(x_s, y_s; 0)$ plane. On the other hand, as for the sampling interval of the wavefield in TFB, there is no clear clue to determine it.

We should pay attention to the spectrum of the surface function to obtain a guideline for the sampling interval in TFB. Figure 10.13 shows sample points in the Fourier space of the parallel coordinates in (a), and tilted coordinates in (b) and (c). Here, spectral remapping is not used in (b) but used in (c). It is assumed that the normal vector $\mathbf{N}_j$ of the polygon is given by $\theta_y = 60°$ and $\theta_x = 0$. The manner of rotation used here is shown in Fig. 10.14. The rotation is performed in order of the x-axis and then y-axis. Therefore, the rotation matrix is

$$\mathbf{T}^{-1} = \mathbf{R}_y(\theta_y)\mathbf{R}_x(\theta_x). \tag{10.36}$$

To calculate a polygon field in PFB, we have to perform the inverse Fourier transform in (10.33) using FFT. Thus, the sample points must be arrayed uniformly and the sampling window must be rectangular in PFB, as shown in Fig. 10.13a. Here, sampling position (u_p, v_q) is given by (4.92). This sampling position is projected to (u_s, v_s) coordinates shown in (c). The value of a sample point in PFB; $H_j[p, q; z_{\mathrm{obj}} - z_j^{(0)}] = H_j(u_p, v_q; z_{\mathrm{obj}} - z_j^{(0)})$ is obtained by resampling the output of FFT of the surface function, where the sample points is uniformly arrayed in a rectangular shape.

As in Fig. 10.13c, the sample points projected from those in PFB are located around the origin of (u_s, v_s) coordinates, because of the spectral remapping. Thus, we need the sampled spectrum $H_{s,j}[p', q']$ of the surface function only within the rectangular area of $\delta u_s \times \delta v_s$ depicted in (c). According to the property of DFT, it is found that the following sampling intervals are sufficient for performing the resampling to obtain $H_j[p, q; z_{\mathrm{obj}} - z_j^{(0)}]$.

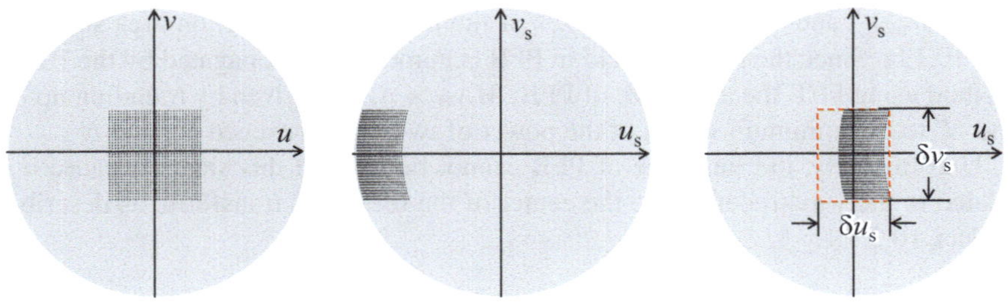

(a) Sampling in parallel plane (b) No spectral remapping (c) Spectral remapping

Fig. 10.13 Distribution of sample points in the Fourier space: **a** Sample points in the parallel local coordinates, corresponding sample points in the tilted local coordinates without **b** and with **c** spectral remapping. $\theta_y = 60°$, $\theta_x = 0$, $\Delta x = \Delta y = 1$ [μm], $M = N = 32$, and $\lambda = 633$ [μm]

Fig. 10.14 The normal vector used for examples. Here, note that rotation on the x-axis is performed prior to rotation on the y-axis

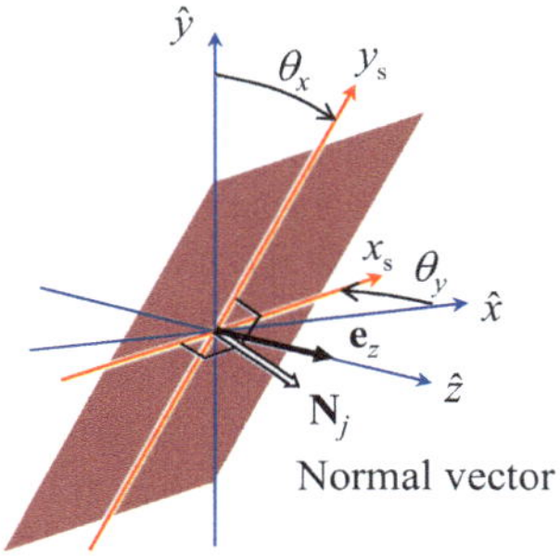

$$\Delta x_s = \frac{1}{\delta u_s} \quad \text{and} \quad \Delta y_s = \frac{1}{\delta v_s}, \tag{10.37}$$

In the example of Fig. 10.13c, δu_s is smaller than $M\Delta u = \Delta x^{-1}$, and thus, $\Delta x_s > \Delta x$. This leads to decreasing the number of sample points in the x_s direction and reducing the computational load of FFT. Not only for the example of Fig. 10.13 but also for all tilted polygons, the sampling intervals of the surface function should be determined by this technique in order to reduce computation time. The sampling interval Δx_s is roughly twice Δx in the case of Fig. 10.13c. This suggest that the sampling window of the surface function covers the same area as that in the parallel plane in roughly speaking, as shown in Fig. 10.15.

However, the actual shape of distribution of sample points is not simple in the Fourier space of the tilted coordinates (see Fig. 10.23). It is very difficult to expect the shape by the rotation matrix only. Thus, we should measure δu_s and δv_s of actual distribution projected by the given rotation matrix. In practice, we can easily determine these using only a few trial sample points. This probing technique is again discussed in Sect. 10.3.10 after consideration of back-face culling.

The number of samples in TFB varies dependently on the center of the rotational transform, as described in the next section.

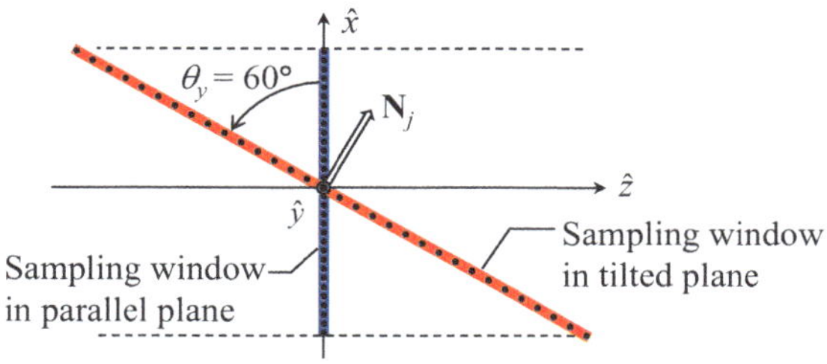

Fig. 10.15 A rough estimate of the size and number of the sampling window in the tilted plane

10.3.7 How to Determine Sizes of PFB and TFB

There is a constraint in determination of the sizes of the sampling window in PFB and TFB. The center of these sampling window, i.e., the origin of the tilted and parallel local coordinates must agree with the center of the rotational transform. Thus, we have to discuss the method for determination of the center of rotation prior to the array sizes of PFB and TFB.

10.3.7.1 Center of Rotational Transform

There are mainly two manners to determine the center of the rotational transform. The center of rotation is given by:

(i) The center of the polygon.
(ii) Orthogonal projection of the center of the maximum diffraction area to the polygon plane.

Figure 10.16 shows arrangement of the sampling windows in these two methods. The rotational transform calculates the field in the rotated plane. Here, it should be emphasized that the center of rotation must agrees with the center of the sampling windows. In other word, when using the sampling manner shown in Fig. 4.6b, the center of rotation agrees with the origin of the local coordinates. This brings advantage and disadvantage to two methods.

When we define the center of rotation as the center of the polygon according to the method (i), the sampling window of the surface function stored in TFB may be minimized but that in PFB exceeds the maximum diffraction area of the polygon, as shown in Fig. 10.16a. This is because the sampling window in PFB must satisfy two conditions; it must cover the maximum diffraction area and the origin must agree with the center of rotation, i.e., the sampling window must spread symmetrically to the center of rotation. This most likely expand the PFB unnecessarily, as in Fig. 10.16a.

The method (ii) gives the opposite advantage and disadvantage. In this method, we take the center of rotation at the center of the maximum diffraction area of the polygon. To be more precise, the center of rotation is a point given by orthogonal projection of the center of the maximum diffraction area to the polygon. Thus, we can

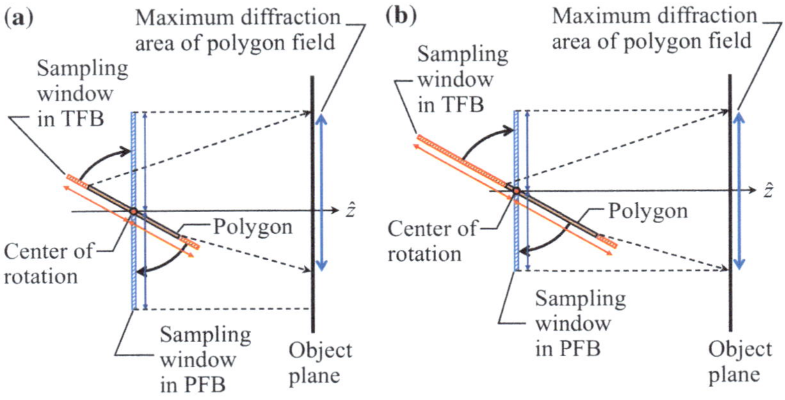

Fig. 10.16 Two methods for determination of the center of the rotational transform. **a** and **b** are corresponding to the methods (i) and (ii) in the text, respectively

minimize the size of PFB in this case. However the size of TFB most likely increases unnecessarily, as shown in Fig. 10.16b. This is also because the sampling window of the surface function stored in TFB must spread symmetrically to the center of rotation.

As a result, it seems that both methods have their merit and demerit evenly. However, the method (ii) is recommended for rendering a simple object, because we have to carry out resampling of spectrum of the surface function at every sample point in PFB. Resampling using interpolation is the most time-consuming process in the polygon-based method, as mentioned in Sect. 10.5. Therefore, it is better to reduce the number of sample points in PFB by use of the method (ii).

However, if using the polygon-by-polygon silhouette method for occlusion processing of a complex object model (see Chap. 11), note that occlusion errors may be caused in the method (ii) because the center of rotation does not agree with that of the polygon. In this case, the field in PFB should be propagated to the plane that includes the center of rotation to block the background field by the silhouette mask or aperture of the polygon. Thus, the total computation time in the method (ii) may be longer than that in the method (i). In this case the method (i) is the better choice.

Here, it should be noted that there are several ways to define the center of a polygon. Figure 10.17 shows typical examples in case of a triangle that is the most important polygon shape in general. In (a), the center of the triangle is given by the centroid. When we produce a rectangular sampling window whose center agrees with the centroid of the triangle, it looks imbalance. In (b), the center of the sampling window agrees with that of the bounding box of the same triangle. This definitely looks better than (a). Thus, the center of a polygon should be given by the center of the bounding box.

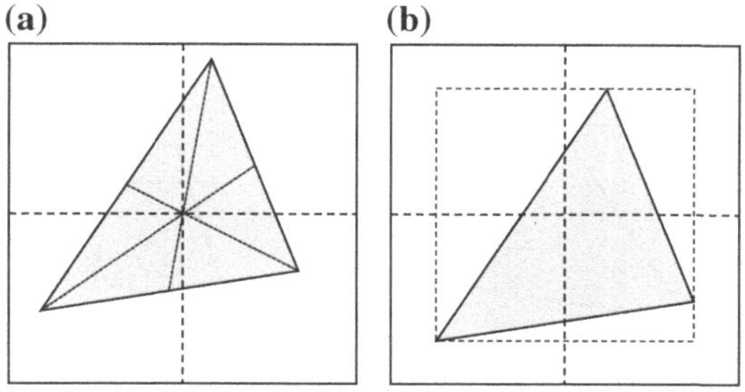

Fig. 10.17 Examples of rectangular sampling windows whose centers agree with **a** the centroid and **b** center of the bounding box

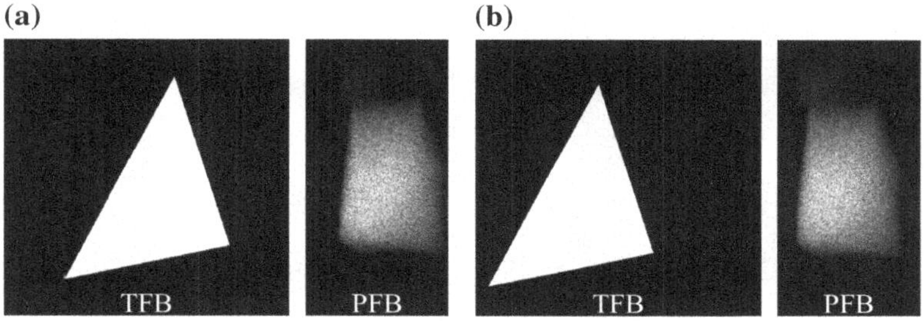

Fig. 10.18 Practical examples of the intensity image of the field in TFB and PFB. Here, the polygon fields in PFB are in the object plane after short propagation at a distance of 5 mm, and the scale is different from that in TFB. **a** and **b** are corresponding to the methods (i) and (ii) for determination of the rotation center. $M_{\text{TFB}} = N_{\text{TFB}} = 8192$, $M_{\text{PFB}} = 8192$, $N_{\text{PFB}} = 16384$, $\theta_x = 0$, $\theta_y = 60$ [°], $\lambda = 633$ [nm], $\Delta x = \Delta y = 1$ [μm], $\Delta x_s = 1.27$ [μm], and $\Delta y_s = 1$ [μm]

10.3.7.2 Practical Examples of Field in PFB and TFB

Figure 10.18 shows practical examples of the intensity image of the field stored in TFB and PFB whose parameters are obtained from the above technique. Here, the sampling intervals of the object field are 1 μm in both the x and y directions. The sample polygon is a triangle that has a square bounding box whose one side is 6 mm in length, as shown in Fig. 10.19a. Orthogonal projection of the triangle that is rotated with an angle of 60° around the y-axis is shown in (b).

When the method (i) is used for determination of the center of rotational transform, the size of TFB is easily determined by the size of the polygon and the sampling interval in TFB. The example is shown in Fig. 10.18a. The sampling window must have the number of samples enough to cover the whole of the polygon face at the given sampling interval; that is 1.27 μm in this case. If we use an ordinary FFT package, M_{TFB} and N_{TFB} are the least number of the power of two but greater than the size sufficiently for the polygon face. Since the center of the sampling window is identical to the center of rotation, the center agrees with the center of the polygon.

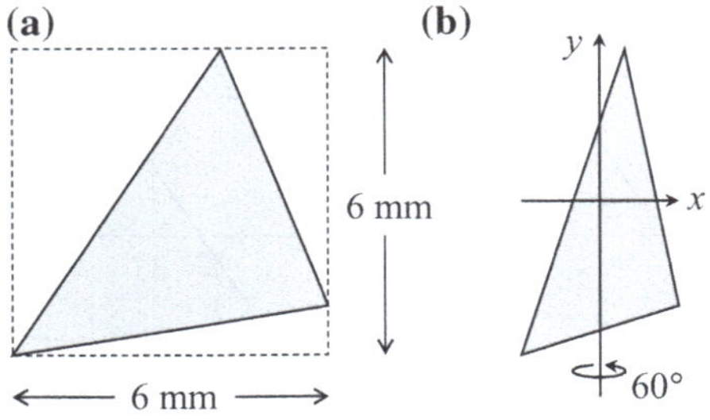

Fig. 10.19 **a** The sample triangle. **b** Orthogonal projection of the triangle rotated with $\theta_y = 60°$

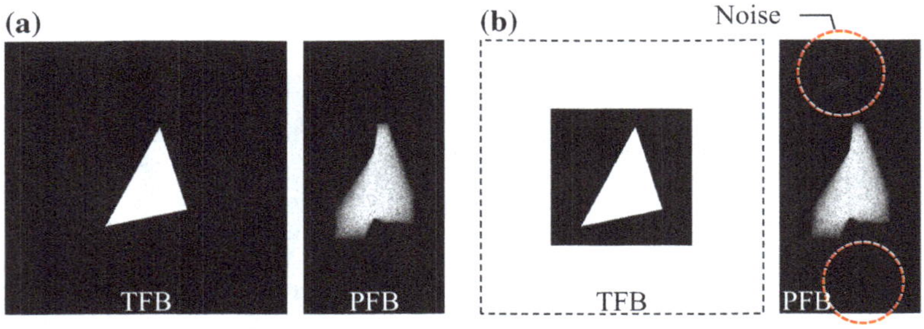

Fig. 10.20 Amplitude images of the field with **a** and without **b** the quadruple extension of the surface function

Determination of M_{PFB} and N_{PFB} is, in contrast, not easy in this case. The sampling window in PFB must satisfy two constraints; the sampling window must have the sufficient size to cover the whole of the maximum diffraction area of the polygon in the object plane, and the center of the sampling window must agree the center of rotation, i.e., the polygon, as shown in Fig. 10.16a. Further, M_{PFB} and N_{PFB} must be the least number of the power of two when using an ordinary FFT.

The above relation is inverted if the method (ii) is adopted for determination of the center of rotation. The example is shown in Fig. 10.18b. In this case, the size of PFB is straightforwardly obtained from the maximum diffraction area of the polygon. As mentioned in Sect. 10.3.5, the sizes M_{PFB} and N_{PFB} are determined by rounding up M' and N' in (10.35) to the least value of the power of two but greater than or equal to M' and N'.

In contrast, M_{TFB} and N_{TFB} must satisfy two constraints. The center of the sampling window must agree with the rotation center, while the size is large enough to cover the polygon surface up at the given sampling interval, as shown in Fig. 10.16b. The numbers of sample points are rounded up to fit the used FFT.

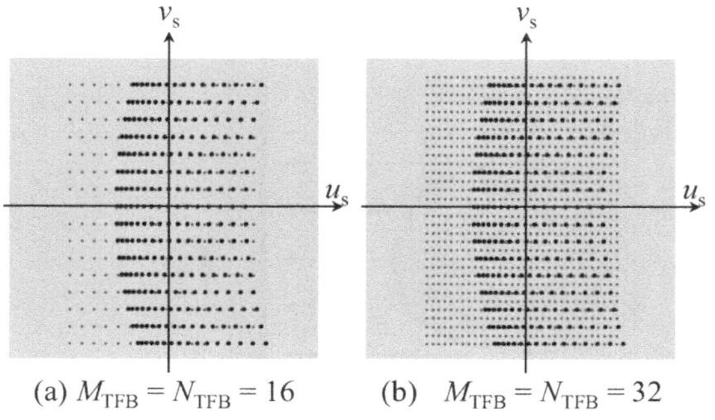

Fig. 10.21 Examples of spectral sample points (red) in TFB without **a** and with **b** the quadruple extension of the surface function. Black dots are also spectral sample points projected from PFB. $M_{PFB} = N_{PFB} = 16$ and $\theta_y = 60°$ in common

10.3.7.3 Quadruple Extension of Surface Function

The quadruple extension of the surface function is recommended to reduce noise in the polygon field. Figure 10.20 shows comparison of the polygon field calculated with and without the quadruple extension. Here, note that the image is depicted as amplitude images to enhance the noise. The polygon field calculated without the quadruple extension is definitely nosy in almost all cases, as shown in (b).

Figure 10.21 shows examples of spectral sample points in the tilted coordinates. Here, the polygon is rotated only around the y-axis at an angle of 60°. The black dots are sample points projected from those in the parallel coordinates. The sampling density increases in the tilted coordinates because of non-linear projection by coordinates rotation. As a result, there is several black points between red points that are spectral sample points of the surface function by FFT, as in (a). Resampling of spectrum of the surface function does not work effectively at black sample points in this case. This is most likely the reason of the noise in Fig. 10.20b. Resampling works better by use of the quadruple extension of the surface function. Spectral sampling density of the extended surface function increases as high as that of the projected sample points, as in Fig. 10.21b.

10.3.8 Back-Face Culling

Back-face culling is a terminology in computer graphics (CG). The back-face culling means a process to remove polygons that face a backward direction. Since back-face polygons are not seen in CG, we can improve rendering speed by removing them from the process list. However, the back-face culling in computer holography is a more difficult problem than what we have expected. This is because the back-face polygons are not always invisible in computer holography.

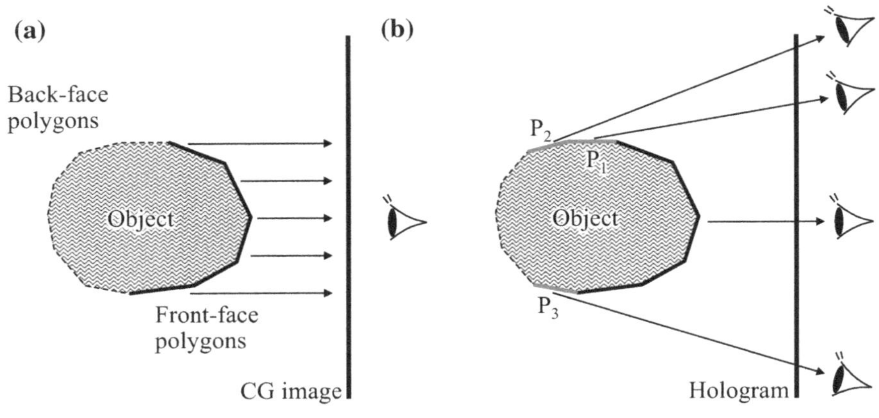

Fig. 10.22 Back-face culling in **a** orthogonal projection of computer graphics, and **b** computer holography

10.3.8.1 Back-Face Culling in Computer Holography

Figure 10.22 explains the difference of the back-face culling between CG and computer holography. In CG, light always travels perpendicularly to the image plane in orthogonal projection,[2] as in (a). Thus, polygons whose normal vector is parallel to the image plane or has a negative z-component are not seen, and thus not depicted in the resultant CG image. It is easy to discriminate between visible and invisible polygons.

Unlike CG, light travels in a variety of direction within the viewing angle in computer holography. The viewers can change their viewing-point freely, as in (b). For example, the surface of polygon P_1 is perpendicular to the hologram plane, i.e., the normal vector is parallel to the hologram. This polygon is commonly invisible in static CG images, but most likely visible from upper viewpoints in HD-CGHs, even if the CGH reconstructs a still 3D image. Even in polygons P_2 and P_3 that are facing back a little, the polygon surface may be seen from viewpoints at the edge of the viewing zone. Thus, we need some unified criterion to perform back-face culling in computer holography. This issue is related to distribution of the sample points of the spectrum stored in TFB, as described in the next section.

10.3.8.2 Problem of Large-Angle Rotation

Let us get back to the problem of TFB, because the problem of back-face culling in computer holography is closely related to the spectral sampling window in the tilted plane, which is projected from the rectangular window in the parallel plane.

---

[2]Light converges into the viewpoint in perspective projection of CG.

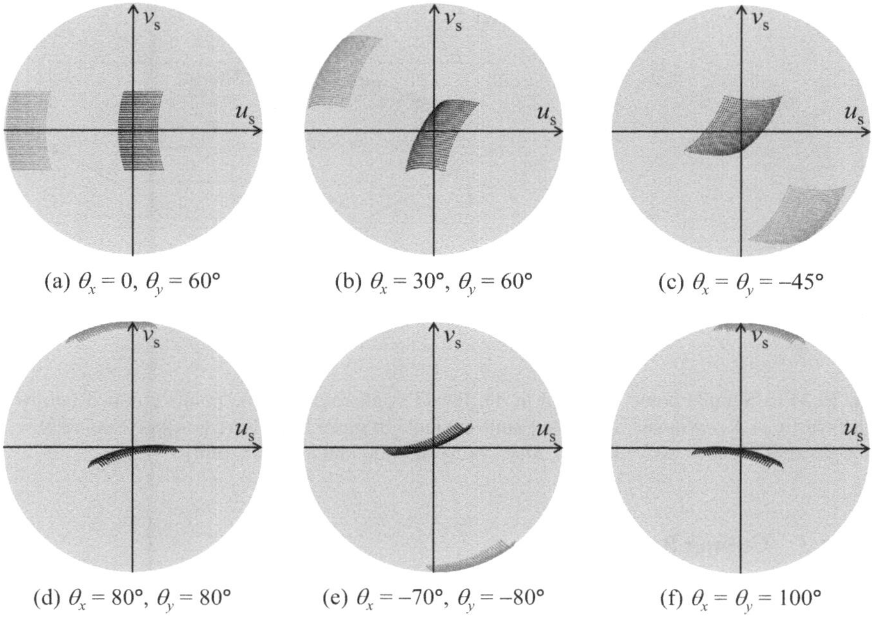

Fig. 10.23 Examples of sample points depicted for a variety of normal vectors of the polygon. Spectral remapping is not used in gray sample points. $\Delta x = \Delta y = 1$ [μm], $M_{\text{PFB}} = N_{\text{PFB}} = 32$ and $\lambda = 633$ [nm]

Sample points in the tilted coordinates is shown for various normal vectors of the polygon in Fig. 10.23. Angles of the normal vector, given by Fig. 10.14, are not so large in (a)–(c). In contrast, (d)–(f) shows examples of sampling distribution in cases of large-angled normal vectors. Here, the rotation matrix is again given by (10.36), and gray dots represent sample points without spectral remapping.

We can find strange shapes that look like a folding structure in projected sample points of (d)–(f). Figure 10.24 explains why the folding structure is generated in the coordinate rotation. Sample points in PFB are mapped on Ewald's sphere in (a), where $\mathbf{T} = \mathbf{R}_y(80°)\mathbf{R}_x(80°)$ correspondingly to Fig. 10.23d. The sample points span over the front side as well as back side of the Ewald's sphere. Therefore, some of sample points have a negative value of w_s. These sample points are depicted in green color in (b) that is magnified sampling distribution of Fig. 10.23d. It is clearly founded out that the folding structure comes from curvature of the Ewald's sphere. How to treat the sample points located in the other side?

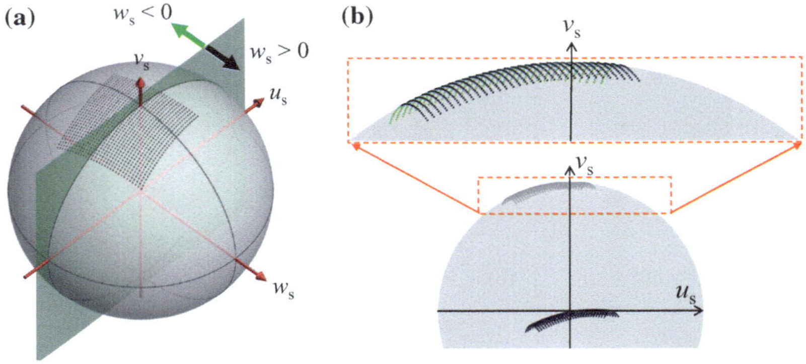

Fig. 10.24 **a** Sample points depicted in the Ewald's sphere. **b** Sample points grouped into two colors in (u_S, v_S) coordinates. Sample points depicted in green color have a negative value of w_S. $\theta_x = \theta_y = 80°$, $\Delta x = \Delta y = 1$ [μm], $M_{\text{PFB}} = N_{\text{PFB}} = 32$ and $\lambda = 633$ [nm]

10.3.8.3 Culling Rate

The sample points located in the back side of the Ewald's sphere represent fields that travel in the backward direction, as shown in Fig. 10.25a. Here, the backward field is depicted by green arrows in (a). A CGH reconstructs light only within the maximum diffraction angle given by the sampling intervals of the object field, as shown in (b)–(e). This constraint is represented by the rectangular sampling window in PFB shown in Fig. 10.13a.

The rectangular sampling window and sample points in PFB are projected to the tilted coordinates in TFB. A sample point projected to the tilted coordinates gives the field component observed in the tilted coordinate system of the polygon. If the projected point is located in the back side of Ewald's sphere, i.e., one of the green points in Fig. 10.24b, the fact means that the field component travels in the backward direction.

If a polygon is approximately parallel to the hologram and object plane, the backward fields do not reach to the object plane properly, as shown in Fig. 10.25b. Therefore, these fields do not affect the object field. The fields whose spectra are depicted in Fig. 10.23a–c agree with this situation.

However, as shown in Fig. 10.25c, if a polygon has the surface nearly perpendicular to the object plane, a part of the backward fields is emitted within the maximum diffraction zone. This backward field component has a negative value of w_S. In the case of (c), almost all projected sample points have a positive w_S, while a few of sample points have a negative w_S.

When the polygon face is right perpendicular, i.e., the normal is parallel to the hologram plane, the ratio of projected sample points with positive w_s to negative w_s is exactly even, as shown in (d). If a polygon face has an even larger angle, number of positive w_s sample points is less than that of negative sample points, as in (e). If the polygon has a "single-side property", which means that only the front side of the

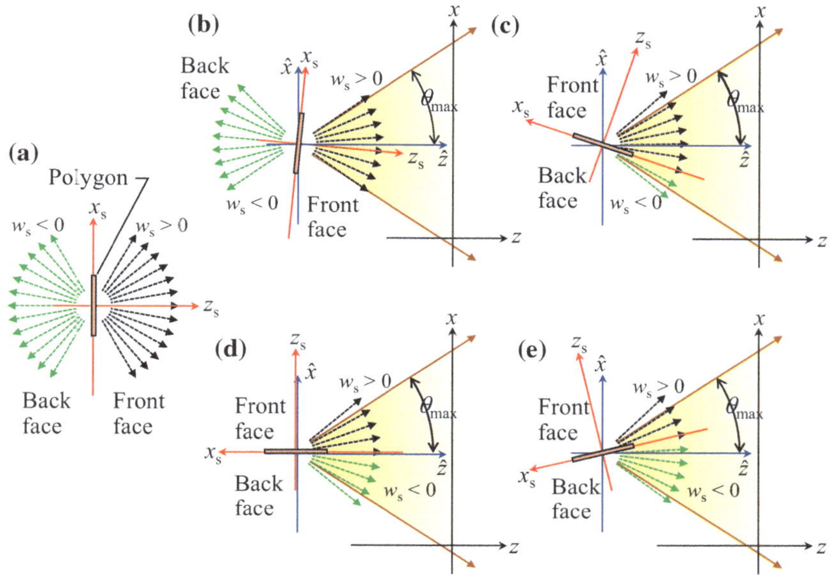

Fig. 10.25 Backward light (green arrows) and forward light (black arrows) of polygons with various tilt-angles

polygon is visible, the backward field should not be calculated; the amplitude must be zero. This is performed by substituting zero into the value of the sample points having negative w_s. In contrast, in the polygon having a "both-sides" property, all projected sample points have the value by resampling of the spectrum of the surface function.

In Fig. 10.25, positive w_s field components decrease from (b) to (e) in order. The polygons corresponding to (d) and (e) are not depicted in the CG image if the polygons have single-side property. However, even these polygons are visible in computer holography, as mentioned in Sect. 10.3.8.1. In that case, should all polygons that have only one positive w_s sample point be contributed to the object field? Polygons that have a few positive points may not be seen in the practical CGH. Therefore, this is an issue of demanded quality of the 3D image reconstructed by the CGH.

In the algorithm of the polygon-based method described above, some trial sample points in (u, v) of the parallel coordinates, which are uniformly distributed in the rectangular region of $1/\Delta x \times 1/\Delta y$, must be projected to the tilted coordinates (u_s, v_s) in order to obtain the shape of distribution of the projected point and determine the sampling interval Δx_s and Δy_s of the surface function. Thus, let N_{all} and N_{front} represent the numbers of all trial sample points and sample points having positive w_s in all trial points, respectively. In addition, we introduce a controllable parameter r_{cull}, called *culling rate*, into the algorithm.

A polygon with the single-side property is processed if

$$\frac{N_{\text{front}}}{N_{\text{all}}} \geq r_{\text{cull}}, \tag{10.38}$$

and not processed if

$$\frac{N_{\text{front}}}{N_{\text{all}}} < r_{\text{cull}} \text{ or } N_{\text{front}} = 0. \tag{10.39}$$

The culling rate has a range of $0 \leq r_{\text{cull}} \leq 1$, and generally described as follows:

$r_{\text{cull}} = 1$: The polygon is processed if the whole light is expected to be emitted within the maximum diffraction angle and contributed to the object field. Thus, polygons whose light is partially contributed to the object field are culled and abandoned, i.e., not processed.

$r_{\text{cull}} = 0.5$: Similarly to CG, all back-face polygons and perpendicular polygons to the hologram plane are culled and abandoned.

$r_{\text{cull}} = 0$: Only polygons that do not have any spectral sample point having positive w_{s}, i.e., $N_{\text{front}} = 0$, are culled and abandoned. All polygons are processed if they have an influence to the object field, even just a little.

As expected by above description, a higher culling rate leads to lower computational effort and quality of the reconstructed 3D image. If a polygon has the both-sides property, the polygon is always processed in any case. Here, note that the single-side and both-sides properties are commonly given by the modeling software used for design of the object model, and thus, designated by the designer.

10.3.9 Overall Algorithm for Rendering Diffused Surface

Algorithm 10.2 summarizes above techniques as an algorithm that includes back-face culling and determination of the parameters of TFB and PFB. Trial sample points arrayed within $\Delta x^{-1} \times \Delta y^{-1}$ rectangular region of (u, v) coordinates are converted into $(u_{\text{s}}, v_{\text{s}})$ coordinates, and then, N_{front}, which is the number of trial sample points that satisfy $w_{\text{s}} > 0$, is counted at lines 2–4. Line 5 is the back-face culling. If $N_{\text{front}}/N_{\text{all}} < r_{\text{cull}}$ or $N_{\text{front}} = 0$, we skip the polygon unless the polygon has the both-sides property. Here, "one-sided" and "both-sided" denote that the polygon has one-side and both-sides properties, respectively.

After removing the back-face trial sample points having negative w_{s} if one-sided, the size of the bounding box of the remaining trial points is measured in $(u_{\text{s}}, v_{\text{s}})$ coordinates, i.e., the maximum and minimum values of the position are measured for the remaining trial sample points at line 9. Here, both-sided polygons are always processed and the back-face trial points are not removed in the measurement. The sampling interval of the surface function are determined by (10.37).

Lines 12–18 show the selection of the center of the rotational transform and the corresponding way to determine the size of TFB and PFB. The maximum diffraction area of the polygon is obtained using the technique shown in Fig. 10.12. The sampling

Algorithm 10.2 An overall procedure to calculate the polygon field of polygon P_j in the object plane

Require: values of λ, Δx, Δy and z_{obj}, and vertex positions of polygon
Require: values of r_{dif} and r_{cull} // controllable parameters
calculate matrix elements $a_1, \ldots, a_6$ from vertex positions through normal vector
generate N_{all}-points trial sample points within rectangular region of $\Delta x^{-1} \times \Delta y^{-1}$ in (u, v) coordinates
project trial sample points to (u_s, v_s) coordinates using rotation matrix
count N_{front} that is number of trial sample points having positive w_s
if (one-sided **and** $N_{\text{front}} \neq 0$ **and** $N_{\text{front}}/N_{\text{all}} \geq r_{\text{cull}}$) **or** both-sided **then** // back-face culling
  if one-sided **then**
    remove trial points having negative w_s
  end if
  measure size of bounding box of remaining trial sample point: $\delta u_s \times \delta v_s$
  $\Delta x_s \leftarrow 1/\delta u_s$
  $\Delta y_s \leftarrow 1/\delta v_s$
  if select center of polygon as center of rotational transform **then** // method (i)
    determine $M_{\text{TFB}} \times N_{\text{TFB}}$ for TFB to cover over polygon
    determine $M_{\text{PFB}} \times N_{\text{PFB}}$ for PFB to cover over maximum diffraction area in object plane, and for center of PFB to be center of polygon
  else // method (ii)
    determine $M_{\text{PFB}} \times N_{\text{PFB}}$ for PFB to cover over maximum diffraction area in object plane
    determine $M_{\text{TFB}} \times N_{\text{TFB}}$ for TFB to cover over polygon, and for center of TFB to be center of rotational transform
  end if
  generate surface function $h_{s,j}[m, n; 0]$ in TFB
  $H_{s,j}[p', q'] = \textbf{FFT}\left\{h_{s,j}[m, n; 0]\right\}$ // Execute FFT of surface function in TFB
  for $q = 0$ **to** $N_{\text{PFB}} - 1$ **do**
    for $p = 0$ **to** $M_{\text{PFB}} - 1$ **do**
      $u_p \leftarrow (p - M_{\text{PFB}}/2)\Delta u_j$
      $v_q \leftarrow (q - N_{\text{PFB}}/2)\Delta v_j$
      if $u_p^2 + v_q^2 \leq \lambda^{-2}$ **then**
        $w[p, q] \leftarrow (\lambda^{-2} - u_p^2 - v_q^2)^{1/2}$
        $w_s[p, q] \leftarrow a_7 u_p + a_8 v_q + a_9 w[p, q]$
        if one-sided **and** $w_s[p, q] < 0$ **then** // culling of backward field
          $H_j[p, q; 0] \leftarrow 0$
        end if
        $u_s[p, q] \leftarrow a_1 u_p + a_2 v_q + a_3(w[p, q] - \lambda^{-1})$
        $v_s[p, q] \leftarrow a_4 u_p + a_5 v_q + a_6(w[p, q] - \lambda^{-1})$
        get value of $H_{s,j}(u_s[p, q], v_s[p, q])$ from spectrum $H_{s,j}[p', q']$ by resampling
        $H_j[p, q; 0] \leftarrow H_{s,j}(u_s[p, q], v_s[p, q])$ // $H_j[p, q; 0]$ is stored in PFB
      else
        $H_j[p, q; 0] \leftarrow 0$
      end if
    end for
  end for
  $H_j[p, q; z_{\text{obj}}] \leftarrow H_{\text{BLAS}}[p, q; z_{\text{obj}} - z_j^{(0)}]H_j[p, q; 0]$ // multiply transfer function
  $h_j[m, n; z_{\text{obj}}] \leftarrow \textbf{FFT}^{-1}\left\{H_j[p, q; z_{\text{obj}}]\right\}$ // execute inverse FFT in PFB
  add $h_j[m, n; z_{\text{obj}}]$ to object field taking shift $x_j^{(0)}$ and $y_j^{(0)}$ into account
end if

numbers $M_{\mathrm{TFB}} \times N_{\mathrm{TFB}}$ and $M_{\mathrm{TFB}} \times N_{\mathrm{TFB}}$ are determined according to the center of rotation.

The procedures below line 19 is the same as those in Algorithm 10.1 except for lines 28–30. In those lines, if the sample point in TFB converted from a sample point in PFB has a negative w_s, zero is substituted into the sample point in PFB unless the polygon has the both-sides property. This is an important process to avoid the backside field of the polygon entering and affecting the object field.

Applying above process for each polygon comprising an object model, the object field is calculated in the object plane in PFB. The object field is commonly propagated to the hologram plane after that. This propagation is, in general, necessary to arrange the object model apart from the hologram plane in order to avoid aliasing errors in the fringe pattern, as mentioned in Sect. 8.8. A technique to create image CGHs in which the object is arranged across the hologram is to reduce the diffraction rate r_{dif} so that the generated fringe patten does not produce aliasing errors.

It should be noted that the quadruple extension of the object filed is generally required in the final propagation. This is because the object field has a given diffusibility and spreads over the sampling window, i.e., the hologram area (see Sect. 6.4.3). As a result, this final propagation consumes the biggest resource of computer memory in creation of a CGH in many cases, and imposes a limit of the size of the created CGH. A technique to get over the problem is to use band-limiting of the polygon field described in Sect. 10.4.

10.3.10 Variation of Probing Sample Points

In above techniques, to determine the sampling interval in TFB, a set of trial sample points must be projected to the Fourier space in the tilted coordinates as probes. This probing usually does not apply so heavy load to the computation in high-definition CGHs, because FFT and interpolation of the surface function usually consume much more computational resources. However, the probing may place load on the computation of small CGHs reconstructed by spatial light modulate (SLM) or something similar devices.

Figure 10.26 shows examples of probes with a few sample points. The upper row shows the cases where a normal rectangular array of trial sample points, shown in (a), is used for probing. The number of sample points is 25 ($=5 \times 5$). It may be better to use odd number for the sampling number, because there is no sample points in the axes and origin if the sampling number is even. As shown in (b) and (c), the shape of the array does not change so much even in a few numbers of sampling.

Lower row of Fig. 10.26 shows the case of circularly-arrayed sample points used as the probe. The number of sample points is 17 in this example. This is based on an idea that the corner of the sampling window does not affect the final object field so much. The bounding box of the projected sample points becomes smaller in this case, as shown in (b) and (c). As a result, the sampling interval of the surface

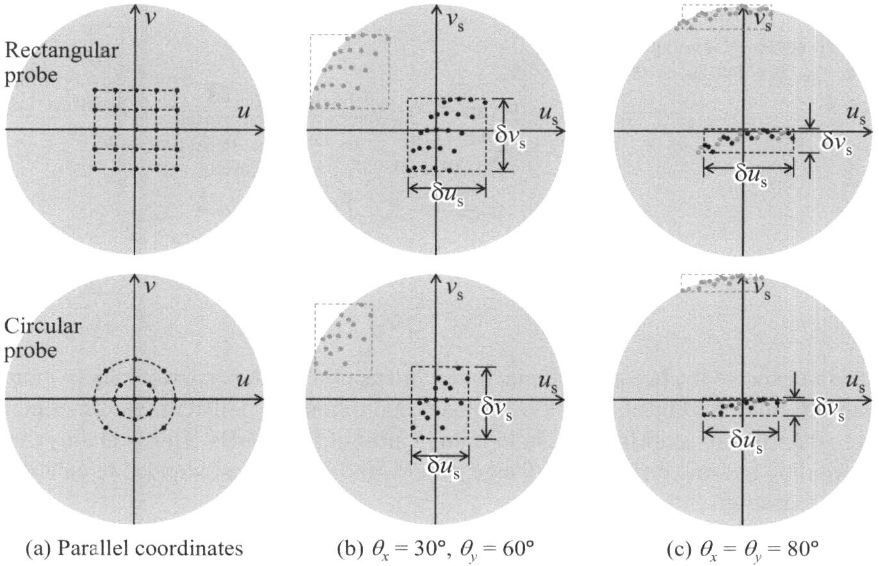

Fig. 10.26 Examples of probes having a few sample points. **a** Trial sample points in parallel coordinates, and sample points projected by rotation matrix **b** $\mathbf{T} = \mathbf{R}_y(30°)\mathbf{R}_y(60°)$ and **c** $\mathbf{T} = \mathbf{R}_y(80°)\mathbf{R}_x(80°)$. The plots in light color show the sample points without spectral remapping. The green plots again show sample points having a negative w_s

function increases in the real space, and resultant reduction of the number of samples commonly leads to decrease of computation time.

10.4 Band Limiting of Polygon Field

In the polygon-based method, one of the most powerful tuning techniques is band limiting of polygon fields [72].

10.4.1 Principle

Figure 10.27 shows the field emitted from a polygon. The polygon field is calculated so as to spread over the maximum diffraction area of the polygon in the technique described in previous sections. However, if the polygon is far from the hologram or is located apart from the optical axis that intersects the hologram at the center, the field may spread so much as to run over the hologram area. The field traveling outside the hologram is definitely superfluous. This situation often occurs in HD-

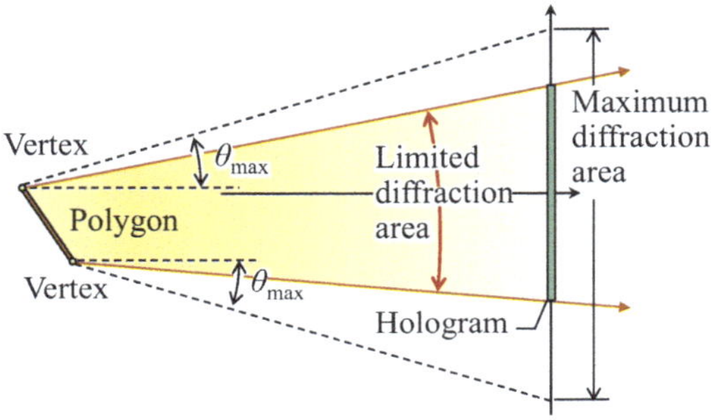

Fig. 10.27 A limited diffraction area of a polygon field. θ_{max} is either $\theta_{max,x}$ or $\theta_{max,y}$

CGH unexpectedly, because the maximum diffraction angle is approximately more than 20°, and the object model is arranged approximately 5–10 cm, or sometimes more than 10 cm, apart from the hologram in most of HD-CGHs. This is mainly due to avoiding aliasing errors in the fringe pattern, and sometimes intended to enhance depth sensation.

Therefore, we may be able to reduce computation time by limiting calculation of the polygon field to the area of the hologram. This technique is also very useful for reduction of memory usage as well as reduction of computation time,[3] because we need the final propagation in order to place the object apart from the hologram. In this final propagation, the quadruple extension of the sampling window is commonly required to avoid the edge effect (see Sect. 6.4.3). The extension makes the computational load much heavier. The reason that the quadruple extension is required in convolution-based numerical propagation is that the diffracted field runs over the sampling window and invades the neighboring region in periodicity of the sampled field, as shown in Fig. 6.10.

If the object field does not spread so much, and thus, does not run over its sampling window, the quadruple extension of the sampling window is no longer required in the final propagation. Since the sampling window of the object field is commonly identical with the hologram area, if all polygon fields of an object spread only within the hologram area, we can omit the quadruple extension and reduce the maximum memory usage in the computation. Accordingly, a larger CGH can be computed by the same computer.

10.4.2 Limit of Bandwidth

Figure 10.28 shows the coordinates and geometry used for consideration of the band limit. The hologram is positioned at the center of $(x, y, 0)$ plane in the global coordinate system as usual. Suppose that the hologram size is $W_{h,x} \times W_{h,y}$. Applying the same way as that in (6.53), the spatial frequency of the field emitted from a point at

[3]In fact, this technique was first devised to reduce memory usage [72].

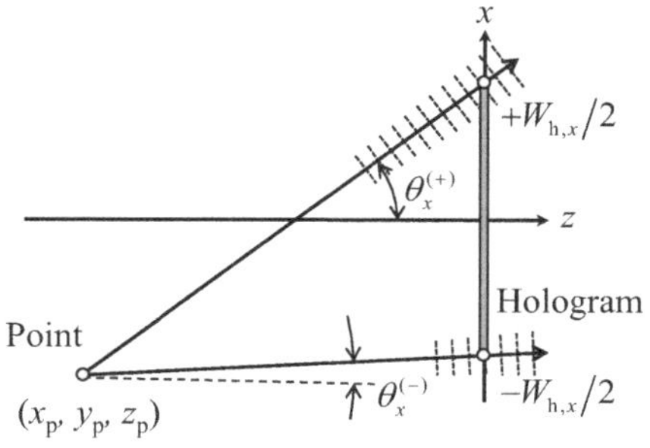

Fig. 10.28 The field passing through the edge of a hologram

$\mathbf{r}_p = (x_p, y_p, z_p)$ is given at the edge of the hologram as follows [72]:

$$
\begin{aligned}
u_{\text{edge}}^{(\pm)}(\mathbf{r}_p) &= \sin\theta_x^{(\pm)}/\lambda \\
&= \frac{-(x_p \mp W_{h,x}/2)}{\lambda\sqrt{(x_p \mp W_{h,x}/2)^2 + z_p^2}} \\
v_{\text{edge}}^{(\pm)}(\mathbf{r}_p) &= \frac{-(y_p \mp W_{h,y}/2)}{\lambda\sqrt{(y_p \mp W_{h,y}/2)^2 + z_p^2}}
\end{aligned}
\tag{10.40}
$$

where $\theta_x^{(\pm)}$ is the angle in $(x, 0, z)$ plane indicated in Fig. 10.28. Using the above edge frequencies, the band limit of the field emitted from the position $\mathbf{r}_p$ is written as

$$
u_{\text{edge}}^{(-)}(\mathbf{r}_p) < u < u_{\text{edge}}^{(+)}(\mathbf{r}_p) \quad \text{and} \quad v_{\text{edge}}^{(-)}(\mathbf{r}_p) < v < v_{\text{edge}}^{(+)}(\mathbf{r}_p). \tag{10.41}
$$

As a result, the limited band of a point forms a rectangle in the Fourier space.

As for the band limit of a polygon, there may be several candidates to be consider. Three typical methods are shown in Fig. 10.29. The individual band limit on the vertex V_n or the center point of a polygon are depicted with blue triangles. The band limit of the polygon is indicated by the red rectangle. Here, the spatial frequency of a polygon field is limited to:

(i) The minimum band that includes all band limits of the polygon vertexes.
(ii) Only the band limit of the center point of the polygon.
(iii) The common region of all band limits of the polygon vertexes.

Since the bandwidth is reduced in order of (i) to (iii), the computation time is also reduced in the same order. The noise produced by the final propagation without the quadruple extension also decrease in the same order. Therefore, the method (iii) is most likely the best technique. However, strange phenomenon, such as disappearance of polygons seen from the edge of the viewing angle, may be observed in the method (iii).

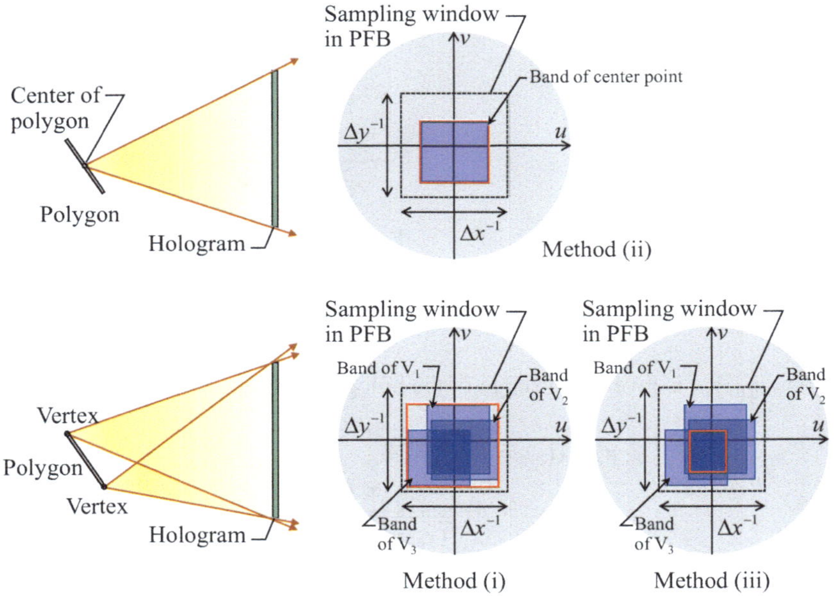

Fig. 10.29 Three typical methods for limiting the spatial bandwidth of a polygon field. V_n is vertex n of a polygon. The blue rectangles indicate the band-limit for each vertex, while the red rectangles indicate the band-limit for the polygon

10.4.3 Modification of Algorithm

To make use of the band limit for reduction of computation time, we have to modify Algorithm 10.2. As shown in the next section, resampling by interpolation is most time-consuming in the polygon-based method. Therefore, we should reduce the number of performing the resampling in line 33 of Algorithm 10.2.

In practice, it is most appropriate to revise the line 25 of Algorithm 10.2, because it controls invoking the steps for resampling. Thus, the line

25: **if** $u_p^2 + v_q^2 \leq \lambda^{-2}$ **then**

should be modified into the following.

25: **if** $u_p^2 + v_q^2 \leq \lambda^{-2}$ **and** $(u_{\mathrm{PBL}}^{(-)} < u_p < u_{\mathrm{PBL}}^{(+)}$ **and** $v_{\mathrm{PBL}}^{(-)} < v_q < v_{\mathrm{PBL}}^{(+)})$ **then**

Here, $u_{\mathrm{PBL}}^{(\pm)}$ and $v_{\mathrm{PBL}}^{(\pm)}$ are upper and lower limit frequencies for the current polygon. Because these limit frequencies depend on the position and shape of the polygon, we must obtain the limit frequencies for each polygon from the vertex positions. More specifically, $u_{\mathrm{edge}}^{(\pm)}$ and $v_{\mathrm{edge}}^{(\pm)}$ are first calculated for each vertex or for the center point

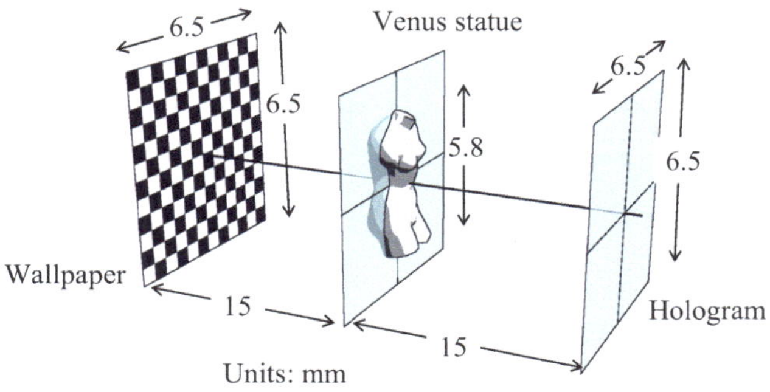

Fig. 10.30 The 3D model of high-definition CGH "The Venus" [67]

Table 10.2 Parameters of "The Venus" CGH

Parameters	Value	Units
Number of pixels (sample points)	65,536 × 65,536	
Pixel pitches (sampling intervals)	1.0 × 1.0	μm
Design wavelength	633	nm
CGH size	6.5 × 6.5	cm^2
Viewing angle	37 × 37	°
Number of polygons (front-face polygons)	1396 (718)	
Size of Venus statue ($W \times D \times H$)	26.7 × 21.8 × 57.8	mm
Culling rate (r_{cull})	0.5	
Diffraction rate (r_{dif})	1.0	

of the current polygon. Limit frequencies $u_{\mathrm{PBL}}^{(\pm)}$ and $v_{\mathrm{PBL}}^{(\pm)}$ for the current polygon are then determined according to the used method, shown in Fig. 10.29.[4]

10.5 Computation Time of Object Field

Actual computation time for an object field definitely depends on the object model and the calculation environment used for measurement, such as CPU, the number of cores, and installed memory, FFT package and so on. Therefore, there may be no sense in measuring and showing computation time peculiar to a given environment and object model. However, the breakdown of processing time is worth showing, and actual computation time at the current stage may also be valuable for recording.

Actual computation time was measured in the object field of HD-CGH "The Venus", introduced in Sect. 1.3. The 3D scene and parameters are shown in Fig. 10.30

[4]The three methods shown in Fig. 10.29 are just examples. There may be another better techniques to provide the band limit for a polygon.

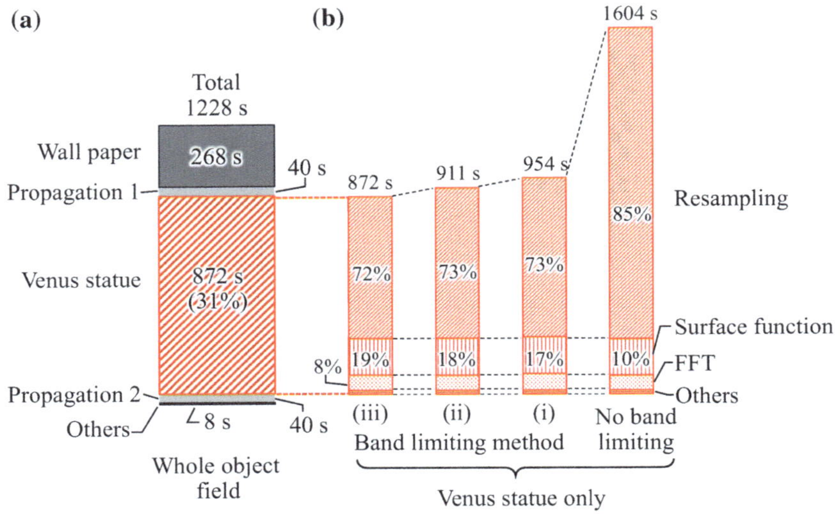

Fig. 10.31 Computation time of the object field of "The Venus": **a** computation time for the whole 3D scene, and **b** only for the Venus statue. CPU: Intel i9-9920X (3.5 GHz/12 Cores), Memory: 128 GB, FFT: Intel MKL11.2

and Table 10.2, respectively. The number of sample points of the object field is approximately 4 billion (=64K × 64K). The main object is a statue of Venus, composed of 1396 polygons. Because $r_{\text{cull}} = 0.5$, only 718 front-face polygons are processes by the polygon-based method.

Figure 10.31a shows the breakdown of measured computation time of the whole 3D scene. Total computation time was 1228 s; it is about 20 min.[5] This computation time is not only for the Venus statue but also for synthesis of the wall paper and occlusion processing by the silhouette method described in Sect. 11.2.2. Computation time only for the Venus statue is 872 s. Here, the bandwidths of polygons are limited using the method (iii) mentioned in Sect. 10.4.2. In this measurement, 72% of computation time of the polygon-based method is consumed by resampling. Generation of surface functions occupies 19% of the total computation time. This is most likely because Mersenne twister algorithm is used for the random generator. Computation time used for FFT is only 8% in this example.

Computation time tends to be longer as the bandwidth increases. When not using any band limiting, computation time is roughly twice as long as that in the case of method (iii). Here note that the quadruple extension is not used in double field propagation. Thus, the object field cannot be obtained properly when the bandwidths are not limited because of the edge effect of the field propagation.

[5]Computation time of the first HD-CGH "The Venus" was approximately 48 h in 2009 [67]. The 3D model of the Venus statue is the same as that used for the measurement in this section.

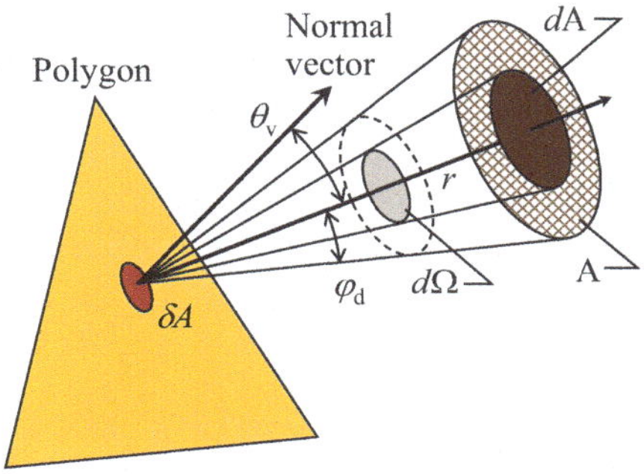

Fig. 10.32 The model of brightness of a polygon expressed by the surface function sampled at an equidistant grid

10.6 Shading and Texture-Mapping of Diffused Surface

Shading and Texture-mapping are very important to reconstruct realistic 3D images by HD-CGHs. In diffused surfaces, the techniques of shading and texture-mapping are essentially the same as those in CG, and there are a lot of shading models and techniques in CG. Although many shading models may also be possible in computer holography, only basic techniques are introduced in this book. It is expected for the reader to invent their new techniques.

10.6.1 Brightness of Reconstructed Surface

In the polygon-based method, brightness of a reconstructed surface varies dependently on the angle of the surface. As a result, objects are shaded as if an unexpected illumination throws light. To compensate for unexpected shading it is necessary to investigate which parameters govern the brightness of the surface in reconstruction.

10.6.1.1 Radiometric Analysis of Polygon Field

Figure 10.32 is a theoretical model that predicts the brightness of a surface represented by the sampled surface function $h_s(x_s, y_s)$.

Suppose that the amplitude of a surface function is constant i.e., $a(x_s, y_s) \equiv a$, and suppose that a^2 provides optical intensity on the surface. In such cases, radiant flux Φ of small area δA on the surface is simply

$$\begin{aligned}\Phi &= \iint_{\delta A} |h_s(x_s, y_s)|^2 \mathrm{d}x_s \mathrm{d}y_s \\ &\cong \delta A a^2. \end{aligned} \tag{10.42}$$

However, this consideration is not enough rigorous to represent the radiant flux in sampled surface functions. If the sampling density decreases, the number of sample points properly decreases within the small area δA. The radiant flux should be decreases in this case. Therefore, we introduce relative sampling density $\sigma\,(\leq 1)$ that is proportional to the surface density of sample points. Suppose that a continuous surface function gives $\sigma = 1$. Consequently, we represent the radiant flux in a sampled surface function as follows.[6]

$$\Phi_s \cong \delta A \sigma a^2. \tag{10.43}$$

Assuming that the small area emits light within a diffusion angle in a direction at θ_v to the normal vector, the solid angle corresponding to the diffusion corn is given as $\Omega = A/r^2$, where $A = \pi(r \tan\varphi_d)^2$ is the section of the diffusion corn at a distance of r and φ_d is the diffusion angle of light, which depends on the diffuser function $\phi(x_s, y_s)$ of (10.1). According to photometry, brightness of the surface, observed in the direction at the angle of θ_v, is given by

$$L = \frac{d\Phi_s/d\Omega}{\cos\theta_v \delta A}. \tag{10.44}$$

Assuming that light is diffused almost uniformly:

$$\frac{d\Phi_s}{dA} \simeq \frac{\Phi_s}{A}, \tag{10.45}$$

the brightness is rewritten by substituting $d\Phi_s \simeq (\Phi_s/A)dA$, $d\Omega = dA/r^2$, and (10.43) into (10.44) as follows:

$$L \simeq \frac{\sigma a^2}{\pi \tan^2\varphi_d \cos\theta_v}. \tag{10.46}$$

As a result, the brightness of the surface depends on the surface density of sampling, the diffusiveness of the diffuser function, and the amplitude of the surface function. In addition, the brightness of the surface is governed by the observation angle θ_v shown in Fig. 10.33. In other words, if several surfaces with the same surface function are reconstructed from a hologram, the brightness varies according to the direction of the normal vector of the surface. This phenomenon causes unexpected shading.

10.6.1.2 Compensation of Brightness

Since only the simple theoretical model has been discussed so far, (10.46) is partially appropriate for the brightness of optically reconstructed surfaces of real holograms. The brightness given in (10.46) diverges in limit $\theta_v \to \pi/2$, but an actual hologram

---

[6] As for sampling density σ, description of the radiant flux was not sufficiently accurate in the original paper [59].

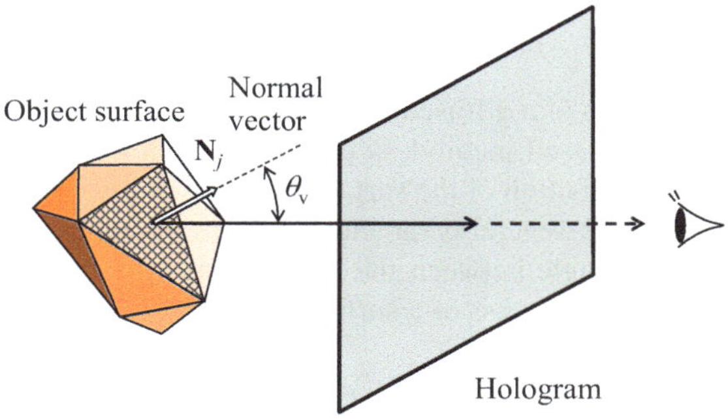

Fig. 10.33 The angle between the normal vector of a polygon and observer's line of sight

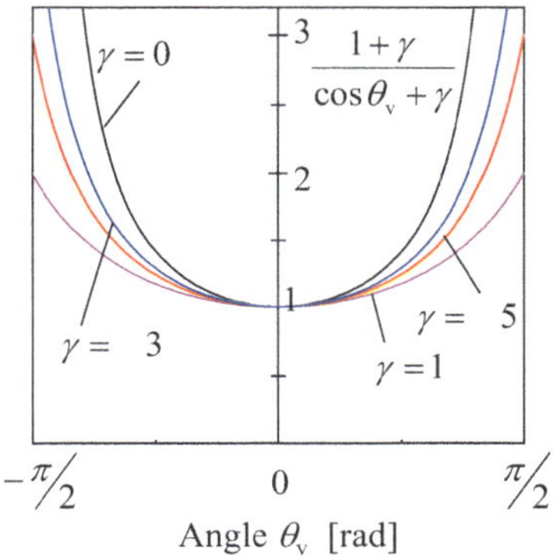

Fig. 10.34 Curves of the angle factor for several values of γ

cannot produce infinite brightness for the reconstructed surface. Thus, (10.46) is not sufficient to compensate for the brightness. To avoid the divergence of brightness in (10.46), an angle factor $(1 + \gamma)/(\cos\theta_v + \gamma)$, shown in Fig. 10.34, should be introduced instead of $1/\cos\theta_v$ a priori. This angle factor is unity in $\theta_v = 0$ and $1 + 1/\gamma$ in $\theta_v = \pi/2$.

Consequently, the brightness is given as

$$L = \frac{\sigma a^2}{\pi \tan^2 \varphi_d} \frac{(1 + \gamma)}{(\cos\theta_v + \gamma)}, \tag{10.47}$$

where γ is a parameter that plays a role to avoid the divergence of brightness and prevents overcompensation. Since γ is dependent on actual methods to fabricate holograms such as encoding of the field or the property of recording materials, it should be determined experimentally.

10.6.2 Amplitude of Surface Function

Brightness of a diffused surface, which is reconstructed by a CGH created using the polygon-based method, is primarily given by $a_j(x_s, y_s)$ in (10.2), which is amplitude distribution of the surface function of polygon j. However, actual brightness is affected by several parameters, such as the sampling density of the surface function and the angle between the surface and the observer's line of sight. According to (10.46), brightness of a surface has dependency:

$$L_j(x_s, y_s) \propto \frac{\sigma_j}{\cos\theta_v} a_j^2(x_s, y_s), \tag{10.48}$$

where σ_j is a relative sampling density of the surface function of polygon j. Here, $(x_s, y_s) = (x_s, y_s, 0)$ is the tilted local coordinates in the surface of polygon j. Although we should append the suffix j to x_s and y_s, it is omitted accordingly to the convention mentioned in Sect. 10.2.2.

Equation (10.48) is not suitable for practical shading because we cannot know the value of θ_v in creating the CGH. In addition, the brightness diverges in $\theta_v = \pi/2$ as described in the preceding section. Therefore, we adopt the non-divergent brightness model in (10.47), and replace θ_v by the angle between the normal vector of the polygon and the optical axis:

$$\theta_v \approx \theta_N. \tag{10.49}$$

In addition, we adopt the following as the relative sampling density.

$$\sigma_j = \frac{\Delta x \Delta y}{\Delta x_s \Delta y_s}, \tag{10.50}$$

where Δx_s, Δy_s, Δx, and Δy are sampling intervals of the surface function and object field again. Although Δx_s and Δy_s are peculiar to the polygon j, the suffix is omitted again. Equation (10.50) does not exactly correspond with the original description of σ in Sect. 10.6.1.1, but is sufficient for the purpose of shading because only the relative brightness is important in this case.

As a result, suppose that dimensionless brightness is represented by

$$I_j(x_s, y_s) = \left(\frac{1+\gamma}{\cos\theta_N + \gamma}\right) \frac{\Delta x \Delta y}{\Delta x_s \Delta y_s} a_j^2(x_s, y_s), \tag{10.51}$$

where

$$\cos\theta_N = \begin{cases} z_N & z_N \geq 0 \\ 0 & \text{otherwise} \end{cases}. \tag{10.52}$$

Here, $z_N = \mathbf{N}_j \cdot \mathbf{e}_z$ is the z-component of the normal vector.

On the other hand, the brightness distribution must shape the polygon primarily and brightness varies depending on shading of the polygon. Thus, $I_j(x_s, y_s)$ should

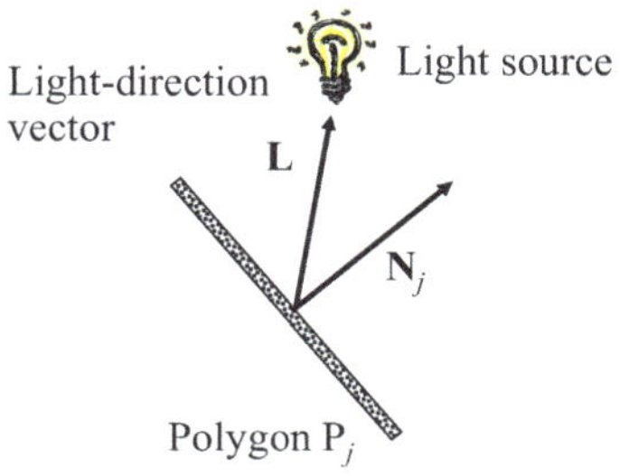

Fig. 10.35 Lambertian reflectance

be represented as

$$I_j(x_s, y_s) = I_{\text{shade},j}(x_s, y_s) I_{\text{shape},j}(x_s, y_s), \tag{10.53}$$

Here, $I_{\text{shape},j}(x_s, y_s)$ gives the shape of the polygon j and is defined as

$$I_{\text{shape},j}(x_s, y_s) = \begin{cases} 1 & \text{inside polygon } j \\ 0 & \text{otherwise} \end{cases}, \tag{10.54}$$

while $I_{\text{shade},j}(x_s, y_s)$ shades the polygon, as described in the following section. By substituting (10.53) into (10.51), the amplitude distribution is generally written as

$$a_j(x_s, y_s) = a_{\text{shape},j}(x_s, y_s) \sqrt{I_{\text{shade},j}(x_s, y_s)}, \tag{10.55}$$

where

$$a_{\text{shape},j}(x_s, y_s) \equiv \sqrt{\left(\frac{\cos\theta_N + \gamma}{1+\gamma} \right) \frac{\Delta x_s \Delta y_s}{\Delta x \Delta y} I_{\text{shape},j}(x_s, y_s)}. \tag{10.56}$$

10.6.3 Shading of Diffused Surfaces

Suppose that the surface function of a diffused polygon is represented by

$$h_{\text{dif},j}(x_s, y_s) = a_{\text{dif},j}(x_s, y_s) g_{\text{dif}}(x_s, y_s), \tag{10.57}$$

$$g_{\text{dif}}(x_s, y_s) = \exp[i\phi_{\text{dif}}(x_s, y_s)], \tag{10.58}$$

where $\phi_{\text{dif}}(x_s, y_s)$ is a random phase distribution to spread the polygon field isotropically.

10.6.3.1 Lambertian Reflectance

In ordinary CG, Lambertian reflectance is commonly used as a model for diffuse reflection. We use the same model in computer holography. Using the model, brightness of polygon j is represented by

$$I_j = k_{\mathrm{d},j}(\mathbf{L}\cdot\mathbf{N}_j)I_{\mathrm{d}}, \tag{10.59}$$

where $k_{\mathrm{d},j}$ and I_{d} are a reflection constant of the polygon and intensity of the incoming light, respectively. $\mathbf{N}_j$ and $\mathbf{L}$ are the polygon's normal vector and a normalized light-direction vector pointing from the surface to the light source, as shown in Fig. 10.35. If $\mathbf{L}\cdot\mathbf{N}_j < 0$, we take $I_j \equiv 0$ because no light illuminates the surface.

In general, ambient light is introduced into the model to avoid the non-illuminated surfaces being perfect black like objects laid on the moon surface:

$$I_{\mathrm{dif},j} = k_{\mathrm{d},j}(\mathbf{L}\cdot\mathbf{N}_j)I_{\mathrm{d}} + k_{\mathrm{a},j}I_{\mathrm{a}}, \tag{10.60}$$

where $k_{\mathrm{a},j}$ and I_{a} are a reflection constant and intensity of the ambient light, respectively. Here, note that the constants $k_{\mathrm{d},j}$, $k_{\mathrm{a},j}$, I_{d}, and I_{a} should depend on the color. Thus, these should be treated as a function of the wavelength in creation of a color CGH.

10.6.3.2 Flat Shading

In flat shading, the brightness $I_{\mathrm{shade},j}(x_{\mathrm{s}}, y_{\mathrm{s}})$ in (10.53) is a constant given by (10.60):

$$I_{\mathrm{shade},j}(x_{\mathrm{s}}, y_{\mathrm{s}}) \equiv I_{\mathrm{dif},j}. \tag{10.61}$$

Amplitude distribution $a_{\mathrm{dif},j}(x_{\mathrm{s}}, y_{\mathrm{s}})$ in flat shading is obtained by substituting (10.60) and (10.61) into (10.55):

$$a_{\mathrm{dif},j}(x_{\mathrm{s}}, y_{\mathrm{s}}) = a_{\mathrm{shape},j}(x_{\mathrm{s}}, y_{\mathrm{s}})\left[k_{\mathrm{d},j}(\mathbf{L}\cdot\mathbf{N}_j)I_{\mathrm{d}} + k_{\mathrm{a},j}I_{\mathrm{a}}\right]^{1/2}. \tag{10.62}$$

We can omit several parameters to avoid redundancy in actual creation of a polygon field. Choosing $k_{\mathrm{d},j} = I_{\mathrm{d}} = I_{\mathrm{a}} \equiv 1$ and $k_{\mathrm{a}} \equiv k_{\mathrm{a},j}$ (assuming the ambient constant is independent of the polygon), the above equation is

$$a_{\mathrm{dif},j}(x_{\mathrm{s}}, y_{\mathrm{s}}) = a_{\mathrm{shape},j}(x_{\mathrm{s}}, y_{\mathrm{s}})\left[\mathbf{L}\cdot\mathbf{N}_j + k_{\mathrm{a}}\right]^{1/2}. \tag{10.63}$$

In this case, only k_{a} determines the rate of ambient light.

An example of the surface function in flat shading is shown in Fig. 10.36a. Figure 10.37a is a close-up photograph of the surface of an object that is optically reconstructed by high-definition CGH "Moai-II" (see Appendix A.1). This is an example of flat shading of diffused surfaces.

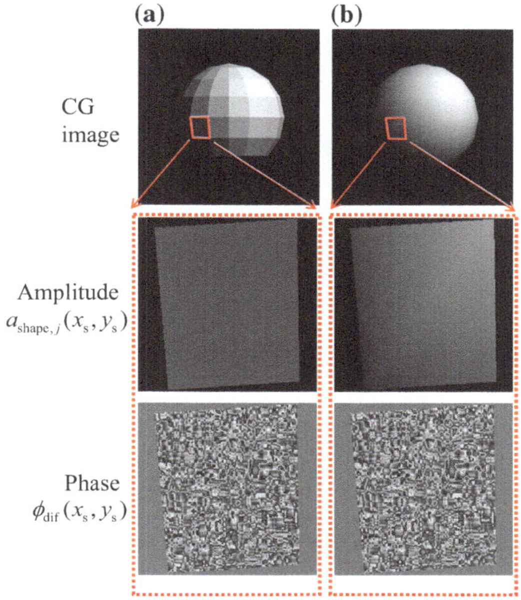

Fig. 10.36 Comparisons between surface functions in **a** flat shading, and **b** smooth shading [77]

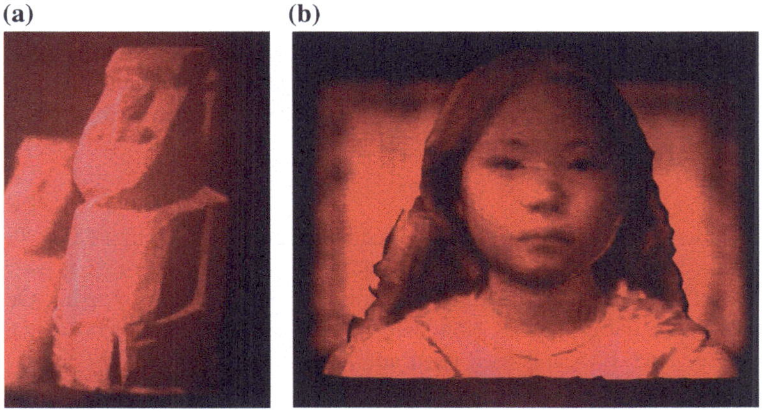

Fig. 10.37 Examples of optical reconstruction of high-definition CGHs with diffused surfaces: **a** "Moai II", and **b** "Shion" (see Appendix A.1 for the parameters). Moai-II is rendered using simple flat shading of diffused surfaces, while Shion is rendered using Gouraud shading and texture-mapping

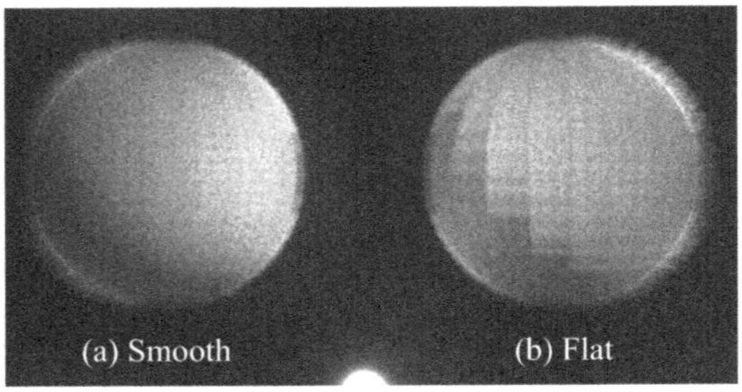

Fig. 10.38 Simulated reconstruction of two spheres rendered with different shading techniques

10.6.3.3 Smooth Shading

Smooth shading of diffused surfaces is also achieved by the same technique as that in CG. In this case, amplitude distribution $I_{\text{shade},j}(x_s, y_s)$ is not constant but varies so that the polygon surfaces have gradation to remove the border between polygons. Notable techniques for smooth shading are Gouraud and Phong shadings in CG.

In Gouraud shading, the normal vector for each vertex of a polygon is obtained by averaging normal vectors of polygons that share the vertex. Then, brightness for each vertex is calculated using (10.60), and $I_{\text{shade},j}(x_s, y_s)$ is determined by using bilinear interpolation of the vertex brightnesses. Figure 10.36b shows an example of the surface function in Gouraud shading.

Unlike Gouraud shading, normal vectors for each sample point of $I_{\text{shade},j}(x_s, y_s)$ are first calculated by interpolation of the vertex's normal vectors in Phong shading. Brightness for each sample point of $I_{\text{shade},j}(x_s, y_s)$ is, then, calculated by (10.60).

When a CGH is created using smooth shading, the amplitude distribution is simply given by

$$a_{\text{dif},j}(x_s, y_s) = a_{\text{shape},j}(x_s, y_s) I_{\text{shade},j}(x_s, y_s)^{1/2}. \tag{10.64}$$

Here, parameters k_d, I_d, k_a, and I_a are reflected to rendering through $I_{\text{shape},j}(x_s, y_s)$. Comparison between flat and smooth shading of diffused surfaces is shown in Fig. 10.38. The photograph of a high-definition CGH rendered by using Gouraud shading is shown in Fig. 10.37b. Note that not only Gouraud shading but also texture mapping (see Sect. 10.6.4) is used for rendering in this CGH.

Both techniques of Gouraud and Phong shading can be applicable to the polygon-based computer holography. It is known in CG that Phong shading works better for rendering of surface highlight. However, it should be noted that the position of surface highlight, produced in specular surfaces, must shift properly as moving the observer's viewpoint. We cannot realize this effect by the simple Phong shading. To achieve highlight shift in specular surfaces, we need special rendering techniques explained in Sect. 10.7.

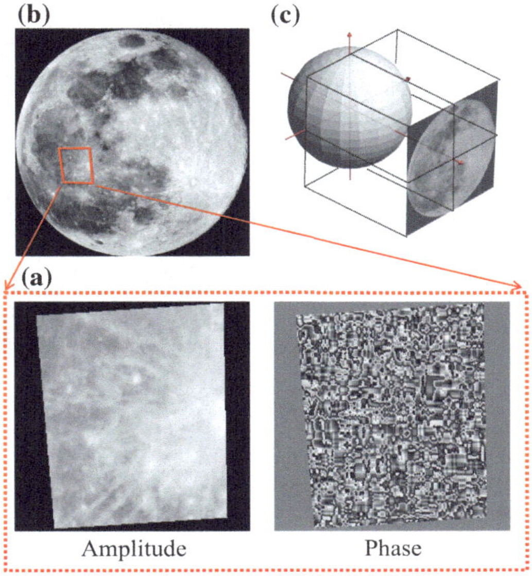

Fig. 10.39 An example of texture-mapping by orthogonal projection: **a** The surface function, **b** an astrophotograph of the real moon, and **c** illustration of orthogonal projection

10.6.4 Texture-Mapping

Texture-mapping as well as shading can also be introduced into the polygon-based computer holography by using the same technique as that in CG. To perform texture-mapping, (10.62) and (10.64) are simply extended to

$$a_{\text{dif},j}(x_s, y_s) = a_{\text{shape},j}(x_s, y_s) a_{\text{tex},j}(x_s, y_s) \left[k_{\text{d},j}(\mathbf{L} \cdot \mathbf{N}_j) I_\text{d} + k_{\text{a},j} I_\text{a}\right]^{1/2}, \quad (10.65)$$

and

$$a_{\text{dif},j}(x_s, y_s) = a_{\text{shape},j}(x_s, y_s) a_{\text{tex},j}(x_s, y_s) I_{\text{shade},j}(x_s, y_s)^{1/2}, \quad (10.66)$$

in flat and smooth shading, respectively. Here,

$$a_{\text{tex},j}(x_s, y_s) = I_{\text{tex},j}(x_s, y_s)^{1/2}, \quad (10.67)$$

and $I_{\text{tex},j}(x_s, y_s)$ is the brightness of the texture.

As in CG, various mapping techniques can be applicable to producing $I_{\text{tex},j}(x_s, y_s)$. Figure 10.39 is an example of texture-mapping by orthogonal projection. An astrophotograph of the real moon is mapped on the polygon-meshed sphere. Figure 10.40 is the photograph of optical reconstruction of HD-CGH "The Moon", where the astrophotograph is mapped on the sphere with flat shading. Here, the background stars is not depicted by a 2D image but produced by 300 point sources of light. Another HD-CGH named "Shion", whose optical reconstruction is shown in Fig. 10.37b, is created by texture-mapping the photograph of a real face to the polygon mesh with

Fig. 10.40 Optical reconstruction of HD-CGH "The Moon". An astrophotograph of the real moon is texture-mapped to the polygon-meshed sphere with flat shading (see Appendix A.1), Video link, https://youtu.be/DdOveIue3sc

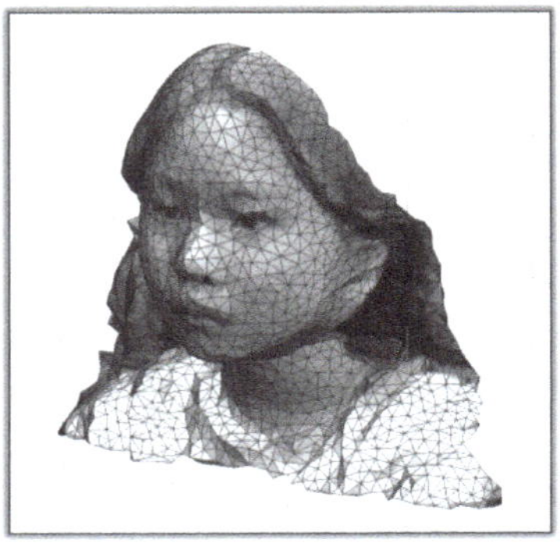

Fig. 10.41 The 3D model of Shion. The photograph of a real face is texture-mapped to the polygon-mesh measured using a 3D laser scanner

Gouraud shading. The CG model, shown in Fig. 10.41, is essentially the same 3D model as that in "Brothers" in Figs. 1.6 and 11.5, and is produced by measuring the shape of the live face using a 3D laser scanner.[7]

Not only orthogonal projection but also another technique of texture-mapping such as UV-mapping can be used for the polygon-based computer holography.

10.7 Rendering Specular Surfaces

Unlike diffused surfaces, rendering of specular surfaces is very different from that in CG. In diffused surfaces, brightness of a surface is nearly constant; it does not change dependently on the observer's viewpoint. In contrast, as the viewpoint moves, brightness of specular polygons must vary under flat shading [95], and the position of highlight must move in smooth specular surfaces [93]. These effects can be realized by modification of the spatial spectrum of the surfaces.

[7] A girl appearing in the 3D images in common is the author's daughter: Shion. She was 12 years old when the author measured the 3D shape of her face using a 3D scanner.

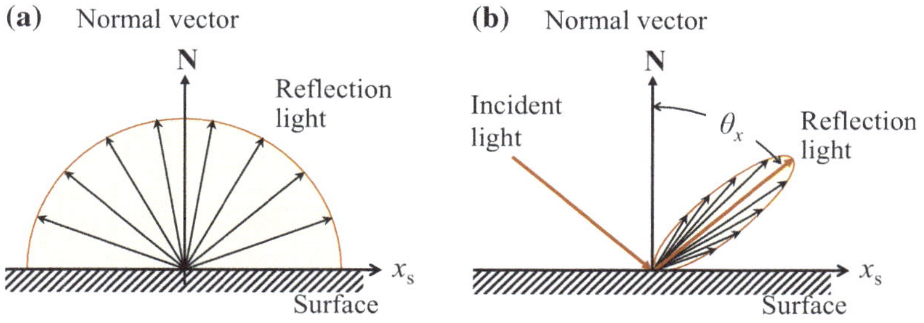

Fig. 10.42 Schematic illustration of **a** diffuse and **b** specular reflection

Fig. 10.43 Schematic illustration of the spatial spectrum of diffuse and specular reflection

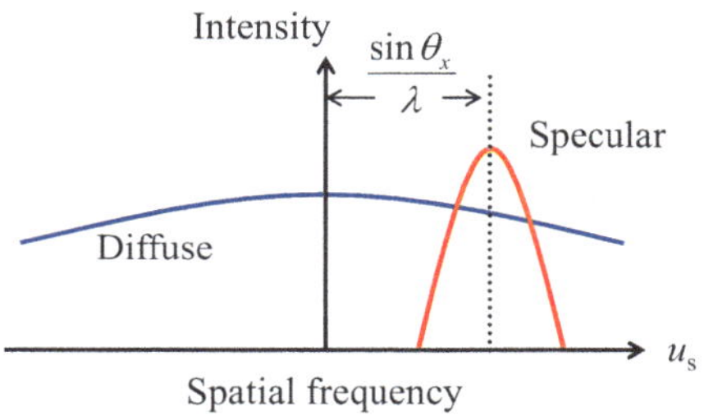

10.7.1 Spectrum of Diffuse and Specular Reflection

Figure 10.42 schematically illustrates two types of surface reflection: diffuse and specular reflection. The spatial spectra of these two types of reflection are also schematically depicted in Fig. 10.43. Because the spectrum represents the far-field pattern of the light, diffuse reflection has a broadband spectrum. A random pattern is thus provided to the phase distribution $\phi(x_s, y_s)$ in rendering of diffused surfaces, as mentioned in Sect. 10.6.3.

In specular reflection, energy of the reflected light is concentrated around the reflection direction. The spectrum is therefore narrow as compared with diffuse reflection, and is shifted from the origin, as shown in Fig. 10.43. When the reflection angle is θ_x in the $(x_s, 0, z_s)$ plane of the tilted local coordinates, the shift magnitude is obtained from the far-field diffraction in (5.47) and the direction angles[8] of field propagation in (5.8). Thus, the frequency shift corresponding to the angle θ_x is

$$u_{\text{shift}} = \frac{\sin\theta_x}{\lambda}, \tag{10.68}$$

where λ is the wavelength of light as usual. The amount of the frequency shift is found easily using the far-field consideration, whereas the shape of the spectrum cannot be determined without a reflection model.

[8]Note that the relation between the direction and reflection angles is $\theta_x = \pi - \alpha$.

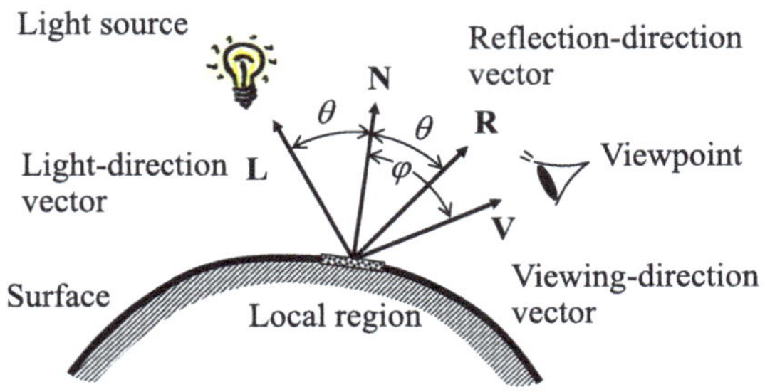

Fig. 10.44 Vectors in the Phong reflection model

10.7.2 Phong Reflection Model

The Phong reflection model is a well-known model to provide local brightness of a surface having the specular component in CG [102].[9] Brightness of a local region of a surface, illuminated by a single light source, is represented in the Phong model as follows:

$$I_{\text{Phong}}(\mathbf{V}, \mathbf{L}, \mathbf{N}) = k_{\text{d}}(\mathbf{L} \cdot \mathbf{N})I_{\text{d}} + k_{\text{s}}(\mathbf{R} \cdot \mathbf{V})^{\alpha} I_{\text{s}} + k_{\text{a}}I_{\text{a}}, \tag{10.69}$$

where k_d, k_s, and k_a are reflection constants for diffuse, specular, and ambient components, respectively. I_d, I_s, and I_a are intensity components that are also corresponding to the diffuse, specular, and ambient lights, respectively. In creating color models, these parameters depend on the color, or are treated as a function of the wavelength.

Vectors **N** and **L** are again a normal vector of the local region and normalized light-direction vector, respectively. **R** is a unit vector of the direction of regularly reflected light, and thus, given by

$$\mathbf{R} = \frac{2(\mathbf{L} \cdot \mathbf{N})\mathbf{N} - \mathbf{L}}{|2(\mathbf{L} \cdot \mathbf{N})\mathbf{N} - \mathbf{L}|}. \tag{10.70}$$

Vector **V** is also a unit vector of the viewing direction, which points toward the viewpoint.

The first term of (10.69) is the component of diffused light, and given by Lambertian reflection in (10.59). The third term is of ambient light. These are the same as those in rendering of diffused surfaces, and are not affected by the viewing direction **V**. The third term represents the specular components of reflection. We can write down the specular component as

$$\begin{aligned} I_{\text{spec}}(\varphi; \alpha) &= (\mathbf{R} \cdot \mathbf{V})^{\alpha} \\ &= [\cos(\varphi - \theta)]^{\alpha}, \end{aligned} \tag{10.71}$$

where α is a parameter that provides degree of specularity of the surface, and is called a *shininess constant*. φ is an angle formed between the normal and view-direction

[9] The Phong reflection model is a different idea from the Phong shading mentioned in Sect. 10.6.3.3.

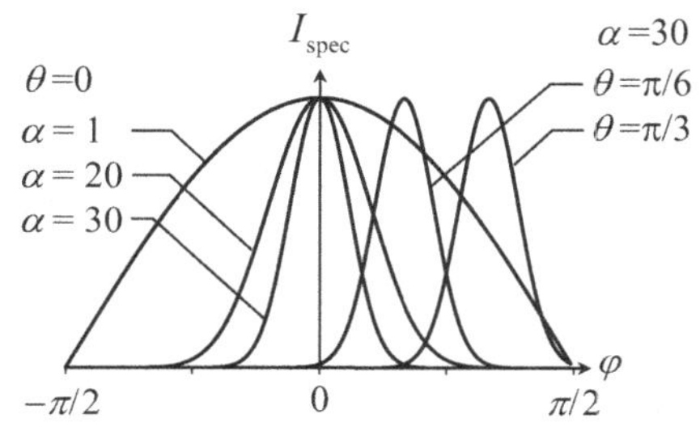

Fig. 10.45 Brightness change of the specular component of the Phong reflection model

vectors. Figure 10.45 shows brightness variation given by (10.71) when moving the viewpoint. The specular component is large when the view direction almost agree with regular reflection. When α is large, the surface causes a nearly mirror-like reflection; brightness rapidly reduces with increasing $|\varphi - \theta|$.

It should be noted that the Phong reflection model is not a physical model, i.e., (10.71) does not have any physical base. Nevertheless, the Phong model is useful for rendering specular surfaces because of its simplicity.

10.7.3 Spectral Envelope of Specular Component

The fact that specular surfaces have a limited diffusibility leads to narrowing the bandwidth of the reflected light. Here, the reflected light is represented by the surface function in the polygon-based method. This means that the spectrum of the surface function must be band-limited for creation of specular surfaces. We use the specular component of the Phong model to determine the shape of the spectrum of the phase distribution $\phi(x_s, y_s)$.

In far-fields, the light traveling in the viewing direction can be regarded as a plane wave. We therefore interpret the viewing direction as a unit wave vector as follows:

$$\mathbf{V} = \mathbf{k}/k, \tag{10.72}$$

where $\mathbf{k}$ and k are again a wave vector and wave number, respectively. As mentioned in Sect. 5.2.1.1, the wave vector can be written by the spatial frequencies. Using (5.5), the wave vector is represented as follows:

$$\mathbf{k} = 2\pi(u, v, w), \tag{10.73}$$

where u and v are again Fourier frequencies in the tilted coordinates and limited in $u^2 + v^2 \leq 1/\lambda^2$. The frequency w with respect to z is not independent of u and v, and is given by $w = (\lambda^{-2} - u^2 - v^2)^{1/2}$ as in (5.7).

By substituting (10.73) into (10.72), the viewing vector is

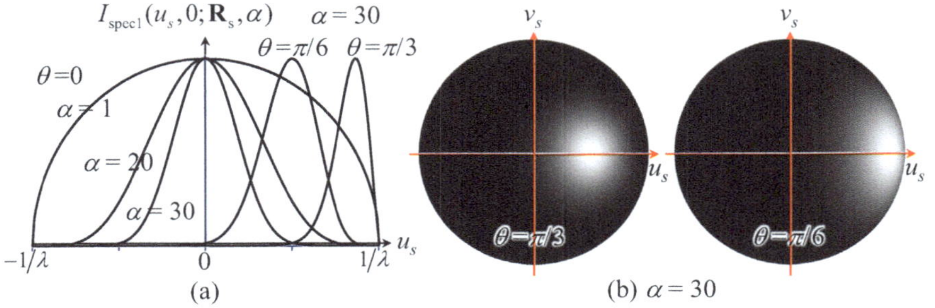

Fig. 10.46 Examples of the spectral envelope based of the Phong specular reflection: **a** the shape of $I_{\mathrm{spec1}}(u_s, 0; \mathbf{R}_s, \alpha)$, and **b** grayscale images of $I_{\mathrm{spec1}}(u_s, v_s; \mathbf{R}_s, 30)$

$$\mathbf{V}(u, v) = \lambda \left(u, v, \sqrt{\lambda^{-2} - u^2 - v^2} \right). \tag{10.74}$$

The brightness of specular reflection can be rewritten as a function of the spatial frequencies by substituting (10.74) into (10.71):

$$\begin{aligned} & I_{\mathrm{spec1}}(u, v; \mathbf{R}, \alpha) \\ & = [\mathbf{R} \cdot \mathbf{V}(u, v)]^{\alpha} \\ & = \begin{cases} \lambda^{\alpha} \left(R_x u + R_y v + R_z \sqrt{\lambda^{-2} - u^2 - v^2} \right)^{\alpha} & \mathbf{R} \cdot \mathbf{V}(u, v) \geq 0 \\ 0 & \text{otherwise} \end{cases}, \end{aligned} \tag{10.75}$$

where

$$\mathbf{R} = \left(R_x, R_y, R_z \right). \tag{10.76}$$

Equation (10.75) provides the first candidate of the spectral envelope corresponding to the specular component of the Phong reflection model.

Examples of the spectral envelope are shown in Fig. 10.46. Here, the reflection vector is defined by $\mathbf{R}_s = (\sin\theta, 0, \cos\theta)$ in the same manner as that in Figs. 10.44 and 10.45. The shape of the spectral envelops is similar to those in the Phong shading. When $\theta = 0$, the shapes are symmetrical with respect to the center frequency and resemble those in the Phong model, but a little deformation appears in the case of a large refection angle.

10.7.4 Generation of Specular Diffuser for Surface Function

To create the specular surfaces using the polygon-based method, the phase factor of the surface function $\exp[i\phi(x_s, y_s)]$ must be modified so that its spectral envelope is fitted to (10.75). Here, we have to change the coordinates system because the surface functions are defined in the tilted local coordinates. Also, the reflection-direction vector must be expressed in component form in the tilted coordinates:

$$\begin{aligned}\mathbf{R}_{s,j} &= \mathbf{T}_j^{-1}\mathbf{R}_j \\ &= \left(R_{s,x}, R_{s,y}, R_{s,z}\right),\end{aligned} \tag{10.77}$$

where $\mathbf{T}_j$ and $\mathbf{R}_j$ are obtained from normal vector $\mathbf{N}_j$ of polygon j using (10.18) and (10.70), respectively.

Although the reflection vector is dependent on the polygon, we omit the suffix j in the following sections unless confusion is caused in the formulation.

10.7.4.1 One Step Generation

The square root of $I_{\text{spec1}}(u_s, v_s; \mathbf{R}_s, \alpha)$ give the spectral envelope because the brightness of a surface corresponds to the intensity of the reflection field. The shaped spectrum of the diffuser is therefore given by:

$$G_{\text{spec1}}(u_s, v_s; \mathbf{R}_s, \alpha) = G_{\text{dif}}(u_s, v_s)\sqrt{I_{\text{spec1}}(u_s, v_s; \mathbf{R}_s, \alpha)}, \tag{10.78}$$

$$G_{\text{dif}}(u_s, v_s) = \mathscr{F}\left\{g_{\text{dif}}(x_s, y_s)\right\}, \tag{10.79}$$

where $g_{\text{dif}}(x_s, y_s)$ and $G_{\text{dif}}(u_s, v_s)$ are a random phase factor in (10.58) and its spectrum, respectively. Figure 10.47a shows an example of $G_{\text{dif}}(u_s, v_s)$. Examples of the spectral envelope $I_{\text{spec1}}(u_s, v_s; \mathbf{R}_s, 30°)$ and shaped spectrum $G_{\text{spec1}}(u_s, v_s; \mathbf{R}_s, 30°)$ are also shown in (b) and (c), respectively.

The phase distribution for the specular diffuser is obtained by the Fourier transform of the shaped spectrum $G_{\text{spec1}}(u_s, v_s; \mathbf{R}_s, \alpha)$. However, unfortunately the amplitude distribution $|\mathscr{F}^{-1}\left\{G_{\text{spec1}}(u_s, v_s; \mathbf{R}_s, \alpha)\right\}|$ is not constant but similar to a random pattern. We cannot use $|\mathscr{F}^{-1}\left\{G_{\text{spec1}}(u_s, v_s; \mathbf{R}_s, \alpha)\right\}|$ itself for the diffuser-phase because it most likely causes unnecessary amplitude modulation, i.e., noise. Thus, we define the diffuser-phase distribution as

$$g_{\text{spec1}}(x_s, y_s; \mathbf{R}_s, \alpha) \equiv \exp[i\phi_{\text{spec1}}(x_s, y_s; \mathbf{R}_s, \alpha)], \tag{10.80}$$

$$\phi_{\text{spec1}}(x_s, y_s; \mathbf{R}_s, \alpha) \equiv \arg\left\{\mathscr{F}^{-1}\left\{G_{\text{spec1}}(u_s, v_s; \mathbf{R}_s, \alpha)\right\}\right\}, \tag{10.81}$$

where $\arg\{\xi\}$ is again the argument of ξ.

10.7.4.2 Possibility of Improving Specular Diffuser

It is ensured that the amplitude of the phase factor is properly unity:

$$|g_{\text{spec1}}(x_s, y_s; \mathbf{R}_s, \alpha)| \equiv 1, \tag{10.82}$$

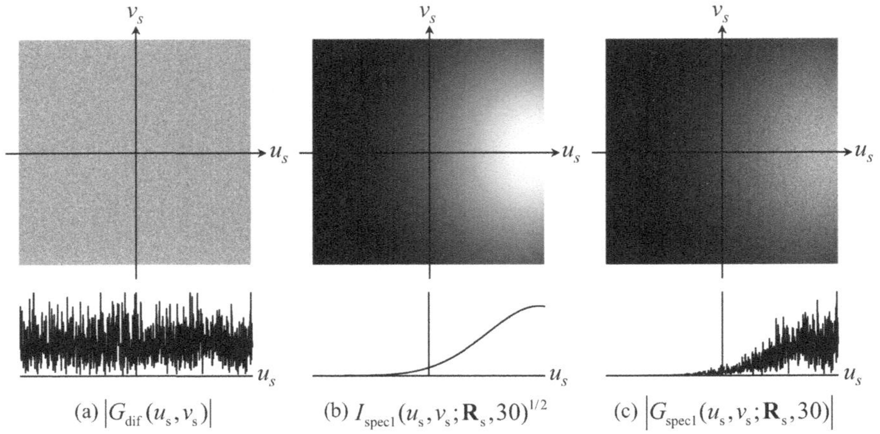

Fig. 10.47 An example of spectrum-shaping based on the Phong reflection model. $M = N = 1024$, $\Delta x = \Delta y = 0.6$ [μm], $\lambda = 633$ [nm], and $\theta = 30°$. The grayscale images are depicted with the standard encoding gamma of 1/2.2

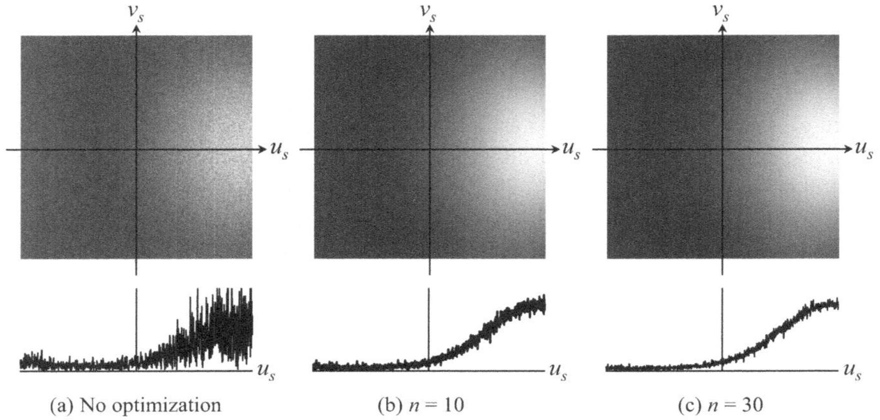

Fig. 10.48 Examples of the spectrum of $g_{\mathrm{specl}}(x_s, y_s; \mathbf{R}_s, \alpha)$. The amplitude images of $|\mathcal{F}^{-1}\{g_{\mathrm{specl}}(x_s, y_s; \mathbf{R}_s, 30°)\}|$ is depicted **a** without optimization, and with optimization by **b** 10 and **c** 30 times iteration of the GS algorithm. $M = N = 1024$, $\Delta x = \Delta y = 0.6$ [μm], $\lambda = 633$ [nm], and $\theta = 30°$

but the spectrum of $g_{\mathrm{specl}}(x_s, y_s; \mathbf{R}_s, \alpha)$ is not identical to $G_{\mathrm{specl}}(u_s, v_s; \mathbf{R}_s, \alpha)$, as shown in Fig. 10.48a, and is even noisier than that in Fig. 10.47c. The shaped spectrum $G_{\mathrm{specl}}(u_s, v_s; \mathbf{R}_s, 30°)$ has almost zero-value in the region where the spectral envelope $I_{\mathrm{specl}}(u_s, v_s; \mathbf{R}_s, 30°)$ nearly takes zero as in Fig. 10.47c. However, noise is produced in the same region in Fig. 10.48a. This spectral noise may degrade the reconstructed image of the CGH.

Ideally speaking, the specular diffuser must keep a constant amplitude as in (10.82) in the real domain, and besides the envelope surface must fits with the curved surface of $I_{\mathrm{spec1}}(u_s, v_s; \mathbf{R}_s, \alpha)$ in the spectral domain. To satisfy both conditions as much as possible, we can make use of very famous Gerchberg-Saxton (GS) algorithm [18].

The flow of the GS algorithm for shaping the diffuser spectrum is shown in Fig. 10.49. In this algorithm, constraints are imposed on the amplitude distribution for each domain. Equation (10.82) presents the constraint in the real domain, i.e., the amplitude of the phase factor must be unity. Otherwise, unnecessary textures, which is not given by (10.65), is produced in the polygon surface. Thus, we only keep the phase of the field and replace the amplitude by unity in the real domain.

$$g'(x_s, y_s) = \exp\left[i \arg\{g(x_s, y_s)\}\right].$$

Here, $g(x_s, y_s) = \mathscr{F}^{-1}\left\{G_{\mathrm{spec1}}(u_s, v_s; \mathbf{R}_s, \alpha)\right\}$ for the first time of iteration. In the spectrum domain, the constraint is written by

$$G'(u_s, v_s) = \sqrt{I_{\mathrm{spec1}}(u_s, v_s; \mathbf{R}_s, \alpha)} \exp\left[i \arg\{G(u_s, v_s)\}\right]. \tag{10.83}$$

The phase distribution is also kept in the complex spectrum, but the amplitude distribution is replaced by that of the specular envelope.

The envelope of the spectral amplitude is gradually converging to the target shape by iterating the loop many times. Finally, the diffuser factor for the specular component of the Phong reflection model is given by

$$g_{\mathrm{spec1}}(x_s, y_s; \mathbf{R}_s, \alpha) = g'(x_s, y_s). \tag{10.84}$$

Figure 10.48b and c show the spectra of $g_{\mathrm{spec1}}(x_s, y_s; \mathbf{R}, 30°)$ after 10 and 30 times iterations of the GS algorithm, respectively. The spectrum becomes not only more similar to the envelope in Fig. 10.47b but also less noisy.

In the last of this section, let us emphasize that the validity of above improvement is not confirmed in actual reconstruction of HD-CGHs. The spectrum apparently becomes less noisy by use of the GS algorithm. However, unfortunately, we cannot recognize the improvement in the reconstructed surfaces. Optimization of the specular diffuser is a future work.

10.7.5 Fast Generation of Specular Diffuser by Shifting Spectrum

Each field of specular polygons is created by shaping the spectral envelope. However, actual computation is most likely time-consuming when calculating the whole wavefield of an object because each of the polygons comprising the object has its own reflection vector $\mathbf{R}_j$, as shown in Fig. 10.50. The diffuser must be generated for

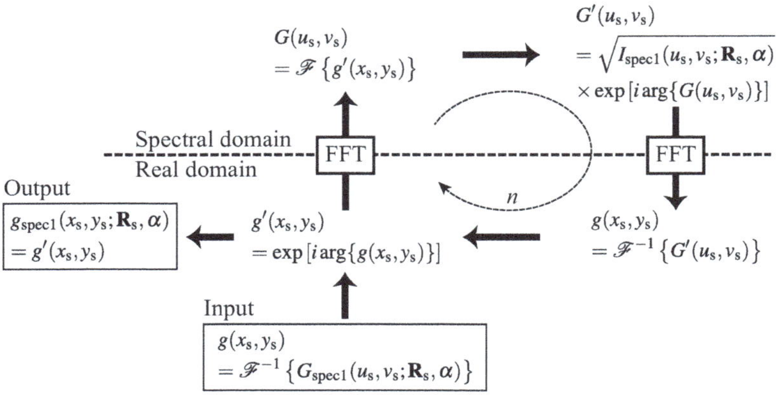

Fig. 10.49 The Gerchberg-Saxton algorithm for shaping the specular spectrum [18]. n denotes the number of iterations

Fig. 10.50 Variations of the reflection vector

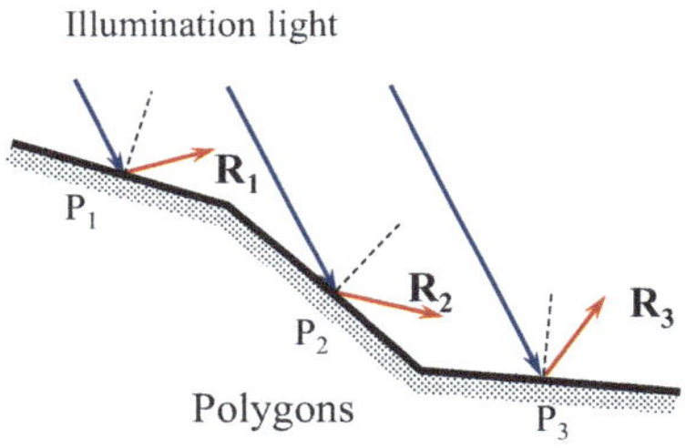

individual polygon if using $g_{\text{spec1}}(u_s, v_s; \mathbf{R}_{s,j}, \alpha)$. Here, $\mathbf{R}_{s,j} = \mathbf{T}_j^{-1}\mathbf{R}_j$ again. To generate $g_{\text{spec1}}(u_s, v_s; \mathbf{R}_{s,j}, \alpha)$ using (10.78)–(10.80), at least a double FFT is necessary for each specular polygon.

To reduce the computational effort, i.e., remove FFTs, we can adopt a spectrum-shifting technique. In this technique, we consider a specular light emitted in the normal direction as a first step. In this case, the specular diffuser based on the Phong model is written as

$$g_{\text{spec0}}(x_s, y_s; \alpha) \equiv g_{\text{spec1}}(x_s, y_s; \mathbf{N}_s, \alpha), \tag{10.85}$$

where $\mathbf{N}_s$ is a normal vector whose components are given in the tilted local coordinates. The components are simply constants:

$$\begin{aligned} \mathbf{N}_s &= \mathbf{T}_j^{-1}\mathbf{N}_j \\ &\equiv (0, 0, 1). \end{aligned} \tag{10.86}$$

Light emitted by a surface function with the diffuser $g_{\text{spec0}}(x_s, y_s; \alpha)$ always travels perpendicularly to the surface. However, we can change the light direction by multiplying the wavefield of a plane wave into the diffuser. The phase factor to emit the light in the direction of $\mathbf{R}_{s,j}$ is given by

$$g_{\text{spec2}}(x_s, y_s; \mathbf{R}_{s,j}, \alpha) = g_{\text{spec0}}(x_s, y_s; \alpha) \exp[ik\mathbf{R}_{s,j} \cdot \mathbf{r}_{s,j}], \tag{10.87}$$

where $\mathbf{r}_{s,j}$ is a position vector represented in the tilted local coordinates of polygon j, and $\exp[ik\mathbf{R}_{s,j} \cdot \mathbf{r}_{s,j}] = \exp[ik(R_{s,x}x_s + R_{s,y}y_s)]$ is the plane wave traveling in the $\mathbf{R}_{s,j}$ direction. Here, the suffix j is omitted in the components according to the convention.

In Fourier space, the spectrum of the modified diffuser is written as:

$$\begin{aligned} G_{\text{spec2}}(u_s, v_s; \mathbf{R}_{s,j}, \alpha) &= \mathscr{F}\left\{g_{\text{spec2}}(x_s, y_s; \mathbf{R}_{s,j}, \alpha)\right\} \\ &= G_{\text{spec0}}\left(u_s - \frac{R_{s,x}}{\lambda}, v_s - \frac{R_{s,y}}{\lambda}; \alpha\right), \end{aligned} \tag{10.88}$$

$$G_{\text{spec0}}(u_s, v_s, \alpha) = \mathscr{F}\left\{g_{\text{spec0}}(x_s, y_s; \alpha)\right\}. \tag{10.89}$$

In this case, the FFT is executed only once for each polygon, because the phase factor $g_{\text{spec0}}(x_s, y_s; \alpha)$ and the spectrum $G_{\text{spec0}}(u_s, v_s, \alpha)$ are independent of the reflection direction, and thus can be precomputed. The phase factor for the direction $\mathbf{R}_{s,j}$ is calculated using the precomputed $G_{\text{spec0}}(u_s, v_s, \alpha)$:

$$g_{\text{spec2}}(x_s, y_s; \mathbf{R}_{s,j}, \alpha) = \mathscr{F}^{-1}\left\{G_{\text{spec0}}\left(u_s - \frac{R_{s,x}}{\lambda}, v_s - \frac{R_{s,y}}{\lambda}; \alpha\right)\right\}. \tag{10.90}$$

As a result, we can generate the specular diffuser by carrying out the single FFT.

By using the same procedure as (10.87) and (10.88), the spectral envelope of $G_{\text{spec2}}(u_s, v_s; \mathbf{R}_s, \alpha)$ is given by:

$$\begin{aligned} I_{\text{spec2}}(u_s, v_s; \mathbf{R}_s, \alpha) &= \mathscr{F}\left\{\mathscr{F}^{-1}\left\{I_{\text{spec1}}(u_s, v_s; \mathbf{N}_s, \alpha)\right\} \exp[ik\mathbf{R}_s \cdot \mathbf{r}_s]\right\} \\ &= I_{\text{spec1}}\left(u_s - \frac{R_{s,x}}{\lambda}, v_s - \frac{R_{s,y}}{\lambda}; \mathbf{N}_s, \alpha\right), \end{aligned} \tag{10.91}$$

$$= \left[1 - (\lambda u_s - R_{s,x})^2 - (\lambda v_s - R_{s,y})^2\right]^{\alpha/2} \tag{10.92}$$

Examples of $I_{\text{spec2}}(u_s, v_s; \mathbf{R}, \alpha)$ are shown in Fig. 10.51b, as compared with $I_{\text{spec1}}(u_s, v_s; \mathbf{R}, \alpha)$ in (a). The spectrum fits well with $I_{\text{spec1}}(u_s, v_s; \mathbf{R}, \alpha)$, especially in the case of the small reflection angle. Although a little difference is found in the large reflection angle, we can accelerate the computation of the whole field of specular objects by using the shape $I_{\text{spec2}}(u_s, v_s; \mathbf{R}, \alpha)$.

Unfortunately, the technique described in this section also have a drawback; that is, we cannot change the sampling window size of the surface function. When the sampling window size changes, we have to reset the process and regenerate $g_{\text{spec0}}(x_s, y_s; \alpha)$. Thus, the advantage of the above algorithm is lost in practice. This means that the technique proposed in this section cannot be used with the tech-

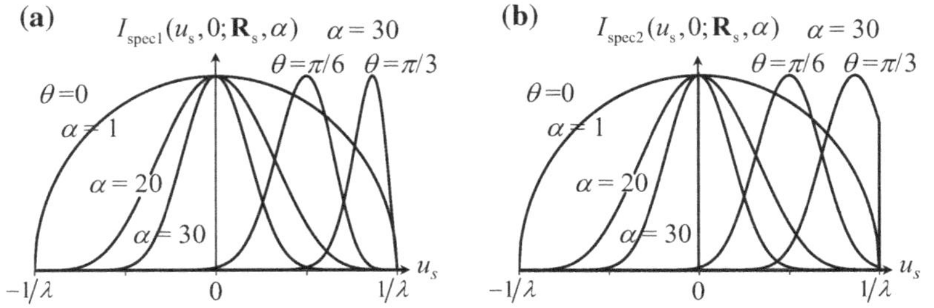

Fig. 10.51 Comparison between the spectrum envelops based of the Phong specular reflection. **a** $I_{\text{spec1}}(u_s, v_s; \mathbf{R}_s, \alpha)$, and **b** $I_{\text{spec2}}(u_s, v_s; \mathbf{R}_s, \alpha)$

nique mentioned in Sects. 10.3.6 and 10.3.10, in which the sampling window is optimized for each polygon. The sampling window in TFB must be constant if using $g_{\text{spec2}}(x_s, y_s; \mathbf{R}_{s,j}, \alpha)$ for the specular diffuser.

10.7.6 Flat Specular Shading

The surface function corresponding to Phong's specular component is given by

$$h_{\text{spec},j}(x_s, y_s) = a_{\text{spec},j}(x_s, y_s) g_{\text{spec2}}(x_s, y_s; \mathbf{R}_j, \alpha), \tag{10.93}$$

$$a_{\text{spec},j}(x_s, y_s) \equiv a_{\text{shape},j}(x_s, y_s) a_{\text{tex},j}(x_s, y_s). \tag{10.94}$$

Here, $g_{\text{spec2}}(x_s, y_s; \mathbf{R}_j, \alpha)$ is adopted as the specular diffuser. The amplitude distribution $a_{\text{spec},j}(x_s, y_s)$ is defined by removing the factor related to Lambertian reflectance from $a_{\text{dif},j}(x_s, y_s)$ in (10.65).

10.7.6.1 Surface Function Based on Phong Model

To realize flat shading based on the Phong reflection model, it seems to be better to configure the surface function by the weighted sum of the surface functions for diffuse, ambient and specular reflections, as in the original Phong model given in (10.69). However, it is impossible to compose the surface function by exactly the same manner as the original Phong model in (10.69). The reason is clear; CG images only reproduce brightness of the 3D model, whereas CGHs must reconstruct the phase as well as brightness, especially in specular models. Therefore, let us define a Phong-like surface function by adding the specular surface function (10.93)–(10.57) a priori.

$$h_{\text{Phong1},j}(x_s, y_s) = h_{\text{dif},j}(x_s, y_s) + (k_{s,j} I_s)^{1/2} h_{\text{spec},j}(x_s, y_s; \mathbf{R}_j, \alpha), \tag{10.95}$$

where $k_{s,j}$ and I_s are a reflection constant and light intensity of the specular component. Substituting (10.65) and (10.93) into (10.95), the Phong-like surface function is written down as

$$h_{\text{Phong1},j}(x_s, y_s) = a_{\text{shape},j}(x_s, y_s)a_{\text{tex},j}(x_s, y_s) \times \left[\left(k_{d,j} I_d \mathbf{L} \cdot \mathbf{N}_j + k_{a,j} I_a\right)^{1/2} g_{\text{dif}}(x_s, y_s) + (k_{s,j} I_s)^{1/2} g_{\text{spec2}}(x_s, y_s; \mathbf{R}_j, \alpha) \right]. \tag{10.96}$$

Alternatively, we can adopt a simple weighted sum as the surface function:

$$h_{\text{Phong2},j}(x_s, y_s) = a_{\text{shape},j}(x_s, y_s)a_{\text{tex},j}(x_s, y_s) \times \left[\left\{ K_{d,j} \left(\mathbf{L} \cdot \mathbf{N}_j\right)^{1/2} + K_{a,j} \right\} g_{\text{dif}}(x_s, y_s) + K_{s,j} g_{\text{spec2}}(x_s, y_s; \mathbf{R}_j, \alpha) \right], \tag{10.97}$$

where the coefficients $K_{d,j} = (k_{d,j} I_d)^{1/2}$, $K_{a,j} = (k_{a,j} I_a)^{1/2}$ and $K_{s,j} = (k_{s,j} I_s)^{1/2}$ are the weights of the components.

The surface function $h_{\text{Phong2},j}(x_s, y_s)$ is not equivalent to $h_{\text{Phong1},j}(x_s, y_s)$ but seems to match with the original Phong model well. However, the CGH created by the surface function $h_{\text{Phong1},j}(x_s, y_s)$ may fit with the CG image better than that by $h_{\text{Phong2},j}(x_s, y_s)$.

10.7.6.2 Procedure for Rendering Surfaces

To simplify the following explanation, rewrite the surface functions in the preceding section as follows:

$$h_{\text{Phong}}(x_s, y_s) = K'_d h_{\text{dif}}(x_s, y_s) + K'_s h_{\text{spec}}(x_s, y_s; \mathbf{R}_s, \alpha). \tag{10.98}$$

where K'_d and K'_s are weights of the diffuse surface function $h_{\text{dif}}(x_s, y_s)$ and specular surface function $h_{\text{spec}}(x_s, y_s; \mathbf{R}_s, \alpha)$, respectively. In addition, let K'_d and K'_s represent all factors that are not depending on x_s, y_s, $\mathbf{R}_s$, and α. We also omit the suffix j here.

The weighted sum can be carried out in the spectral domain as well as the real domain:

$$H_{\text{Phong}}(u_s, v_s) = K'_d H_{\text{dif}}(u_s, v_s) + K'_s H_{\text{spec}}(u_s, v_s; \mathbf{R}_s, \alpha), \tag{10.99}$$

where $H_{\text{dif}}(u_s, v_s) = \mathcal{F}\{h_{\text{dif}}(x_s, y_s)\}$ and $H_{\text{spec}}(u_s, v_s; \mathbf{R}_s, \alpha) = \mathcal{F}\left\{h_{\text{spec}}(x_s, y_s; \mathbf{R}_s, \alpha)\right\}$. Supposing that $g_{\text{spec2}}(x_s, y_s; \mathbf{R}_s, \alpha)$ in (10.87) is used for generating the specular spectrum $H_{\text{spec}}(u_s, v_s; \mathbf{R}_s, \alpha)$, the first step is to generate the following surface function:

$$h_{\text{spec0}}(x_s, y_s; \alpha) = a_{\text{spec}}(x_s, y_s) g_{\text{spec0}}(x_s, y_s; \alpha). \tag{10.100}$$

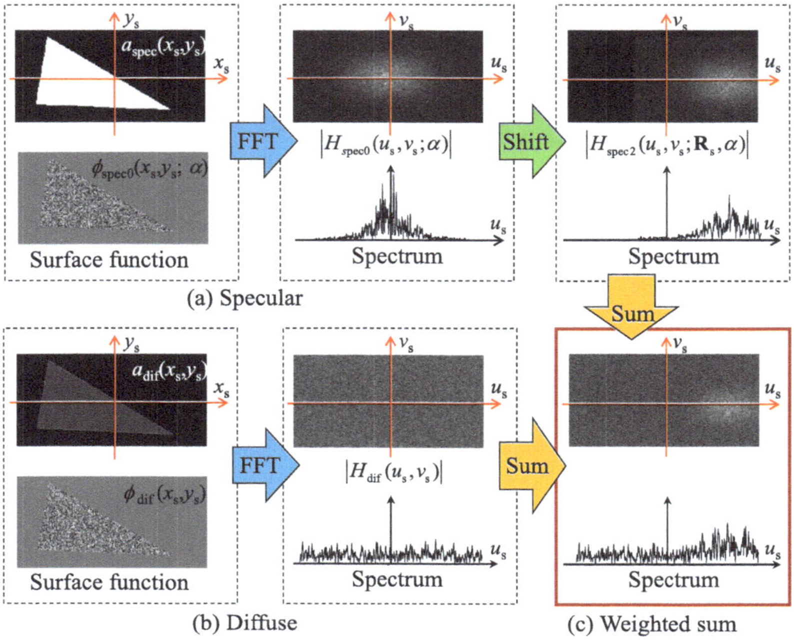

Fig. 10.52 Example of the procedure for generating a specular surface [95]

Here, remember that $g_{\text{spec0}}(x_s, y_s; \alpha)$ in (10.85) is independent of the reflection vector $\mathbf{R}_s$ and can be precomputed. Using the same technique as that in (10.88), the specular spectrum is given by

$$\begin{aligned} H_{\text{spec2}}(u_s, v_s; \mathbf{R}_s, \alpha) &= \mathscr{F}\left\{h_{\text{spec0}}(x_s, y_s; \alpha)\exp[ik\mathbf{R}_s \cdot \mathbf{r}_s]\right\} \\ &= H_{\text{spec0}}\left(u_s - \frac{R_{s,x}}{\lambda}, v_s - \frac{R_{s,y}}{\lambda}; \alpha\right), \end{aligned} \tag{10.101}$$

where

$$H_{\text{spec0}}(u_s, v_s; \alpha) = \mathscr{F}\left\{h_{\text{spec0}}(x_s, y_s; \alpha)\right\}. \tag{10.102}$$

Figure 10.52 shows an example of generation of the surface function. The specular surface function $h_{\text{spec0}}(x_s, y_s; \alpha)$ with the specular diffuser $g_{\text{spec0}}(x_s, y_s; \alpha)$ of (10.100) is Fourier-transformed and then shifted as in (a). Here, the amplitude distribution for the specular reflection $a_{\text{spec}}(x_s, y_s)$ gives the shape and texture of the polygon but does not present shading, because the shading of the object comes from the diffuse component. The diffuse surface function is also Fourier-transformed and superposed into the spectrum of the specular surface function.

Note that the example in Fig. 10.52 only shows the basic procedure for specular rendering using the specular spectrum $H_{\text{spec2}}(u_s, v_s; \mathbf{R}_s, \alpha)$. In actual specular rendering by the polygon-based method, we must take the spectral remapping into account, as described in the next section.

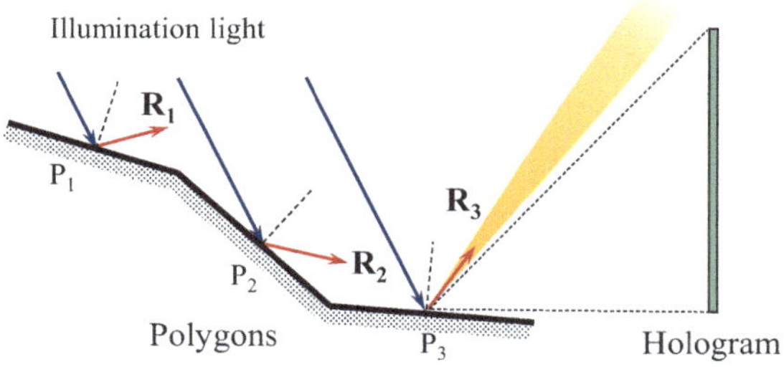

Fig. 10.53 Various specular polygons reflecting the illumination light

10.7.6.3 Cancellation of Spectrum Remapping

As mentioned in Sect. 10.2.2, the spectrum of a surface function is commonly shifted in order to reduce the computational load in rendering diffused surfaces. This spectrum remapping is also used in the rendering based on the Phong model because the Phong model has the diffuse component as well as the specular component. The spectrum remapping changes the direction of the field emitted by the polygon. This direction change has little effect on the diffuse light because the diffuse light is emitted in every direction originally. However, specular light is strongly affected by the spectrum remapping, i.e., the light direction is unexpectedly changed by the remapping. To avoid the direction change, the specular light must be shifted in the opposite direction to the spectrum remapping in order to cancel the effect.

The spectrum of the Phong-like surface function (10.99) is shifted by the spectrum remapping as follows:

$$H'_{\mathrm{Phong}}(u_s, v_s) = H_{\mathrm{Phong}}(u_s - u_0, v_s - v_0). \tag{10.103}$$

According to the spectrum remapping described in Sect. 10.2.2, the shift amount is

$$u_0 = \frac{a_3}{\lambda} \quad \text{and} \quad v_0 = \frac{a_6}{\lambda}. \tag{10.104}$$

To cancel the shift, the specular spectrum in (10.101) must be modified into

$$\begin{aligned} H_{\mathrm{spec}}(u_s, v_s; \mathbf{R}_s, \alpha) &= H_{\mathrm{spec2}}(u_s + u_0, v_s + v_0; \mathbf{R}_s, \alpha) \\ &= H_{\mathrm{spec0}}\left(u_s - \frac{R_x}{\lambda} + u_0, v_s - \frac{R_y}{\lambda} + v_0; \alpha\right). \end{aligned} \tag{10.105}$$

This is the actual specular spectrum used in the polygon-based method to render specular surfaces. The bandwidth of $H_{\mathrm{spec}}(u_s, v_s; \mathbf{R}_s, \alpha)$ is commonly narrowed by cancellation of the spectrum remapping. As a result, the computational load can be reduced in the practical rendering.

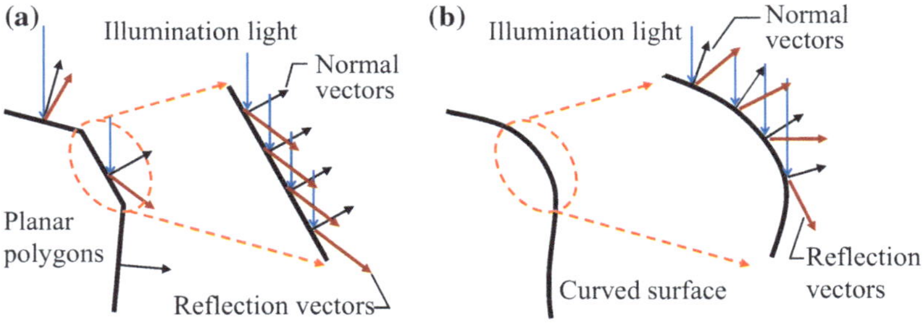

Fig. 10.54 Reflection from **a** planar polygons and **b** a curved surface

10.7.6.4 Removal of Unreachable Light

Unlike diffused surfaces, it is sometimes difficult to handle specular polygon fields. For example, the specular field reflected by polygon P_3 does not reach the hologram, as shown in Fig. 10.53. Thus, the specular field should not be calculated to avoid unnecessary noise. The same technique as the band limiting in Sect. 10.4 should be introduced into computation of the specular filed.

In another case, a reflection vector $\mathbf{R}_j$ may have a negative z component. This means that reflection light travels backward and never reach the hologram. If light of the diffuse component reach the hologram, the polygon is not removed by the back-face culling described in Sect. 10.3.8. A technique to cull unreachable specular light is required to avoid unnecessary calculation and the resultant noise.

10.7.7 Smooth Specular Shading

Light reflected by a planar polygon has the same direction anywhere in the polygon surface, as shown in Fig. 10.54a. This gives flat specular shading, and causes reconstruction of angled facets. In contrast, the normal vector smoothly changes depending on the surface position in curved surfaces, as in Fig. 10.54b. Thus, when illumination light is reflected by the curved surface, the direction of regular reflection from the surface smoothly changes. Reducing the size of individual polygons and increasing their number can make the surface look smoother. However, increasing the number of polygons usually increases the required computational effort and calculation time.

10.7.7.1 Principle of Smooth Shading Based on Phong Shading

Phong shading is a well-known technique for specular smooth shading in CG [102]. Here, note that the Phong shading is a different technique from the Phong reflection

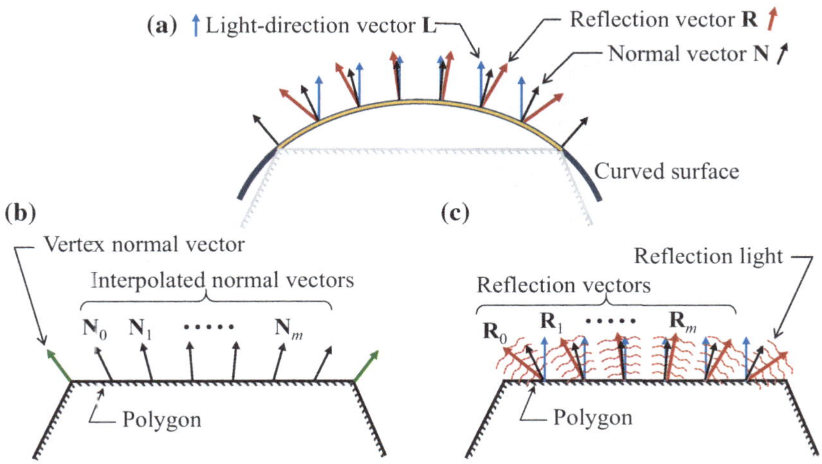

Fig. 10.55 Schematic illustrations of **a** normal vectors of a curved surface, **b** interpolated normal vectors in the Phong shading technique, and **c** field emission to imitate the curved surface

model mentioned in Sect. 10.7.2. This is a kind of interpolation technique to remove the border of shade at polygon edges, and is also called Phong interpolation.

Figure 10.55a schematically shows local normal vector **N** and reflection vector **R** in a curved surface. In the Phong shading, to imitate a curved surface, the normal vector of a polygon is not constant in the polygon surface, but varies dependently on the position, as shown in Fig. 10.55b. In practice, normal vectors $\mathbf{N}_m$ are obtained using interpolation between the vertex normal vectors of the polygon, which are given by the average of the normal vectors of neighboring polygons. The surface brightness corresponding to each normal vector is determined using (10.69) and (10.70). As a result, if the normal vectors are interpolated sufficiently densely, then the change in brightness is sufficiently smooth that the polygon appears to be a curved specular surface in CG rendering.

The original Phong shading technique in CG does not work well in holography. This is because it changes only the surface brightness, and does not change the direction of light emitted by the polygon surface. Here, let us emphasize again that viewers of a hologram can change their viewpoints, unlike in CG. Thus, the appearance of the polygon surface should smoothly change as the viewpoint moves in computer holography. To imitate a curved surface, each portion of a polygon surface must emit light in different directions represented by $\mathbf{R}_m$, as shown in Fig. 10.55c. Here, the reflection vector $\mathbf{R}_m$ is given by (10.70) and the interpolated normal vectors $\mathbf{N}_m$.

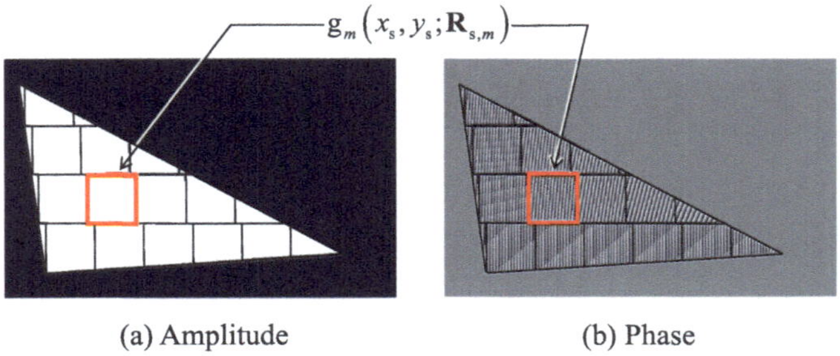

Fig. 10.56 An example of fragmentary plane waves. Each fragment corresponds to a segment of the specular surface function

10.7.7.2 Specular Surface Function for Smooth Shading

To change the field direction for each portion of a polygon, we divide the surface function into rectangular segments corresponding to the reflection vector $\mathbf{R}_m$. We then change the direction of the light by multiplying each segment of the surface function by a plane wave traveling in the direction of $\mathbf{R}_m$. The fragmentary plane wave limited inside the rectangle of segment m is given by

$$g_m(x_s, y_s; \mathbf{R}_{s,m}) = \text{rect}_m(x_s, y_s) W(\mathbf{R}_{s,m}) \exp[ik\mathbf{R}_{s,m} \cdot \mathbf{r}_s], \tag{10.106}$$

where $\mathbf{r}_s$ is again a position vector in the tilted local coordinates. $\text{rect}_m(x_s, y_s)$ is a rectangular function for segment m:

$$\text{rect}_m(x_s, y_s) = \begin{cases} 1 & \text{inside segment } m \\ 0 & \text{otherwise} \end{cases}. \tag{10.107}$$

Factor $W(\mathbf{R}_{s,m})$ is also defined as

$$W(\mathbf{R}_{s,m}) = \begin{cases} 1 & (\mathbf{T}_j \mathbf{R}_{s,m}) \cdot \mathbf{e}_z \geq 0 \\ 0 & \text{otherwise} \end{cases}, \tag{10.108}$$

where $\mathbf{e}_z$ and $\mathbf{T}_j$ are again a unit vector of the $\hat{z}$-axis and rotation matrix of polygon j, respectively. This factor avoids backward reflection, where reflected light travels in the opposite direction to the hologram. An example of fragmentary plane waves is shown in Fig. 10.56.

In ambient light and diffuse reflection, light emitted from a surface inherently spreads over all directions. Therefore, the direction of light does not need to be changed in these fields. The original Phong shading technique in CG, i.e., change in brightness of the surface, achieves smooth shading with regard to ambient and diffuse

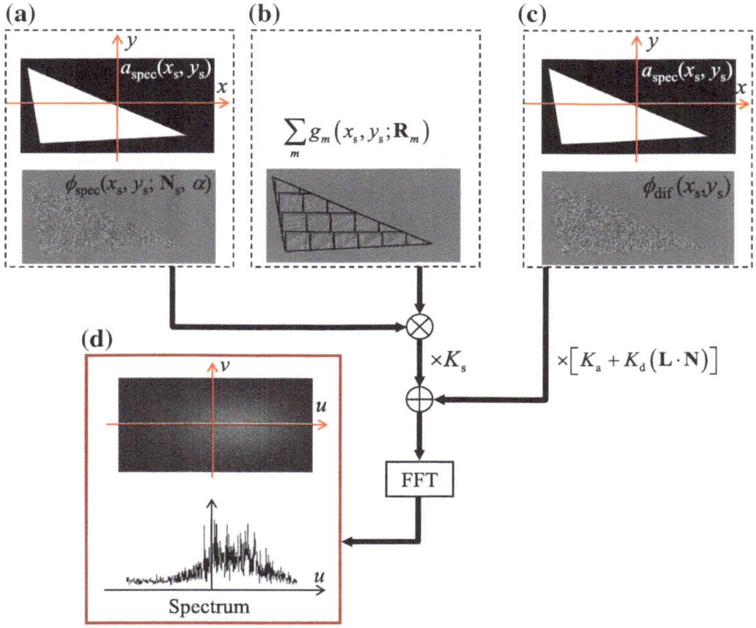

Fig. 10.57 Procedure for rendering a specular curved polygon [93]. **a** Specular surface function for flat shading, **b** fragmentary plane waves, **c** diffuse surface function, and **d** spectrum of the surface function for the specular curved surface

components even in computer holography [77]. Only specular reflection requires the fragmentary plane waves to render a curved surface.

The specular surface function is properly given by the inverse Fourier transform of the specular spectrum (10.105):

$$h_{spec}(x_s, y_s; \mathbf{R}_s, \alpha) = \mathcal{F}^{-1}\left\{H_{spec}(u_s, v_s; \mathbf{R}_s, \alpha)\right\} \tag{10.109}$$

This specular surface function emits light to the direction of $\mathbf{R}_s$. However, this is not required in rendering a specular curved surface because the field direction is changed by the multiplying fragmentary plane waves of (10.106). Therefore, instead of $h_{spec}(x_s, y_s; \mathbf{R}_s, \alpha)$, the surface function $h_{spec}(x_s, y_s; \mathbf{N}_s, \alpha)$ which emits light in a direction perpendicular to the polygon should be used in this case. As a result, a smooth specular surface function is written as

$$h_{smooth}(x_s, y_s; \alpha) = h_{spec}(x_s, y_s; \mathbf{N}_s, \alpha) \sum_m g_m(x_s, y_s; \mathbf{R}_{s,m}). \tag{10.110}$$

Here, note that the specular surface function $h_{spec}(x_s, y_s; \mathbf{N}_s, \alpha)$ sometimes has very high spatial frequency depending on the direction of the polygon. The frag-

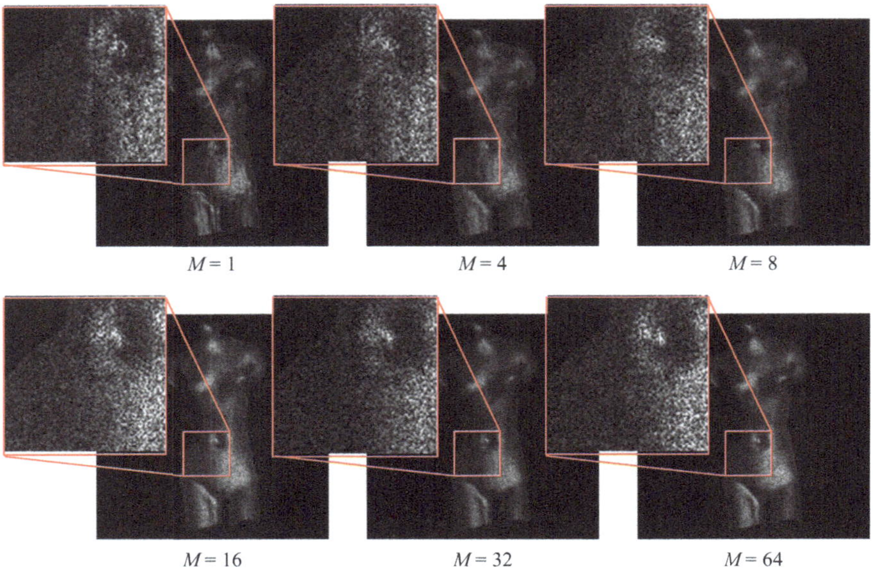

Fig. 10.58 Simulated reconstruction of object fields calculated using different segment sizes [93]

mentary plane waves $\sum_m g_m(x_s, y_s; \mathbf{R}_{s,m})$ can also have high frequency component in some cases. As a result, we sometimes need very small sampling intervals nearly equal to one half wavelength to generate the surface function $h_{\text{smooth}}(x_s, y_s; \alpha)$.

10.7.7.3 Procedure for Rendering of Specular Curved Surfaces

Figure 10.57 shows the procedure for generating the surface function of a specular curved surface. The diffuse surface function shown in (c) gives ambient light and diffuse reflection. Although this is the diffuse component, the amplitude distribution is also used for specular reflection, i.e., $a_{\text{spec}}(x_s, y_s) = a_{\text{shape}}(x_s, y_s)a_{\text{tex}}(x_s, y_s)$. The Phong interpolation is not appropriate to generate the amplitude distribution in this case, because highlight of the surface reflection is brought by the phase distribution. Flat shading by Lambertian reflectance is good enough for the diffuse component, but if the border of polygons appear in the reconstruction, the use of Gouraud shading may be effective to remove it.

The specular surface function shown in Fig. 10.57a has a narrow band spectrum, whose bandwidth is specified by a shininess constant α, and whose peak is at the origin. To change the direction of light inside the polygon surface, the field (a) is multiplied by the fragmentary plane waves shown in (b). The direction of the plane waves is determined by (10.70), using the light-direction vector $\mathbf{L}$ and interpolated normal vectors based on the Phong interpolation. The final surface function is given

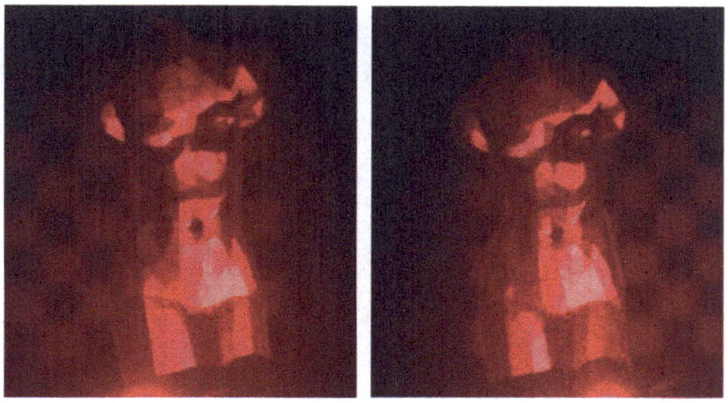

Fig. 10.59 Optical reconstruction of "The Metal Venus I" created with specular flat shading [95]. The photographs are taken from left and right angles. The illumination light source is a He–Ne laser. Video link, https://doi.org/10.1364/AO.50.00H245.m001

by the simple weighted sum of the diffuse and specular surface functions using the weights K_a, K_d, and K_s in Fig. 10.57. It is also possible to use another style of the weighted sum by $k_d I_d$, $k_a I_a$, and $k_s I_s$, as in (10.96). Figure 10.57d shows the spectrum of the final surface function.

10.7.7.4 Segment Size

It is important to choose an appropriate segment size in specular surface functions. The field direction is controlled using fragmentary plane waves in the proposed method. If the segment size is too small, the fragmentary plane waves may not be able to properly control the field direction. In contrast, a segment size that is too large will most likely result in a decrease in the smoothness of the curved surface. We calculated some object fields with different segment sizes, to investigate the influence of segment size on the reconstruction of curved surfaces. The object model used in this experiment is that of the "The Venus" introduced in Chap. 1 (see also Sect. 10.5). This model is composed of 718 front face polygons. To make clear the effect of segment size, the object field was calculated by specular reflection only, i.e., $K_a = K_d = 0$. The wavelength is 633 nm, and the sampling interval of the surface function is 320 nm. The object field was calculated with 16K $\times$ 16K (1K = 1024) samplings and 1.0 μm sampling intervals for the object having a height of 1.4 cm. The reconstruction is simulated using virtual imaging (see Chap. 13).

Figure 10.58 shows the simulated reconstruction. Here, the sampled surface function is divided into $M \times M$ sampling square segments. It is verified that the curved surface is degraded for segment sizes of less than $M = 8$. The reconstruction is almost unchanged for segment sizes of $M \geq 16$.

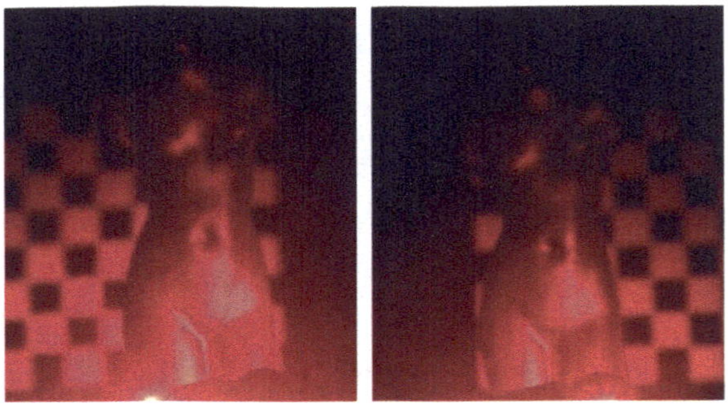

Fig. 10.60 Optical reconstruction of "The Metal Venus II" created with specular smooth shading [93]. The photographs are taken from left and right angles. The illumination light source is a He–Ne laser. Video link, https://doi.org/10.1364/AO.56.000F37.v001

Table 10.3 Parameters used for creating The Metal Venus I and II

The Metal Venus	I	II
Number of pixels	4.3×10^9 (65, 536 × 65, 536)	
Pixel pitches	1.0 μm × 1.0 μm	
Reconstruction wavelength	633 nm	
Hologram coding	Binary amplitude	
Number of front-face polygons	718	
Dimension of Venus model ($W \times H \times D$)	26.7 mm × 57.8 mm × 21.8 mm	
Light-direction vector L in global coordinates	(0.75, 0.25, 1.0)	
Shininess constant (α)	10	
Weight of specular light (K_s)	0.8	0.4
Weight of diffuse light (K_d)	1.0	1.0
Weight of ambient light (K_a)	0	0.1
Sampling interval of diffuse surface function	1 μm × 1 μm	320 nm × 320 nm
Sampling interval of specular surface function	250 nm × 250 nm	320 nm × 320 nm

10.7.8 Examples of High-Definition CGHs with Specular Shading

Figures 10.59 and 10.60 show optical reconstruction of HD-CGHs with specular shading. These are a kind of twins, and named "The Metal Venus I" and "The Metal Venus II". These have the same 3D scene, but the surface is rendered with specular flat shading and specular smooth shading in The Metal Venus I and II, respectively. Several major parameters are summarized in Table 10.3.

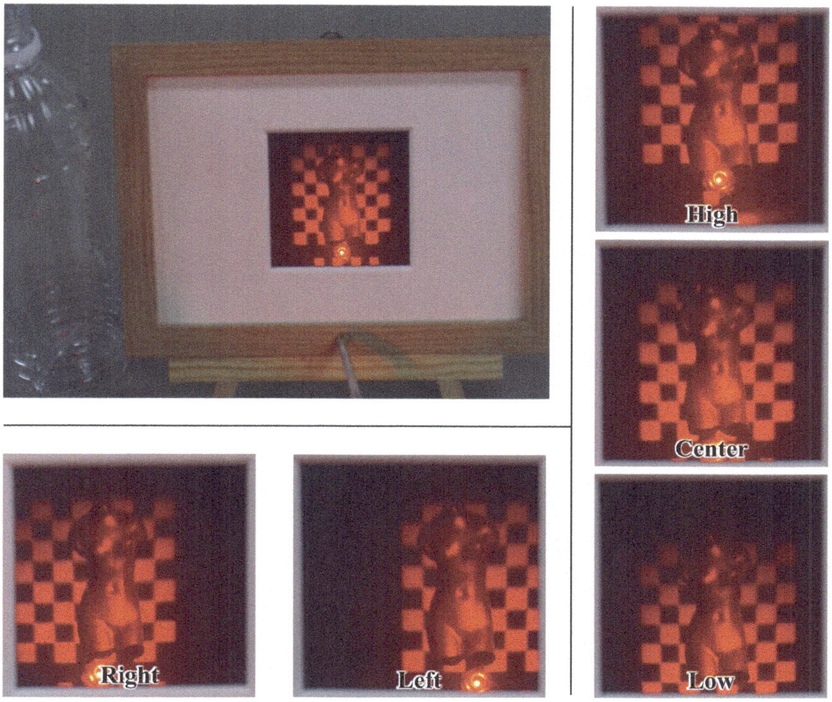

Fig. 10.61 Optical reconstruction of The Metal Venus II by illumination of an ordinary red LED [93]. The photographs are taken from different angles. Video link, https://doi.org/10.1364/AO.56.000F37.v002

Photographs of optical reconstruction in Figs. 10.59 and 10.60 clearly show the difference between flat and smooth specular shadings. Figure 10.61 shows the optical reconstruction of The Metal Venus II using a red LED as the light source. Motion parallax as well as specular curved surfaces are verified in these photographs and videos.

Chapter 11
The Silhouette Method

Technique for Occlusion Processing in Field Rendering

Abstract Occlusion is one of the most important cues in depth perception, and it is also the most difficult process in computer holography. The silhouette method is a technique to process occlusion in computer holography. In this chapter, the simplest silhouette method to process occlusion between separate objects is discussed in the beginning. The Babinet's principle is then introduced to extend the silhouette method to processing of occlusion between polygons. This type of silhouette method, called the switch-back technique, is the basis of rendering complicated 3D models.

11.1 Occlusion

Occlusion is well known as one of the most important cues in depth perception. Thus, reconstruction of occlusion is highly important in computer holography and many other 3D techniques. To reconstruct occlusion, one object in front of another object must hide the object to the rear. In other words, the light on the rear object must be shielded by the front object. The technique is therefore similar to *hidden surface removal* in conventional CG. However, while the viewpoint cannot be changed when seeing a still CG image, the viewpoint in computer holography can be moved even in still images, as repeatedly mentioned in this book. This makes hidden surface removal in computer holography much more difficult than that in CG.

Hidden surface removal in computer holography, which is also called *occlusion processing* or *occlusion culling*, is necessarily associated with the numerical techniques used to generate the object field. Most researchers prefer point-based methods, in which surfaces or sometimes wire-framed objects are expressed as large collections of point light sources. These techniques are generally ray-oriented; occlusion processing is commonly interpreted as a visibility test in ray casting, which examines the existence of obstacles between a sampling point in the hologram plane and a point source of the object. If something interrupts the ray between the sampling point and the object point, the ray is removed from the computation. The visibility test is suitably efficient for horizontal-parallax-only (HPO) holograms. However, the visibility test is commonly too time-consuming to perform occlusion processing in full-parallax CGHs.

K. Matsushima, *Introduction to Computer Holography*,
Series in Display Science and Technology,
https://doi.org/10.1007/978-3-030-38435-7_11

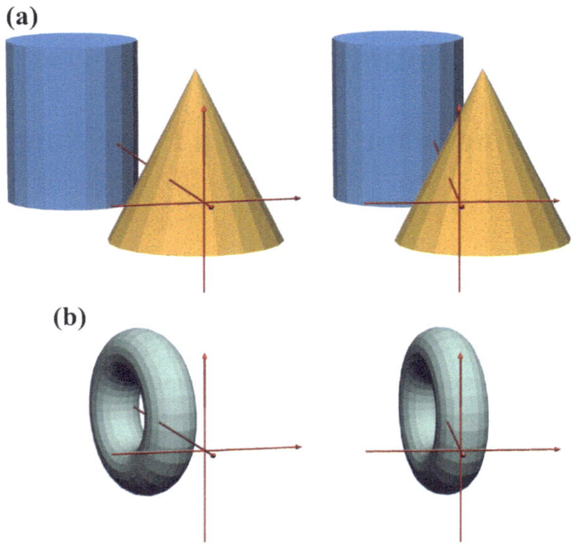

Fig. 11.1 Classification of occlusion: **a** mutual occlusion, and **b** self-occlusion

Occlusion is classified into two categories: mutual occlusion and self-occlusion, as shown in Fig. 11.1. Mutual occlusion means that an object is hidden by another separate object or objects, while self-occlusion refers to the situation where part of a surface is hidden by another surface of the same object. The HD-CGHs created in the early days commonly reconstruct mutual occlusions only. The original silhouette masking technique was proposed to process both mutual and self-occlusions [45]. However, to process self-occlusions, the masking process must be performed for each polygon. This is very time-consuming, especially in HD-CGHs, because the original silhouette method requires the same number of numerical propagations of the wavefield as the number of masking. Therefore, the silhouette masking process was performed for each object rather than for each polygon when creating the HD-CGHs. Because this type of silhouette method cannot deal with self-occlusion, self-occluded objects are reconstructed as partially transparent objects. We had therefore carefully chosen object models without any self-occlusions for HD-CGH creation.

The situation was completely changed by invention of the *switch-back technique* [75],[1] described in Sect. 11.3. This technique can speedup the masking process for each polygon very much, and makes it possible to convert a CG model into the CGH automatically.

[1] The switch-back technique was intuitively invented by M. Nakamura when he was a masters student. However, anybody including himself had not been able to explain how the technique works well for a long time. The formulation and interpretation were finished several years later [75].

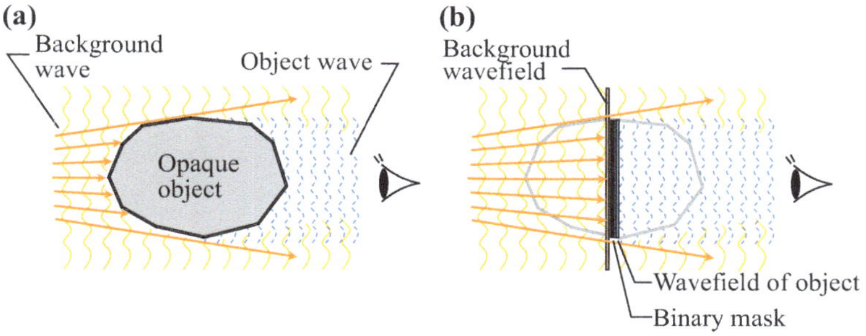

Fig. 11.2 **a** Schematic illustration of light-shielding of a real opaque object, and **b** emulation of light-shielding by the silhouette method

11.2 Processing of Mutual Occlusion

Opaque objects hide their background, i.e., any part of the background that overlaps the object is shielded by that object, as shown in Fig. 11.2a. If we see the object as shown in (a), we can see the near-side front surface of that object and the part of the background that does not overlap the object. If we change the viewpoint, then the visible portion of the background changes with the viewpoint.

11.2.1 Silhouette Method

The *silhouette method* provides a simple and low-cost computer-based emulation of the physical phenomenon mentioned above. Consider the background light to be expressed using the wavefield $g(x, y)$. Here, the (x, y) plane perpendicular to the optical axis is placed at a position where the cross section of the object takes the maximum, as shown in Fig. 11.2b.

Part of the background field, which is overlapping the object, cannot pass through the object. The field amplitude must vanishes in the area of the overlap. This is expressed by multiplying $g(x, y)$ by a binary function $M(x, y)$, which has values of zero inside the object silhouette and unity outside the silhouette. This binary function, which is simply generated by orthogonal projection of the object, is called a *silhouette mask*. Consider the wavefield of the object to be given by $O(x, y)$; the final wavefield that the viewer sees is then presented by $M(x, y)g(x, y) + O(x, y)$.

If the viewpoint moves, then the viewed background changes with the viewpoint, because the background field $g(x, y)$ includes off-axis light in a specific range given by the maximum diffraction angle in (3.54) of the background field.

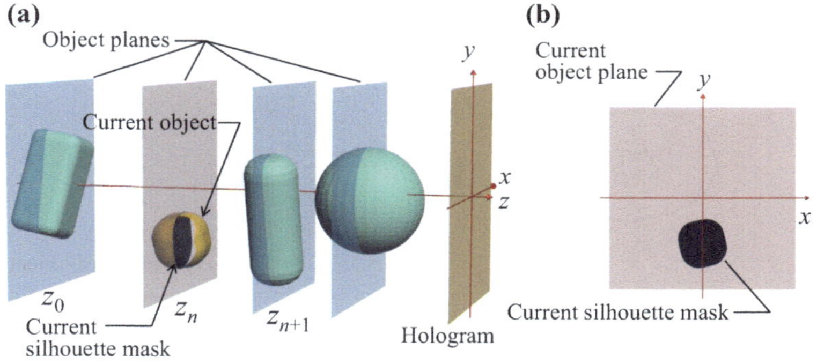

Fig. 11.3 **a** Schematic illustration of object-by-object light-shielding, and **b** the silhouette mask of the current object

11.2.2 Formulation of Object-by-Object Light-Shielding for Multiple Objects

Consider multiple objects in the 3D scene of a hologram, as shown in Fig. 11.3a. Here, the origin of the coordinate system is placed in the plane of the hologram as usual, and the objects are numbered in order of depth. The object furthest from the hologram is numbered $n = 0$. The light in the 3D scene travels approximately along the z-axis toward the hologram. Let us define the wavefield $g_n(x, y) \equiv g_n(x, y; z_n)$ in the plane (x, y, z_n). We choose the depth position z_n of the plane such that the cross-section of the object approximately reaches a maximum in the plane. These planes are called the *object planes*.

In the object plane (x, y, z_n), the shielded wavefield is given as $g_n(x, y)M_n(x, y)$, as described above. Here, $M_n(x, y)$ is again the silhouette mask, which is defined in the object plane (x, y, z_n) as follows:

$$M_n(x, y) = \begin{cases} 0 & \text{inside orthogonal projection of object } n \\ 1 & \text{otherwise} \end{cases}. \tag{11.1}$$

The object n also emits light. If we consider the wavefield to be given by $O_n(x, y)$ in the object plane, then the total wavefield is given by $g_n(x, y)M_n(x, y) + O_n(x, y)$. This field propagates to the next object plane, (x, y, z_{n+1}), as defined for object $n + 1$. Therefore, the procedure for the silhouette method is expressed using a recurrence formula:

$$g_{n+1}(x, y) = \mathscr{P}_{n+1,n}\left\{M_n(x, y)g_n(x, y) + O_n(x, y)\right\}, \tag{11.2}$$

where $\mathscr{P}\{\cdot\}$ is a propagation operator introduced in Sect. 5.4. To be accurate, the notation $\mathscr{P}_{n+1,n}\{\xi(x, y)\}$ stands for the numerical propagation of the wavefield $\xi(x, y)$ from the plane at z_n to that at z_{n+1}.

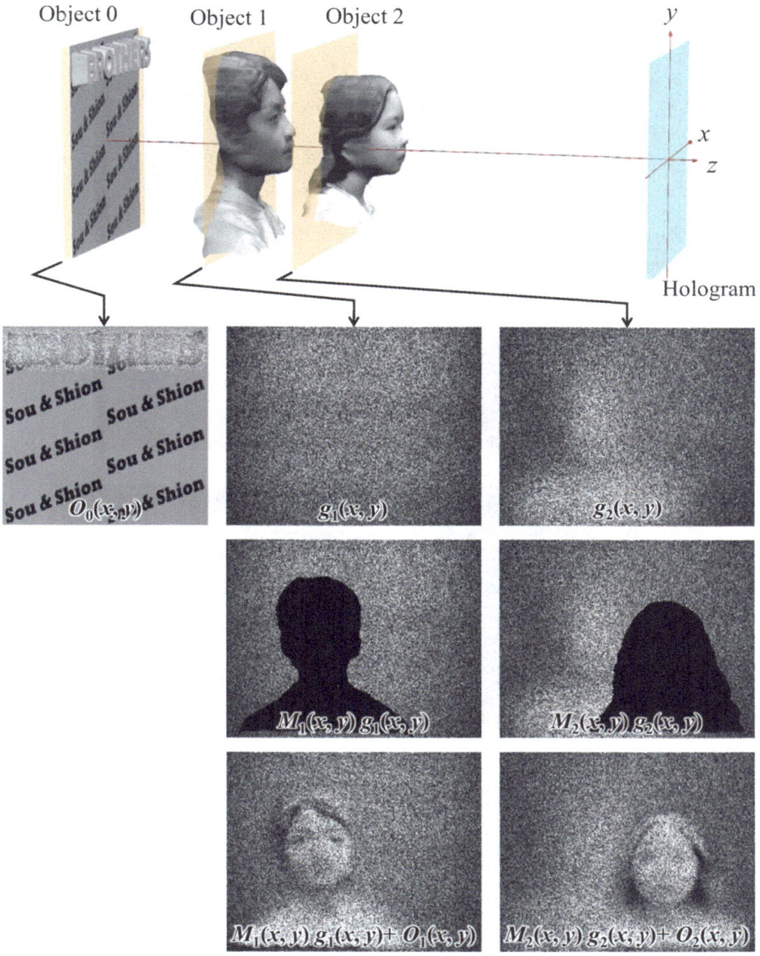

Fig. 11.4 The process to calculate the object field of Brothers by using O-O light-shielding. Amplitude distribution of wavefields is depicted for each step of light-shielding

We start the light-shielding process from the plane (x, y, z_0), where $g_0(x, y) \equiv 0$. The field $g_1(x, y)$ is calculated from the silhouette mask $M_0(x, y)$ and object field $O_0(x, y)$ of the first object placed at the farthest position from the hologram. According to the recurrence formula (11.2), the wavefield is then sequentially calculated toward the closest object to the hologram in order of the z-position. The propagation is required N times to process an entire 3D scene composed of N objects. This type of silhouette method is referred to as *object-by-object (O-O) light-shielding* in this book.

Fig. 11.5 Optical reconstruction of HD-CGH "Brothers". The pictures are taken from different viewpoints. Video link, https://youtu.be/RCNyPNV7gHM

Here, note that the O-O light-shielding by the silhouette method is independent of the technique to generate the object field $O_n(x, y)$. We can use the point-based method as well as the polygon-based method to calculate the object field. Because the silhouette method uses the technique of numerical field propagation described in Chap. 6 or Chap. 12, the method is considered to be fit better with the field-oriented polygon-based method than the ray-oriented point-based method, as described in Sect. 11.3.10.

11.2.3 Actual Example of Object-by-Object Light-Shielding

Figure 11.4 shows the actual process to calculate the object field of the HD-CGH "Brothers" mentioned in Chap. 1. The 3D scene is composed of three objects. The object 0 is an image and polygon-meshed 3D letters, while the 2nd and 3rd objects are

Table 11.1 Parameters of "Brothers" CGHs. Brothers V2 was actually on display at Massachusetts Institute of Technology (MIT) Museum

Parameters	Value		Units
	Brothers V1	Brothers V2	
Number of pixels	21,474,836,480	25,769,803,776	
$(M \times N)$	$(163{,}840 \times 131{,}072)$	$(196{,}608 \times 131{,}072)$	
Pixel pitches $(\Delta x_h \times \Delta y_h)$	0.8×0.8	0.64×0.8	μm
CGH Size $(W_x \times W_y)$	131.1×104.9	125.8×104.9	mm
Design wavelength	633		nm
	Object 1	Object 2	
Total number of polygons	2,865	3,654	
Object size (W × H × D)	$90.0 \times 86.2 \times 38.7$	$80.0 \times 70.3 \times 44.9$	mm
Center position (x, y, z)	$(-15, 0, -200)$	$(20, 0, -150)$	mm

the polygon-meshed objects of live faces, which were measured by using a 3D laser scanner. The wavefield $g_1(x, y)$ is calculated from the object 0 in a plane that slices the object 1. Here, note that all amplitude images in Fig. 11.4 are not realistic because the original CGH is composed more than 25 billion pixels with sub-micron pixel pitches. The masked wavefield $M_2(x, y)g_2(x, y) + O_2(x, y)$ is finally propagated to the hologram plane and gives the object field of the whole 3D scene.

Figure 11.5 shows optical reconstruction of the actual "Brothers" CGH (see also Sect. 1.3). The parameters are summarized in Table 11.1. There are actually two versions; photographs in Figs. 11.5 and 1.6 are of V1. We can verify that mutual occlusion among objects are reconstructed successfully. One more HD-CGH is "Aqua 2" shown in Fig. 11.6. The 3D scene of this CGH is composed of 10 objects and the occlusion is processed by O-O light-shielding.

11.2.4 Translucent Object

Simple rendering of a translucent CG model is realized by adjusting transmittance of silhouette masks. Suppose t_n is amplitude transmittance of object n. The translucent object partially transmits the background field inside the silhouette of the object; the background field is $t_n g_n(x, y)$ inside the silhouette. Therefore, the silhouette mask of a translucent object is defined as

$$M_n(x, y; t_n) = \begin{cases} t_n & \text{inside orthogonal projection of object } n \\ 1 & \text{otherwise} \end{cases}. \tag{11.3}$$

Using the binary mask $M_n(x, y)$, the translucent mask function is given by

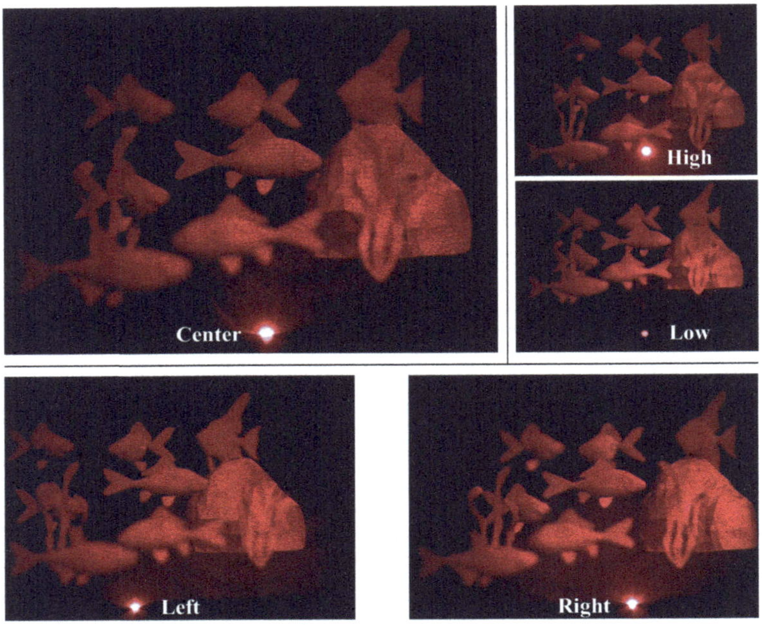

Fig. 11.6 Optical reconstruction of HD-CGH "Aqua 2" (see Appendix A.1 for the parameters). Video link, https://youtu.be/DdOveIue3sc

$$M_n(x, y; t_n) = t_n + (1 - t_n)M_n(x, y). \tag{11.4}$$

The recurrence formula (11.2) is in this case

$$g_{n+1}(x, y) = \mathscr{P}_{n+1,n}\left\{M_n(x, y; t_n)g_n(x, y) + O_n(x, y)\right\}. \tag{11.5}$$

Needless to say, this is a simplified technique to reconstruct a translucent object because real objects commonly have a different refractive index from the atmosphere. This means that the background field is always refracted by the object in the real world.

It is easy to emulate refraction in numerical propagation. By replacing the wavelength, we can simulate the refraction phenomenon occurring at a planar interface between two different optical media. However, imitation of refraction in a 3D object is very difficult to realize at this stage. Equation (11.5) only provides a low-cost alternative to the regular physical simulation.

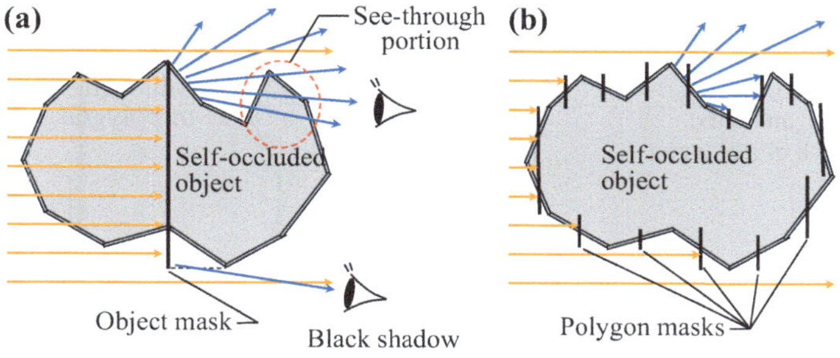

Fig. 11.7 Comparison between **a** object-by-object and **b** polygon-by-polygon light-shielding methods

11.3 Switch-Back Technique for Processing Self-Occlusion by the Silhouette Method

If an object has a complicated shape or concave surfaces, as shown in Fig. 11.7a, then O-O light-shielding does not work well. In this case, the reconstructed images may have a transparent portion or sometimes a black shadow that occurs when the viewer sees the mask directly.

11.3.1 Principle of Polygon-by-Polygon Light-Shielding and Associated Problem

The occlusion errors involved in processing self-occluded objects can be avoided by light-shielding using masks for each polygon, as shown in Fig. 11.7b. In this case, the light is shielded using many masks that are much smaller than the mask for the whole object, as shown in Fig. 11.8b. The masked field is numerically propagated for a short distance between masks, as shown in Fig. 11.8a. Only the unit of processing differs from that of O-O light-shielding. We call this type of light shielding *polygon-by-polygon (P-P) light-shielding*.

The procedure and the formulation used for P-P light-shielding are exactly the same as those of O-O light-shielding. We can simply apply the recurrence formula of (11.2) to each polygon. The concept of P-P light-shielding is very simple. However, this simple P-P shielding is not useful in practice, because we need the same number of field propagations of the whole wavefield as that of the polygons to complete the P-P shielding. This can be very time-consuming, particularly for HD-CGHs that are composed of more than a billion pixels.

P-P shielding has features that distinguish it from O-O shielding. The silhouette mask of a single polygon is, in general, very small when compared with the wavefield. The ratio of silhouette area to field area is commonly less than 1%, and is sometimes

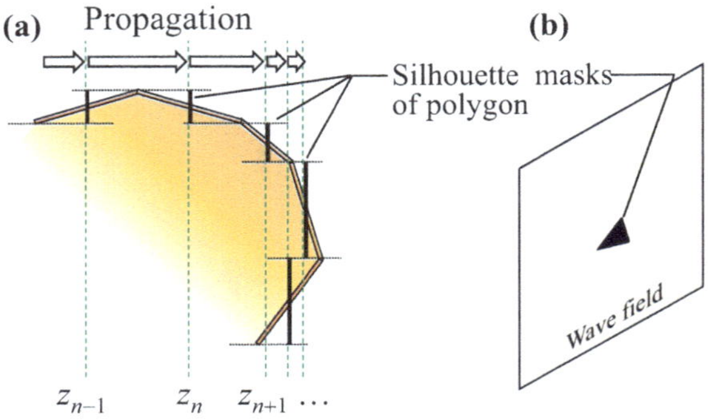

Fig. 11.8 Schematic illustration of **a** polygon-by-polygon light-shielding, and **b** the silhouette mask of a polygon

less than 0.1% for objects formed using many polygons. This means that the use of silhouette-shaped apertures rather than silhouette-shaped masks offers a considerable advantage, as detailed in the following section.

11.3.2 The Babinet's Principle

According to *Babinet's principle*, it is possible to calculate a masked field using an aperture that is exactly complementary to the mask. Let us use Fresnel propagation for the following discussion on the Babinet's principle. The propagation of a masked wavefield is given by a double integral in (5.27):

$$\begin{aligned} g(x, y; z) &= \mathcal{P}_d \{g(x, y; z_0)\} \\ &= \iint g(x_s, y_s; z_0) h_{FR}(x - x_s, y - y_s; d) dx_s dy_s, \end{aligned} \tag{11.6}$$

where $h_{RF}(x, y)$ is given by (5.29) and $d = z - z_0$ as usual. We omit the trivial coefficient $i\lambda d A_{FR}(d)$ here.

As shown in Fig. 11.9a, suppose that the domain of integration is divided into two regions S_M and S_A as follows.

$$\begin{aligned} \iint & g(x_s, y_s; z_0) h_{FR}(x - x_s, y - y_s) dx_s dy_s \\ &\equiv \iint_{S_M} g(x_s, y_s; z_0) h_{FR}(x - x_s, y - y_s; d) dx_s dy_s \\ &+ \iint_{S_A} g(x_s, y_s; z_0) h_{FR}(x - x_s, y - y_s; d) dx_s dy_s. \end{aligned} \tag{11.7}$$

We discuss light-shielding using the Babinet's principle on the basis of (11.7).

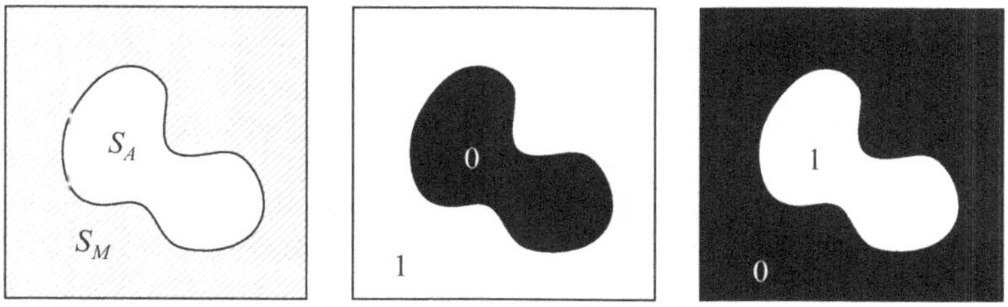

(a) (b) $M(x, y)$ (c) $A(x, y)$

Fig. 11.9 **a** Two regions of the domain of integration, **b** the binary mask and **c** aperture functions corresponding to the regions in (**a**)

11.3.2.1 Light-Shielding by Opaque Mask

Suppose that the region S_A is covered with an opaque mask, i.e., $g(x, y; z_0)$ only has non-zero values outside S_A. The first integral in the right-hand side of (11.7) can be represented as

$$\begin{aligned}
&\iint_{S_M} g(x_s, y_s; z_0) h_{\mathrm{FR}}(x - x_s, y - y_s; d) \mathrm{d}x_s \mathrm{d}y_s \\
&= \iint M(x_s, y_s) g(x_s, y_s; z_0) h_{\mathrm{FR}}(x - x_s, y - y_s; d) \mathrm{d}x_s \mathrm{d}y_s. \\
&= \mathscr{P}_d \{M(x, y) g(x, y; z_0)\}.
\end{aligned} \tag{11.8}$$

The binary mask function $M(x, y)$ in this case is shown in Fig. 11.9b. Also, suppose that the second integral can be represented by

$$\begin{aligned}
&\iint_{S_A} g(x_s, y_s; z_0) h_{\mathrm{FR}}(x - x_s, y - y_s; d) \mathrm{d}x_s \mathrm{d}y_s \\
&= \iint A(x_s, y_s) g(x_s, y_s; z_0) h_{\mathrm{FR}}(x - x_s, y - y_s; d) \mathrm{d}x_s \mathrm{d}y_s, \\
&= \mathscr{P}_d \{A(x, y) g(x, y; z_0)\}.
\end{aligned} \tag{11.9}$$

Substituting (11.7)–(11.9) into (11.6), we can obtain the following.

$$\mathscr{P}_d \{M(x, y) g(x, y; z_0)\} = \mathscr{P}_d \{g(x, y; z_0)\} - \mathscr{P}_d \{A(x, y) g(x, y; z_0)\}. \tag{11.10}$$

This means that the binary function $A(x, y)$ is complimentary to the mask function $M(x, y)$:

$$M(x, y) + A(x, y) \equiv 1. \tag{11.11}$$

The example of $A(x, y)$ is shown in Fig. 11.9c. As a result, we can conclude that propagation of the masked field: $\mathscr{P}_d \{M(x, y) g(x, y; z_0)\}$ is obtained by subtracting

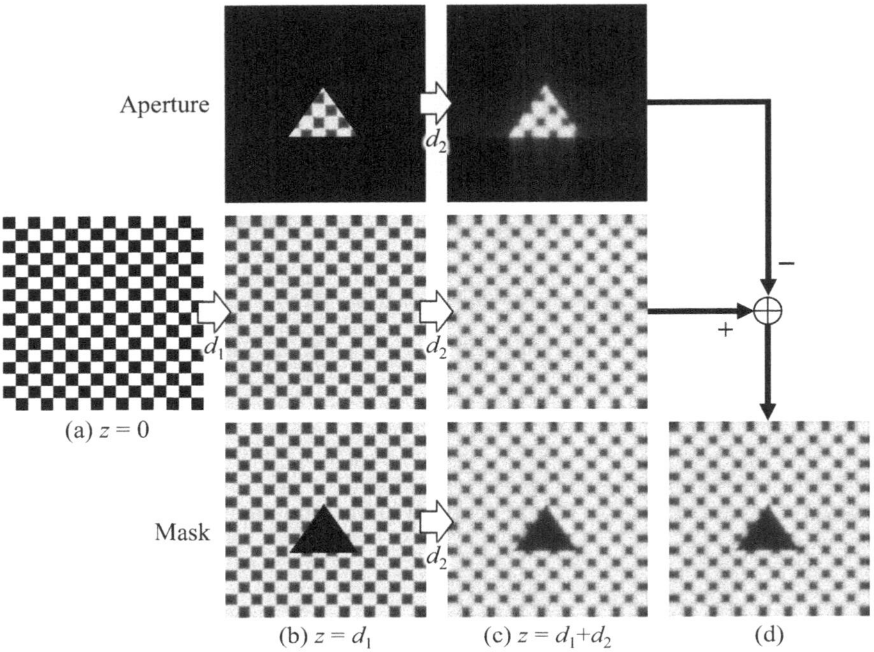

Fig. 11.10 A numerical experiment of the Babinet's principle. **a** Background field at $z = 0$, amplitude image of propagated field at **b** $z = d_1$ and **c** $z = d_1 + d_2$, and **d** the subtracted field. $\lambda = 633$ μm, $M = N = 1024$, $\Delta x = \Delta y = 10$ μm, $d_1 = 2$ [mm], and $d_2 = 3$ [mm]

the field $\mathscr{P}_d\{A(x, y)g(x, y; z_0)\}$, shielded by the complimentary aperture, from the whole propagated field $\mathscr{P}_d\{g(x, y; z_0)\}$.

Figure 11.10 shows a numerical experiment of the Babinet's principle. The image in (a) is a background pattern. The phase distribution is randomized to diffuse the background pattern. This background field is numerically propagated at a distance of d_1 using the BLAS method (see Sect. 6.4). The images depicted in the column (b) are amplitude distributions of the propagated field (middle row), and the fields multiplied by a mask (lower row) and aperture (upper row). The images in the column (c) also show amplitude patterns of the fields that are again propagated at $d_2 = 3$ [mm] by the BLAS method. The image in the lower row of the column (c) shows diffraction of the masked field, i.e., the field in the left-hand side of (11.10), while (d) is the result of subtraction in the right-hand side. These two fields perfectly correspond with each other.

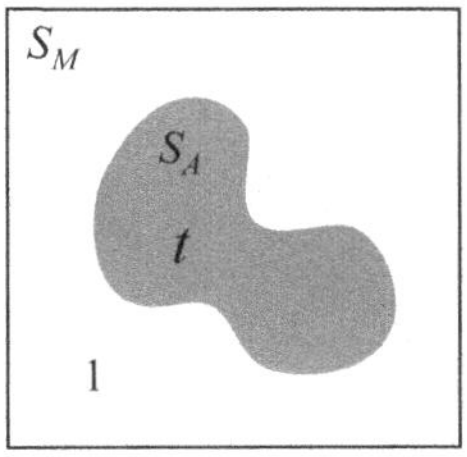

Fig. 11.11 A translucent mask $M(x, y; t)$. The amplitude transmittance is t in the region S_A and 1 in the region S_M

11.3.2.2 Light-Shielding by Translucent Mask

A translucent mask defined by (11.3) and (11.4) is schematically illustrated in Fig. 11.11. Because amplitude transmittance is t and 1 in the region S_A and S_M, respectively, propagation of the field shielded by the translucent mask is represented by

$$\mathscr{P}_d\{M(x, y; t)g(x, y; z_0)\} \equiv \iint_{S_M} g(x_s, y_s; z_0)h_{FR}(x - x_s, y - y_s; d)\mathrm{d}x_s\mathrm{d}y_s + t\iint_{S_A} g(x_s, y_s; z_0)h_{FR}(x - x_s, y - y_s; d)\mathrm{d}x_s\mathrm{d}y_s. \tag{11.12}$$

Using expressions in (11.8) and (11.9), the propagated field is rewritten as

$$\mathscr{P}_d\{M(x, y; t)g(x, y; z_0)\} = \mathscr{P}_d\{M(x, y)g(x, y; z_0)\} + t\mathscr{P}_d\{A(x, y)g(x, y; z_0)\}. \tag{11.13}$$

By substituting (11.10) into (11.13), we obtain the following.

$$\mathscr{P}_d\{M(x, y; t)g(x, y; z_0)\} = \mathscr{P}_d\{g(x, y; z_0)\} - \mathscr{P}_d\{(1 - t)A(x, y)g(x, y; z_0)\}. \tag{11.14}$$

Therefore, the translucent mask-function is associated with the aperture-function as follows.

$$M(x, y; t) = 1 - (1 - t)A(x, y). \tag{11.15}$$

As a result, we can also calculate the propagated field shielded by the translucent mask by subtraction between the fully-propagated field and the field shielded by the aperture.

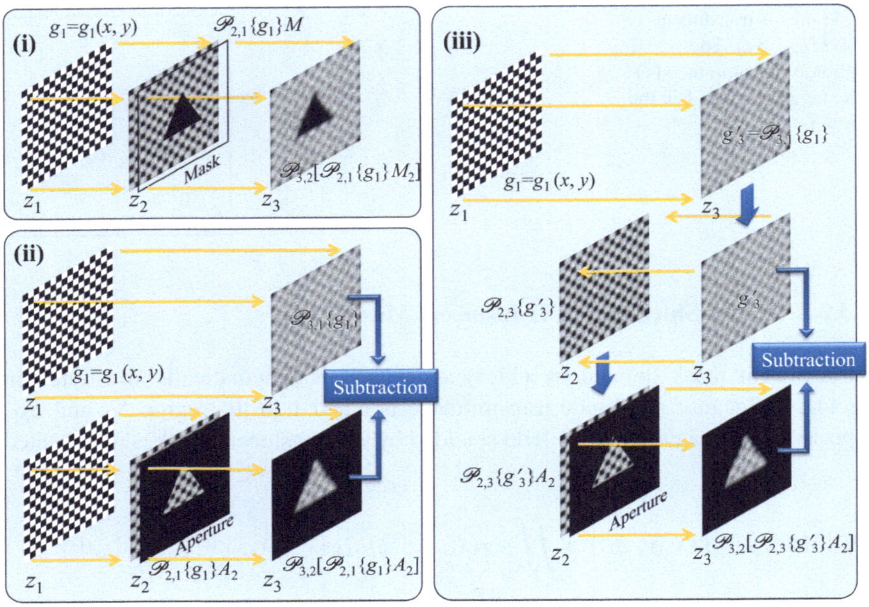

Fig. 11.12 Three types of masking procedure. The final masked field is given by the subtraction of two fields in procedures (ii) and (iii)

11.3.3 Light-Shielding by Use of Aperture Instead of Mask

Figure 11.12 schematically shows three possible procedures for performing the single light-shielding step in the silhouette method. Here, the object fields are ignored for simplicity.

Procedure (i) shows the original masking technique given directly by (11.2). Consider a wavefield $g_1 = g_1(x, y)$ given in the z_1 plane. The wavefield is propagated to the z_2 plane and is then masked using the mask function $M_2 = M_2(x, y)$. The masked field, given by $\mathscr{P}_{2,1}\{g_1\}M_2$, is then propagated again to the z_3 plane. The field in the z_3 plane is thus given by:

$$g_3(x, y) = \mathscr{P}_{3,2}\left\{\mathscr{P}_{2,1}\{g_1(x, y)\}M_2(x, y)\right\}. \tag{11.16}$$

Procedure (ii) is equivalent to procedure (i) according to Babinet's principle. In this case, we use an inverted mask function, i.e., an aperture function in (11.11), which is here defined as:

$$A_n(x, y) = 1 - M_n(x, y). \tag{11.17}$$

The field $\mathscr{P}_{2,1}\{g_1\}$ in the z_2 plane is masked by the aperture and is then propagated to the z_3 plane. The field is thus given by $\mathscr{P}_{3,2}\left\{\mathscr{P}_{2,1}\{g_1\}A_2\right\}$ in the z_3 plane. Babinet's principle ensures that the subtraction of this field from $\mathscr{P}_{3,1}\{g_1\}$ gives the same g_3 as that given in (11.16). In fact, by substituting $M_2 = 1 - A_2$ into (11.16), the field in the z_3 plane can be written as:

$$\begin{aligned} g_3(x, y) &= \mathscr{P}_{3,2}\left\{\mathscr{P}_{2,1}\{g_1(x, y)\} - \mathscr{P}_{2,1}\{g_1(x, y)\}A_2(x, y)\right\} \\ &= \mathscr{P}_{3,1}\{g_1(x, y)\} - \mathscr{P}_{3,2}\left\{\mathscr{P}_{2,1}\{g_1(x, y)\}A_2(x, y)\right\}, \end{aligned} \tag{11.18}$$

where the propagation $\mathscr{P}_{3,2}\left\{\mathscr{P}_{2,1}\{\xi(x, y)\}\right\}$ is equivalent to $\mathscr{P}_{3,1}\{\xi(x, y)\}$, according to the definition of the propagation operator.

Procedure (iii) also gives the same result as those found using procedures (i) and (ii). In this case, the field g_1 is propagated to the z_3 plane once without any masking. This temporal field is written as:

$$g'_3(x, y) = \mathscr{P}_{3,1}\{g_1(x, y)\}. \tag{11.19}$$

We retain this temporal field. The field before masking in the z_3 plane is obtained from back-propagation of the retained field $g'_3(x, y)$. The masked field $\mathscr{P}_{2,3}\left\{g'_3(x, y)\right\}$ $A_2(x, y)$ is again forward-propagated to the z_3 plane and is subtracted from the retained field $g'_3(x, y)$. Thus, the final field g_3 is given by:

$$g_3(x, y) = g'_3(x, y) - \mathscr{P}_{3,2}\left\{\mathscr{P}_{2,3}\left\{g'_3(x, y)\right\}A_2(x, y)\right\}. \tag{11.20}$$

Procedures (ii) and (iii), which use the aperture function rather than the mask function, require propagation three times, while procedure (i) uses the mask function and needs to be performed only twice. Therefore, procedures (ii) and (iii) seem to be redundant when compared with procedure (i). However, as shown in Fig. 11.13, use of the aperture offers a major advantage in that the full field is no longer required for two of the three propagations, because only a small portion of the field passes through the aperture. Therefore, the propagation of only the small part of this field is necessary to perform light-shielding. Here, we must again emphasize that the masked area is much smaller than the area of the whole wavefield in the case of P-P light-shielding.

In addition, procedure (iii) has the feature that both the second and third propagations are performed between the z_2 and z_3 planes. The only difference is the direction. Also, the final field g_3 is given by the difference between the temporal field g'_3 and the resultant of the "round-trip" propagation. As shown in the next section, this has a useful effect for the processing of multiple polygons; all intermediate information is accumulated in a single plane.

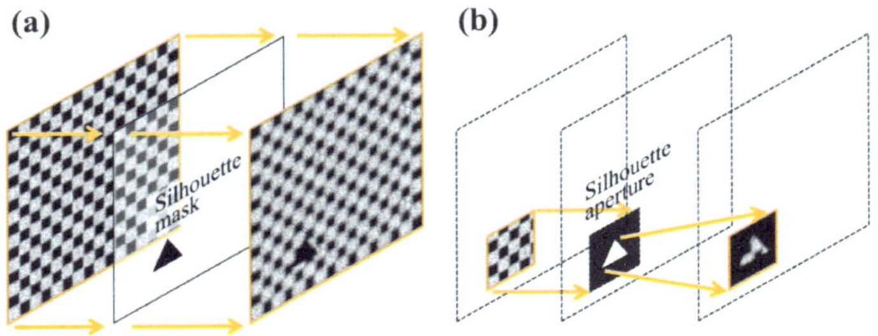

Fig. 11.13 Schematic illustration of the advantages of the silhouette aperture over the silhouette mask

11.3.4 Formulation for Multiple Polygons

The examples described above are for a single polygon only. To formulate multiple shielding, we must redefine the object plane. This object plane is similar to that in O-O light-shielding; it is perpendicular to the optical axis and is located near or across the object. However, the role of the object plane in P-P light-shielding is quite different from the role of the plane in O-O light-shielding. We intend to accumulate the temporal field in the object plane, which is the intermediate state of P-P light-shielding before completion of the occlusion processing. We can retain every object field under computation in the object plane using a formulation based on procedure (iii), as follows.

Suppose that the operation $\mathscr{P}_{\text{obj},n}\{g_n(x,y)\}$ propagates the field $g_n(x,y)$ in the z_n plane to the object plane. By using operator $\mathscr{P}_{\text{obj},n+1}\{\}$ for both sides of (11.2), the recurrence formula can be rewritten as

$$\begin{aligned}\mathscr{P}_{\text{obj},n+1}\{g_{n+1}(x,y)\} &= \mathscr{P}_{\text{obj},n+1}\left\{\mathscr{P}_{n+1,n}\{M_n(x,y)g_n(x,y)+O_n(x,y)\}\right\} \\ &= \mathscr{P}_{\text{obj},n}\{M_n(x,y)g_n(x,y)+O_n(x,y)\}, \end{aligned} \tag{11.21}$$

where $\mathscr{P}_{\text{obj},n+1}\left\{\mathscr{P}_{n+1,n}\{\xi(x,y)\}\right\}$ is equivalent to $\mathscr{P}_{\text{obj},n}\{\xi(x,y)\}$, according to the definition of the operator. By substituting $M_n(x,y)$ from (11.17) into (11.21), the recurrence formula can then be rewritten as

$$\mathscr{P}_{\text{obj},n+1}\left\{g_{n+1}(x,y)\right\} = \mathscr{P}_{\text{obj},n}\{g_n(x,y)\} - \mathscr{P}_{\text{obj},n}\{A_n(x,y)g_n(x,y) - O_n(x,y)\}. \tag{11.22}$$

We then introduce a new symbol:

$$g_n^{\text{obj}}(x,y) \equiv \mathscr{P}_{\text{obj},n}\{g_n(x,y)\}, \tag{11.23}$$

where $g_n^{\text{obj}}(x,y)$ is the temporally-accumulated wavefield in the object plane, i.e., the sub-total object field from polygon 0 to polygon $n-1$ with P-P light-shielding. Using this new symbol, (11.22) can be rewritten as follows:

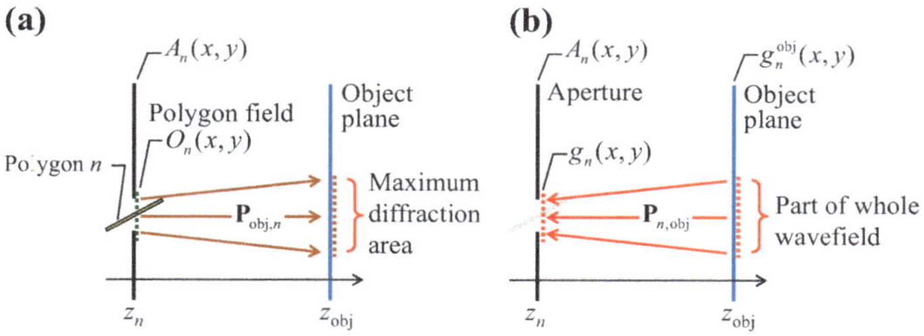

Fig. 11.14 **a** Forward propagation of the localized field to the object plane, and **b** backward propagation of the small part of the accumulated field in the object plane to the current aperture

$$g_{n+1}^{\text{obj}}(x, y) = g_n^{\text{obj}}(x, y) + \mathscr{P}_{\text{obj},n} \{O_n(x, y) - A_n(x, y)g_n(x, y)\}. \tag{11.24}$$

This recurrence formula states that the field $g_n(x, y)$ is required to advance by a single step of the calculation, i.e., to obtain $g_{n+1}^{\text{obj}}(x, y)$ from $g_n^{\text{obj}}(x, y)$. From the definition of $g_n^{\text{obj}}(x, y)$ in (11.23), this is provided by backward propagation as follows:

$$g_n(x, y) = \mathscr{P}_{n,\text{obj}} \left\{g_n^{\text{obj}}(x, y)\right\}. \tag{11.25}$$

As a result, (11.24) and (11.25) comprise a simultaneous recurrence formula that offers another expression of the original recurrence formula of (11.2).

Computation based on (11.24) and (11.25) requires double numerical propagation for each polygon, while the process was singular in the original recurrence formula. This seems to be ineffective. However, we should emphasize again that the propagation of the whole field is not necessary in this case. The polygon field $O_n(x, y)$ is localized around the polygon and the aperture in the z_n plane, as shown in Fig. 11.14a. In addition, $g_n(x, y)$ is required only around the aperture, as shown in (b). Therefore, we rewrite the simultaneous recurrence formulas as:

$$g_n(x, y) = \mathbf{P}_{n,\text{obj}} \left\{g_n^{\text{obj}}(x, y)\right\}, \tag{11.26}$$

$$g_{n+1}^{\text{obj}}(x, y) = g_n^{\text{obj}}(x, y) + \mathbf{P}_{\text{obj},n} \{O_n(x, y) - A_n(x, y)g_n(x, y)\}, \tag{11.27}$$

where the new notation $\mathbf{P}_{n,\text{obj}}\{g_n^{\text{obj}}(x, y)\}$ stands for the backward propagation of only a small part of the accumulated field $g_n^{\text{obj}}(x, y)$ to the aperture plane at z_n. The other notation, $\mathbf{P}_{\text{obj},n}\{O_n(x, y) - A_n(x, y)g_n(x, y)\}$, stands for forward propagation of the field that is localized around the polygon to the object plane. Greatly reduced computational effort is required for the two propagations when compared with that required for the original whole field propagation.

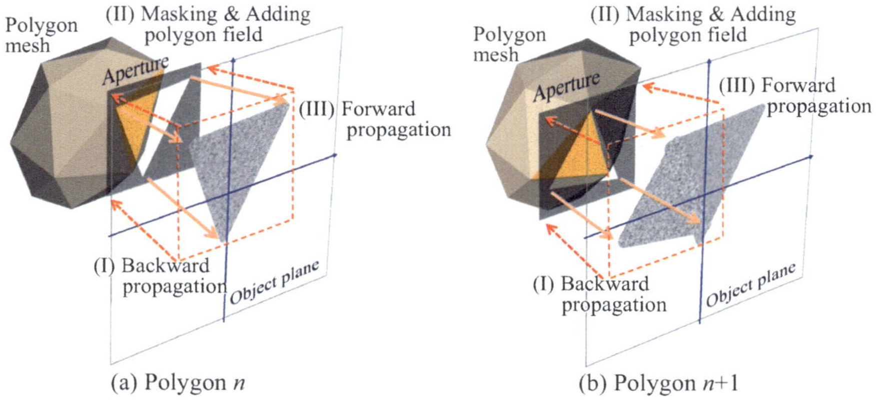

Fig. 11.15 Schematic explanation of the procedure for the switch-back technique: **a** switch-back for the current polygon n, and **b** for the next polygon $n + 1$

11.3.5 Practical Procedure for Computation of Object Field with P-P Shielding

In the case where there is no background field behind the object, we can set $g_0^{\mathrm{obj}}(x, y) \equiv 0$. Thus, backward propagation in (11.26) gives $g_0(x, y) \equiv 0$. By forward propagation in (11.27), the wavefield of polygon 0 is given by $g_1^{\mathrm{obj}}(x, y) = \mathbf{P}_{\mathrm{obj},0}\{O_0(x, y)\}$ in the object plane. The three steps described below are essentially repeated after this, as shown in Fig. 11.15.

Step (I). A small part, marked by a red-dashed line in (a), is extracted from the accumulated field $g_n^{\mathrm{obj}}(x, y)$ in the object plane. The extracted partial field is kept in a separate small sampling array (frame buffer) and propagated backward to the plane for polygon n, i.e., $g_n(x, y)$ is computed in the small frame buffer according to (11.26). Here, the small part required for the propagation is the region that is affected by the field of polygon n and by the light-shielding. The affected region is obtained by the technique explained in the following section.

Step (II). The field $g_n(x, y)$ is multiplied by the silhouette aperture function, $A_n(x, y)$, and then the resultant is subtracted from the polygon field $O_n(x, y)$, as per (11.27). Note that all operations can be performed in the small frame buffer through this step.

Step (III). The resultant field, still kept in the small frame buffer, is propagated forward to the object plane and then added to the current accumulated field, $g_n^{\mathrm{obj}}(x, y)$. This in turn gives the next accumulated field $g_{n+1}^{\mathrm{obj}}(x, y)$, as shown in (11.27).

If the object is composed of N polygons, then the accumulated field $g_N^{\mathrm{obj}}(x, y)$ gives the final object field. Thus, after performing step (III) when $n = N - 1$, we

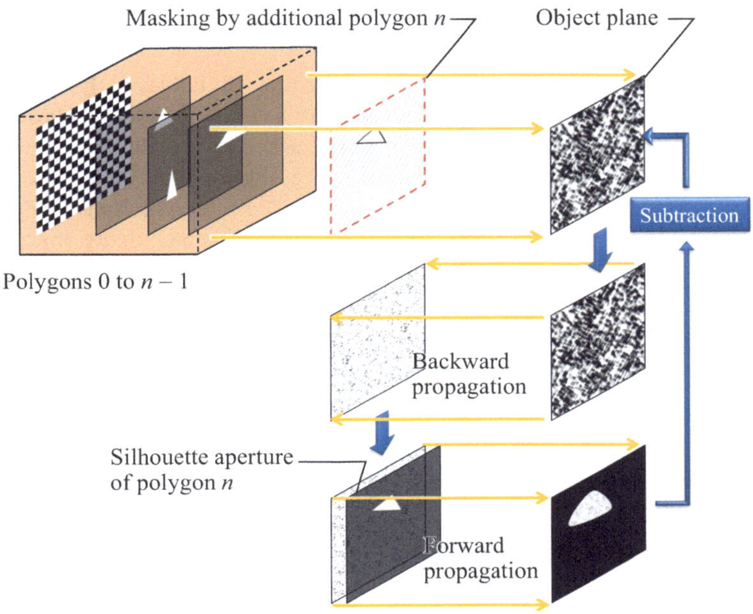

Fig. 11.16 Inductive explanation of P-P light-shielding by the switch-back propagation. It is assumed that the object field of the polygon 0 to $n - 1$ exists in the object plane from the beginning. Light-shielding by the additional polygon n is performed by the switch-back propagation

can obtain the final object field. As a result, the three steps must be performed $N - 1$ times in the case where $g_0^{\text{obj}}(x, y) \equiv 0$. The number of propagations is $2N - 1$. If $g_0^{\text{obj}}(x, y) \neq 0$, i.e., if there is a non-zero background field, then the three steps must be started from $g_0^{\text{obj}}(x, y)$. In this case, the number of repetitions required and the propagation are N and $2N$, respectively.

As mentioned earlier, the local wavefields are propagated back and forth to process each polygon in this technique. We therefore refer to this technique as the *switch-back technique*.

11.3.6 Inductive Explanation of the Switch-Back Technique

The switch-back technique may look a magical method because occlusion processing is performed by the simple round-trip propagation. The principle is not intuitive; it is only proved by mathematical formulas. Let us try to explain the principle by an inductive method.

Assume that the object field of polygons 0 to $n - 1$ has been already calculated with proper P-P light-shielding in the object plane, as shown in Fig. 11.16. Here, polygon fields are omitted and only the light-shielding is performed for simplicity.

Consider a situation where we add one more polygon n to the object field with P-P light-shielding.

To perform the light-shielding, using the procedure (iii) in Fig. 11.12, the existing wavefield that includes the fields of polygons 0 to $n-1$ is backward propagated to the mask position of the polygon n. The field is masked by the silhouette aperture of the polygon n and then forward propagated to the object plane. By subtracting the field from the existing object field, the new object field, in which the background field is shielded by the silhouette of the polygon n, is calculated in the object plane.

Light-shielding by polygon $n+1$ is also processed using the same procedure as that for the polygon n. This is the reason that the switch-back technique achieves P-P light-shielding properly.

11.3.7 Numerical Technique and Sampling Window for Switch-Back Propagation

It is very important to avoid aliasing errors in numerical propagations of the step (I) and (III). If the aliasing error occurs in each propagation, the accumulated error is considerable because N is usually very large. Therefore, the method of the numerical propagation should be carefully chosen in the practical implementation. The BLAS method in Sect. 6.4 should be used for the numerical propagation, because the BLAS method solves the problem of aliasing errors of the sampled transfer function in short distance propagation. However, if the destination field is expanded by diffraction over the sampling window, the quadruple extension of the sampling window is required to avoid degradation by field invasion. But, the extension increases the computational load. In practice, it is commonly better than the use of the quadruple extension to give a larger sampling window than the maximum diffraction area from the beginning.

The maximum diffraction area of the polygon field $O_n(x, y)$ is easily estimated in the object plane using the technique described in Sect. 10.3.5; it is given by the rectangular region including all maximum diffraction areas of point sources placed at the vertexes of the polygon. This is schematically shown in Fig. 11.17,

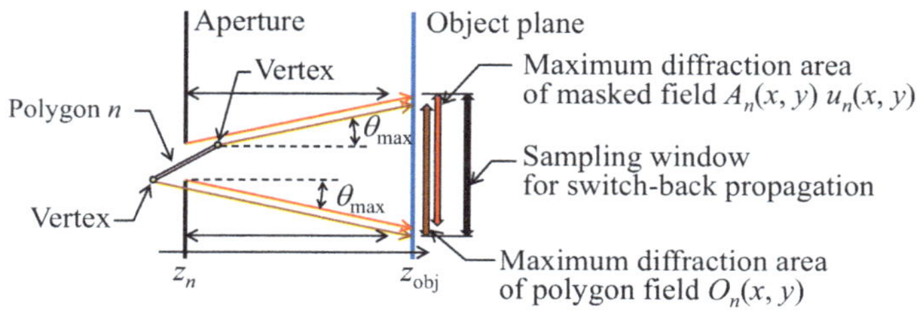

Fig. 11.17 Schematic illustration of maximum diffraction areas and the minimum sampling window required for avoiding arising errors in switch-back propagation

where θ_{max} is again maximum diffraction angle. The maximum diffraction area of the masked field $A_n(x, y)g_n(x, y)$ is also estimated by the same technique, as in Fig. 11.17. The sampling window for switch-back propagation must include both areas that are slightly different from each other. If this condition is satisfied, any aliasing error caused by convolution-based propagation can be avoided. As a result, the quadruple extension is unnecessary in this case, because it is ensured that the field is not expanded to the outside of the sampling window.

11.3.8 Emulation of Alpha Blend of CG

Alpha blend is a technique in CG to treat translucency of models and to blend colors of multiple objects. Polygons composing a CG model usually have color codes including alpha channel. An alpha value α_{bl} ranges from 0 to 1; $0.0 \le \alpha_{\text{bl}} \le 1.0$, where $\alpha_{\text{bl}} = 0.0$ represents completely transparent, and $\alpha_{\text{bl}} = 1.0$ represents fully opaque. Since the alpha value represents opacity of the polygon, optical transmittance should be associated with the alpha value as

$$T = 1 - \alpha_{\text{bl}}. \tag{11.28}$$

Thus, amplitude transmittance is

$$\begin{aligned} t &= \sqrt{T} \\ &= \sqrt{1 - \alpha_{\text{bl}}}. \end{aligned} \tag{11.29}$$

In computer holography, opaqueness is strongly associated with occlusion processing. As mentioned in Sect. 11.2.4, translucency of an object can be treated by translucent mask $M_n(x, y; t_n)$ defined in (11.3) and (11.4). However, translucency in O-O light-shielding is a vague concept due to lack of refraction occurring in a volumetric object. On the other hand, translucency is meaningful in P-P light-shielding because a polygon is considered as an object without thickness. We can consider translucency of a polygon to be a technique corresponding to the alpha blend in CG.

According to discussion on Babinet's principle for a translucent mask in Sect. 11.3.2.2, The translucent mask of polygon n is given by (11.15):

$$M_n(x, y; t_n) = 1 - (1 - t_n)A_n(x, y), \tag{11.30}$$

where t_n is amplitude transmittance of the polygon, obtained from the alpha value using (11.29). By substituting (11.30) into (11.5) and applying operator $\mathscr{P}_{\text{obj},n}\{\}$, we can derive a recurrence formula corresponding to (11.24):

$$g_{n+1}^{\text{obj}}(x, y) = g_n^{\text{obj}}(x, y) + \mathscr{P}_{\text{obj},n}\left\{O_n(x, y) - (1 - t_n)A_n(x, y)g_n(x, y)\right\}. \tag{11.31}$$

This formula is identical with (11.24) in the case where $t_n = 0$, while no light-shielding is performed in $t_n = 1$.

Rendering of translucent objects using (11.31) is not exactly the same as that by the alpha blend of CG. When the alpha value of a polygon is zero, the polygon disappears from the 3D scene in CG. However, although the background field is not shielded by the polygon having $t_n = 1$, the polygon itself does not vanish when using (11.31). Therefore, the following formula is most likely suitable for reproducing the same effect as that by the alpha blend in CG.

$$g_{n+1}^{\text{obj}}(x, y) = g_n^{\text{obj}}(x, y) + \mathscr{P}_{\text{obj},n} \left\{ \sqrt{\alpha_{\text{bl},n}} O_n(x, y) - (1 - t_n) A_n(x, y) g_n(x, y) \right\}. \tag{11.32}$$

Therefore, the recurrence formula used for actual rendering is

$$\begin{aligned} g_{n+1}^{\text{obj}}(x, y) = g_n^{\text{obj}}(x, y) \\ + \mathbf{P}_{\text{obj},n} \left\{ \sqrt{\alpha_{\text{bl},n}} O_n(x, y) - (1 - \sqrt{1 - \alpha_{\text{bl},n}}) A_n(x, y) g_n(x, y) \right\}, \end{aligned} \tag{11.33}$$

where $\alpha_{\text{bl},n}$ is the alpha value of the polygon n.

11.3.9 Acceleration by Dividing Object

The computation time required for the object field in the switch-back technique depends on the average distance between the polygons and the object plane, because the maximum diffraction area extends with increasing propagation distance, and thus the number of sampling points in the numerical propagation also increases with distance. Therefore, if the object extends for a long distance in the z-direction, i.e., the object has a large depth, then the technique will require a long computation time.

It is easy to predict that division of an object into multiple sub-objects and the use of an object plane for each sub-object, i.e., using multiple object planes, are valid for computation time reduction. In this case, the switch-back procedure presented in Sect. 11.3.5 is applied sequentially for each sub-object, as shown in Fig. 11.18.

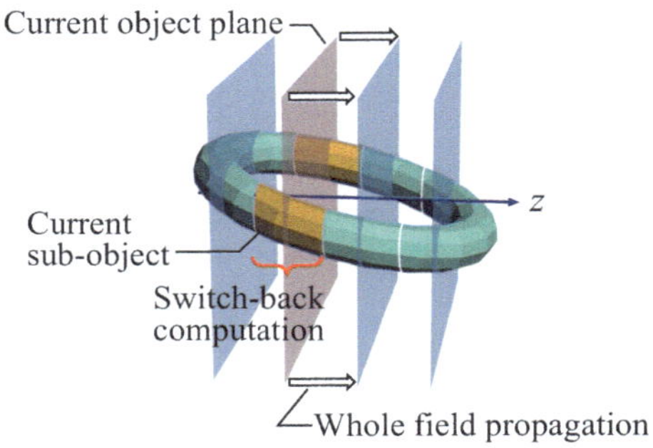

Fig. 11.18 Schematic illustration of the division of an object and the multiple object planes

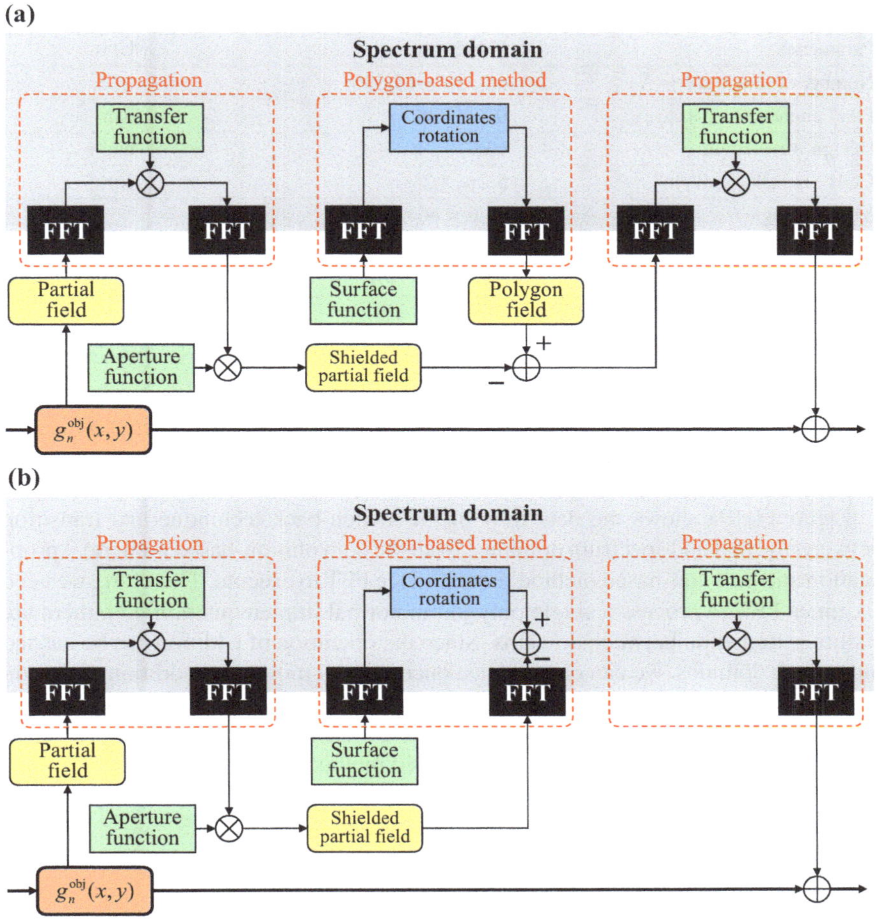

Fig. 11.19 The data-flow of the switch-back technique used with the polygon-based method in **a** normal and **b** integrated implementation

After the switch-back procedure is complete in the current object plane, the resultant object field is fully propagated to the next object plane. This field is the background field $g_0^{\text{obj}}(x, y)$ for the switch-back computation of the next sub-object.

11.3.10 Integration with the Polygon-Based Method

As mentioned before, the silhouette method is independent of the technique to generate the polygon fields. We can use the switch-back technique with the point-based

Table 11.2 Parameters of "Five Rings" CGH

Parameters	Value	Units
Number of pixels ($M \times N$)	65,536 × 65,536	
Pixel pitches ($\Delta x_h \times \Delta y_h$)	0.8 × 0.8	μm
Design wavelength	633	nm
CGH Size ($W_x \times W_y$)	5.2 × 5.2	cm
Viewing angle ($\theta_{max,x} \times \theta_{max,y}$)	46.6 × 46.6	°
Number of polygons	5,000	
Object size (W × H × D)	(40 × 26 × 39)	mm
Center position of reference spherical field (x, y, z)	(0, −35, −250)	mm

method as well as the polygon-based method. However, it is usually better to use the switch-back technique with the polygon-based method because both techniques are on the basis of wave-optics.

Figure 11.19a shows the data-flow of the switch-back technique and transition between the real and spectrum domains. Both the convolution-based numerical propagation and polygon-based method require twice FFT to execute. Therefore, we need six times FFT to process a single polygon in normal implementation, i.e., there are six times transition between domains. Since the operation of addition can be carried out in both domains, we can omit at least once FFT by moving an addition operation from the real domain to the spectrum domain, as shown in Fig. 11.19b. As a result, the number of FFT-executions is reduced to five for each polygon by integrating the switch-back technique with the polygon-based method.

11.3.11 Actual Examples of P-P Light-Shielding and Computation Time

Let us measure computation time of field rendering by combination of the switch-back technique and the polygon-based method. The 3D model is "Five Rings". Because this model is used in several sections, e.g., Sect. 8.7.2, the 3D scene is already shown in Fig. 8.15. The 3D model is composed of just 5,000 polygons and has a lot of self-occlusion. The number of samples of the object field is approximately 4 billion (64K × 64K), and the viewing angle is nearly 45° in full-parallax. Thus, the object field cannot be calculated by any other method than the switch-back technique. The detailed parameters of the actual HD-CGH are summarized in Table 11.2.

Measurement of computation time is performed for different numbers of sub-models to verify acceleration by dividing the object. Here, bandwidth of the polygon fields is limited using the method (iii) in Sect. 10.4.2. Thus, the quadruple extension is not used even in the whole field propagation. Moreover, the switch-back computation is integrated with the polygon-based method as described in the preceding section. Accordingly, five is the number of FFT execution necessary for each polygon.

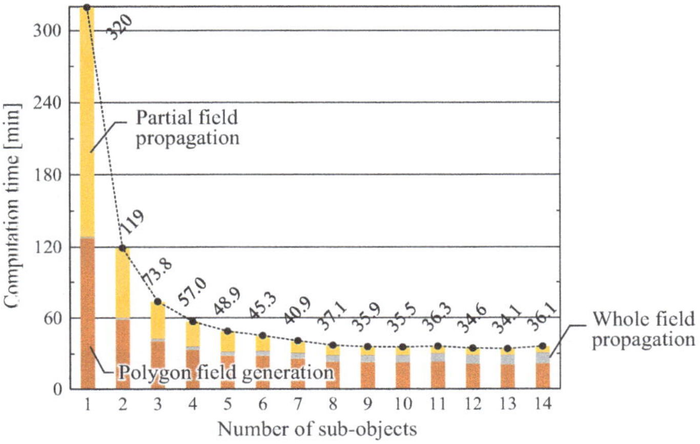

Fig. 11.20 Computation time of the object field of "FiveRings". CPU: Intel i9-9920X (3.5 GHz/12 Cores), Memory: 128 GB, FFT: Intel MKL11.2

Fig. 11.21 Optical reconstruction of Five Rings. The illumination light source is a He–Ne laser [75]

Measured computation times are shown in Fig. 11.20. The results confirm that computation time decreases with increasing the number of sub-objects as predicted in the previous section. However, because the number of whole field propagations agrees with the number of sub-models, computation time is almost constant or increase a little when the object model is divided into approximately more than 10 sub-models in this example. The shortest computation time was about 34 min at 13 sub-models.

Photographs of optical reconstruction of "Five Rings" are shown in Figs. 11.21 and 11.22 The close-up image in Fig. 11.21 is reconstructed by a He–Ne laser, while photographs and movies in Fig. 11.22 are taken using illuminating light by an ordinary red LED. In both cases, we can confirm a clear reproduction of self-occlusion.

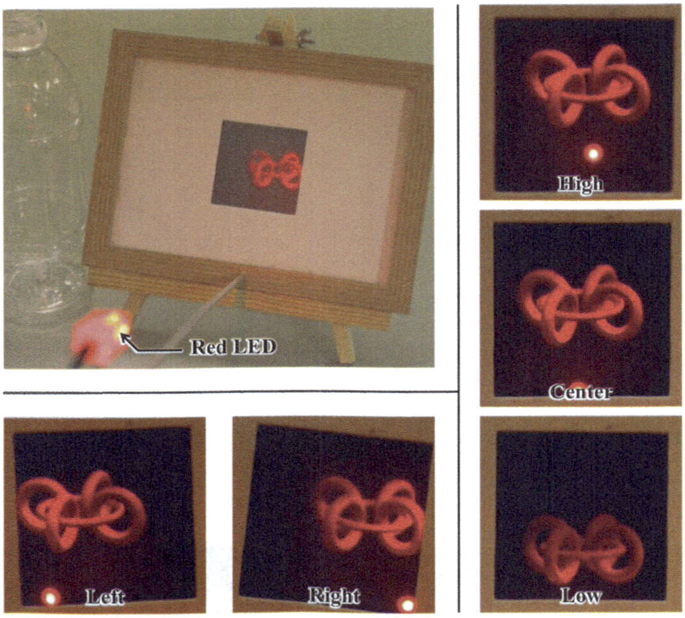

Fig. 11.22 Optical reconstruction of Five Rings [75]. The illumination light source is an ordinary red LED. Video link, https://doi.org/10.1364/OE.22.024450.m001

Figure 11.23a is a photograph of optical reconstruction of another HD-CH "Tri-plane IIb". The size of this HD-CGH is 18 cm $\times$ 12 cm, and the total number of pixels is just 45 billion. The 3D model, composed of 13034 polygons, has very complicated shape, as shown in the wireframe of (b). We can verify that the 3D object is faithfully converted to the holographic 3D image. Besides, because the fringe oversampling technique is used for the coding (see Sect. 8.8.3), the conjugate image is hardly seen in the optical reconstruction.

11.4 Limitation of the Silhouette Method

Even in P-P light shielding, the silhouette method is an approximate technique of true light-shielding after all; the light-shielding is therefore not perfect. Figure 11.24a is a photograph of "Sailing Warship II" CGH again. This photograph is taken from a front viewpoint, while close-up photograph in Fig. 11.24b is taken from a very left angle. Many cracks are perceived in the hull and sails in photograph (b). These cracks are caused by *occlusion errors* inherent in the silhouette method.

The cause of the occlusion error is schematically illustrated in Fig. 11.25. Here, the silhouette mask is arranged at the back end of the polygons to prevent the silhouette

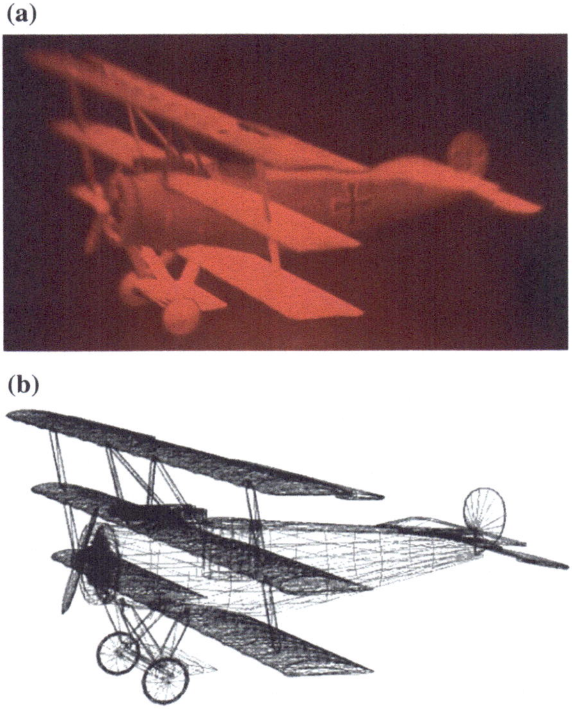

Fig. 11.23 **a** Optical reconstruction of HD-CGH "Triplane IIb", and **b** the wireframe of the 3D model. $M = 225,000$, $N = 200,000$, $\Delta x_h = 0.8$ [μm], and $\Delta y_h = 0.6$ [μm]. Video link, https://youtu.be/yTr8PXL9uZ4, https://youtu.be/rFoGerFXVNs

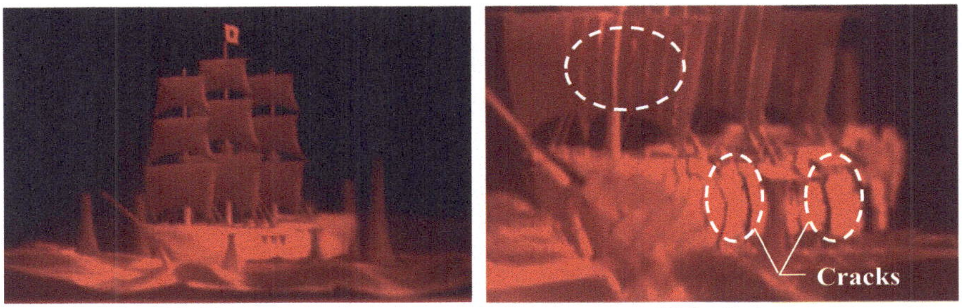

(a) Front view (b) Left view

Fig. 11.24 Photographs of optical reconstruction of "Sailing Warship II" Video link, https://youtu.be/8USLC6HEPsQ

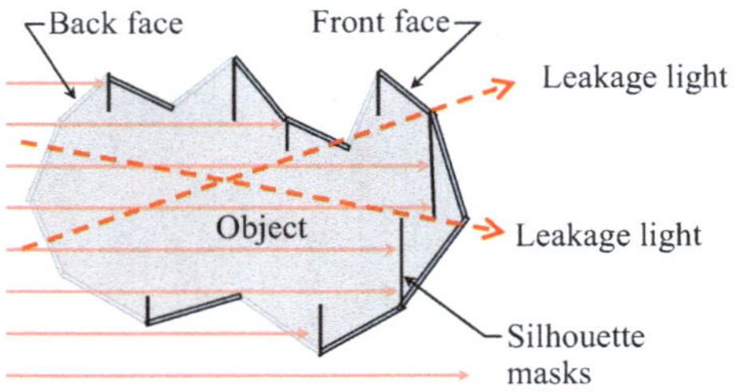

Fig. 11.25 Schematic illustration of the origin of occlusion errors

masks from being perceived directly. The figure shows paraxial fields are blocked properly but angled fields cannot be shield and leak out. This is most likely the origin of the cracks detected in Sailing Warship II. Masking background fields along the polygon surfaces is an effective technique to avoid the problem [60].

Chapter 12
Shifted Field Propagation

Abstract The shifted field propagation is a sort of numerical propagation between parallel planes. However, the technique makes it possible to propagate the source field to the destination sampling window that is laterally shifted from the source sampling window. Using this propagation technique, it is possible to propagate very large-scale fields that cannot be loaded in the computer memory. In this chapter, we discuss several techniques to realize the shifted field propagation: shifted far-field and Fresnel propagation, and the shifted angular spectrum method. The technique of scaled FFT used for the shifted Fresnel propagation is also described in this chapter.

12.1 Introduction

Off-axis or shifted field propagation is also a technique to propagate a wavefield in the source sampling window to the destination sampling window that is parallel to the source sampling window but laterally shifted.

12.1.1 What is Shifted Field Propagation

In numerical propagation of wavefields, the destination field is sometimes required in a region apart from the optical axis. This situation is depicted in Fig. 12.1a. The source sampling window is placed around the origin of the coordinate system, whereas the region of interest in the destination plane is located apart from the origin. In this case, the destination field is generally calculated by the following procedure:

Step 1. Extend the source sampling window so that the sampling window includes the region of interest after numerical propagation.
Step 2. Pad the extended source sampling window with zeros.
Step 3. Numerically propagate the source wavefield onto the destination plane.
Step 4. Cut out the region of interest from the destination sampling window.

K. Matsushima, *Introduction to Computer Holography*,
Series in Display Science and Technology,
https://doi.org/10.1007/978-3-030-38435-7_12

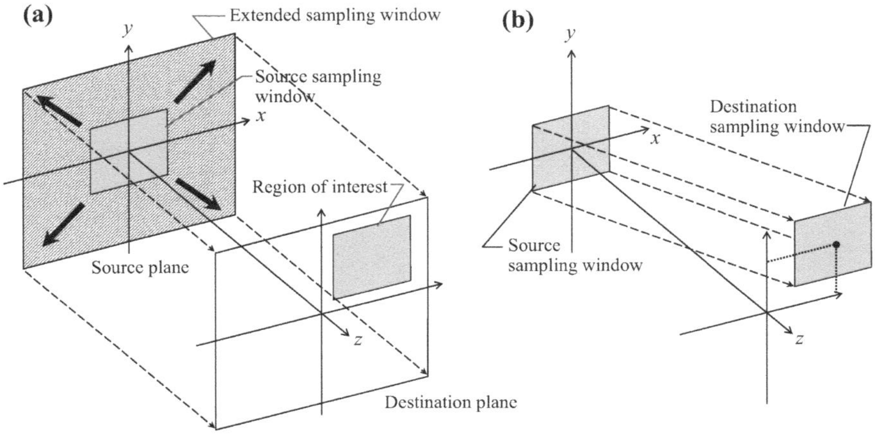

Fig. 12.1 Schematic illustration of **a** parallel and **b** shifted field propagation to calculate the field far from the center of the source field

Here, in Step 3 the source wavefield is propagated using conventional numerical methods that do not change the sampling interval, such as convolution-based techniques. The diffracted field can be calculated using this procedure. However, the procedure usually requires a huge computational effort, especially in cases where the region of interest is very far from the optical axis.

The shifted field propagation technique is one of the technique that propagates a wavefield between parallel planes, but this technique allow us to calculate a destination field shifted from that of the input source field, as shown in Fig. 12.1b. This is very useful in cases where the field is not paraxial and travels in off-axis directions.

12.1.2 Rectangular Tiling

The shifted field propagation is also essential to handling a large-scale field that cannot be loaded on the main memory of a computer. In this case, a technique called *rectangular tiling* makes it possible to propagate the large field. Figure 12.2 shows three typical patterns of the rectangular tiling.

In the case of (a), the field is so diffusive that a large sampling window is necessary in the destination plane. We can calculate the destination field in this case by dividing the destination sampling window into some sub-windows. Here, let us emphasize that we need only the memory for the source field and a destination sub-field in the computer. The final result can be obtained by scanning the destination sub-window and patching the sub-field together. In the opposite situation, i.e., in the case of converging light, the source field is divided into sub-fields as shown in Fig. 12.2b.

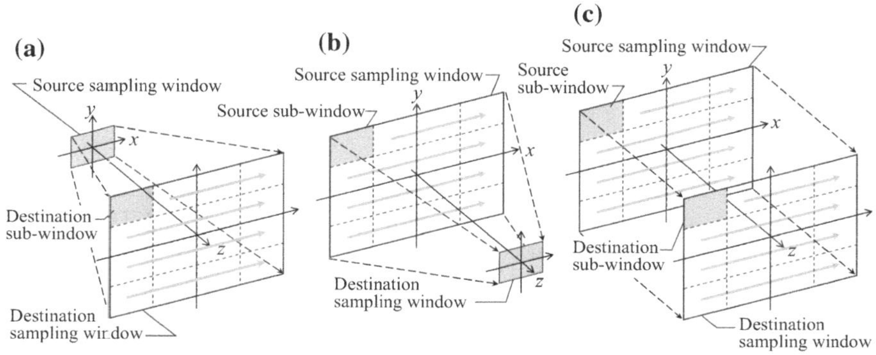

Fig. 12.2 Three typical patterns of rectangular tiling

All destination fields calculated from the source sub-fields is summed up in the destination plane in this case.

In computer holography, we often encounter the situation depicted in Fig. 12.2c, where both the source and destination fields are too large to load them into the memory of a computer. It is impossible to load any one of the source field or destination field when we are creating a large-scale CGH. Even in this case, we can carry out the propagation using the technique of shifted field propagation by loading sub-fields of the source field and the destination field one by one. Note that both the sub-windows in the source and destination field must be scanned in this case, and thus, we need considerable computational effort that includes I/O operation of the field data. However, it is much better to be able to carry out the large-scale field propagation than unable to do it.

12.2 Mathematical Preliminary

In ordinary DFT/FFT, the sampling interval in the frequency domain is automatically determined by (4.78). This sometimes imposes severe restrictions on wave-optical computation. The technique of fractional DFT and scaled FFT allows us to remove the limitation on the sampling interval of DFT/FFT.

12.2.1 Fractional DFT

Fractional DFT[1] based on Bluestein's algorithm is proposed to remove restriction on the sampling interval of ordinary DFT [3]. The fractional DFT is defined as

[1] This technique is called "fractional Fourier transform" in the original paper [3]. However, we call the technique "fractional DFT" in this book in order to avoid confusion with a "fractional Fourier transform" that is widely used but has different definition in mathematics [87]. In addition, a FFT operation applying fractional DFT to symmetrical sampling is referred to as a "scaled FFT" in this book.

$$F[p] = \mathbf{FDFT}^{(s)}\{f[m]\} \equiv \sum_{m=0}^{M-1} f[m]\exp(-i2\pi mps). \ (p = 0, \ldots, M-1) \tag{12.1}$$

The scaling parameter s may be any complex number but usually a real number. When $s = 1/M$, above definition agrees with the ordinary raw DFT/FFT in (4.79). The sampling interval of the Fourier frequency is given by $\Delta u = s/\Delta x$.

Substituting $2mp = m^2 + p^2 - (m-p)^2$ into (12.1), we get

$$F[p] = \sum_{m=0}^{M-1} f[m]\exp\left[-i\pi\{m^2 + p^2 - (m-p)^2\}s\right] = \exp[-i\pi p^2 s]\sum_{m=0}^{M-1} g_1[m]g_2[m-p], \tag{12.2}$$

where

$$g_1[m] = f[m]\exp(-i\pi m^2 s), \tag{12.3}$$

$$g_2[m] = \exp(i\pi m^2 s). \tag{12.4}$$

Since (12.2) has the form of discrete convolution in (4.96), we can calculate it using DFT/FFT. Here, note that $g_2[m]$ is evaluated with $-(M-1) \leq m \leq M-1$. Because FFT-based convolution gives circular convolution, $g_2[m]$ must have a periodicity $g_2[m-p] = g_2[m-p+M]$. To ensure this, $g_1[m]$ and $g_2[m]$ are extended to length $2M$ and evaluated as follows:

$$g_1[m] = \begin{cases} f[m]\exp(-i\pi m^2 s) & 0 \leq m < M \\ 0 & M \leq m < 2M \end{cases}, \tag{12.5}$$

$$g_2[m] = \begin{cases} \exp(i\pi m^2 s) & 0 \leq m < M \\ \exp[i\pi(m-2M)^2 s] & M \leq m < 2M \end{cases}. \tag{12.6}$$

Using this technique, the discrete convolution is calculated by

$$g[m] = g_1[m] \otimes g_2[m] = \mathbf{RFFT}^{-1}\{\mathbf{RFFT}\{g_1[m]\} \times \mathbf{RFFT}\{g_2[m]\}\}, \quad (m = 0, \ldots, M-1). \tag{12.7}$$

Here, it should be emphasized that $g[m]$ holds proper values only for $0 \leq m < M$ and the remaining results for $M \leq m < 2M$ must be discarded.

As a result, the fractional DFT in (12.1) of a sampled function $f[m]$ ($m = 0, \ldots, M-1$) is given by

$$\begin{aligned} F[p] &= \mathbf{FDFT}^{(s)}\{f[m]\} \\ &= \exp(-i\pi p^2 s) \times g[p], \quad (p = 0, \ \ldots, \ M-1), \end{aligned} \tag{12.8}$$

where the index m of $g[m]$ in (12.7) is changed to p.

In 2D sampled function $f[m, n]$ ($m = 0, \ldots, M-1$ and $n = 0, \ldots, N-1$), the fractional DFT can be evaluated by the same procedure as above:

$$\begin{aligned} F[p, q] &= \mathbf{FDFT}^{(s,t)}\{f[m, n]\} \\ &= \sum_{m=0}^{M-1}\sum_{n=0}^{N-1} f[m, n]\exp[-i2\pi(mps + nqt)] \\ &= \exp[-i\pi(p^2 s + q^2 t)] \times g[p, q], \\ &\quad (p = 0, \ldots, M-1 \text{ and } q = 0, \ldots, N-1), \end{aligned} \tag{12.9}$$

where t is a scaling parameter with respect to y coordinates. The sampling intervals in the Fourier space are given by

$$\Delta u = \frac{s}{\Delta x} \quad \text{and} \quad \Delta v = \frac{t}{\Delta y}. \tag{12.10}$$

12.2.2 *Scaled FFT for Symmetric Sampling*

Fractional DFT is defined with positive indexes as shown in the preceding section. However, it is inconvenient that the origin of the coordinate system is not placed at the center of the sampling window. To apply symmetrical sampling, we define symmetric fractional DFT as follows:

$$\begin{aligned} F[p] &= \mathbf{SFFT}^{(\alpha)}\{f[m]\} \\ &= \sum_{m=-M/2}^{M/2-1} f[m]\exp\left(\frac{-i2\pi mp}{M}\alpha\right), \end{aligned} \tag{12.11}$$

where α is a scaling parameter. This is referred to as *scaled FFT* in this book. The scaled FFT can also be written as

$$\begin{aligned} F(u'_p) &= \mathbf{SFFT}^{(\alpha)}\left\{f(x'_m)\right\} \\ &= \sum_{m=-M/2}^{M/2-1} f(x'_m)\exp(-i2\pi u'_p x'_m), \end{aligned} \tag{12.12}$$

where $x'_m = m\Delta x$ and $u'_p = p\Delta u$. The scaling parameter is given by

$$\alpha = M\Delta u \Delta x. \tag{12.13}$$

Substituting $m - M/2$ for m in (12.12) and using the sampling manner represented in (4.91) and (4.92), the following is equivalent to (12.12):

$$\begin{aligned} F(u_p) &= \mathbf{SFFT}^{(\alpha)}\{f(x_m)\} \\ &= \sum_{m=0}^{M-1} f(x_m)\exp(-i2\pi u_p x_m). \end{aligned} \tag{12.14}$$

Therefore, to apply symmetrical sampling as in Fig. 4.6b, substituting (4.91) and (4.92) into above, we get

$$\begin{aligned} F[p] = \exp&\left[i\pi\alpha\left(p - \frac{M}{2}\right)\right] \\ &\times \sum_{m=0}^{M-1} f[m]\exp(i\pi m\alpha)\exp\left(\frac{-i2\pi pm\alpha}{M}\right), \\ &(p = 0, \ldots, M-1), \end{aligned} \tag{12.15}$$

where it is assumed that M is a multiple of 4. Here, note that (12.15) results in (4.94) when $\alpha = 1$.

Using the fractional DFT, (12.15) is represented as

$$\mathbf{SFFT}^{(\alpha)}\{f[m]\} = \exp\left[i\pi\alpha\left(p - \frac{M}{2}\right)\right]\mathbf{FDFT}^{(\alpha/M)}\{f[m]\exp(i\pi\alpha m)\}. \tag{12.16}$$

This shows how to evaluate scaled FFT in practice. In 2D sampled function $f[m, n]$ ($m = 0, \ldots, M-1$ and $n = 0, \ldots, N-1$), scaled FFT is also evaluated using the same manner:

$$\begin{aligned} F[p, q] &= \mathbf{SFFT}^{(\alpha,\beta)}\{f[m, n]\} \\ &\equiv \exp\left[i\pi\left\{\alpha\left(p - \frac{M}{2}\right) + \beta\left(q - \frac{N}{2}\right)\right\}\right] \\ &\quad \times \mathbf{FDFT}^{(\alpha/M,\ \beta/N)}\{f[m, n]\exp[i\pi(\alpha m + \beta n)]\}, \\ &\quad (p = 0, \ldots, M-1 \text{ and } q = 0, \ldots, N-1), \end{aligned}$$

where β is a scaling parameter in the y direction and defined by

$$\beta \equiv N\Delta u \Delta x. \tag{12.17}$$

Examples of the scaled FFT are shown in Fig. 12.3. Input is a 2D rectangle function with the size of 2×2. In (a), because both scaling parameters α and β are set to

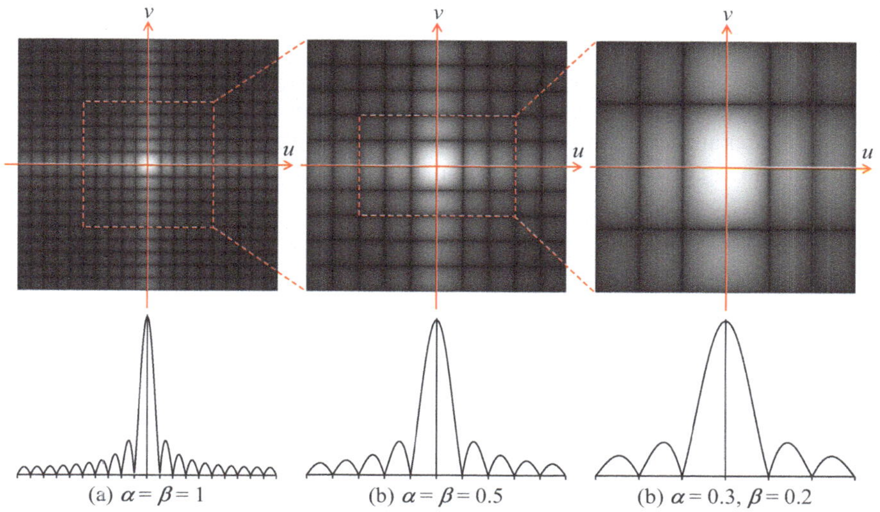

Fig. 12.3 Examples of scaled FFT. $|F(u_p, v_q)|$ is depicted as grayscale images. $f(x_m, y_n) = \text{rect}\,(x_m/2, y_n/2)$, $M = N = 512$, $\Delta x = \Delta y = 0.1$

Fig. 12.4 Examples of scaled FFT with scaling parameters more than unity. $\alpha = \beta = 3$. Other parameters and the input function are the same as those in Fig. 12.3

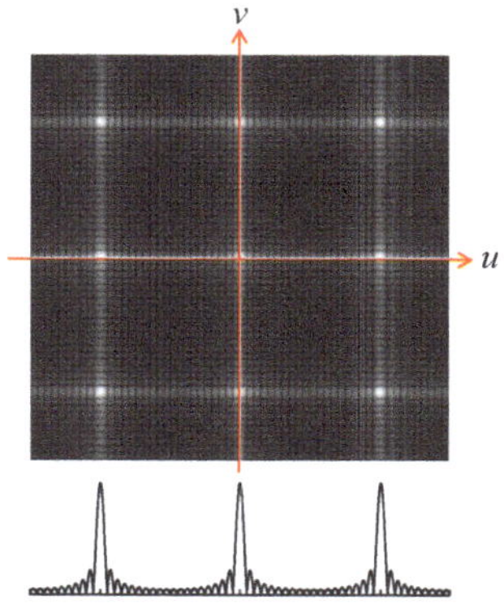

unity, the result agrees with the ordinary FFT. In contrast, (b) and (c) show that scaled FFT overcomes the limitation of ordinary FFT. We can freely enlarge the spectrum. Here, note that every spectrum has the same number of sampling. The scaled FFT changes only the sampling interval.

An example of scaled FFT with scaling parameters more than unity is shown in Fig. 12.4, where $\alpha = \beta = 3$. Here, let us emphasize that scaled FFT can change the sampling interval of the Fourier spectrum but cannot change the nature of sampled functions. As mentioned in Sect. 4.5, the Fourier spectrum of a function sampled at an interval of Δx is a periodic function whose period is $1/\Delta x$. Figure 12.4 truly shows this periodic nature of the sampled function.

The scaled FFT gains freedom represented by the scaling parameters α and β at the cost of three times FFT of $4MN$ sample points. However, note that $g_2[m]$ in (12.6) is independent of $f[m]$. This means that **RFFT** $\{g_2[m, n]\}$ can be precomputed when scaled FFT of different data array $f[m, n]$ is executed in the same sampling number and scaling parameters. This is useful for reducing computational effort in rectangular tiling in large-scale field propagation using the shifted Fresnel propagation, as mentioned in Sect. 12.4.

12.3 Shifted Far-Field Propagation

Figure 12.5 shows the geometry used for formulation of the shifted far-field propagation and shifted Fresnel propagation. The sampled source wavefield is placed at $z = z_0$, while the destination field is at $z = z_0 + d$ and the center of the sampling window is located at $(x_0, y_0, z_0 + d)$.

12.3.1 Formulation

In the 1D form, the far-field propagation in (5.43) is represented by

$$\begin{aligned} g_{\text{far}}(x'_p; z_0 + d) &= A_{\text{far}}(x'_p; d) \int_{-\infty}^{\infty} g(x_s; z_0) \exp\left[-i\frac{2\pi x'_p x_s}{\lambda d}\right] dx_s \\ &\cong A_{\text{far}}(x'_p; d) \sum_{m=-M/2}^{M/2-1} g(x'_{s,m}; z_0) \exp\left(-i\frac{2\pi x'_p x'_{s,m}}{\lambda d}\right), \end{aligned} \tag{12.18}$$

where the position of sample points is given by

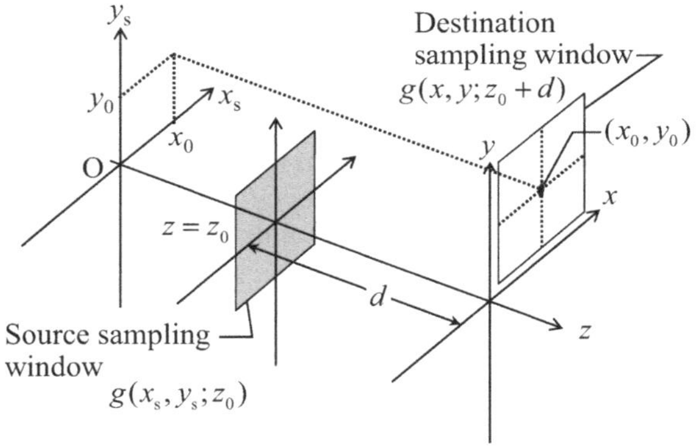

Fig. 12.5 Sampling windows used in shifted far-field and Fresnel propagation

$$\begin{aligned} x'_{s,m} &= m\Delta x_s, \\ x'_p &= p\Delta x + x_0. \end{aligned} \tag{12.19}$$

Δx_s and Δx are again sampling intervals of the source and destination field, respectively. Substituting (12.19) into (12.18), we get

$$\begin{aligned} g_{\text{Sfar}}(x'_p; z_0 + d) &= A_{\text{far}}(x'_p; d) \\ &\times \sum_{m=-M/2}^{M/2-1} g(x'_{s,m}; z_0) \exp\left(-i\frac{kx_0 x'_{s,m}}{d}\right) \exp\left(-i2\pi mp\frac{\Delta x \Delta x_s}{\lambda d}\right). \end{aligned} \tag{12.20}$$

When we use the symmetrical FFT represented in (4.83), the above equation can be rewritten as

$$g_{\text{Sfar}}[p; z_0 + d] = A_{\text{far}}[p; d]\mathbf{FFT}\left\{g[m; z_0] \exp\left(-i\frac{kx_0 x_{s,m}}{d}\right)\right\}, \tag{12.21}$$

where

$$\frac{\Delta x \Delta x_s}{\lambda d} = \frac{1}{M}. \tag{12.22}$$

It is definitely better to apply scaled FFT instead of ordinary symmetrical FFT from the perspective of the usefulness. Using scaled FFT in (12.11), (12.20) is rewritten as

$$g_{\text{Sfar}}[p; z_0 + d] = A_{\text{far}}[p; d]\mathbf{SFFT}^{(\alpha)}\left\{g[m; z_0] \exp\left(-i\frac{kx_0 x_{s,m}}{d}\right)\right\}, \tag{12.23}$$

where

$$\frac{\Delta x \Delta x_s}{\lambda d} = \frac{\alpha}{M}. \tag{12.24}$$

As a result, the shifted far-field propagation is formulated in the 2D form as follows:

$$\begin{aligned} g_{\text{Sfar}}[p, q; z_0 + d] &= A_{\text{far}}[p, q; d] \\ &\times \mathbf{SFFT}^{(\alpha,\beta)}\left\{g[m, n; z_0] \exp\left[-i\frac{k}{d}(x_0 x_{s,m} + y_0 y_{s,n})\right]\right\}, \end{aligned} \tag{12.25}$$

where the scaling parameters of scaled FFT are given by

$$\begin{aligned} \alpha &= \frac{\Delta x \Delta x_s}{\lambda d} M, \\ \beta &= \frac{\Delta x \Delta x_s}{\lambda d} N. \end{aligned} \tag{12.26}$$

Here, note that when using the symmetrical FFT or the scaled FFT the sampling manner is the same as that shown in Fig. 4.6b, (4.91), and (4.92), i.e., the sampling position of the source field is given by (6.1). Let us present the sampling position

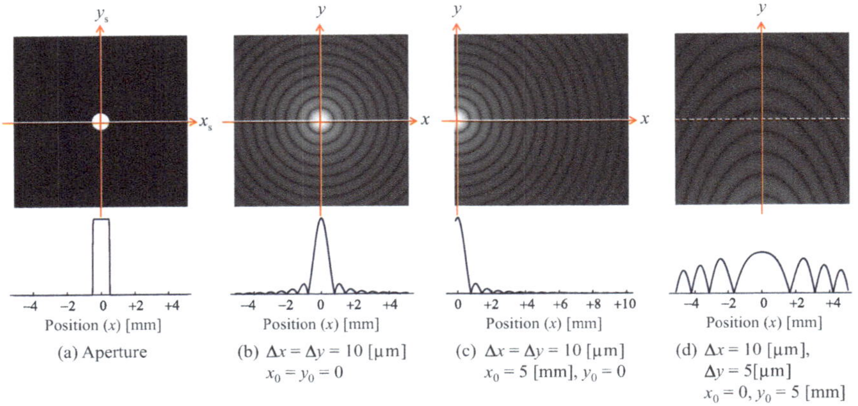

Fig. 12.6 Examples of the shifted far-fieald propagation. **a** Source field $|g(x_{s,m}, y_{s,n}; z_0)|$, $M = N = 1024$, $\Delta x_s = \Delta y_s = 10$ [μm], $\lambda = 633$ [nm]. **b–d** Amplitude image of the destination field $|g_{\mathrm{Sfar}}(x_p, y_q; z_0 + d)/(\lambda d)^2|$ for different shifts and sampling intervals. $d = 1$ [m]. Note that the graph of (**d**) shows amplitude distribution along the broken line and has a different vertical scale from others

again for the reader's convenience.

$$
\begin{aligned}
x_{s,m} &= \Delta x_s \left(m - \frac{M}{2} \right), \quad (m = 0, \ldots, M - 1), \\
y_{s,n} &= \Delta y_s \left(n - \frac{N}{2} \right), \quad (n = 0, \ldots, N - 1),
\end{aligned} \tag{12.27}
$$

where Δx_s and Δy_s are again the sampling interval of the source field. The sampling position of the destination field is also given by

$$
\begin{aligned}
x_p &= \Delta x \left(p - \frac{M}{2} \right) + x_0, \quad (p = 0, \ldots, M - 1), \\
y_q &= \Delta y \left(q - \frac{N}{2} \right) + y_0, \quad (q = 0, \ldots, N - 1),
\end{aligned} \tag{12.28}
$$

where Δx and Δy are sampling intervals of the destination field.

The technique of scaled far-field propagation has a big advantage of variability of the destination sampling interval as well as shifting the position of the sampling window.

12.3.2 Numerical Example

Actual examples of the shifted far-field propagation are shown in Fig. 12.6. Here, the source field is provided by a circular aperture of 1 mm in diameter. The normalized amplitude of the destination field is depicted in grayscale images. When compared with Fig. 6.2, it is verified that the shifted far-field propagation is more useful than the far-field propagation using a simple FFT. In this technique, we can keep the same sampling interval of the destination field as the source field, unlike the simple far-field propagation. Besides, we can freely change the sampling interval as well as the position of the destination window, as shown in Fig. 6.2d. Here, the image is depicted with the same pixel pitch in both vertical and horizontal directions, but the signal is sampled at non-isotropic sampling intervals.

12.3.3 Sampling Problem

Three factors mainly cause sampling problems or numerical errors in the shifted far-field propagation.

(i) Periodicity inherited from the scaled FFT
(ii) Aliasing errors before the Fourier transform
(iii) Aliasing errors after the Fourier transform

The factors (i) and (iii) are common to the shifted Fresnel propagation described in the next section. Thus, we treat only the factor (ii) in this section. The factors (i) and (iii) are discussed in Sect. 12.4.3, because the shifted Fresnel propagation is more widely used than the shifted far-field propagation in general. Actual examples of sampling problems are also provided in the next section.

As for factor (ii), the exponential factor in (12.25) before the Fourier transform may cause a sampling problem. When defining the factor as

$$\exp\left[-i\frac{k}{d}(x_0 x_s + y_0 y_s)\right] = \exp[i\phi(x_s, y_s)],$$

$$\phi(x_s, y_s) \equiv -\frac{k}{d}(x_0 x_s + y_0 y_s), \tag{12.29}$$

the local frequency with respect to x_s is given by

$$f_x = \frac{1}{2\pi}\frac{\partial\phi(x_s, y_s)}{\partial x_s} = -\frac{x_0}{\lambda d}. \tag{12.30}$$

Since the local frequency must satisfy $2|f_x| < \Delta x_s^{-1}$, we can find that there are limits on the amount of the shift as follows:

$$|x_0| < \frac{\lambda}{2\Delta x_s}d \quad \text{and} \quad |y_0| < \frac{\lambda}{2\Delta y_s}d. \tag{12.31}$$

These are also rewritten using the destination window sizes in (6.8) in the far-field propagation.

$$|x_0| < \frac{W_{\text{far},x}}{2} \quad \text{and} \quad |y_0| < \frac{W_{\text{far},y}}{2}. \tag{12.32}$$

12.4 Shifted Fresnel Propagation

Historically, the technique of the shifted Fresnel propagation based on the fractional DFT is proposed prior to the shifted far-field propagation [87]. This technique gains freedom to change the sampling interval as well as shift the sampling window by using the scaled FFT,[2] and thus is very useful in numerical propagation.

12.4.1 *Formulation*

As with formulation of the shifted far-field propagation, the Fresnel propagation in (6.21) is represented in the 1D form as follows:

$$\begin{aligned} g_{\text{FR}}(x'_p; z_0 + d) &= A_{\text{far}}(x'_p; d) \int_{-\infty}^{\infty} g(x_{\text{s}}; z_0) \exp\left[i\frac{\pi x_{\text{s}}^2}{\lambda d}\right] \exp\left[-i\frac{2\pi x'_p x_{\text{s}}}{\lambda d}\right] \text{d}x_{\text{s}} \\ &\cong A_{\text{far}}(x'_p; d) \\ &\quad \times \sum_{m=-M/2}^{M/2-1} g(x'_{\text{s},m}; z_0) \exp\left[i\frac{\pi x'^2_{\text{s},m}}{\lambda d}\right] \exp\left(-i\frac{2\pi x'_p x'_{\text{s},m}}{\lambda d}\right), \end{aligned} \tag{12.33}$$

where sampling positions are $x'_{\text{s},m} = m\Delta x_{\text{s}}$ and $x'_p = p\Delta x + x_0$. We can rewrite the above as

$$\begin{aligned} g_{\text{SFR}}(x'_p; z_0 + d) &= A_{\text{far}}(x'_p; d) \\ \times \sum_{m=-M/2}^{M/2-1} g(x'_{\text{s},m}; z_0) &\exp\left[i\frac{k}{d}\left(\frac{x'^2_{\text{s},m}}{2} - x_0 x'_{\text{s},m}\right)\right] \exp\left(-i2\pi mp\frac{\Delta x \Delta x_{\text{s}}}{\lambda d}\right). \end{aligned} \tag{12.34}$$

In comparison with (12.20), it is found that the shifted Fresnel propagation is formulated by simply multiplying the integrand of the far-field propagation by $\exp[ikx'^2_{\text{s},m}/(2d)]$. Thus, we can derive the formula of the shifted Fresnel propagation

[2] The shifted Fresnel propagation is originally formulated by using fractional DFT [87]. However, in this book we formulate the technique using scaled FFT that is more useful for our purpose.

from that of the shifted far-field propagation in (12.25) as follow:

$$g_{\mathrm{SFR}}[p, q; z_0 + d] = A_{\mathrm{far}}[p, q; d] \times \mathbf{SFFT}^{(\alpha,\beta)} \left\{ g[m, n; z_0] \exp\left[i\frac{k}{d} \left(\frac{x_{\mathrm{s},m}^2 + y_{\mathrm{s},n}^2}{2} - x_0 x_{\mathrm{s},m} - y_0 y_{\mathrm{s},n} \right) \right] \right\}. \tag{12.35}$$

The scaling parameters α and β are the same as those of the shifted far-field propagation, and thus, given by (12.26). Sampling positions $x_{\mathrm{s},m}$, $y_{\mathrm{s},n}$, x_p and y_q are given by (12.27) and (12.28).

12.4.2 Numerical Example

Figure 12.7 shows actual examples of destination fields obtained by the shifted Fresnel propagation in the case of no shift. All parameters except for the destination sampling interval are the same as those in Fig. 6.4, which is of the single-step Fresnel propagation. In Fig. 12.7, all examples have the same sampling interval of 3 [μm], which is chosen to give the similar results in Fig. 6.4. Here, note that a rectangular region of 3 mm are clipped from the whole sampling window in Fig. 6.4, because the destination sampling intervals change proportionally to the propagation distance in the single-step Fresnel propagation. But we can directly calculate the destination fields of the same area without clipping in this technique.

The second example is also diffraction by a circular aperture whose diameter is 5 mm, but an off-axis plane wave irradiates the aperture, as shown in Fig. 12.8. In this case, the diffraction pattern shifts as the field advances. Figure 12.9 shows the examples. The upper row shows the field diffraction calculated by the single-step Fresnel propagation for comparison. The number of sampling is 2048 × 2048. This is chosen for the diffracted field to keep within the sampling window. The lower row shows the destination fields calculated by the shifted Fresnel propagation. The destination sampling window shifts with $x_0 = 5$ [mm] in (b) and 10 [mm] in (c). Since we can shift the destination sampling window, the number of sample points required for the whole region of interest is always 1024 × 1024 in this technique.

12.4.3 Sampling Problem

As mentioned in Sect. 12.3.3, there are three factors that may cause sampling problems or numerical errors in the shifted Fresnel propagation.

(i) Periodicity inherited from the scaled FFT

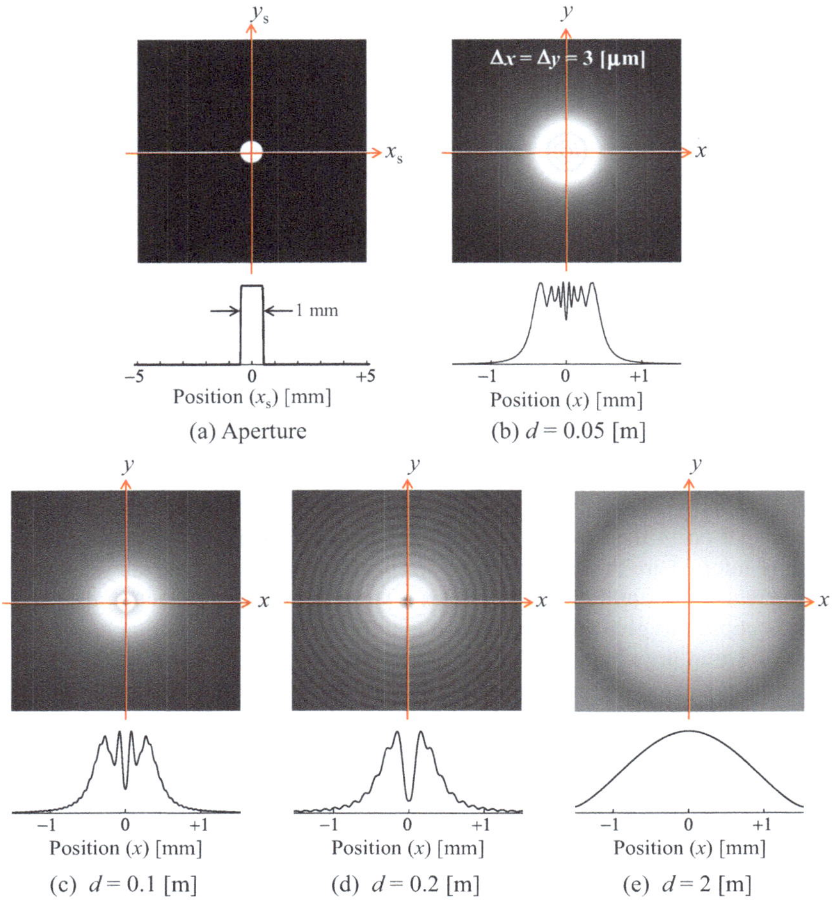

Fig. 12.7 Amplitude images calculated by the shifted Fresnel propagation. **a** Source field $|g(x_{s,m}, y_{s,n}; z_0)|$, $M = N = 1024$, $\Delta x_s = \Delta y_s = 10$ [μm], $\lambda = 633$ [nm]. **b–d** The destination field $|g_{SFR}(x_p, y_q; z)/(\lambda d)^2|$ for different propagation distances $d = z - z_0$. $\Delta x = \Delta y = 3$ [μm], $x_0 = y_0 = 0$

Fig. 12.8 Plane wave diffracted by a circular aperture

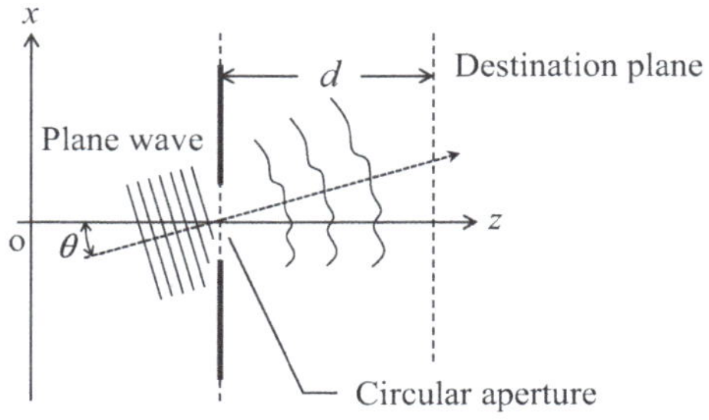

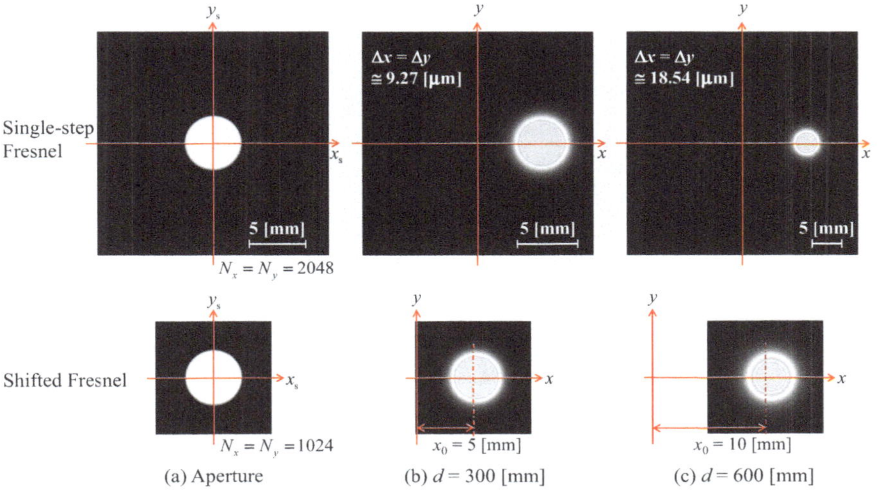

Fig. 12.9 Amplitude images of the destination fields calculated by the single-step Fresnel propagation (upper row) and the shifted Fresnel propagation (lower row) at different propagation distances. The diameter of the aperture is 5 mm and $\theta = 1$ [°]. $\Delta x_s = \Delta y_s = 10$ [μm] and $\lambda = 633$ [nm] in common with both methods. The destination sampling interval does not change in the shifted Fresnel propagation and $\Delta x = \Delta y = 10$ [μm]

As mentioned in Sect. 12.2.2, when scaling parameter α or β is more than unity, the periodic structure of sampled functions appears in the Fourier spectrum. To avoid this effect, the scaling parameters must not exceed unity, i.e., $\alpha \leq 1$ and $\beta \leq 1$. Therefore, the minimum propagation distance for avoiding the periodicity problem is derived from (12.26).

$$d \geq \frac{\Delta x}{\lambda} W_{sx} \quad \text{and} \quad \frac{\Delta x}{\lambda} W_{sy}. \tag{12.36}$$

where W_{sx} and W_{sy} are again the sizes of the source sampling window.

There is also an alternative interpretation on the limitation. When $\alpha = \beta = 1$, the size of the destination sampling window obtained from (12.26) agrees with that in the far-field propagation, i.e.,

$$W_{\text{far},x} = \frac{\lambda}{\Delta x_s} d \quad \text{and} \quad W_{\text{far},y} = \frac{\lambda}{\Delta y_s} d. \tag{12.37}$$

These give the period of repetition of the destination field. In other words, $W_{\text{far},x} \times W_{\text{far},y}$ gives the size of a cell or window. If the destination field keeps inside the cell, it does not experience any periodicity. In contrast, if the destination field leaves the cell, multiple images or aliased images appear in the destination field.

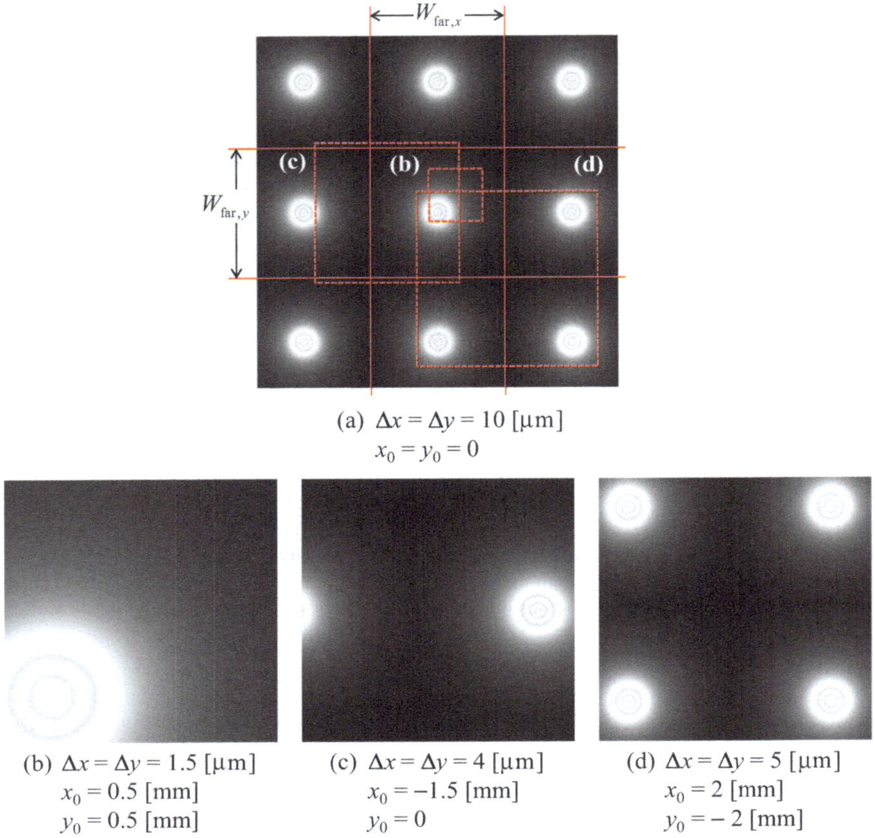

Fig. 12.10 Examples of the shifted Fresnel propagation. **a**–**d** Amplitude images of the destination field $|g_{\text{SFR}}(x_p, y_q; z_0 + d)/(\lambda d)^2|$ for different shifts and sampling intervals. The source field is a circular aperture of 1 [mm] in diameter. $M = N = 1024$, $\Delta x_s = \Delta y_s = 10$ [μm], $\lambda = 633$ [nm]. $d = 60$ [mm], and $W_{\text{far},x} = W_{\text{far},y} \cong 3.8$ [mm]

Figure 12.10 shows the destination field propagated at $d = 60$ [mm] by the shifted Fresnel propagation. In (a), the parameters violate the relation (12.36), because $W_{sx}\Delta x/\lambda \cong 161$ [mm] in this case. Thus, multiple cells appear in the amplitude image. The parameters in (b) satisfies the relation (12.36) and the destination window keeps inside the center cell, whereas the parameters in (c) and (d) violate the relation (12.36) a little and the destination window go across the cell border.

(ii) Aliasing errors before the Fourier transform

To simplify the problem, let us concentrate only on the x direction. Using the same procedure as that in Sect. 12.3.3, i.e.,

$$\phi(x_s) \equiv \frac{k}{d}\left(\frac{x_s^2}{2} - x_0 x_s\right), \tag{12.38}$$

$$f_x = \frac{1}{2\pi}\frac{\partial \phi(x_s)}{\partial x_s} = \frac{x_s - x_0}{\lambda d}, \tag{12.39}$$

we find the following aliasing-free condition of the propagation distance:

$$|x_s - x_0| < \frac{\lambda}{2\Delta x_s} d. \tag{12.40}$$

Here, assume that source field $g(x_s; z_0)$ has non-zero value in a range $|x_s| \le D_x/2$. For example, the source field given by a rectangular aperture whose width is D_x fits with this condition. In this case, because the relation (12.40) must be satisfied in a range $-D_x/2 \le x_s \le D_x/2$, the amount of the shift must satisfy the following relation to avoid the sampling problem:

$$|x_0| \le \frac{1}{2}\left|W_{\mathrm{far},x} - D_x\right| \quad \text{and} \quad |y_0| \le \frac{1}{2}\left|W_{\mathrm{far},y} - D_y\right| \tag{12.41}$$

When the source field spreads over the whole sampling window, the above relations are given by

$$|x_0| < \frac{1}{2}\left|W_{\mathrm{far},x} - W_{sx}\right| \quad \text{and} \quad |y_0| < \frac{1}{2}\left|W_{\mathrm{far},y} - W_{sy}\right|. \tag{12.42}$$

Here, W_{sx} and W_{sy} are again the sizes of the source sampling window. The parameters in Fig. 12.10c, d violate the condition (12.41) a little.

(iii) Aliasing errors after the Fourier transform
Factor $A_{\mathrm{far}}(x, y; d)$ also causes another sampling problem for the phase distribution of the destination field as in ordinary far-field propagation, as mentioned in Sect. 6.1.4. According to relation (6.14), the destination field is limited to a range

$$-\frac{\lambda d}{2\Delta x} < x < \frac{\lambda d}{2\Delta x} \quad \text{and} \quad -\frac{\lambda d}{2\Delta y} < y < \frac{\lambda d}{2\Delta y}, \tag{12.43}$$

to avoid aliasing errors of the phase distribution. This means that the destination sampling window must be placed inside the range imposed by the above relations, as shown in Fig. 12.11. Results of Fig. 12.10a, d violate this condition, but the amplitude images are not affected by this error.

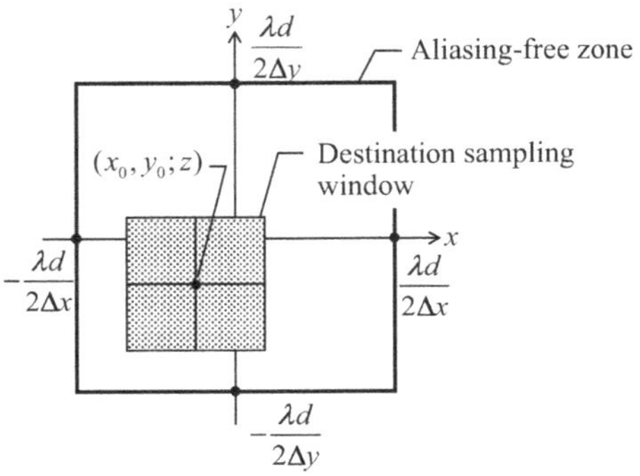

Fig. 12.11 The aliasing-free zone of phase distribution of the destination sampling window in the shifted far-field and Fresnel propagation

12.5 Shifted Angular Spectrum Method

The shifted Fresnel method is very useful technique, which can change the destination sampling intervals as well as the position of the sampling window. However, the technique has several restrictions as mentioned in the preceding section. In particular, the limit of the distance in (12.36) makes it impossible to propagate wavefields at a short distance. This problem results from use of the scaled FFT. Thus, we need another technique for short-distance shifted propagation.

The shifted angular spectrum method, based on fast convolution using FFT, removes the restriction of the shifted Fresnel propagation and propagate large-scale wavefields even at a short distance [62].

12.5.1 Coordinate System

The coordinate system used for the formulation is shown in Fig. 12.12. The source field is represented using source coordinate system (x_s, y_s, z) and placed in plane (x_s, y_s, z_0). On the other hand, the destination field is represented in different coordinate system (x, y, z) and placed in destination plane $(x, y; z_0 + d)$. Unlike the propagation technique mentioned so far, these coordinate systems do not share their origin. Every coordinate axis is parallel to each other, but coordinates x and y of the destination is shifted from x_s and y_s of the source as follows:

$$\begin{aligned} x &= x_s - x_0, \\ y &= y_s - y_0, \end{aligned} \tag{12.44}$$

where x_0 and y_0 represent lateral shifts of the coordinate origin of the destination field.

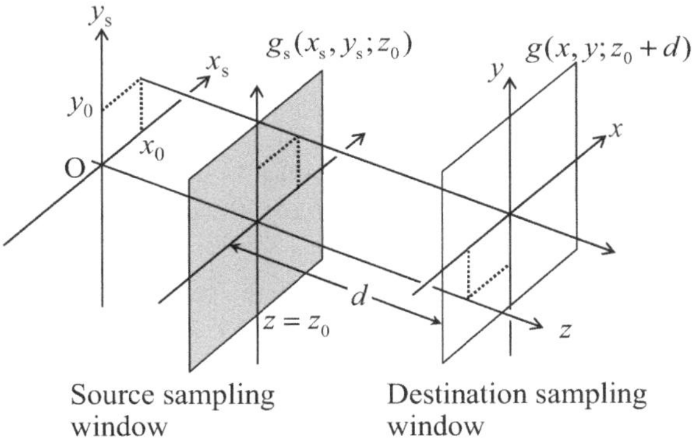

Fig. 12.12 Coordinate systems used in formulation of the shifted angular spectrum method

12.5.2 Formulation

Supposing that the source field is $g_s(x_s, y_s; z_0)$, the destination field in the (x_s, y_s, z) coordinates is obtained by convolution integral. Using (5.19) in the angular spectrum method, we get

$$g_s(x_s, y_s; z_0 + d) = \iint_{-\infty}^{+\infty} g_s(x', y'; z_0) h_{AS}(x_s - x', y_s - y'; d) \mathrm{d}x' \mathrm{d}y'. \quad (12.45)$$

When the destination field is represented by $g(x, y; z_0 + d)$ in the shifted coordinate system shown in Fig. 12.12, it can be simply written as a shift of the source field:

$$g(x, y; z_0 + d) = g_s(x + x_0, y + y_0; z_0 + d). \quad (12.46)$$

Therefore, substituting $x + x_0$ and $y + y_0$ for x_s and y_s in (12.45) respectively, the destination field is written as

$$\begin{aligned} g(x, y; z_0 + d) &= g_s(x + x_0, y + y_0; z_0 + d) \\ &= \iint_{-\infty}^{+\infty} g_s(x', y'; z_0) h_{AS}(x + x_0 - x', y + y_0 - y'; d) \mathrm{d}x' \mathrm{d}y' \\ &= \iint_{-\infty}^{+\infty} g_s(x', y'; z_0) h_{SAS}(x - x', x - y'; d) \mathrm{d}x' \mathrm{d}y', \end{aligned} \quad (12.47)$$

where the propagation kernel is defined by

$$h_{SAS}(x, y; d) \equiv h_{AS}(x + x_0, y + y_0; d). \quad (12.48)$$

We can easily obtain the transfer function using the shift theorem of the Fourier transform, i.e.,

$$
\begin{aligned}
H_{\mathrm{SAS}}(u, v; d) &= \mathscr{F}\{h_{\mathrm{SAS}}(x, y; d)\} \\
&= H_{\mathrm{AS}}(u, v; d)\exp[i2\pi(x_0 u + y_0 v)] \\
&= \exp[i2\pi(x_0 u + y_0 v + dw)],
\end{aligned} \tag{12.49}
$$

where (5.18) is used. w is again a Fourier frequency with respect to z coordinates and defined in (5.7). Conclusively, the destination field is calculated by

$$
\begin{aligned}
g(x, y; z_0 + d) &= g_s \otimes h_{\mathrm{SAS}}(x, y; d) \\
&= \mathscr{F}^{-1}\{\mathscr{F}\{g_s(x_s, y_s; d)\} H_{\mathrm{SAS}}(u, v; d)\}.
\end{aligned} \tag{12.50}
$$

Since the destination field is calculated by convolution, the sampling intervals are the same as those of the source field in this technique, i.e., $\Delta x = \Delta x_s$ and $\Delta y = \Delta y_s$.

12.5.3 Band Limiting

When we actually apply this technique to a sampled wavefield, the sampled transfer function $H_{\mathrm{SAS}}(u, v; d)$ causes aliasing errors very easily. Thus, a band limit is always necessary to obtain the destination field without noise.

For simplicity, we first discuss one-dimensional wavefields. In this case, the transfer function in (12.49) is redefined as follows:

$$
\begin{aligned}
H_{\mathrm{SAS}}(u; d) &= \exp[(i\phi(u; d)], \\
\phi(u; d) &\equiv 2\pi\left[x_0 u + d\sqrt{\lambda^{-1} - u^2}\right]
\end{aligned} \tag{12.51}
$$

The local signal frequency of the transfer function is given by

$$
\begin{aligned}
f_u &= \frac{1}{2\pi}\frac{\partial\phi(u; d)}{\partial u} \\
&= x_0 - \frac{u}{\sqrt{\lambda^{-1} - u^2}} d.
\end{aligned} \tag{12.52}
$$

The following condition is necessary to satisfy $\Delta u^{-1} > 2|f_u|$.

$$
x_0 - \frac{1}{2\Delta u} < \frac{ud}{\sqrt{\lambda^{-1} - u^2}} < x_0 + \frac{1}{2\Delta u}, \tag{12.53}
$$

where Δu is a sampling interval of Fourier frequency u as usual. However, it should be noted that the sampling interval is not given by W_{sx}^{-1} unlike the shifted far-field and Fresnel method. Since we must use the quadruple extension in Sect. 4.7.3 to convert circular convolution to linear convolution, the number of source sample points is always doubled in this technique. Therefore, the sampling interval is given by

$$\Delta u = \frac{1}{2W_{sx}}. \tag{12.54}$$

Using the size of the source sampling window, relations (12.53) are rewritten as

$$\frac{x_0 - W_{sx}}{d} < \frac{u}{\sqrt{\lambda^{-1} - u^2}} < \frac{x_0 + W_{sx}}{d}. \tag{12.55}$$

These conditions must be satisfied to avoid aliasing errors in the sampled transfer function $H_{\mathrm{SAS}}(u; d)$. Regarding to frequency u, the conditions are summarized in the following three cases.

Case (i) $W_{sx} < x_0$

$$u_{\mathrm{BL}}^{(-)} < u < u_{\mathrm{BL}}^{(+)} \tag{12.56}$$

Case (ii) $-W_{sx} < x_0 \leq W_{sx}$

$$-u_{\mathrm{BL}}^{(-)} < u < u_{\mathrm{BL}}^{(+)} \tag{12.57}$$

Case (iii) $x_0 \leq -W_{sx}$

$$-u_{\mathrm{BL}}^{(-)} < u < -u_{\mathrm{BL}}^{(+)} \tag{12.58}$$

The constant $u_{\mathrm{BL}}^{(\pm)}$ is given by

$$u_{\mathrm{BL}}^{(\pm)} \equiv \frac{1}{\lambda}\left[\left(\frac{d}{x_0 \pm W_{sx}}\right)^2 + 1\right]^{-1/2}. \tag{12.59}$$

When $x_0 = 0$, the constant $u_{\mathrm{BL}}^{(\pm)}$ is exactly the same as u_{BL} in (6.44).

In two-dimensional cases, the Nyquist condition is given as two-dimensional regions in the (u, v) plane. The local signal frequencies of the transfer function $H_{\mathrm{SAS}}(u, v; d)(= \exp[i\phi(u, v; d)])$ are given by

$$\begin{aligned} f_u &= \frac{1}{2\pi}\frac{\partial \phi(u, v; d)}{\partial u} = x_0 - \frac{u}{w}d, \\ f_v &= \frac{1}{2\pi}\frac{\partial \phi(u, v; d)}{\partial v} = y_0 - \frac{u}{w}d, \end{aligned} \tag{12.60}$$

where $\phi(u, v; d) = \exp[2\pi(x_0 u + y_0 v + dw)]$ and $w = (\lambda^{-2} - u^2 - v^2)^{1/2}$. The sampling interval with respect to v is also given by

$$\Delta v = \frac{1}{2W_{sy}}, \tag{12.61}$$

because we must perform the quadruple extension.

To avoid sapling problems in the 2D transfer function, the following Nyquist conditions must be satisfied simultaneously:

$$\Delta u^{-1} > 2|f_u|, \tag{12.62}$$

$$\Delta v^{-1} > 2|f_v|. \tag{12.63}$$

Since these conditions are very similar, let us formulate the region that satisfies only condition (12.62) in the (u, v) space as follows:

Case (i) $W_{sx} < x_0$

$$u > 0 \quad \text{and} \quad \frac{u^2}{\left[u_{\mathrm{BL}}^{(-)}\right]^2} + \frac{v^2}{\lambda^{-2}} > 1 \quad \text{and} \quad \frac{u^2}{\left[u_{\mathrm{BL}}^{(+)}\right]^2} + \frac{v^2}{\lambda^{-2}} < 1 \tag{12.64}$$

Case (ii) $-W_{sx} < x_0 \leq W_{sx}$

$$\left(u \leq 0 \quad \text{and} \quad \frac{u^2}{\left[u_{\mathrm{BL}}^{(-)}\right]^2} + \frac{v^2}{\lambda^{-2}} < 1 \right) \quad \text{or} \quad \left(u > 0 \quad \text{and} \quad \frac{u^2}{\left[u_{\mathrm{BL}}^{(+)}\right]^2} + \frac{v^2}{\lambda^{-2}} < 1 \right) \tag{12.65}$$

Case (iii) $x_0 \leq -W_{sx}$

$$u < 0 \quad \text{and} \quad \frac{u^2}{\left[u_{\mathrm{BL}}^{(+)}\right]^2} + \frac{v^2}{\lambda^{-2}} > 1 \quad \text{and} \quad \frac{u^2}{\left[u_{\mathrm{BL}}^{(-)}\right]^2} + \frac{v^2}{\lambda^{-2}} < 1 \tag{12.66}$$

Here, the region satisfying the other condition (12.63) is simply obtained by switching the symbols x and u to y and v in the above relations, respectively. In this case, the constant $v_{\mathrm{BL}}^{(\pm)}$ is defined by

$$v_{\mathrm{BL}}^{(\pm)} \equiv \frac{1}{\lambda}\left[\left(\frac{d}{y_0 \pm W_{sy}}\right)^2 + 1\right]^{-1/2}. \tag{12.67}$$

Note that the region under the conditions $x_0 = 0$ or $y_0 = 0$, that is, the special case of (ii) is given by

$$\frac{u^2}{u_{\mathrm{BL}}^2} + \frac{v^2}{\lambda^{-2}} < 1 \quad \text{and} \quad \frac{u^2}{\lambda^{-2}} + \frac{v^2}{v_{\mathrm{BL}}^2} < 1. \tag{12.68}$$

This agrees with the relation (6.48) of the angular spectrum method.

Relations (12.64)–(12.66) for shifting x_0 give a region specified by the combination of vertical ellipsoidal regions with a major diameter $2\lambda^{-1}$ in the (u, v) plane, while the relations for shifting y_0 give horizontal ellipsoidal regions. Figure 12.13

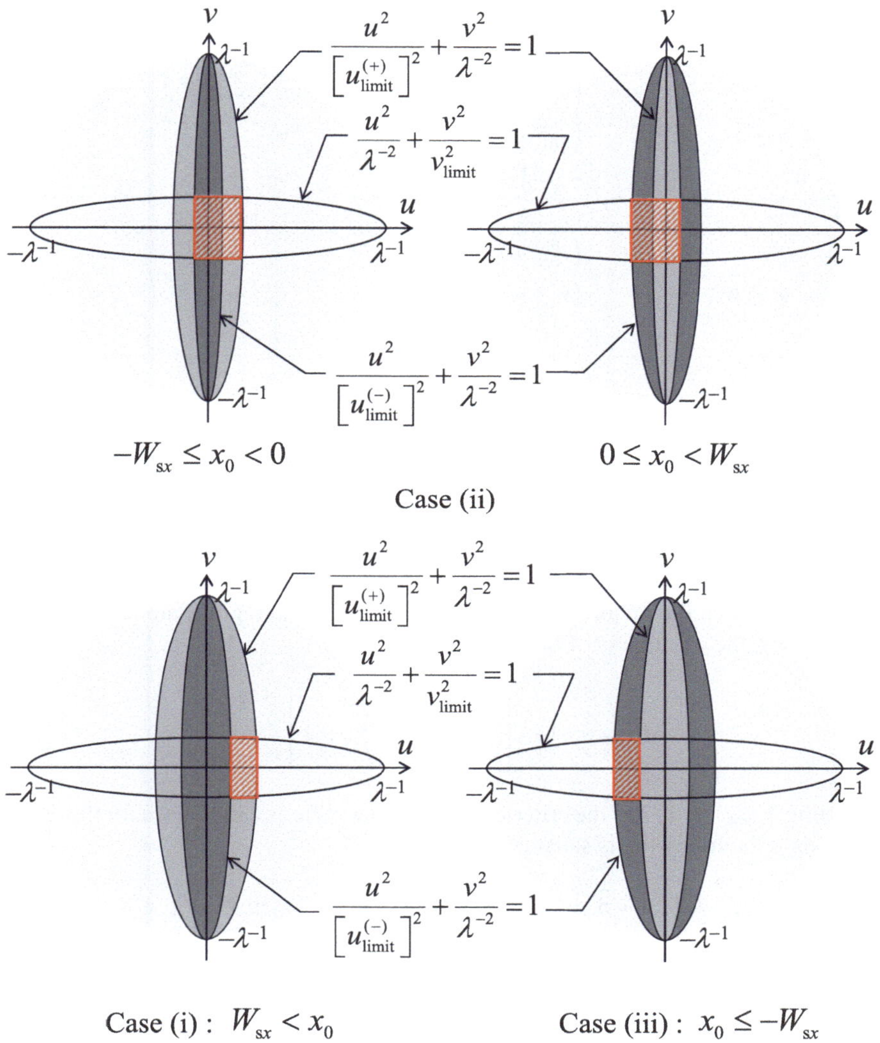

Fig. 12.13 Schematic illustrations of frequency regions to avoid aliasing errors of the sampled transfer function. Here, $x_0 \neq 0$ and $y_0 = 0$

shows these ellipsoidal regions in cases where $x_0 \neq 0$ and $y_0 = 0$. In this figure, the overlap regions that satisfy both Nyquist conditions (12.62) and (12.63) are schematically depicted as red-hatched areas. The sampled transfer function should be limited in the overlap region to avoid sampling problems. This is equivalent to limiting the bandwidth of the source field within the overlap region.

Here, note that the inside ellipse in Case (i) is switched to the outside ellipse in Case (iii). This switching occurs when x_0 changes its sign in Case (ii). Although

Table 12.1 Constants used for the band-limit to avoid aliasing errors

Case	u_0	u_{width}
$W_{sx} < x_0$	$\left(u_{\text{BL}}^{(+)} + u_{\text{BL}}^{(-)}\right)/2$	$u_{\text{BL}}^{(+)} - u_{\text{BL}}^{(-)}$
$-W_{sx} < x_0 \leq W_{sx}$	$\left(u_{\text{BL}}^{(+)} - u_{\text{BL}}^{(-)}\right)/2$	$u_{\text{BL}}^{(+)} + u_{\text{BL}}^{(-)}$
$x_0 \leq -W_{sx}$	$-\left(u_{\text{BL}}^{(+)} + u_{\text{BL}}^{(-)}\right)/2$	$u_{\text{BL}}^{(-)} - u_{\text{BL}}^{(+)}$
Case	v_0	v_{width}
$W_{sy} < y_0$	$\left(v_{\text{BL}}^{(+)} + v_{\text{BL}}^{(-)}\right)/2$	$v_{\text{BL}}^{(+)} - v_{\text{BL}}^{(-)}$
$-W_{sy} < y_0 \leq W_{sy}$	$\left(v_{\text{BL}}^{(+)} - v_{\text{BL}}^{(-)}\right)/2$	$v_{\text{BL}}^{(+)} + v_{\text{BL}}^{(-)}$
$y_0 \leq -W_{sy}$	$-\left(v_{\text{BL}}^{(+)} + v_{\text{BL}}^{(-)}\right)/2$	$v_{\text{BL}}^{(-)} - v_{\text{BL}}^{(+)}$

relation (12.65) seems, in a sense, to be written as a separate form, the mathematical expression of Case (ii) is the same, irrespective of the sign of x_0.

It should be noted that band limiting regions in (12.64)–(12.66) are not required if the ellipses shown in Fig. 12.13 is sufficiently oblate, as mentioned in Sect. 6.4.1. Instead the elliptic regions, a simple rectangular region can be used to limit the bandwidth. Since the major and minor radii of the elliptic regions are λ^{-1} and $u_{\text{BL}}^{(\pm)}$ respectively, the oblateness of the ellipses is given by

$$f_{\text{oblate}} = 1 - \left[\left(\frac{d}{x_0 \pm W_{sx}}\right)^2 + 1\right]^{-1/2}. \tag{12.69}$$

Adopting $f_{\text{oblate}} \gg 1/2$ as the criterion, rectangular regions can be used for the band-limit when the following is satisfied:

$$|d| \gg \sqrt{3} \times |x_0 \pm W_{sx}| \quad \text{and} \quad \sqrt{3} \times |y_0 \pm W_{sy}|. \tag{12.70}$$

In this case, the band-limit can be approximated as a simple rectangular region by combining the one-dimensional relations of (12.56)–(12.59). As a result, the band-limited transfer function is represented by

$$H'_{\text{SAS}}(u, v; d) = H_{\text{SAS}}(u, v; d)\text{rect}\left(\frac{u - u_0}{u_{\text{width}}}\right)\text{rect}\left(\frac{v - v_0}{v_{\text{width}}}\right), \tag{12.71}$$

where constants u_0, v_0, u_{width} and v_{width} are listed in Table 12.1. An example of the approximated rectangular region is shown in Fig. 12.14.

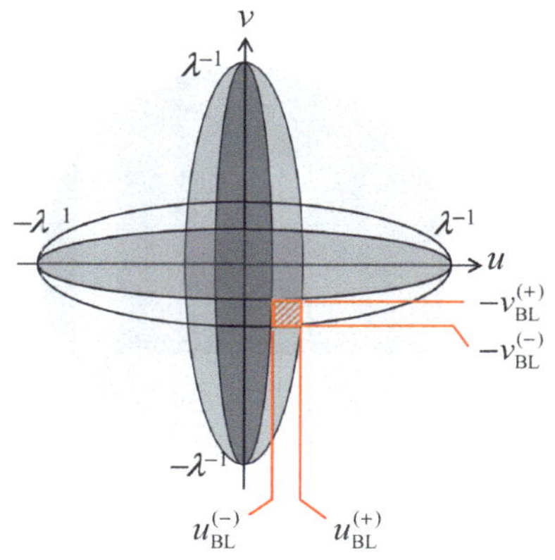

Fig. 12.14 Schematic illustration of the approximated rectangular region for the band-limit. Here, the region is depicted with $x_0 > +W_{sx}$ and $y_0 < -W_{sy}$ as an example

12.5.4 Actual Procedure for Numerical Calculation

In practical numerical process, we first calculate the spectrum of the source field using FFT. The input data is represented by $g_s[m, n; z_0] = g_s(x_{s,m}, y_{s,n}; z_0)$ ($m = 0, \ldots, M-1$ and $n = 0, \ldots, N-1$), where the sampling manner in (6.1) is used for the source field. Here, the sizes of the source sampling window are properly

$$W_{sx} = M\Delta x_s \quad \text{and} \quad W_{sy} = N\Delta y_s. \tag{12.72}$$

where Δx_s and Δy_s are the sampling intervals of the source field again. Note that the sampling intervals do not change in the destination field because this technique is a convolution-based method. The constants $u_{BL}^{(\pm)}$ and $v_{BL}^{(\pm)}$ are calculated by (12.59) and (12.67) using W_{sx} and W_{sy}.

The input data is first embed in a zero-padded sampling grid with $2M \times 2N$ sample points to apply the quadruple extension. Then, the sampled spectrum is calculated using FFT:

$$G_s[p, q; z_0] = \mathbf{FFT}\{g_s[m, n; z_0]\}, \tag{12.73}$$

where sampling positions of the spectrum are given by

$$u_p = \frac{1}{2\Delta x}(p - M) \quad \text{and} \quad v_q = \frac{1}{2\Delta y}(q - N). \tag{12.74}$$
$$(p = 0, \ldots, 2M-1 \quad \text{and} \quad q = 0, \ldots, 2N-1)$$

If the propagation distance d satisfies the criterion (12.70), the sampled transfer function is generated by

$$H_{SAS}[p, q; d] = H'_{SAS}(u_p, v_q; d). \tag{12.75}$$

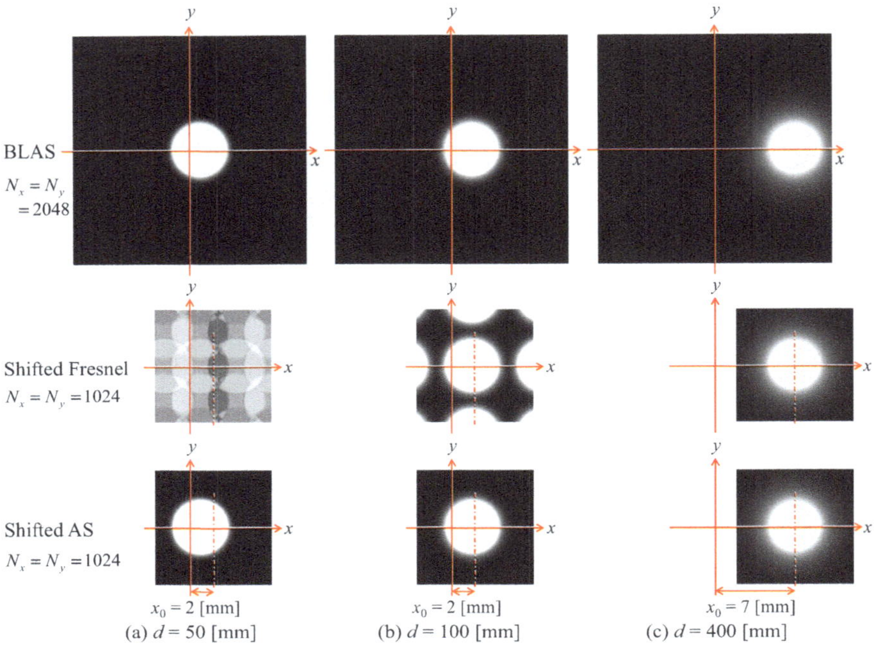

Fig. 12.15 Amplitude images of the destination fields calculated by the band-limited angular spectrum method (upper row), shifted Fresnel propagation (middle row), and shifted angular spectrum method (lower row) at different propagation distances. The source field is an off-axis plane wave diffracted by a circular aperture. $\Delta x_s = \Delta y_s = 10$ [μm] and $\lambda = 633$ [nm]

Here, note that the sampling position (u_p, v_q) is given by (12.74). The constants describing the rectangular region are given in Table 12.1 and calculated dependently on the cases. If the criterion is not satisfied, the band-limit region must be generated using Fig. 12.13.

The destination field is obtained using inverse FFT:

$$g[m, n; z_0 + d] = \mathbf{FFT}^{-1} \{G_s[p, q; z_0] H_{SAS}[p, q; d]\} . \tag{12.76}$$

Finally, the $M \times N$ points region is cut out from the center of the sampling grid of $g[m, n; z_0 + d]$, as shown in Fig. 4.14. Here, the sampling positions of the destination field $g(x_m, y_n; z_0 + d)(= g[m, n; z_0 + d])$ are given by (4.91), where $m = 0, \ldots, M - 1$ and $n = 0, \ldots, N - 1$, again.

12.5.5 Numerical Example

Numerical examples of the destination field calculated by the shifted angular method are shown in the lower row of Fig. 12.15. Here, the source field is the same as that in Fig. 12.9, i.e., a plane wave traveling in an inclined direction of $\theta = 1[°]$ is diffracted by a circular aperture of 5 mm in diameter. The destination fields calculated by the shifted Fresnel propagation using the same parameters as those of the shifted angular spectrum method are also shown in the middle row for comparison. The amplitude images in the upper row are also the destination fields but calculated not by off-axis techniques, i.e., the band-limited angular spectrum method is used for calculation.

When the propagation distance is sufficiently long, as in (c), both the off-axis techniques give the same result. However, in short-distance propagation, the parameters of the shifted Fresnel method no longer satisfy the aliasing-free conditions mentioned in Sect. 12.4.3, and thus, sampling problems are caused as shown in (a) and (b).

In contrast to the shifted Fresnel propagation, the shifted angular method does not have any apparent sampling problem in Fig. 12.15. But, the bandwidth u_{width} and v_{width} decrease with increasing the propagation distance. This leads to lack of sample points within the band-limit region. As a result, numerical errors increase in long-distance propagation. This can be avoided by extending the sampling grid more than the ordinary quadruple extension at the sacrifice of computation time and resources.

12.5.6 Discussion on the Limit Frequency

We limit the spatial frequencies of the source field to the band specified by $u_{\text{BL}}^{(\pm)}$. We can discuss the limit frequency by the same manner as that in BLAS in Sect. 6.4.4. The limit frequency can be physically interpreted as the local spatial frequency of the field emitted from a point within the source sampling window, as shown in Fig. 12.16. The highest spatial frequency is given by the field emitted from the point at the lower end of the source sampling window to the point at the upper end of the destination window. The lowest frequency is given in the same manner. Here, angles θ_{max} and θ_{min} in Fig. 12.16 are given by the geometry of the source and destination window as follows:

$$\begin{aligned} \sin\theta_{\text{max}} &= \frac{|x_0 + (W_{sx} + W_x)/2|}{\sqrt{d^2 + [x_0 + (W_{sx} + W_x)/2]^2}}, \\ \sin\theta_{\text{min}} &= \frac{|x_0 - (W_{sx} + W_x)/2|}{\sqrt{d^2 + [x_0 + (W_{sx} + W_x)/2]^2}}, \end{aligned} \tag{12.77}$$

where W_x is again a size of the destination sampling window. Note that θ_{max} and θ_{min} are switched dependently on the sign of x_0. Since $W_x = W_{sx}$ in this technique, the

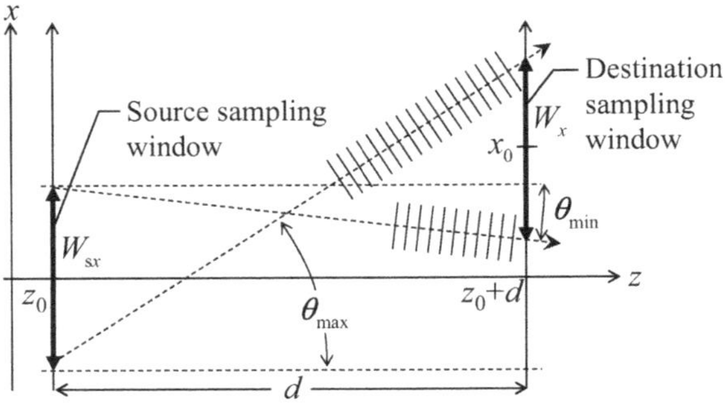

Fig. 12.16 A model for discussion of the highest and lowest spatial frequencies required for shifted field propagation

highest and lowest spatial frequencies are given by:

$$u_{\text{high}} = \frac{\sin \theta_{\text{max}}}{\lambda} = u_{\text{BL}}^{(+)},$$
$$u_{\text{low}} = \frac{\sin \theta_{\text{min}}}{\lambda} = u_{\text{BL}}^{(-)}. \tag{12.78}$$

As a result, the limits imposed on the frequencies by the procedure to avoid sampling problems are in agreement with the highest and lowest frequencies. Therefore, we can interpret the condition specified by $u_{\text{BL}}^{(\pm)}$ as the frequency bandwidth that is required to physically propagate the source wavefield onto the shifted destination area.

Chapter 13
Simulated Reconstruction Based on Virtual Imaging

Abstract Many images shown in this book are produced by simulated reconstruction based on virtual imaging. This is a very powerful technique because it emulates the image formation by cameras and eyes. Unnecessary light and field in holography, i.e., non-diffraction light and conjugate images are also reconstructed exactly like real holograms. Thus, we can confirm reconstruction of a CGH before its fabrication. We start from a simple technique based on back propagation, and then proceed to imaging by use of a virtual lens. Full-color simulated reconstruction as well as monochromatic reconstruction is discussed in this chapter.

13.1 Need for Simulated Reconstruction

Some simulation technique is absolutely needed for creating HD-CGHs. A LD-CGH is usually reconstructed by SLMs or something like those. We can check the reconstructed image promptly in this case. In contrast, a long time is generally required to confirm the resultant 3D image from a HD-CGH. This is because printing a HD-CGH is very time-consuming. We need at least several hours to print a HD-CGH, as mentioned in Chap. 15. Some large HD-CGHs composed of nearly 0.1 trillion pixel, such as "Sailing Warship II" (Fig. 1.7) and "Toy Train" (Fig. 1.10), required more than a whole day to print them. Therefore, some method to confirm the reconstructed image prior to printing is inevitably required to create a large-scale CGH.

As mentioned in Chaps. 7 and 8, thin holograms reconstruct not only the true image but also the conjugate image and non-diffraction light, unless the thin hologram is produced as the ideal phase pattern (see Sect. 8.7.1). These unnecessary images and light disturb observation of the 3D image. Besides, HD-CGHs can reconstruct a very deep 3D seen and large viewing angle. Accordingly, when observers moves their viewpoint and focus point, the unnecessary images and light overlap the true image in some cases, even if these do not disturb the view from a front viewpoint. In another case, when moving the viewpoint, the true image itself changes and becomes something different from what the CGH creator wants to show. Desirable simulation technique should also be able to produce an image with a large depth of field (DOF),

K. Matsushima, *Introduction to Computer Holography*,
Series in Display Science and Technology,
https://doi.org/10.1007/978-3-030-38435-7_13

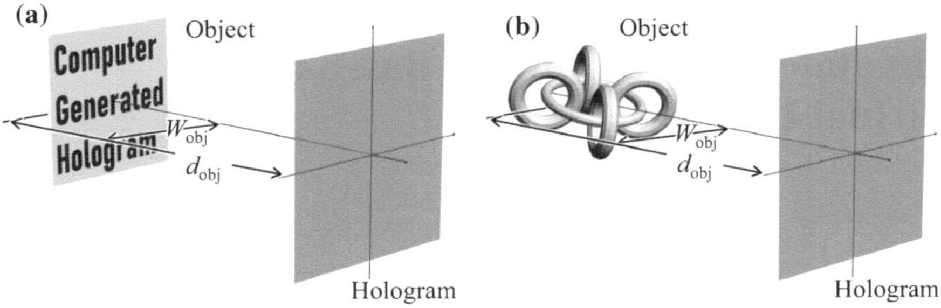

Fig. 13.1 The 3D scenes used for calculating the object field of **a** a 2D planar object and **b** 3D object of the "Five Rings" model

because we want to check the image of a deep 3D scene at a glance. Furthermore, since it is recently possible to create full-color HD-CGHs, the simulation technique should be applicable to reconstruct color CGHs.

As a result, the requirement of an ideal simulation technique is summarized as follows. The simulation technique should be capable of:

- Reproducing the conjugate image and non-diffraction light as well as the true image
- Moving the viewpoint
- Changing the focus
- Changing the depth of field
- Reproducing the color image

These all requirements are provided by the technique based on *virtual imaging*. Thus, the technique is already used everywhere in this book, especially in Chaps. 7 and 8.

13.2 Simulated Reconstruction by Back Propagation

The amplitude images or intensity images of wavefield $g_{obj}(x, y; 0)$, which is the light emitted from diffusive surfaces of an object, cannot convey appearance of the object. This is because the object surface is diffusive and thus the light emitted from the surface promptly spreads over the sampling window as the field propagate in the space. Therefore, the simplest technique of simulated reconstruction is to propagate the wavefield backward to somewhere close to the object. Since the back propagation makes the spread of light smaller than that in the original wavefield, the shape of the object emerges as an amplitude or intensity image of the backward-propagated wavefield $g_{obj}(x, y; -d_{obj})$. We call this technique the *back propagation method*.

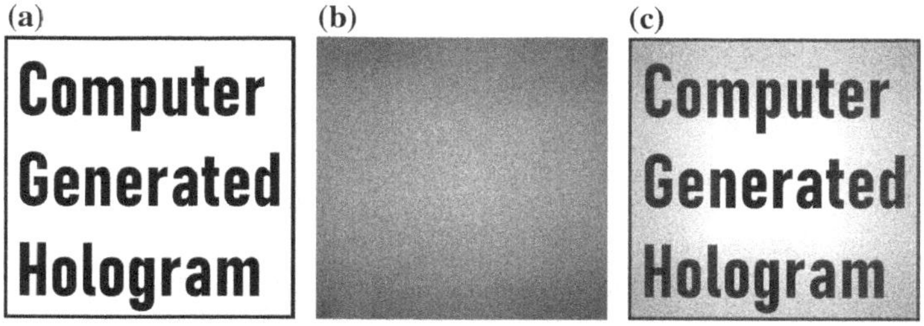

Fig. 13.2 An example of simulated reconstruction of the 2D object by back-propagation. **a** The original 2D image $I_{\text{tex}}(x, y)$, **b** the amplitude image of the object field $|g_{\text{obj}}(x, y; 0)|$, and **c** the simulated reconstruction $|\mathscr{P}_{-d_{\text{obj}}}\left\{g_{\text{obj}}(x, y; 0)\right\}|^2$. $M = N = 4{,}096$, $\Delta x = \Delta y = 5\,[\mu\text{m}]$, $\lambda = 633\,[\text{nm}]$, $W_{\text{obj}} = 20\,[\text{mm}]$, and $d_{\text{obj}} = 100\,[\text{mm}]$

13.2.1 Examples of Reconstruction by Back-Propagation

The 3D scenes used for generating an object field are shown in Fig. 13.1. The object is a 2D model in (a) and 3D model in (b). Here, W_{obj} and d_{obj} are a width of the model and a distance of the center from the CGH. The texture of the 2D model: $I_{\text{tex}}(x, y)$ is shown in Fig. 13.2a. The object field is simply generated by randomizing the phase distribution and then propagating the wavefield using the BLAS method with the quadruple extension:

$$g_{\text{obj}}(x, y; 0) = \mathscr{P}_{d_{\text{obj}}}\left\{\sqrt{I_{\text{tex}}(x, y)}\exp[i\phi_{\text{dif}}(x, y)]\right\}, \tag{13.1}$$

where $\phi_{\text{dif}}(x, y)$ is a randomized phase distribution to emulate the diffused surface (see Sect. 10.1.1). The amplitude image of generated object field $|g_{\text{obj}}(x, y; 0)|$ is shown in (b). The parameters are of a low definition CGH; the field size is approximately 20 mm, and the viewing angle is only 7.3°.

Figure 13.2c shows an example of simulated reconstruction by the back-propagation method: $|\mathscr{P}_{-d_{\text{obj}}}\left\{g_{\text{obj}}(x, y; 0)\right\}|^2$. The same field propagation technique as that used in generating the object field is used for the back-propagation. Here, note that the simulated reconstruction is not calculated from the fringe pattern but object wavefield, i.e., complex amplitudes. Accordingly, the conjugate image and non-diffraction light do not occur in the first place.

This simulated reconstruction reproduces the original image very well because generation of the object field and its back-propagation just compose round-trip propagation in this case:

$$g'_{\text{obj}}(x, y; -d_{\text{obj}}) = \mathscr{P}_{-d_{\text{obj}}}\left\{\mathscr{P}_{d_{\text{obj}}}\left\{\sqrt{I_{\text{tex}}(x, y)}\exp[i\phi_{\text{dif}}(x, y)]\right\}\right\} \tag{13.2}$$

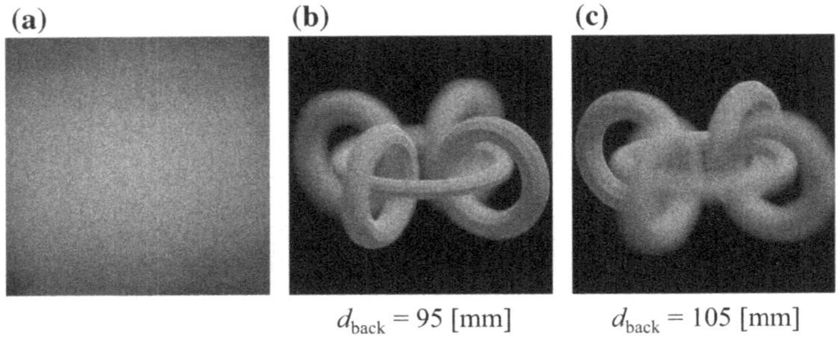

Fig. 13.3 An example of simulated reconstruction of the 3D object by back-propagation: **a** Amplitude images of object field $|g_{\text{obj}}(x, y; 0)|$, and **b**, **c** intensity images of simulated reconstruction $|\mathscr{P}_{-d_{\text{back}}}\{g_{\text{obj}}(x, y; 0)\}|^2$. $M = N = 4{,}096$, $\Delta x = \Delta y = 5\,[\mu\text{m}]$, $\lambda = 633\,[\text{nm}]$, $W_{\text{obj}} = 20\,[\text{mm}]$, and $d_{\text{obj}} = 100\,[\text{mm}]$. The encoding gamma for grayscale images is 1/2.2

Fig. 13.4 Field extent of a back-propagated field

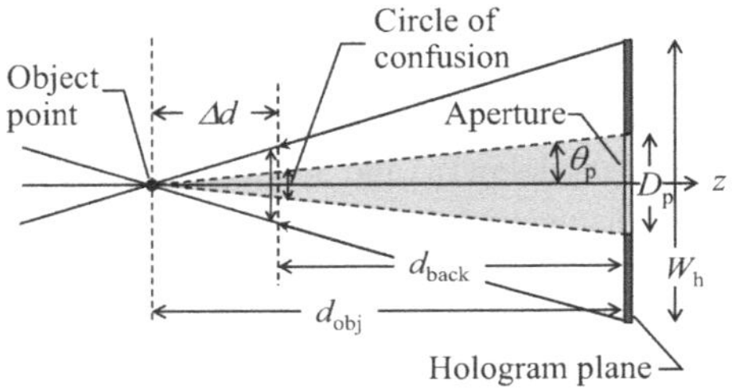

Nevertheless, a little brightness change is produced at the edge of the reconstructed image. This is because the object field overflows the sampling window in the forward propagation and the edge of the field is lost in the generated object field.[1]

Figure 13.3b and c shows simulated reconstruction of the 3D object shown in Fig. 13.1b. Here, the object field is calculated using the switch-back technique (see Sect. 11.3). The depth of the 3D model is approximately 20 mm when $W_{\text{obj}} = 20\,[\text{mm}]$. The simulated reconstruction is performed with different distances of the back-propagation:

$$g'_{\text{obj}}(x, y; -d_{\text{back}}) = \mathscr{P}_{-d_{\text{back}}}\{g_{\text{obj}}(x, y; 0)\}, \tag{13.3}$$

where d_{back} is a distance of the back-propagation. This is because the object has a thickness in this case.

Unlike the case of the 2D object, the reconstructed image is not very clear in both $d_{\text{back}} = 95$ [mm] and 105 [mm], as in (b) and (c). The portion close to image plane $(x, y, -d_{\text{back}})$ is clearly reconstructed but other portion far from the plane is out of

[1]The switch-back technique also uses the round-trip propagation, as described in Sect. 11.3, However, this kind of numerical errors is not caused by the round-trip propagation in the switch-back technique because the sampling window is extended in advance in order to prevent the field from overflowing the sampling window.

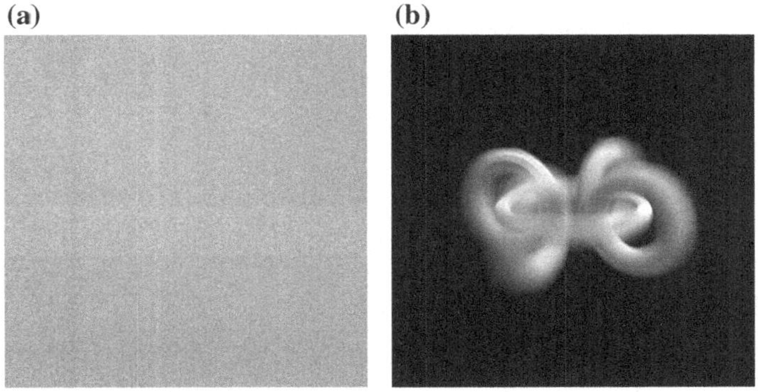

Fig. 13.5 The example of back-propagation in a quasi HD-CGH. Amplitude images of **a** the object field and **b** back-propagated field. $M = N = 32{,}768$, $\Delta x = \Delta y = 1\,[\mu\text{m}]$, $\lambda = 633\,[\text{nm}]$, $W_{\text{obj}} = 20\,[\text{mm}]$, and $d_{\text{obj}} = d_{\text{back}} = 100\,[\text{mm}]$

focus. In other words, depth of field (DOF) is small in the simulated reconstruction by the back-propagation. The reason is very clear. Suppose the field emitted from an object point spreads over the whole hologram and the hologram size is W_{h}, as shown in Fig. 13.4. The field size at $z = -d_{\text{back}}$ is

$$W_{\text{f}} = \frac{W_{\text{h}}}{d_{\text{obj}}} \Delta d, \tag{13.4}$$

where $\Delta d = |d_{\text{obj}} - d_{\text{back}}|$. Thus, if the CGH has a large size, i.e., a large ratio of $W_{\text{h}}/d_{\text{obj}}$, a little thickness of the object leads to strong blurring of the simulated image in this technique.

For example, $W_{\text{h}}/d_{\text{obj}} \cong 0.2$ in the case of Fig. 13.3. Object points at the forefront and back-end of the 3D object have $\Delta d \simeq 10\,[\text{mm}]$ around the center of the object. When we propagate the object field backward to the center of the object, the fields emitted from these longitudinal edge points approximately extends over 2 mm at the center of the object. This is the reason that the DOF is quite small in the back-propagation method.

Figure 13.5 shows another example of a back-propagated object field in a quasi high-definition CGH whose sampling interval and number of samples are 1 μm and $32\text{K} \times 32\text{K}$ (1K = 1024), respectively. $W_{\text{h}}/d_{\text{obj}} \cong 0.33$ in this example. In actual HD-CGHs, the rate $W_{\text{h}}/d_{\text{obj}}$ usually exceeds 0.5. Thus, the back-propagation method is commonly not suitable for simulated reconstruction of a HD-CGH. However, in LD-CGHs or ordinary digital holography (see Sect. 14.2), the back-propagation method offers an easy way to obtain the simulated reconstruction.

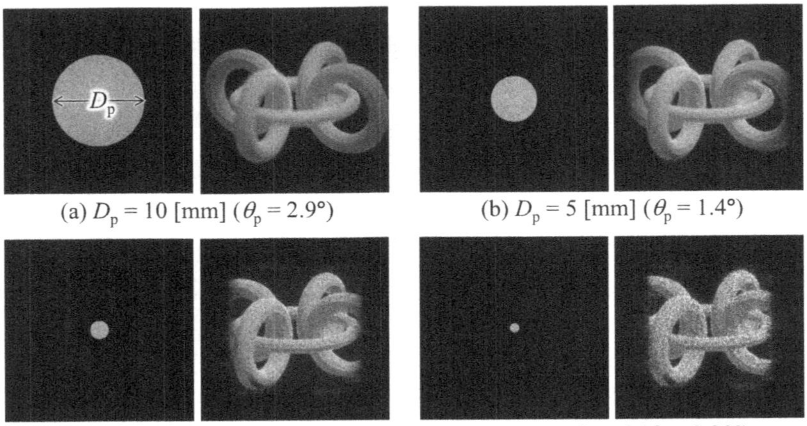

(a) $D_p = 10$ [mm] ($\theta_p = 2.9°$) (b) $D_p = 5$ [mm] ($\theta_p = 1.4°$)

(c) $D_p = 2$ [mm] ($\theta_p = 0.57°$) (d) $D_p = 1$ [mm] ($\theta_p = 0.29°$)

Fig. 13.6 The effect of an aperture on the back-propagation. $M = N = 4{,}096$, $\Delta x = \Delta y = 5\,[\mu\text{m}]$, $\lambda = 633\,[\text{nm}]$, $W_{\text{obj}} = 20\,[\text{mm}]$, and $d_{\text{obj}} = d_{\text{back}} = 100\,[\text{mm}]$

13.2.2 Control of DOF Using Aperture

A simple idea to increase the small DOF in the back-propagation method is to limit the area of the object field using an aperture. As shown in Fig. 13.4, when limiting the area of the object field by an aperture whose diameter is D_p, the field extent given in (13.4) is rewritten as

$$W_f = \frac{D_p}{d_{\text{obj}}}\Delta d. \tag{13.5}$$

Thus, using a small aperture, we can increase the DOF. If the aperture is circular,[2] the value of W_f is corresponding to a diameter of the *circle of confusion* that is well-known idea in photography.

Angle θ_p indicated in Fig. 13.4 is introduced according to a concept of the numerical aperture (NA). The numerical aperture in the atmosphere, where the refractive index is regarded as unity, is written as

$$\text{NA} = \sin\theta_p. \tag{13.6}$$

The angle θ_p is a useful index to DOF and speckle noise, as shown below.

Figure 13.6 shows examples of the effect of an aperture. Here, the object field of Fig. 13.3a is used for the numerical experiment. The reconstructed image becomes sharper with reducing the aperture size. Moreover, narrowing FOV is detected in the small aperture. This is because object points placed at the edge of the object do

[2]It is not very important in this technique whether the aperture is circular or not. A rectangular aperture gives almost the same images as those by a circular aperture.

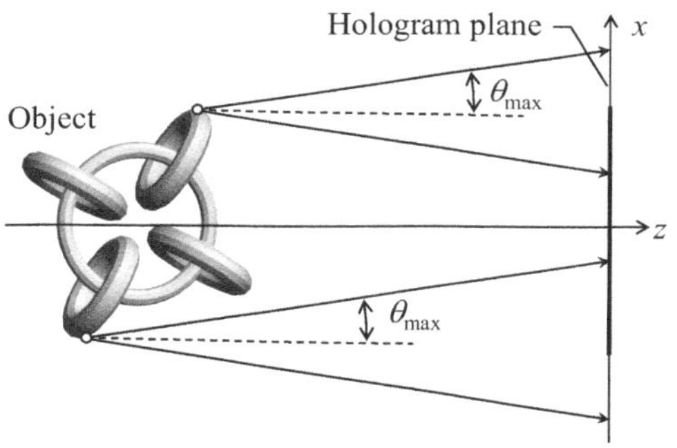

Fig. 13.7 Ranges of the field emitted from object points at the edge of the model

not contribute to the object field around the center. For example, the object filed of Fig. 13.3a has a maximum diffraction angle of 3.6°. Thus, the field emitted from an object point only spread over a range whose size is 6.3 mm when the field propagates at a distance of 100 mm. As a result, light emitted from object points around the edge of the object cannot reach the center of the object field in the hologram plane, as shown in Fig. 13.7.

In addition to FOV narrowing, remarkable speckle noise is caused by the small apertures, as in (d), because only a small portion of the field passes through the aperture. This is a well-known phenomenon in coherent image formation.

13.2.3 Control of View-Direction Using Aperture

Use of an aperture offers not only the increase of DOF but also an ability to change the view-direction in the simulated reconstruction. Figure 13.8 shows the examples. Here, the object field is of the quasi HD-CGH shown in Fig. 13.5. The aperture size is $D_p = 5\,[\text{mm}]$ in common. We can confirm that appearance of the reconstructed image changes dependently on the position of the aperture. The line of sight is considered to be a connecting line between the centers of the aperture and field $\mathcal{P}_{-d_{back}}\left\{g_{obj}(x, y; 0)\right\}$. This is a simplified technique to predict the reconstructed image from different viewpoints.[3] However, it has a drawback that the change of the viewpoint is limited to the inside of the object field.

As mentioned above, the back-propagation method provides a simplified technique to predict the reconstructed images of an object field. However, the technique has several disadvantages such as the FOV narrowing and limited viewpoint. Above all, the technique is difficult to apply it to CGHs whose fringe pattern is encoded by a reference field. We cannot confirm exact effects of the conjugate image and non-diffraction light by the back-propagation method.

---

[3] In practice, this technique was once used for creating free-viewpoint images from digital holograms recorded by an image sensor [91].

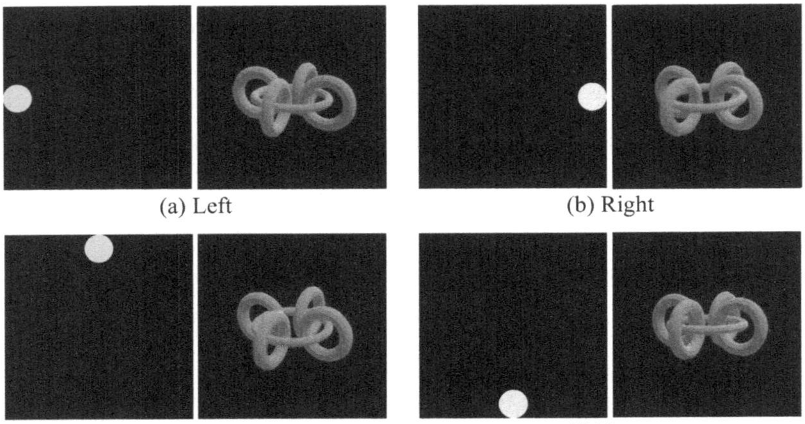

Fig. 13.8 Examples of change of the view-direction using an aperture. $M = N = 32{,}768$, $\Delta x = \Delta y = 1\,[\mu\text{m}]$, $\lambda = 633\,[\text{nm}]$, $W_{\text{obj}} = 20\,[\text{mm}]$, $D_{\text{p}} = 5\,[\text{mm}]$, and $d_{\text{obj}} = d_{\text{back}} = 100\,[\text{mm}]$

13.3 Image Formation by Virtual Lens

Ordinary cameras and human eyes get the image from diffusive light by use of a lens. This can be applicable to simulated reconstruction of an object field and the fringe pattern of a CGH.

13.3.1 Sampling Problem of Virtual Lens

As described in Sect. 5.3.1, a thin lens modulates the incident wavefield by multiplying complex transmittance in (5.55):

$$t_{\text{lens}}(x, y; f) = \exp\left[-i\frac{k}{2f}\left(x^2 + y^2\right)\right] p(x, y), \tag{13.7}$$

where f and $p(x, y)$ are a focal length and pupil function again. The lens function must be sampled to use it for simulated reconstruction. However the sampled lens function may cause aliasing errors because the exponent includes a quadratic term. Thus, we introduce the same technique as that in Sect. 6.1.4 to investigate the property:

$$\begin{aligned} t_{\text{lens}}(x, y; f) &= \exp\left[-i\phi(x, y)\right] p(x, y), \\ \phi(x, y; f) &= \frac{k}{2f}\left(x^2 + y^2\right). \end{aligned} \tag{13.8}$$

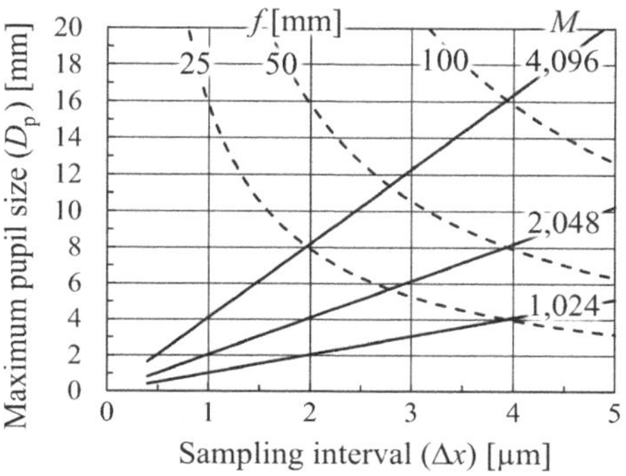

Fig. 13.9 Examples of the maximum diameter of a virtual lens. The solid and broken lines depict limitations by the sampling window and sampling problem, respectively. $\lambda = 633$ [nm]

The local spatial frequency of $t_{\text{lens}}(x, y; f)$ is

$$f_x = \frac{1}{2\pi}\frac{\partial}{\partial x}\phi(x, y; f) = \frac{x}{\lambda f}. \tag{13.9}$$

The local frequency must satisfy $\Delta x^{-1} > 2|f_x|$ to avoid aliasing errors;

$$|x| < \frac{\lambda f}{2\Delta x}. \tag{13.10}$$

Thus, the diameter of the sampled lens function is limited to the maximum size:

$$D_{\max,x} = \frac{\lambda f}{\Delta x} \quad \text{and} \quad D_{\max,y} = \frac{\lambda f}{\Delta y}. \tag{13.11}$$

This meas that pupil diameter D_p must satisfy

$$D_p < D_{\max,x} \quad \text{and} \quad D_p < D_{\max,y}. \tag{13.12}$$

In addition to relations (13.12), the maximum pupil size of a virtual lens is limited by the size of the sampling window of the lens function; we cannot generate the lens larger than the sampling window in virtual imaging:

$$D_p \leq M\Delta x \quad \text{and} \quad D_p \leq N\Delta y. \tag{13.13}$$

Figure 13.9 shows actual examples of the limited pupil size, imposed by the sampling window (solid lines) in (13.13) and the sampling problem (broken lines) in (13.12) at a wavelength of 633 nm. With respect to a sampling interval of the lens function, smaller one of the two limitations gives the maximum pupil size.

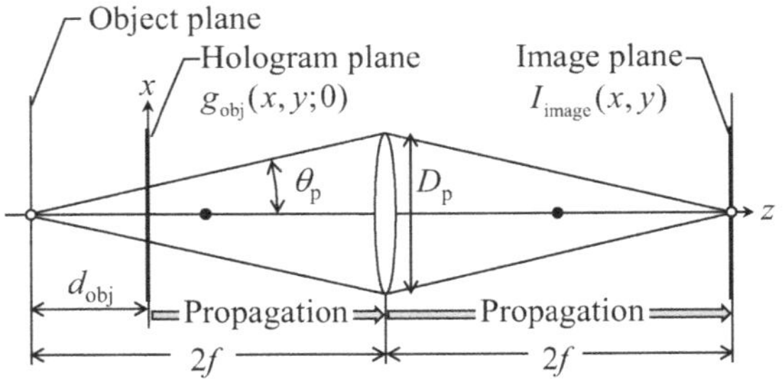

Fig. 13.10 Setup of equal-magnification imaging

On another note, the f-number of the lens is defined as the ratio of the focal length to the diameter of the entrance pupil in photography:

$$N_f = \frac{f}{D_p}. \tag{13.14}$$

The f-number may give an experiential index to DOF and blur to the reader familiar with lens cameras.

13.3.2 *Equal Magnification Imaging by Virtual Lens*

According to the well-known thin lens formula in (5.49), a lens forms the image of a unity magnification if

$$d_1 = d_2 = 2f. \tag{13.15}$$

Using the equal magnification imaging, we can perform simulated reconstruction similar to the back-propagation method in the previous section. The setup of the simulated reconstruction is shown in Fig. 13.10. In this technique, the object field $g_{obj}(x, y; 0)$ is first propagated to the lens plane, and then multiplied by the lens function. The resultant field is again propagated at a distance of $2f$. Therefore, the reconstructed image is expressed by propagation operators (see Sect. 5.4) as follows:

$$I_{image}(x, y) = \left|\mathscr{P}_{2f}\left\{t_{lens}(x, y; f)\mathscr{P}_{(2f-d_{obj})}\left\{g_{obj}(x, y; 0)\right\}\right\}\right|^2. \tag{13.16}$$

Here, we assume a circular lens whose diameter is D_p. Thus the pupil function is given by

$$p(x, y) = \text{circ}(x, y; D_p), \tag{13.17}$$

where circ$(x, y; D)$ is a circular function whose diameter is D.

When we use a convolution method, i.e., the BLAS method for the numerical propagation, the sampling window of the lens function is identical with that of the object field. Because we can freely choose any focal length of the lens in this

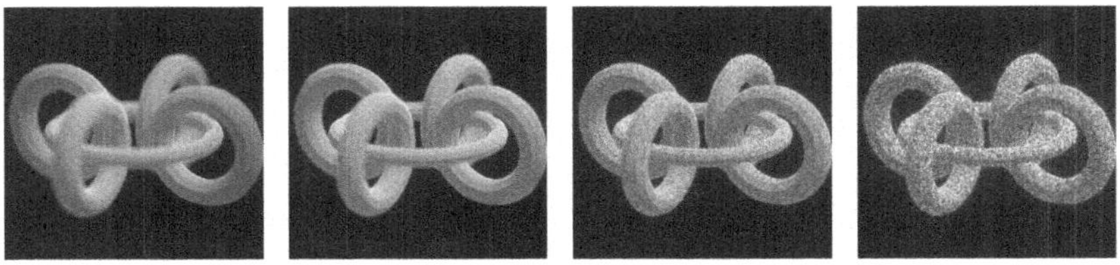

Fig. 13.11 Examples of simulated reconstruction by equal-magnification imaging using a lens. Images **a**–**d** are corresponding to those in Fig. 13.6. $M = N = 4{,}096$, $\Delta x = \Delta y = 5\,[\mu\text{m}]$, $\lambda = 633\,[\text{nm}]$, $W_{\text{obj}} = 20\,[\text{mm}]$, and $d_{\text{obj}} = f = 100\,[\text{mm}]$

technique, limitation of the pupil size is only imposed by the sampling window of the object field. As shown in Fig. 13.10, the angle θ_p is given by

$$\theta_p = \tan^{-1}\left(\frac{D_p}{4f}\right). \tag{13.18}$$

Figure 13.11 shows examples of numerical reconstruction by this technique using a virtual lens whose focal length is 100 mm. Here, the parameters are chosen so that the angle θ_p obtained by (13.18) is matched with the examples of the back-propagation method in Fig. 13.6. We can see reconstructed images similar to those by the back-propagation, but the narrowing of FOV does not occur in this case. As the pupil size is reduced, the DOF becomes larger and the speckle noise is more remarkable, but the edge of the image does not vanished from the reconstructed image unlike the back-propagation. This is because the object field sufficiently spreads over the sampling window by the first propagation to the lens plane.

13.3.3 Reduced Imaging by Virtual Lens

The imaging technique in the previous section, i.e., equal-magnification imaging works well in LD-CGH, but is not suitable for HD-CGHs, because the twice propagation in (13.16) is performed with a large-scale wavefield that usually have more than billion sample points. Besides, the setup shown in Fig. 13.10 does not resemble that in human eyes and ordinary photography at all. If the main purpose of simulated reconstruction is to predict how the 3D image of the CGH is seen by an observer or photographed by a camera, the lens system should resembles eyes and cameras as much as possible.

It is said that the focal length of a human eye is approximately 20 mm. An ordinary camera uses lenses with a focal length of several centimeters for imaging. According to (13.12), a small sampling interval is required to produce this kind of short focus

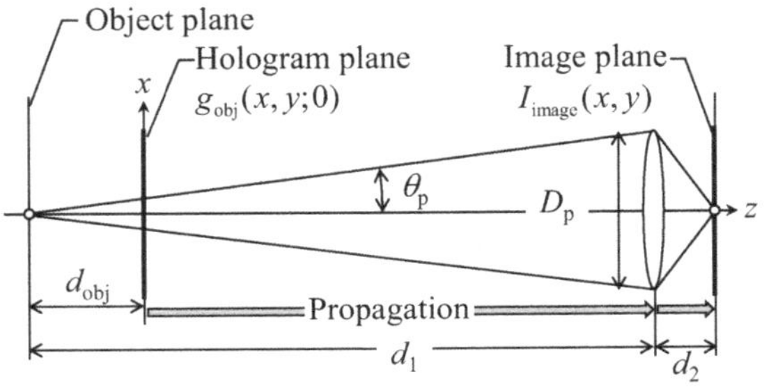

Fig. 13.12 Setup for reduced imaging

virtual lens. In addition, the 3D image of a CGH is seen or photographed roughly at a distance of 20 cm to 1 m. As a result, magnification of the lens system is generally less than 1/10. Therefore, it is better to apply a virtual lens whose sampling interval is determined separately from that of the object field. This can be possible by use of the shifted Fresnel propagation technique (see Sect. 12.4) for the first propagation. Using the setup shown in Fig. 13.12, the reconstructed image is formulated as

$$I_{\text{image}}(x, y) = \left| \mathscr{P}_{d_2} \left\{ t_{\text{lens}}(x, y; f) \mathscr{P}^{(\text{SF})}_{(d_1 - d_{\text{obj}})} \left\{ g_{\text{obj}}(x, y; 0) \right\} \right\} \right|^2, \tag{13.19}$$

where $\mathscr{P}^{(\text{SF})}\{\}$ is a propagation operator using the shifted Fresnel method. The angle θ_{p} is in this case

$$\theta_{\text{p}} = \tan^{-1}\left(\frac{D_{\text{p}}}{2d_1}\right). \tag{13.20}$$

In this technique, we first determine the distance of the viewpoint d_1 in general, and then give the focal length f or the distance d_2. The last parameter d_2 or f is properly given by the thin lens formula to focus the reconstructed image on an object point:

$$\frac{1}{d_1} + \frac{1}{d_2} = \frac{1}{f}. \tag{13.21}$$

Using the shifted Fresnel method for the first propagation, in addition to reduction of the sampling interval, we can decrease the number of sampling using the rectangular tiling technique, as shown in Fig. 12.2b. Thus, after the first propagation, we do not have to treat large-scale wavefields even in HD-CGHs, which sometimes have a gigantic number of sampling. This gives a big advantage for changing the viewpoint, as described in the following section.

Figure 13.13 shows examples of simulated reconstruction using the reduced imaging. The object field is of the quasi HD-CGH shown in Fig. 13.5. Thus, the total number of sample points is over one billion points (32K × 32K). Symbols M_{i} and N_{i} indicated in Fig. 13.13 are the number of pixels of the reconstructed images, i.e., the number of the sample points of the object field after the first propagation, and also

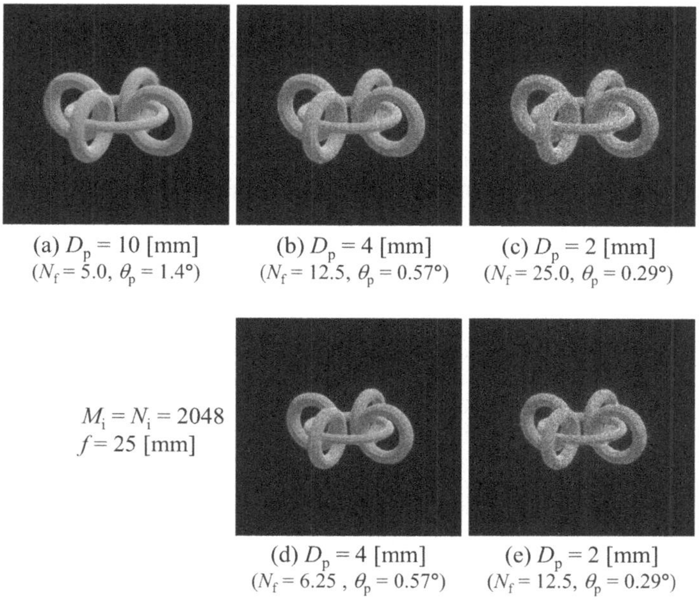

Fig. 13.13 Examples of simulated reconstruction using the reduced imaging. Object field: $M = N = 32{,}768$, $\Delta x = \Delta y = 1\,[\mu m]$, $\lambda = 633\,[nm]$, $W_{obj} = 20\,[mm]$, and $d_{obj} = 100\,[mm]$. Simulated reconstruction: $d_1 = 200\,[mm]$ and $\Delta x_i = \Delta y_i = 2.5\,[\mu m]$

the sample points of the lens function. Let us emphasize the fact that a large-scale wavefield is reconstructed as an ordinary image size.

Here, we set the view distance d_1 to 200 mm in common with all simulations. In this case, using a virtual lens with a focal length of 50 mm, which is a typical focal length of ordinary cameras, the magnification d_2/d_1 is approximately 0.33 and the image size is $0.33 \times W_{obj} \cong 6.67$ [mm]. To set the pupil diameter to at least 10 mm, by consulting the graph in Fig. 13.9, we adopt $M_i = N_i = 4096$ and $\Delta x_i = \Delta y_i = 2.5\,[\mu m]$ as the parameters of the propagated field in Fig. 13.13a–c. Here, Δx_i and Δy_i are sampling intervals of the wave field and lens function. The pupil sizes are chosen so that the angles θ_p agree with those in Figs. 13.6 and 13.11. We find that DOF and speckle noise in Fig. 13.13a–c are similar to Fig. 13.11b–d.

Using a lens with $f = 25$ [mm], the sample points necessary for reconstruction is even less: only 2K $\times$ 2K, as shown in Fig. 13.13d and e. However, unfortunately, we find that the pupil size in this case cannot be set to 10 mm according to Fig. 13.9. The pupil sizes in Fig. 13.13d, e are also chosen for the angles θ_p to match those in Fig. 13.11c and d, respectively. Although the reconstructed images seem to shrink a little, the appearance resembles to Figs. 13.13b, c, and 13.11c, d.

13.3.4 Change of Viewpoint

In the shifted Fresnel propagation technique used for the first propagation, we can change not only the sampling interval of the propagated field but also the lateral position of the destination sampling window. This property of shifted propagation makes it possible to change the viewpoint of simulated reconstruction. The viewpoint is commonly considered as the center of the entrance pupil, i.e., the center of the lens. This is also the center of the destination sampling window. Thus, suppose the destination field propagated by the shifted Fresnel propagation is represented by

$$g'_{\mathrm{obj}}(x', y'; 0) = \mathscr{P}^{(\mathrm{SF})}_{\mathbf{r}_{\mathrm{v}}} \left\{ g_{\mathrm{obj}}(x, y; 0) \right\}, \tag{13.22}$$

where $\mathbf{r}_{\mathrm{v}}$ is a position vector of the viewpoint. The component is

$$\mathbf{r}_{\mathrm{v}} = (x_{\mathrm{v}}, y_{\mathrm{v}}, z_{\mathrm{v}}), \tag{13.23}$$

where

$$z_{\mathrm{v}} = d_1 - d_{\mathrm{obj}}. \tag{13.24}$$

Operation $\mathscr{P}^{(\mathrm{SF})}_{\mathbf{r}} \{\xi(x, y)\}$ denotes propagation of the field $\xi(x, y)$, whose center is at the origin of the coordinates, to position $\mathbf{r}$ presenting the center of the destination sampling window.

As shown in Fig. 13.14a, after the first propagation, we introduce another coordinate system whose origin is placed at $\mathbf{r}_{\mathrm{v}}$:

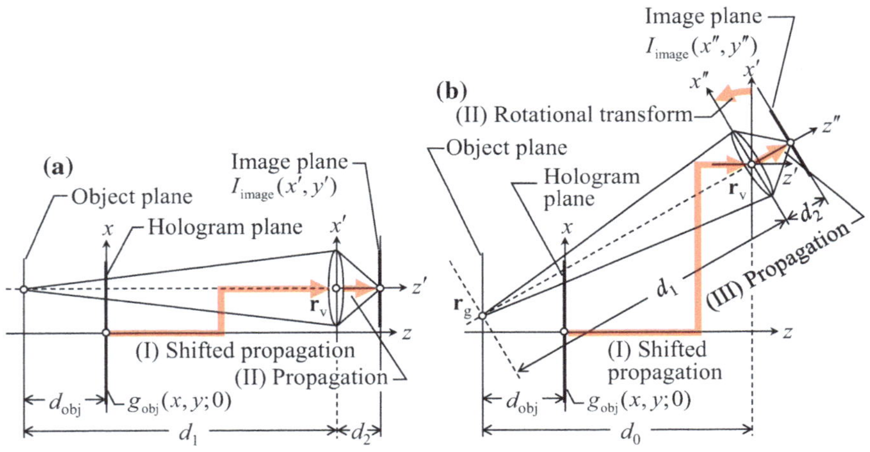

Fig. 13.14 Change of the viewpoint by the shifted propagation **a** without and **b** with rotational transform

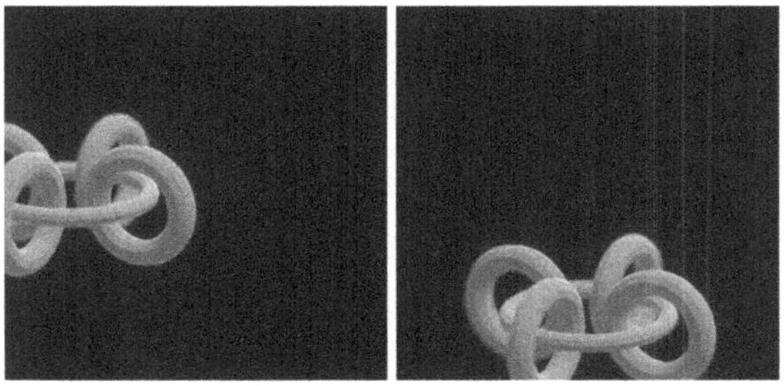

Fig. 13.15 Examples of simulated reconstruction performed without rotational transform of the object field. $D_p = 10\,[\text{mm}]$. Other parameters are the same as those in Fig. 13.13

$$(x', y', z') = (x - x_v, y - y_v, z - z_v). \tag{13.25}$$

The propagated field is then multiplied by the lens function, and again propagated to the image plane. Therefore, the reconstructed image is written as

$$I_{\text{image}}(x', y') = \left| \mathscr{P}_{d_2} \left\{ t_{\text{lens}}(x', y'; f) g'_{\text{obj}}(x', y'; 0) \right\} \right|^2. \tag{13.26}$$

Unfortunately, the image formed by the two steps in (13.22) and (13.26) is just a shifted image of the object, as shown in Fig. 13.14a. Examples of the reconstructed image are shown in Fig. 13.15. To turn look to an arbitrary object point, i.e., direct the line of sight of the virtual eyes or camera to the gazing point, as shown in Fig. 13.14b, we must insert one more step; the rotational transform of wavefields (see Chap. 9) is performed between the first and second propagation. As a result, the reconstructed image is obtained by the following three steps:

$$g'_{\text{obj}}(x', y'; 0) = \mathscr{P}^{(\text{SF})}_{\mathbf{r}_v} \left\{ g_{\text{obj}}(x, y; 0) \right\}, \tag{13.27a}$$

$$g''_{\text{obj}}(x'', y''; 0) = \mathscr{R}_{\mathbf{T}_v} \left\{ g'_{\text{obj}}(x', y'; 0) \right\}, \tag{13.27b}$$

$$I_{\text{image}}(x'', y'') = \left| \mathscr{P}_{d_2} \left\{ t_{\text{lens}}(x'', y''; f) g''_{\text{obj}}(x'', y''; 0) \right\} \right|^2, \tag{13.27c}$$

where $\mathscr{R}_{\mathbf{T}}\{\xi(x, y)\}$ denotes an operator of the rotational transform of wavefield $\xi(x, y)$ with rotation matrix $\mathbf{T}$. Coordinate system (x'', y'', z'') after the rotational transform is properly not parallel to the coordinate systems (x', y', z') and (x, y, z).

Supposing that $\mathbf{r}_g$ is a position vector of the gazing point as in Figs. 13.14b and 13.16, it is easily to found that $d_1 = |\mathbf{r}_v - \mathbf{r}_g|$. Furthermore, the rotation matrix $\mathbf{T}_v$ is obtained by the Rodrigues' rotation formula described in Sect. 10.2.3 using the view vector:

$$\mathbf{v} = \frac{\mathbf{r}_v - \mathbf{r}_g}{|\mathbf{r}_v - \mathbf{r}_g|}, \tag{13.28}$$

and unit vector $\mathbf{e}_z$ of the z-axis as follows.

Fig. 13.16 Vectors used for obtaining the rotation matrix

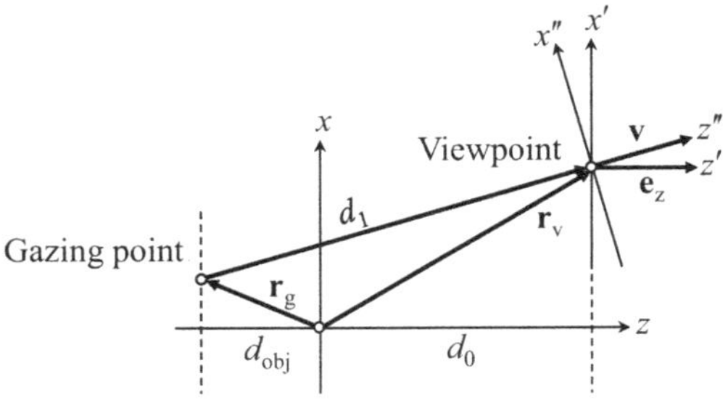

$$\mathbf{T}_v = \cos\theta_v \mathbf{I} + (1 - \cos\theta_v)^t \mathbf{s}_v \cdot \mathbf{s}_v + \sin\theta_v [\mathbf{s}_v]_\times, \tag{13.29}$$

where all symbols for vector and matrix operations are the same as those described in Sect. 10.2.3, and

$$\mathbf{s}_v = \frac{\mathbf{v} \times \mathbf{e}_z}{|\mathbf{v} \times \mathbf{e}_z|}, \quad \cos\theta_v = \mathbf{v} \cdot \mathbf{e}_z, \quad \text{and} \quad \sin\theta_v = |\mathbf{v} \times \mathbf{e}_z|. \tag{13.30}$$

Note that the Rodrigues' rotation formula obtains the rotation matrix to direct a vector toward another vector. However, in the rotational transform, a rotation matrix is necessary for coordinates rotation. Thus, direction of rotation is opposite to that in rotation of a vector.

Here, let us emphasize that the carrier offset cannot be omitted in the rotational transform in this case, as mentioned in Sect. 9.6.2. Therefore, the sampled field after the rotational transform is written as

$$g''(x''_m, y''_n; 0) = \hat{g}(x''_m, y''_n; 0) \exp[i2\pi(u_c x'' + v_c y''], \tag{13.31}$$

where $\hat{g}(x''_m, y''_n; 0)$ is given by renaming x_m and y_n in (9.54). u_c and v_c are also given in (9.46). The carrier factor $\exp[i2\pi(u_c x'' + v_c y'']$ sometimes causes severe aliasing errors especially in cases of a large viewpoint-shift. Thus, it is better to set the sampling intervals $\Delta x''$ and $\Delta y''$ in the lens and image planes to a small value in order to avoid aliasing errors. Accordingly, it is better to use a short focal-length lens to shrink the image formed in the image plane.

Figure 13.17 shows examples of reconstructed images seen from various viewpoints. Here, the sampling interval is 1 μm in the $(x', y', 0)$ and $(x'', y'', 0)$ planes. To put the whole image within the small sampling window, a short focal length; $f = 25$ [mm] is chosen for the lens. Although the viewpoint is changed for each image in Fig. 13.17, the gazing point is always at the center of the object. We can observe the simulated image from various viewpoint by this technique.

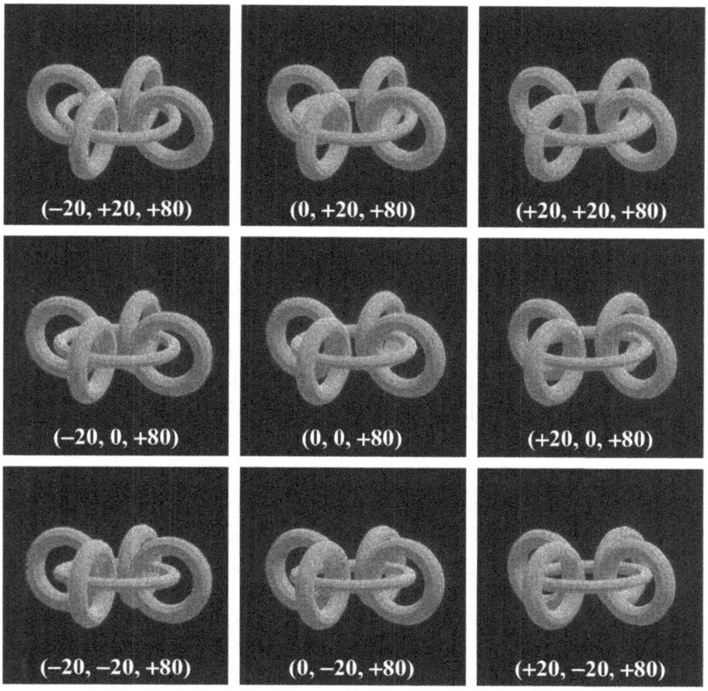

Fig. 13.17 Examples of simulated reconstruction performed with the rotational transform. The position of the viewpoint ($\mathbf{r}_v$) is indicated in reconstructed images. Parameters of simulated reconstruction: $d_0 = 180$ [mm], $f = 25$ [mm], $\mathbf{r}_g = (0, 0, -100$ [mm]), $D_p = 4$ [mm], $\Delta x_i = \Delta y_i = 1$ μm, and $M_i = N_i = 4{,}096$. Object field: $M = N = 32{,}768$, $\Delta x = \Delta y = 1$ [μm], $\lambda = 633$ [nm], $W_{obj} = 20$ [mm], and $d_{obj} = 100$ [mm]

13.4 Simulated Reconstruction from Fringe Pattern

The simulated images are obtained from the object field in the previous sections. Thus, the conjugate image and non-diffraction light are not involved in the reconstruction. These unnecessary image and light are produced in the encoding process for generating a fringe pattern, as mentioned in Chap. 8. Thus, when we reconstruct images from the fringe pattern, these unnecessary fields properly appear in the simulated reconstruction.

13.4.1 Formulation

The simulation technique using a virtual lens, described in the previous section, can be used for the simulated reconstruction of a fringe pattern properly as well. The formulas are simply derived by replacing the object field in (13.27a) by the reconstructed field:

$$g_{\text{obj}}(x, y; 0) \leftarrow \begin{cases} \sqrt{T(x, y)}P(x, y; 0) & \text{Amplitude hologram} \\ t(x, y)P(x, y; 0) & \text{Phase hologram} \end{cases}, \tag{13.32}$$

where $T(x, y)$ represents the fringe pattern of a thin amplitude hologram. An actual fringe pattern is given by (8.16) in a general form. $t(x, y) = \exp[i\phi(x, y)]$ is the phase factor of a thin phase hologram. The common form is given by (8.25).

Wavefield $P(x, y; 0)$ represents the illumination light, and is usually the same as the reference field used for encoding. In the simulated reconstruction, however, a different illumination filed from the reference field is allowed for reconstruction. Even in this case, the simulated image agrees with the actual image, which is optically reconstructed under the same situation; the illumination is not the same as the reference light.

13.4.2 Comparison Between Simulated and Optical Reconstructions

Figure 13.18 shows a comparison between simulated and optical reconstructions in "The Venus". This HD-CGH is optically reconstructed by a plane wave, which is roughly adjusted so as to fit with that used in the numerical encoding. Here, the viewpoints in the simulation are also roughly adjusted so as to fit with the pictures of the optical reconstruction.

Figure 13.19 shows examples of simulated and optical reconstruction of a LD-CGH. The object field is calculated by the technique of the point-cloud in this CGH. The fringe pattern was printed by a home-made printer (see Sect. 15.2.1). Since the incident angle of the illumination light is not large, non-diffraction light sometimes disturbs the view in this CGH. The simulated reconstruction well reproduces the appearance of the optical reconstruction.

Figure 13.20 shows optical and simulated reconstruction of HD-CGH "Moai I". In the 3D scene, one of the moai statues is arranged at $z_{\text{obj}} = -300$ [mm] and the center of the reference spherical wave is placed at $z_R = -250$ [mm]. According to the diagram in Fig. 7.16, an inverted conjugate image is generated before the moai object. This conjugate image is successfully reproduced in Fig. 13.20, and its position and shape well coincide with the optical reconstruction.

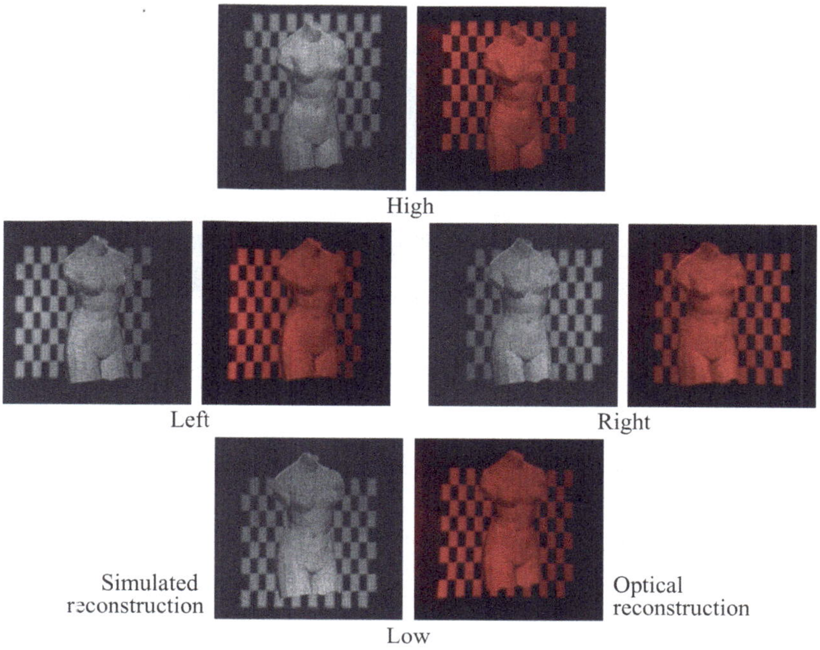

Fig. 13.18 Comparison between simulated and optical reconstructions of "The Venus" CGH. See Table 10.2 for the parameter of the CGH. $d_0 = 450\,[\text{mm}]$, $f = 25\,[\text{mm}]$, $D_\text{p} = 4\,[\text{mm}]$, $\Delta x_\text{i} = \Delta y_\text{i} = 1\,[\mu\text{m}]$, and $M_\text{i} = N_\text{i} = 4{,}096$

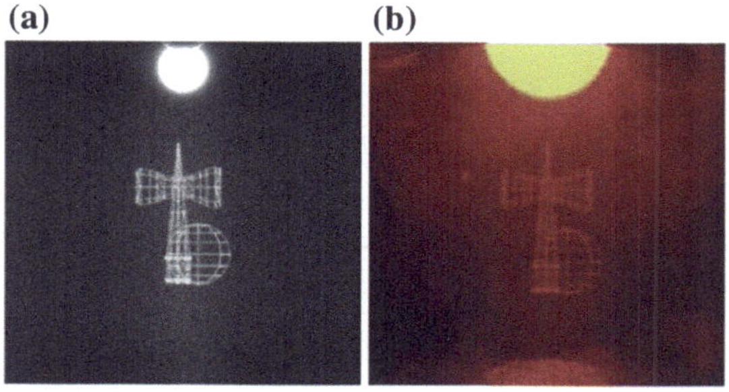

Fig. 13.19 Comparison between **a** simulated and **b** optical reconstructions of a LD-CGH. $M = 16{,}384$, $N = 8{,}129$, $\Delta x = 1.5\,[\mu\text{m}]$, $\Delta y = 3\,[\mu\text{m}]$, $\lambda = 633\,[\text{nm}]$, $W_\text{obj} = 10\,[\text{mm}]$

In this simulation, we set the distance d_2 to 24 mm, which substantially accords with the value in human eyes. The focal length f instead of d_2 is adjusted in order to focus on the gazing point. This behavior resembles to human eyes or zoom cameras, and sometimes gives better results than those in use of a constant focal length.

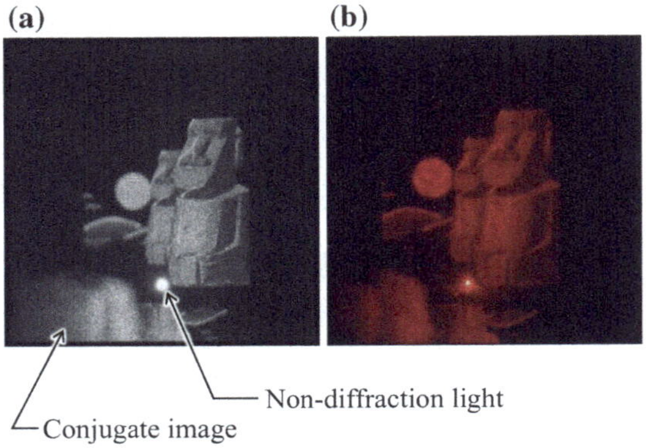

Fig. 13.20 Comparison between **a** simulated and **b** optical reconstructions of a HD-CGH "Moai I". See Appendix A.1 for the parameters of the CGH. $d_0 = 400\,[\text{mm}]$, $d_2 = 24\,[\text{mm}]$, $\Delta x_i = \Delta y_i = 0.5\,[\mu\text{m}]$, and $M_i = N_i = 8{,}192$

13.5 Simulated Reconstruction in Color

Monochrome images were produced from a monochromatic wavefield in the technique of simulated reconstruction so far. This is sufficient to predict the images reconstructed by a monochrome CGH illuminated by a monochromatic light source such as a laser. However, in cases where the illumination light source is not monochromatic; ordinary light sources such as LEDs or light bulbs are used for illumination, the simulated images are not accurate enough to predict the optical reconstruction. Above all, recent progress of high-definition CGHs makes it possible to reconstruct full-color 3D images, as described in Sect. 15.5. In these cases, it is required to produce full-color images from multi-spectral wavefields.

13.5.1 Production of Full-Color Reconstructed Image

Because a wavelength is not explicitly indicated in the formulas of simulated reconstruction of (13.27), we rewrite the formulas as follows:

$$U'(x', y'; \lambda) = \mathscr{P}^{(\text{SF})}_{\mathbf{r}_\text{v}} \{U(x, y; \lambda)\}, \tag{13.33a}$$

$$U''(x'', y''; \lambda) = \mathscr{R}_{\mathbf{T}_\text{v}} \left\{U'(x', y'; \lambda)\right\}, \tag{13.33b}$$

$$I_{\text{image}}(x'', y''; \lambda) = \left|\mathscr{P}_{d_2} \left\{t_{\text{lens}}(x'', y''; f, \lambda) U''(x'', y''; \lambda)\right\}\right|^2, \tag{13.33c}$$

where $\mathbf{r}_\text{v}$ and $\mathbf{T}_\text{v}$ are the same as those in (13.27). Here, note that the lens function is depending on the wavelength.

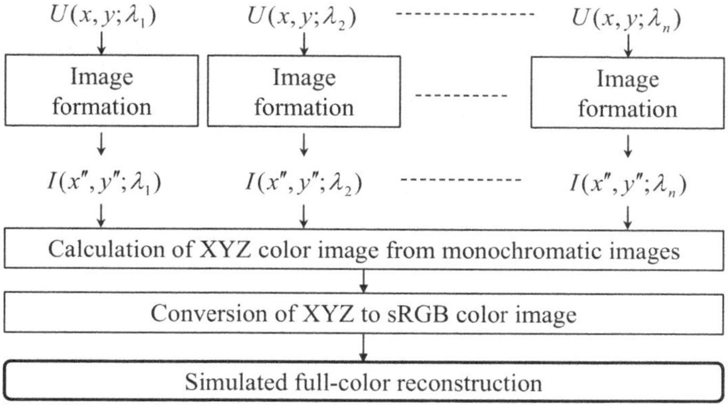

Fig. 13.21 The procedure for producing the color image from multi-spectral intensity images

The input field $U(x, y; \lambda)$ is, in general,

$$U(x, y; \lambda) = \begin{cases} g_{\text{obj}}(x, y; 0) & \text{(Wavefield at wavelength } \lambda) \\ \sqrt{T(x, y; \lambda_c)}P(x, y; \lambda) & \text{(Amplitude hologram)} \\ t(x, y; \lambda_c)P(x, y; \lambda) & \text{(Phase hologram)} \end{cases}, \quad (13.34)$$

where $T(x, y; \lambda_c)$ and $t(x, y; \lambda_c)$ are fringe patterns whose design wavelength is λ_c, i.e., the fringe pattern is generated for reconstruction at a wavelength of λ_c. $P(x, y; \lambda)$ is the wavefield of the illumination light whose wavelength is λ. Here, note that the illumination wavelength λ does not necessarily agree with λ_c, rather these wavelengths are not the same in many cases. This difference is the origin of the chromatic aberration in holography. Here note that (13.34) is a general form. Actual input field depends of the technique to produce the color 3D image.

Figure 13.21 shows the procedure to produce a color image from multi-spectral intensity images. There are some color spaces to treat color quantitatively. Here, we use the most standard one: CIE 1931 XYZ color space. In this color space, three coordinates values representing a color, called the tristimulus values, are given by

$$X = \int_\lambda L(\lambda)\bar{x}(\lambda)d\lambda,$$
$$Y = \int_\lambda L(\lambda)\bar{y}(\lambda)d\lambda,$$
$$Z = \int_\lambda L(\lambda)\bar{z}(\lambda)d\lambda, \quad (13.35)$$

where $L(\lambda)$ is radiance at a wavelength λ. $\bar{x}(\lambda)$, $\bar{y}(\lambda)$ and $\bar{z}(\lambda)$ are the CIE XYZ standard *color matching functions*.

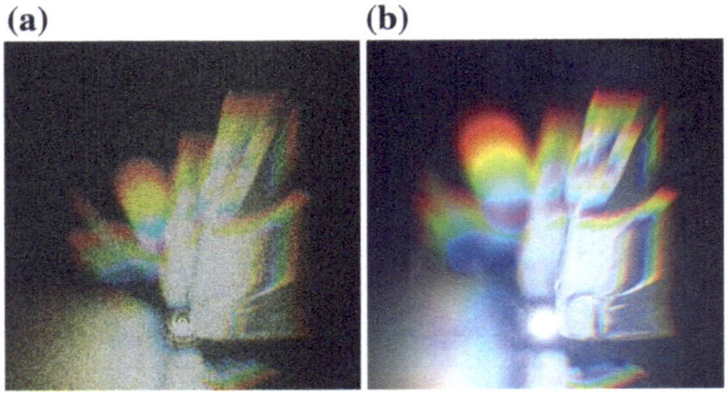

Fig. 13.22 Comparison between **a** simulated and **b** optical reconstructions of Moai I, illuminated by a white LED. $N_c = 6$. The other parameters are the same as those in Fig. 13.20

Because radiance corresponds to brightness, i.e., optical intensity, we can convert the intensity image $I(x'', y''; \lambda)$, generated by the simulated reconstruction in (13.33), into a XYZ color image as follows:

$$X(x'', y'') = \sum_{n}^{N_c} I(x'', y''; \lambda_n)\bar{x}(\lambda_n),$$
$$Y(x'', y'') = \sum_{n}^{N_c} I(x'', y''; \lambda_n)\bar{y}(\lambda_n),$$
$$Z(x'', y'') = \sum_{n}^{N_c} I(x'', y''; \lambda_n)\bar{z}(\lambda_n), \tag{13.36}$$

where λ_n and N_c are the wavelength and number of input wavefields, respectively. The color matching functions $\bar{x}(\lambda)$, $\bar{y}(\lambda)$ and $\bar{z}(\lambda)$ are provided by CIE. The XYZ color image can be easily converted to the sRGB color image by the standard procedure.

13.5.2 Examples of Simulated Reconstruction in Color

We see several actual examples of simulated reconstruction in color.

13.5.2.1 Monochrome CGH Reconstructed by White Light

Optical reconstruction of a monochrome CGH: Moai I is shown in Fig. 13.22b. Here, a white LED is used for the light source. Severe color smear is caused by chromatic aberration because the HD-CGH is a kind of thin hologram. Figure 13.22a shows a simulated reconstruction in color. The simulation is carried out with six wavelengths from 410 to 710 nm at intervals of 50 nm, i.e., the image is reconstructed from the fringe pattern $T(x, y; \lambda_c)$, whose design wavelength a constant 633 nm, using illumination light $P(x, y; \lambda_n)$ ($n = 1, \ldots, 6$). The weighted sum of six monochrome images produces a XYZ color image by (13.36). The resultant color image well reproduces the picture of the optical reconstruction.

13.5.2.2 Full-Color 3D Image Superimposed by Dichroic Mirrors

We can reconstruct a full-color image by superimposing images reconstructed by three HD-CGHs using dichroic mirrors, as mentioned in Sect. 15.5. Figure 13.23c shows pictures of a full-color 3D image reconstructed by this technique. Ideally, the input wave field of the simulated reconstruction is expressed as

$$U(x, y; \lambda) = \sum_{c=R,G,B} \sqrt{T(x, y; \lambda_c) T_c(\lambda)} P(x, y; \lambda), \tag{13.37}$$

where $T(x, y; \lambda_c)$ is a fringe pattern for three primary colors; c = R, G, or B. $T_c(\lambda)$ is a spectral property of the dichroic mirror for color c. We can produce an XYZ image from monochromatic images reconstructed at various wavelengths. However, it is time-consuming to generate monochromatic images at many wavelengths. Besides, we need $T_c(\lambda)$ data to do this. Thus, if accurate color reproduction is not necessary, only three wavelengths used for generating fringe patterns is sufficient to confirm the 3D image prior to printing and reconstructing the three CGHs.

Figure 13.23a shows monochromatic simulated reconstructions of the three CGHs for three primary colors. Here, the wavelengths of the illumination light is the same as those in generating the fringes. A color image obtained by combining the three images using (13.36) is shown in Fig. 13.23b. Although the color is not vivid when compared with the optical reconstruction in (c), the simulated reconstruction is useful enough to confirm the color 3D image.

13.5.2.3 Full-Color CGH Using RGB Color Filters

As mentioned in Sect. 15.6, one of promising techniques to produce full-color HD-CGHs is to make use of RGB color filters. In this type of full-color CGH, the fringe has the structure shown in Fig. 13.24. The fringe pattern given in (15.9) is

$$T(x, y) = \sum_{c=R,G,B} \mathbf{Blk}_c \left\{ T(x, y; \lambda_c) \right\}, \tag{13.38}$$

where λ_R, λ_G, and λ_B are wavelengths corresponding to three primary colors: red, green, and blue. $\mathbf{Blk}_c \{T(x, y; \lambda_c)\}$ denotes clipping the block for color c out of the corresponding fringe pattern $T(x, y; \lambda_c)$. Because RGB color filters are attached to the fringe pattern, the reconstructed field is expressed using the same clipping operator:

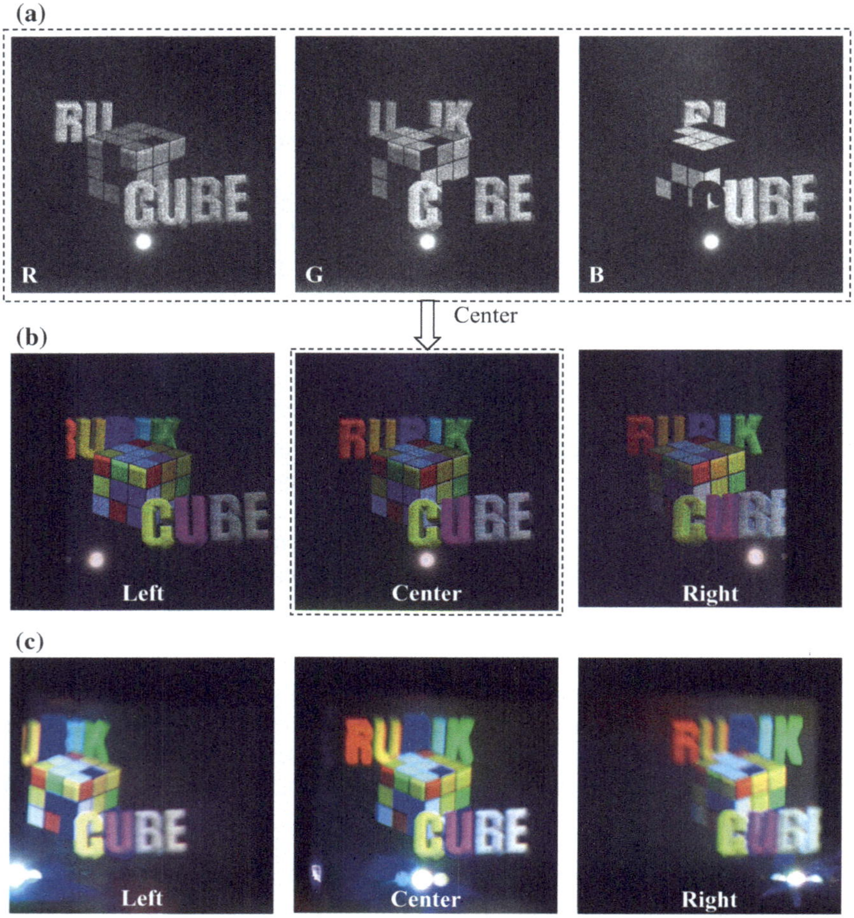

Fig. 13.23 Full-color reconstruction based on superimposing three images reconstructed by three HD-CGHs for primary colors (see Sect. 15.5). **a** Simulated reconstructions of the individual monochrome CGH, **b** full-color simulated images produced from the monochromatic images, and **c** pictures of the optical reconstruction taken from different viewpoints. CGH: $M = N = 65{,}536$, $\Delta x = \Delta y = 0.8\,[\mu\mathrm{m}]$, $\lambda_R = 620\,[\mathrm{nm}]$, $\lambda_G = 540\,[\mathrm{nm}]$, and $\lambda_B = 450\,[\mathrm{nm}]$. Simulated reconstruction: $M_i = N_i = 4{,}096$, $\Delta x_i = \Delta y_i = 1\,[\mu\mathrm{m}]$, and $f = 25\,[\mathrm{mm}]$. Video link, https://youtu.be/0W0Wmh6SW4s

Fig. 13.24 The structure of fringe patterns in full-color CGHs using RGB color filters

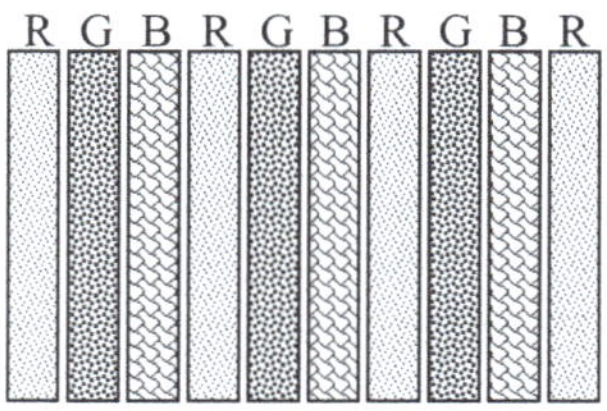

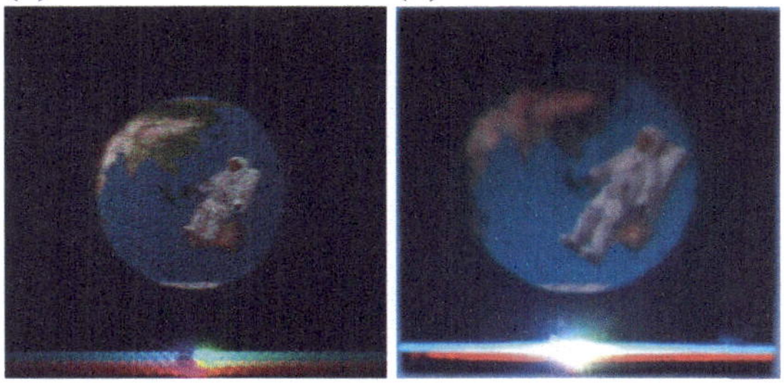

Fig. 13.25 Comparison between **a** simulated and **b** optical reconstruction of a full-color HD-CGH using RGB color filters, named "Zero Gravity". See Appendix A.1 for the parameters. $M = N = 65{,}536$, and $\Delta x = \Delta y = 0.8\,[\mu\text{m}]$. Video link, https://youtu.be/dCLVhMJsGm4

$$U(x, y; \lambda) = \sum_{c=R,G,B} \mathbf{Blk}_c \left\{ \sqrt{T(x, y) T_c(\lambda)} P(x, y; \lambda) \right\}, \tag{13.39}$$

where $T_c(\lambda)$ is spectral transmittance of the filter for color c.

Figure 13.25b shows optical reconstruction of HD-CGH "Zero Gravity", made by using RGB color filters, and (a) shows the simulated reconstruction. Although it is desirable to use many wavelengths and accurate spectral property $T_c(\lambda)$ to reproduce the exact color, only three representative wavelengths of the RGB color filters is used for the color simulated reconstruction here.

Chapter 14
Digitized Holography

Modern Holography for Physical Object

Abstract In this chapter, we discuss a technique to reconstruct real objects by HD-CGHs like traditional optical holography. The object field emitted by a subject is recorded by an image sensor instead of photo-sensitive chemicals unlike traditional holography. The captured object field is arranged in a virtual 3D scene like computer holography mentioned so far, and is optically reconstructed by the HD-CGH. We start from the capturing technique using phase-shifting synthetic aperture lensless-Fourier digital holography, and then proceed to occlusion processing of the physical object and embedding the captured field into a virtual 3D scene. A topic of digital resizing of the captured field is also discussed in this chapter.

14.1 Concept of Digitized Holography

In classical holography, the wavefield emitted from a real object is recorded on photo-sensitive materials in the form of the fringe pattern, as mentioned in Sect. 7.4. The wavefield of the subject is optically reconstructed by diffraction by the fringe pattern after chemical processing of the material. This is the 3D image produced by classical holography. The classical holography makes it possible to reconstruct brilliant 3D images of the physical object. In contrast, computer holography, explained in this book so far, is holography that reconstructs virtual 3D scenes composed of non-physical objects, because HD-CGHs are created from CG models or digital images.

To reconstruct real objects through computer holography, three approaches can be adopted at this time. The easiest approach is to measure the shape of the 3D objects using a rangefinder or 3D scanner and texture-map the photograph of the object onto the obtained polygon mesh. High-definition CGHs "Shion" (Fig. 10.37), "Brothers" (Fig. 11.5), and "Color Shion II" (Fig. 15.42) are the examples of this type of HD-CGH. However, the 3D image obtained by taking this approach may be regarded as a type of synthetic image rather than a real image. The second approach is to use some technique based on multiple-viewpoint images. This may be the most promising approach, but the technique is beyond the scope of this book.

The third approach is to capture the real wavefields of a subject using image sensors. The classical holography mentioned above reconstructs the 3D image of

K. Matsushima, *Introduction to Computer Holography*,
Series in Display Science and Technology,
https://doi.org/10.1007/978-3-030-38435-7_14

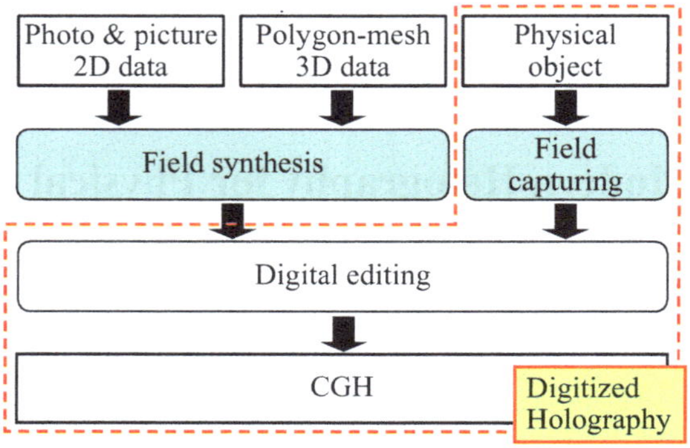

Fig. 14.1 An overall concept of computer holography including reconstruction of physical objects

physical object through two steps: recording and reconstruction. Computer holography handled in this book so far offers only digitization of the reconstruction step. Digitization of the recording step is often referred to as just *digital holography* (DH).[1]

It is theoretically possible to capture any object field using DH, but it is not easy in practice to capture real fields that meet the requirements for creating the high-definition CGHs. The first requirement is that the sampling interval of the captured field is sufficiently small to provide a large viewing angle in the optical reconstruction. In fact, the sampling interval of the captured wavefield should not exceed 1 micron. The second requirement is a large capturing area comparable to the created HD-CGHs. This is needed to extend the viewing angle and to reconstruct as large objects as the hologram itself, whose length is, for example, approximately 10 cm. Both requirements lead to capturing the field with a large number of sample points. This is essentially the same problem as the space bandwidth product in computer holography, as described in Sect. 8.3. Unfortunately, current image sensors available in the market do not meet these requirements.

To resolve the problem, we use lensless-Fourier synthetic-aperture digital holography [5, 91], which is a type of DH using spherical reference waves. This technique resolves the problems of capturing a large field at high density, and makes it possible to reconstruct 3D images of the real object through computer holography. As a result, the whole process of classical holography is replaced by digital counterparts. Thus, we refer to this technique as *digitized holography*.

Digitized holography allows us to digitally edit, archive, transmit, and optically reconstruct the field of really existing 3D objects. In addition, the captured field can be mixed with virtual 3D scenes composed of digital 2D images and polygon-mesh 3D objects. Figure 14.1 shows the extended concept of computer holography, which includes reconstruction of physical objects as well as non-physical objects in a virtual 3D scene.

[1]The word "digital holography" is used in a broad sense in many fields and not restricted to digitized recording of holograms in practice.

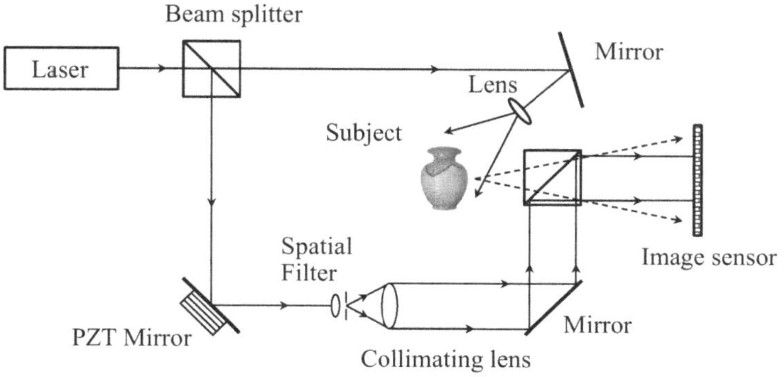

Fig. 14.2 Typical setup for recording the object field in digital holography

14.2 Digital Holography

Digital holography is a digital technique to capture object fields emitted from a subject [20, 109, 110]. Digitization of the recording step in optical holography is simply realized by replacing the photo-sensitive chemical material by an image sensor. There is, however, a problem in practice. Spatial resolution of current image sensors is considerably lower than that of photo-sensitive materials used in traditional holography. This means that the fringe pattern can be captured by DH at a low spatial frequency.

As learned in Sect. 8.8, off-axis holography where the reference wave has a large incident angle generates the fringes at a high spatial frequency. Thus, to avoid aliasing errors in capturing the fringe pattern, an optical setup shown in Fig. 14.2 is commonly used to record the object field by an image sensor. This is often called *inline digital holography*. The inline digital holography can minimizes the fringe frequency, but arises another problem. Because the inline digital holography is equivalent to holography in the cases where $\theta_R = 0$ in Fig. 7.8 and $x_R = 0$ in Fig. 7.12, the conjugate image and non-diffraction light inevitably overlap with the true image. The technique to avoid the problem is *phase-shifting* described in the following section.

14.2.1 Phase-Shifting

In the technique of phase-shifting, a device that shifts the phase of a field is inserted into the optical path of the reference field. A piezoelectric transducer (PZT) is usually used for the phase-shifting device, as shown in Fig. 14.2. Here, suppose the reference field with the phase shifting φ is represented by

$$R(x, y; \varphi) = R(x, y) \exp[i\varphi]. \tag{14.1}$$

Replacing $R(x, y)$ in (7.8) by $R(x, y; \varphi)$, the fringe intensity is rewritten as

$$\begin{aligned} I(\varphi) &= I(x, y, \varphi) \\ &= |O|^2 + |R|^2 + OR^* \exp[-i\varphi] + O^* R \exp[i\varphi] \end{aligned} \tag{14.2}$$

where $O = O(x, y)$ and $R = R(x, y)$.

14.2.1.1 Four-Step Phase-Shifting

When we capture four fringe patterns that have phase shifts $\varphi = 0, \pi/2, \pi, 3\pi/2$, the fringe patterns are

$$I(0) = |O|^2 + |R|^2 + 2\mathrm{Re}\left\{OR^*\right\}, \tag{14.3}$$

$$I\left(\frac{\pi}{2}\right) = |O|^2 + |R|^2 + 2\mathrm{Im}\left\{OR^*\right\}, \tag{14.4}$$

$$I(\pi) = |O|^2 + |R|^2 - 2\mathrm{Re}\left\{OR^*\right\}, \tag{14.5}$$

$$I\left(\frac{3\pi}{2}\right) = |O|^2 + |R|^2 - 2\mathrm{Im}\left\{OR^*\right\}. \tag{14.6}$$

Therefore, after several steps, we get

$$|OR^*| = \frac{\sqrt{[I(0) - I(\pi)]^2 + [I(\pi/2) - I(3\pi/2)]^2}}{4}, \tag{14.7}$$

$$\arg(OR^*) = \tan^{-1}\left[\frac{I(\pi/2) - I(3\pi/2)}{I(0) - I(\pi)}\right]. \tag{14.8}$$

Thus, we can obtain complex field $OR^*(x, y) = O(x, y)R^*(x, y)$. If $R(x, y)$ is a simple wavefield such as plane waves, $O(x, y)$ itself is extracted from the fringe patterns. This is called the *four-step phase-shifting digital holography* [130].

14.2.1.2 Three-Step Phase-Shifting

In fact, four steps are not necessary to obtain $OR^*(x, y)$ from fringe patterns. Three steps are the minimum number to remove unnecessary components and to extract only the object field from digitally captured fringe patterns. The fringe patterns in phase-shifting $\varphi = 0, \alpha, \beta$ are

$$I(0) = |O|^2 + |R|^2 + OR^* + O^* R, \tag{14.9}$$

$$I(\alpha) = |O|^2 + |R|^2 + OR^* \exp[-i\alpha] + O^* R \exp[i\alpha], \tag{14.10}$$

$$I(\beta) = |O|^2 + |R|^2 + OR^* \exp[-i\beta] + O^* R \exp[i\beta]. \tag{14.11}$$

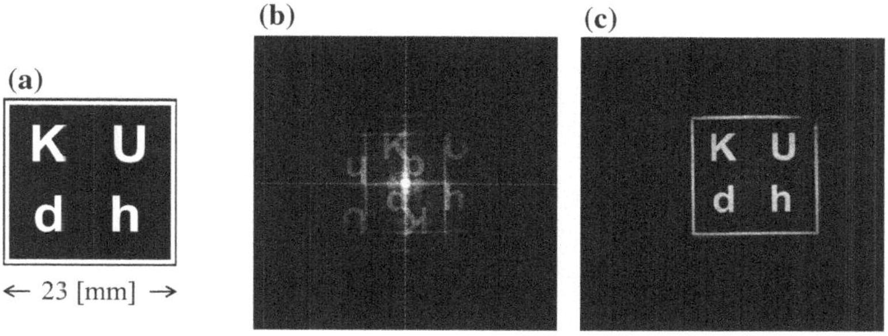

Fig. 14.3 The effect of three-step phase-shifting digital holography [90]. **a** The captured planar object, numerical reconstruction **b** without and **c** with phase-shifting. $\lambda = 532$ [nm], $\delta_x = \delta_y = 6$ [μm], and $z_0 = z_R = 62$ [cm]

The following generalized formula is derived by eliminating $|O|^2$, $|R|^2$, and O^*R from the above equations:

$$OR^* = \frac{i}{2}\frac{[I(0) - I(\alpha)]\exp(i\beta) + [I(\alpha) - I(\beta)] + [I(\beta) - I(0)]\exp(i\alpha)}{\sin(\alpha) - \sin(\beta) - \sin(\alpha - \beta)}. \tag{14.12}$$

When we choose $\alpha = 2\pi/3$ and $\beta = -2\pi/3$, (14.12) is simplified down to [118]

$$OR^* = \frac{1}{3}\left[I(0) - \exp\left(-i\frac{\pi}{3}\right)I\left(\frac{2\pi}{3}\right) - \exp\left(i\frac{\pi}{3}\right)I\left(-\frac{2\pi}{3}\right)\right]. \tag{14.13}$$

This is the technique of *three-step phase-shifting digital holography*.

Figure 14.3 shows the effect of three-step phase-shifting digital holography. The planar object in (a) is captured by lensless-Fourier DH, described in the next section, using an image sensor with a sensor pitch of 6 μm. Numerical reconstruction[2] without phase-shifting is shown in (b). The conjugate image and non-diffraction light are clearly reconstructed in (b). In contrast, only the true image appears in the reconstructed image of (c), which is captured and reconstructed using three-step phase-shifting.

14.2.2 Lensless-Fourier Digital Holography for Converting Sampling Interval

In case of a reference plane waves, it is properly expected that $R(x, y) = 1$ because of the inline configuration of DH. Thus, $O(x, y)$ is directly captured in the sensor

[2]The word "numerical reconstruction" is often used in DH instead of simulated reconstruction.

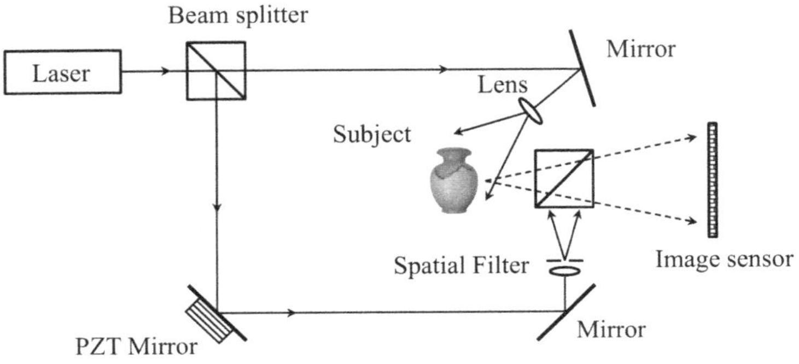

Fig. 14.4 Setup for lensless-Fourier DH

plane. We can obtain the object field close to the subject by back-propagation of $O(x, y)$, as mentioned in Sect. 13.2. If the subject is bigger than the sensor size, extension of the sampling window is required prior to the back-propagation like the situation in Fig. 12.1a.

When a spherical wave is used instead of a plane wave for the reference wave in recording, the object field is simply obtained by the Fourier transform of the captured field [125]. Therefore, the technique is called *lensless-Fourier digital holography*. When lensless-Fourier configuration is applied to DH, the sampling interval and thus size of the object field can be changed by the recording parameters. Because of this big advantage, object fields in digitized holography is commonly recorded using the lensless-Fourier DH.

14.2.2.1 Sampling Intervals

Figure 14.4 shows typical setup for phase-shifting lensless-Fourier DH. The phase-shifting technique in the preceding section works well in a reference spherical wave as well as a reference plane wave. However, the resultant wavefields have considerably different properties.

We have dealt with the theory of holography with a reference spherical wave in Sect. 7.7.2. Inline DH with a reference spherical wave can be analyzed by a similar procedure. Let the object field and reference spherical wave be expressed in (7.44) and (7.41) under Fresnel approximation, respectively. Thus, the field captured by an image sensor using phase-shifting is

$$OR^*(x) = \mathscr{P}_{d_0}\left\{g_{\text{obj}}(x; -d_0)\right\} A_{\text{R}} \exp\left[-ik\left\{d_{\text{R}} + \frac{(x - x_{\text{R}})^2}{2d_{\text{R}}}\right\}\right]. \quad (14.14)$$

Here, we assume that the coordinates origin is at the center of the image sensor and the object field in the sensor plane is given by propagation of the object field

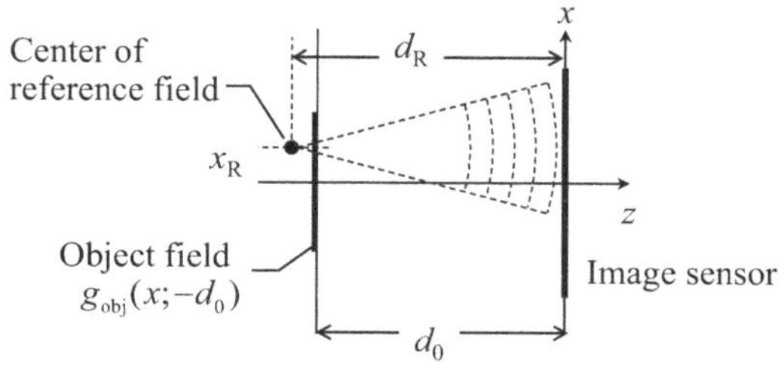

Fig. 14.5 Geometry used for analysis of lensless-Fourier DH

$g_{obj}(x; -d_0)$ placed at $z = -d_0$, as shown in Fig. 14.5. The center of the spherical wave is positioned at $(x_R, -d_R)$. The y-axis is temporarily omitted in formulation.

Using Fresnel propagation in (5.27), the captured field is rewritten as

$$OR^*(x) = A_{FR}(d_0)A_R \exp[-ikd_R] \times \int g_{obj}(x_s; -d_0) \exp\left\{i\frac{\pi}{\lambda}\left[\frac{(x_s - x)^2}{d_0} - \frac{(x_R - x)^2}{d_R}\right]\right\} dx_s. \quad (14.15)$$

Supposing $d_0 = d_R$, the field is simplified to

$$OR^*(x) = \frac{A_R}{i\lambda d_R} \exp\left[i\frac{kx_R}{2d_R}(2x - x_R)\right] \times \int g_{obj}(x_s; -d_R) \exp\left[i\frac{\pi x_s^2}{\lambda d_R}\right] \exp[-i2\pi u x_s]\, dx_s, \quad (14.16)$$

where $u = x/(\lambda d_R)$. Furthermore, we assume $x_R = 0$ and ignore the trivial coefficient:

$$OR^*(x) \cong \mathcal{F}\left\{g_{obj}(x; -d_R) \exp\left[i\frac{\pi x_s^2}{\lambda d_R}\right]\right\}_{u=x/(\lambda d_R)} \quad (14.17)$$

The field is written by the 2D form as follows:

$$OR^*(x, y) = \mathcal{F}\left\{g_{obj}(x, y; -d_R)\phi^*(x, y)\right\}_{u=x/(\lambda d_R), v=y/(\lambda d_R)}, \quad (14.18)$$

$$\phi(x, y) \equiv \exp\left[-i\frac{k}{2d_R}(x^2 + y^2)\right]. \quad (14.19)$$

Conclusively, the object field at a position of $z = -d_R$ is given by

$$g_{obj}(x, y; -d_R) = \mathcal{F}\left\{OR^*(x, y)\right\}_{u=-x/(\lambda d_R), v=-y/(\lambda d_R)} \times \phi(x, y). \quad (14.20)$$

As a result, we can obtain the object field at the $(x, y, -d_R)$ plane by simple Fourier transform[3] of the captured field:

The sampled version of (14.20) is the most interesting for us:

$$g_{\rm obj}(-\lambda d_R u_m, -\lambda d_R v_n; -d_R) = \mathbf{FFT}\left\{OR^*(x_m, y_n)\right\}\phi(x_m, y_n). \tag{14.21}$$

The sampling intervals in the $(x, y, -d_R)$ plane are given by

$$\Delta x = \lambda d_R \Delta u \quad \text{and} \quad \Delta y = \lambda d_R \Delta v. \tag{14.22}$$

Supposing the sensor pitches are δ_x and δ_y in the x and y directions, $\Delta u = 1/(M_0\delta_x)$ and $\Delta v = 1/(N_0\delta_y)$ according to (4.78), and thus

$$\Delta x = \frac{\lambda d_R}{M_0\delta_x} \quad \text{and} \quad \Delta y = \frac{\lambda d_R}{N_0\delta_y}, \tag{14.23}$$

where M_0 and N_0 are numbers of sample points of the complex field $OR^*(x_m, y_n)$. Note that values of M_0 and N_0 are commonly a power of 2 in order to perform FFT, while numbers of pixels $M_{\rm sens}$ and $N_{\rm sens}$ of the image sensor are not power of 2 in general.[4] As a result, we can capture the object field at different sampling intervals from the sensor pitches, i.e., the sampling intervals is controllable in the lensless-Fourier DH.

14.2.2.2 Aliasing-Free Zone

As shown in Fig. 8.19 of Sect. 8.8.2, fringes generated with a reference spherical wave have relatively lower frequencies than those with a plane wave in general. This means that DH using a reference spherical wave has a larger field of view (FOV) than that using a plane wave. Because an object point further from the optical axis generates a higher frequency, as shown in Fig. 8.19e, f, capturing a large subject may cause aliasing errors, i.e., if the subject size exceeds a limit, aliasing errors are produced in the captured field. The limit on the object size increases with increasing the distance from the image sensor like the FOV of cameras. We call the FOV, where aliasing errors do not occur, an *aliasing-free zone*. Figure 14.6a shows the concept of the aliasing-free zone. Estimation of the aliasing-free zone is very important in actual capturing of a subject by lensless-Fourier DH.

Dimensionless spatial frequency $\hat{u}$ is given by (8.32). Here, to estimate the aliasing-free zone roughly, suppose the center of the reference spherical wave is

---

[3]Since $g_{\rm obj}(x, y; -d_R)$ is obtained by the inverse Fourier transform of the both sides of (14.18), "inverse Fourier transform" seems to be more exact expression than simple "Fourier transform". In this case, however, we can choose forward Fourier transform as well as inverse Fourier transform by definition of u and v.

[4]The minimum integers that is a power of 2 and more than sensor pixels are commonly chosen for M_0 and N_0.

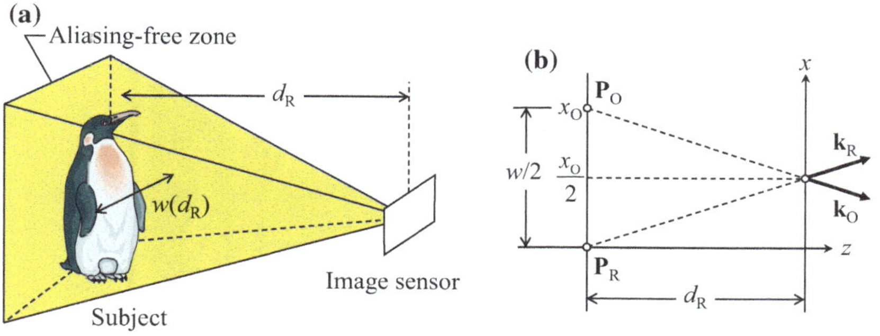

Fig. 14.6 The aliasing-free zone: **a** illustration of the concept, and **b** estimation of the maximum spatial frequency

in the optical axis; $x_R = 0$. In addition, we assume that the position of the subject almost agrees with the center of the spherical wave; $z_O = z_R = -d_R$, as shown in Fig. 14.6b. In this case, the dimensionless frequency is

$$\hat{u}(x) = \frac{x}{\sqrt{x^2 + z_R^2}} - \frac{x - x_O}{\sqrt{(x_O - x)^2 + z_R^2}}. \tag{14.24}$$

We can easily verify that derivative $\partial \hat{u}/\partial x$ takes zero when $x = x_O/2$. Hence, the maximum frequency is

$$\hat{u}\left(\frac{x_O}{2}\right) = \frac{x_O}{\sqrt{(x_O/2)^2 + z_R^2}}. \tag{14.25}$$

The spatial frequency of the fringe must satisfy $|\hat{u}| < \lambda/(2\delta_x)$ as in (8.35) and (8.36) to avoid aliasing errors. Thus, when we take $x_O = w/2$, the following must be satisfied:

$$w < \frac{4\lambda}{\sqrt{16\delta_x^2 - \lambda^2}} d_R. \tag{14.26}$$

$w = w(d_R)$ gives the aliasing-free zone in the x direction. As a result, the aliasing-free zone forms a quadrangular pyramid, as shown in Fig. 14.6.

Figure 14.7 shows examples of numerical reconstruction of large subjects in lensless-Fourier DH. The subject is a 3D object in (a), while a 2D object in (b). The field is captured by an image sensor whose number of pixels is $M_{sens} = N_{sens} = 2{,}000$. The field captured using three-step phase-shifting is embed in an array whose number of sample points is $M_0 = N_0 = 2{,}048$, and then Fourier-transformed. The size of the aliasing-free zone is approximately 53 mm at $d_R \simeq 60$ [cm]. Because the sensor pitch is 6 μm in both directions, the sensor size is just 12 mm × 12 mm. The lensless-Fourier setup makes it possible to capture and reconstruct a larger subject than the image sensor.

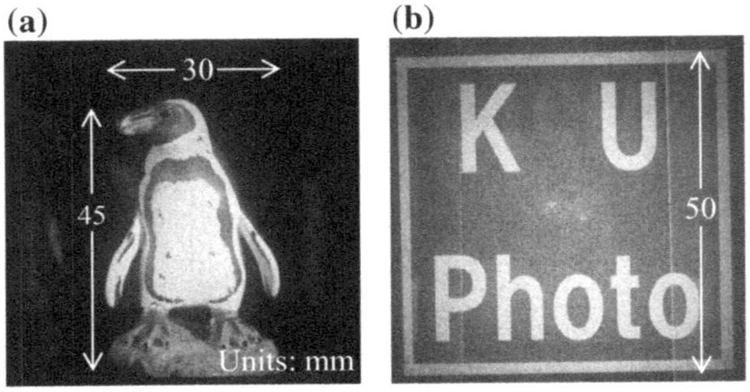

Fig. 14.7 Examples of numerical reconstruction of large subjects captured by lensless-Fourier digital holography with **a** $d_R = 61.7$ [cm] and **b** $d_R = 61.1$ [cm] [90]. $\lambda = 532$ [nm], $\delta_x = \delta_y = 6$ [μm], $M_{sens} = N_{sens} = 2{,}000$, and $M_0 = N_0 = 2{,}048$

14.2.3 Synthetic Aperture Digital Holography for Capturing Large-Scale Wavefield

The sampling interval is increased by extending the distance d_R to capture and reconstruct large subjects in the examples of Fig. 14.7. However, in practice, the sampling intervals of the captured object field must be reduced to meet the requirement in HD-CGHs in order to realize digitized holography. Assuming that $\Delta x = \Delta y = 1$ [μm] is required, we need $d_R \simeq 1.9$ [cm] in the image sensor used in Fig. 14.7. This is an impracticable value in actual setup.

14.2.3.1 Synthetic Aperture in Lensless-Fourier Digital Holography

According to (14.23), the sampling intervals can be reduced not only by decreasing d_R but also by increasing the numbers of samplings M_0 and N_0. Accordingly, to achieve sampling intervals required in HD-CGHs, the number of samplings must be increased inevitably. In addition, the captured area is only 2×2 mm^2 in the object plane even if a sampling interval of 1 μm is achieved by the image sensor used in Fig. 14.7. The technique to increase the number of samplings and the captured area in the object plane is the *synthetic aperture digital holography*. The technique of the synthetic aperture DH was first proposed to improve resolution of DH [5]. However, we use the technique to capture a large area of the object field at a high density.

The idea of the synthetic aperture DH is very simple. Figure 14.8 is a schematic illustration of capturing large-scale wavefields using the synthetic aperture technique. In the synthetic-aperture DH, the image sensor is mechanically translated and captures the wavefield at different positions. Here, the sensor shift is set to be smaller than the sensor area in order to overlap the captured field at each position, as shown in Fig. 14.8. This overlap area is used to avoid translation errors, i.e., sensor positions

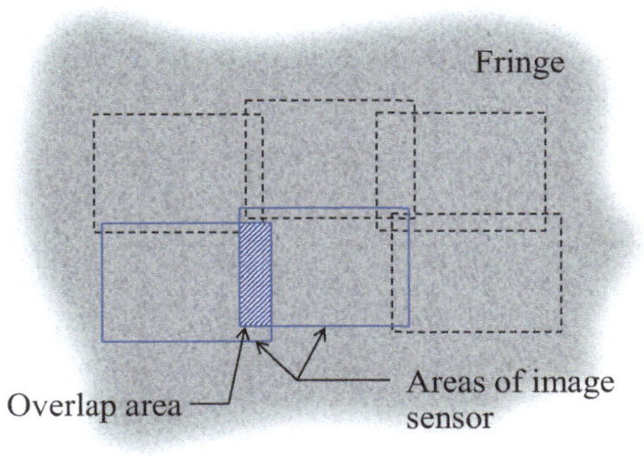

Fig. 14.8 Capture of large-scale wavefields using the synthetic aperture technique

are measured exactly using a cross-correlation function for the captured fields, as described in the following section.

The lensless-Fourier setup is necessary to realize the synthetic aperture DH because it uses a reference spherical wave. When a plane wave is used for the reference field, the spatial frequency of the fringe patterns increases at the edge of the image sensor, as shown in the left column of Fig. 8.19. In this case, extension of the captured area inevitably leads to aliasing errors. In contrast, the increase of the frequency is quite gentle in the fringes generated with a reference spherical wave, as shown in the right column. Therefore, choosing the parameters carefully, we can entirely avoid aliasing errors in synthetic aperture DH.

14.2.3.2 Accurate Alignment Using Cross-Correlation Function

Consider that a large fringe pattern is generated in the sensor plane and we can finally obtain the large complex field $OR^*(x, y)$ from multiple fringes using phase-shifting. Furthermore, suppose that a sampled complex field $g_1(x_m, y_n)$ is captured at a position of the image sensor using phase-shifting. Here, the local position of a sample point is given in the captured field by

$$x_m = \delta_x \left(m - \frac{M_{\text{sens}}}{2}\right) \quad \text{and} \quad y_n = \delta_y \left(m - \frac{N_{\text{sens}}}{2}\right) \tag{14.27}$$

δ_x, δ_y, M_{sens}, and N_{sens} are again sensor pitches and numbers of pixels of the image sensor. The captured complex field $g_1(x_m, y_n)$ can be regarded as a part of the whole field $OR^*(x, y)$, as shown in Fig. 14.9. Therefore,

$$g_1(x_m, y_n) = OR^*(x_m + x_1, y_m + y_1), \tag{14.28}$$

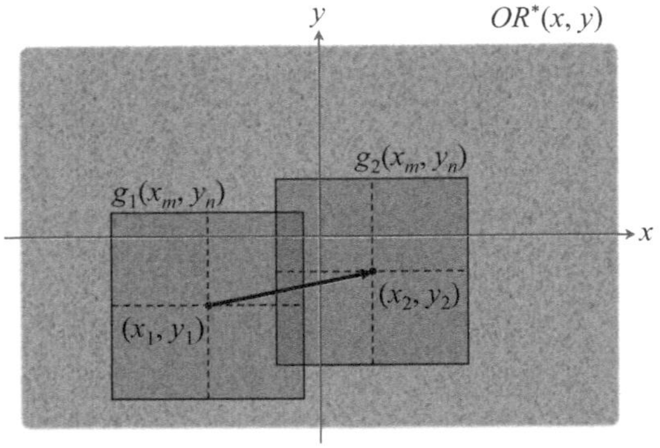

Fig. 14.9 Measurement of the position shift in neighboring complex fields

where (x_1, y_1) is the center position of the captured field. Furthermore, suppose that the second field is then captured at a center position of (x_2, y_2):

$$g_2(x_m, y_n) = OR^*(x_m + x_2, y_m + y_2). \tag{14.29}$$

Here, position (x_2, y_2) is chosen so that the fields overlap each other.

According to (4.52), cross-correlation of sampled fields is represented by

$$g_1 \star g_2(x_m, y_n) = \sum_{m',n'} g_1(x_{m'}, y_{n'}) g_2^*(x_{m'} - x_m, y_{n'} - y_n) \tag{14.30}$$

To simplify formulation, omit the y coordinates temporarily, and substitute (14.28) and (14.29) into (14.30):

$$g_1 \star g_2(x_m) = \sum_{m'} OR^*(x_{m'} + x_1) O^* R(x_{m'} - x_m + x_2) \tag{14.31}$$

Introducing new variable $x_{m''} = x_{m'} + x_1$, the cross-correlation is rewritten as

$$g_1 \star g_2(x_m) = \sum_{m''} OR^*(x_{m''}) O^* R(x_{m''} - x_1 - x_m + x_2) \tag{14.32}$$

When $x_m = x_2 - x_1$, the right-hand side of the above equation obviously takes the maximum value: $\sum_{m''} |OR^*(x_{m''})|^2$. In other words, if the cross-correlation $g_1 \star g_2(x_m)$ takes the maximum value in $x_m = x_{\text{peak}}$, the x_{peak} gives the position shift $x_2 - x_1$.

Accordingly, we first calculate the cross-correlation function for the sampled complex fields $g_1[m, n]$ and $g_2[m, n]$. According to (4.53), the sampled correlation function is calculated by

$$g_1 \star g_2(x_m, y_n) = \mathbf{FFT}^{-1} \left\{ \mathbf{FFT} \{ g_1[m, n] \} \cdot \mathbf{FFT} \left\{ g_2^*[m, n] \right\} \right\}. \tag{14.33}$$

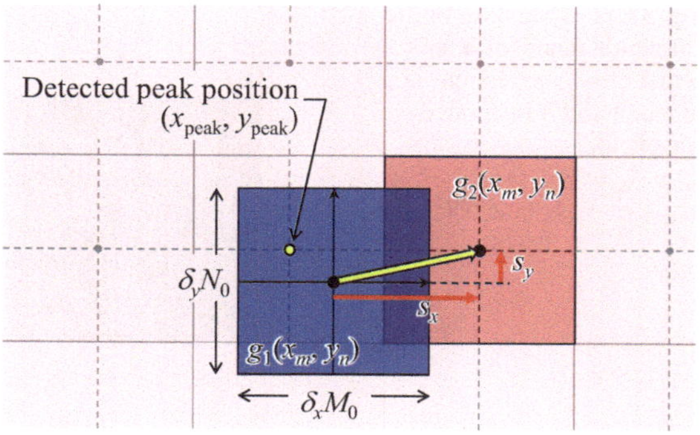

Fig. 14.10 The periodic structure of sampled complex functions in the object plane. Only periodicity of $g_2(x_m, y_n)$ is depicted to avoid confusion. $g_1(x_m, y_n)$ also has periodicity in practice

We then detect the position where the correlation function takes the maximum value. Supposing the peak value is given by $g_1 \star g_2(x_{\text{peak}}, y_{\text{peak}})$, the position shift from $g_1(x_m, y_n)$ to $g_2(x_m, y_n)$ is

$$x_2 - x_1 = x_{\text{peak}}, \quad \text{and} \quad y_2 - y_1 = y_{\text{peak}}. \tag{14.34}$$

Let s_x and s_y represent the position shifts in the x and y directions, as shown in Fig. 14.10. When a small position shift is expected, i.e., it is assumed that $|s_x| < \delta_x M_0/2$ and/or $|s_y| < \delta_y N_0/2$, the position shift is simply given by (14.34):

$$\begin{aligned} s_x &= x_{\text{peak}}, \\ s_y &= y_{\text{peak}}, \end{aligned} \tag{14.35}$$

where M_0 and N_0 are again the numbers of sample points of the complex field in which $g_1(x_m, y_n)$ and $g_2(x_m, y_n)$ are embedded. However, if large motion is expected; $|s_x| \geq \delta_x M_0/2$ and/or $|s_y| \geq \delta_y N_0/2$, the position shifts are given by

$$\begin{aligned} s_x &= \begin{cases} \delta_x M_0 + x_{\text{peak}} & \text{if } x_{\text{peak}} < 0 \\ -\delta_x M_0 + x_{\text{peak}} & \text{if } x_{\text{peak}} > 0 \end{cases} \\ s_y &= \begin{cases} \delta_y N_0 + y_{\text{peak}} & \text{if } y_{\text{peak}} < 0 \\ -\delta_y N_0 + y_{\text{peak}} & \text{if } y_{\text{peak}} > 0 \end{cases} \end{aligned} \tag{14.36}$$

This is because $g_1(x_m, y_n)$ and $g_2(x_m, y_n)$ have periodic structures that originate in sampled functions, as shown in Fig. 14.10.

An actual example of cross-correlation function of neighboring captured fields is shown in Fig. 14.11a. In this example, the command value sent to the translator is 8 mm only in the x direction. Because the size of the sampling window $\delta_x M_0 = \delta_y N_0 = 12.288$ [mm], the position shift definitely corresponds to the case where $s_x \geq \delta_x M_0/2$. As the measured peak position is $(x_{\text{peak}}, y_{\text{peak}}) =$

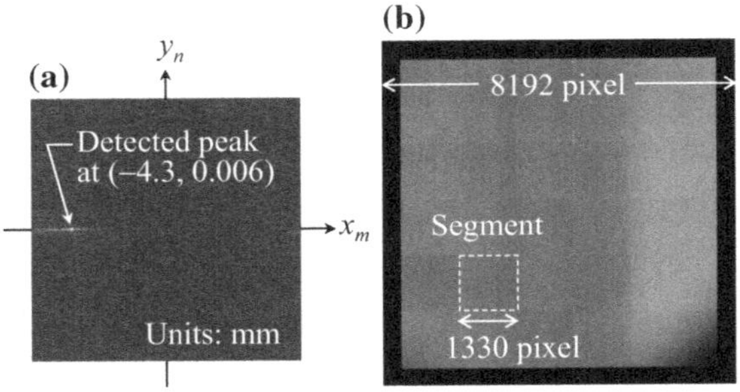

Fig. 14.11 Examples of the amplitude image of **a** an actual cross-correlation function and **b** integrated wavefield. $\delta_x = \delta_y = 6$ [μm], $M_{\text{sens}} = N_{\text{sens}} = 2{,}000$, and $M_0 = N_0 = 2{,}048$

(−4.3 [mm], 0.006 [mm]), we obtain $s_x = 7.988$ [mm] and $s_y = 6$ [μm] from (14.36).

Figure 14.11b shows an example of the amplitude image of the stitched wavefield. As the command value for each shift is 8 mm in both the x and y directions, the number of effective pixels is approximately $1{,}333 \times 1{,}333$ for each position. Since the number of the capturing position is 5×5, the size of the effective captured field is roughly $7{,}330 \times 7{,}330$ pixels.[5] The effective field is kept in an $8{,}192 \times 8{,}192$ array to perform FFT.

14.3 Capture of Object-Field

Using the techniques described in the previous sections, large-scale object fields can be captured in practice.

14.3.1 Monochromatic Object Field

Figure 14.12 shows an example of experimental setup actually used to capture object-fields by lensless-Fourier synthetic aperture DH [63]. There is nothing different from optical holography except for the fact that the scanning image sensor records the fringe pattern. In this setup, mirror M3 is installed on a PZT to shift the reference phase by changing the optical path length. Fringes are captured three times at the same position as shifting the reference phase. It should be noted that beam splitter BS2 must have at least the same size as the subject.

Figure 14.13a shows an example of the integrated field, captured using the optical setup in Fig. 14.12. The integrated field is composed of complex fields captured at 8×12 positions and embedded in a $32{,}768 \times 32{,}768$ array. The Fourier transform

[5]Because the whole sensor pixels are used in the end of the sensor scan, the number of effective pixels is estimated at 7,332 ($=1{,}333 \times 4 + 2{,}000$)

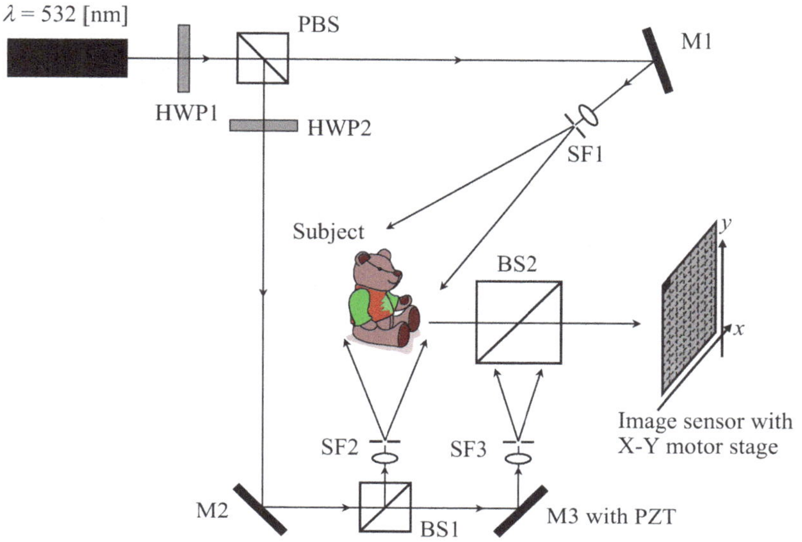

Fig. 14.12 An example of actual experimental setup used to capture object fields by lensless-Fourier synthetic aperture DH at a single wavelength. M, mirror; HWP, half-wavelength plate; BS, beam splitter; PBS, polarizing beam splitter; SF, spatial filter

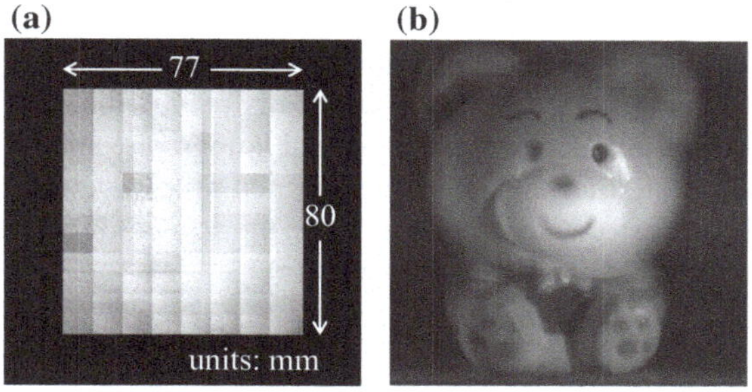

Fig. 14.13 The amplitude images of **a** an integrated field and **b** object field obtained by the Fourier transform [53]. $\lambda = 532$ [nm], $M_{\mathrm{sens}} = 3{,}000$, $N_{\mathrm{sens}} = 2{,}208$, $\delta_x = \delta_y = 3.5$ [μm], $M_0 = N_0 = 32{,}768$, $d_{\mathrm{R}} = 21.5$ [cm], and $\Delta x = \Delta y \simeq 1$ [μm]

of the integrated field is shown in Fig. 14.13b. Distance d_{R} is selected using (14.23) so that sampling intervals of the object field is $\Delta x = \Delta y \cong 1$ [μm] after the Fourier transform. Because the area of the integrated field is 77×88 mm^2, the depth of field (DOF) of the amplitude image is considerably small. We can find out that the amplitude image is nearly in focus on the eyes of the toy bear, i.e., the object field is obtained in the plane that intersects the subject around the eyes.

An actual CGH was created to verify optical reconstruction of the captured field. The picture of the optical reconstruction is shown in Fig. 14.14b. The number of

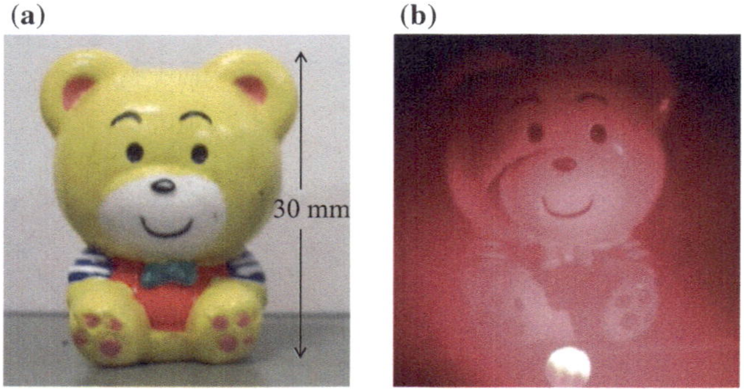

Fig. 14.14 Photographs of **a** the subject and **b** optical reconstruction of the fabricated CGH, whose object field is captured by lensless Fourier synthetic aperture DH. The number of pixels of the CGH is 32,768 × 32,768. Pixel pitch is 1 μm × 1 μm. The object field, whose amplitude image is depicted in Fig. 14.13b, is arranged at $z = -100$ [mm] behind the hologram. The CGH is named "Bear I"

the pixels and pixel pitches are the same as those in the captured field. The object field, whose amplitude image and parameters are shown in Fig. 14.13, is arranged at a position of $z = -100$ [mm], i.e., the object field was propagated at a distance of 100 mm and then numerically interfered with a reference spherical wave. The binary fringe pattern was generated by the technique in Sect. 8.6.3 and printed using laser lithography (see Sect. 15.3). This is the typical digitized holography. The object field of a subject is recorded by DH, and is optically reconstructed by the CGH, i.e., both recording and reconstruction steps in optical holography are replaced by digital technologies. Here, note that the CGH is generated at a wavelength of 532 nm in Fig. 14.14b, but is reconstructed at 633 nm. Thus, the size of the reconstructed image is not exactly the same as the subject shown in (a).

14.3.2 Object Fields in Full-Color

Monochromatic digitized holography can be extended to full-color holography [80]. In this case, the object field is captured at three wavelengths corresponding to RGB primary colors. An example of the optical setup is shown in Fig. 14.15. The outputs of three light sources are switched using electrically controlled mechanical shutters (S). The three beams are combined into a single beam using beam splitters. The combined beam is then divided into two arms to provide reference and illumination light. We use half-wavelength plates (HWP) to control the distribution ratios for each of the wavelengths. Here, note that HWP4 must work at three wavelengths. Although HWPs that feature a broad working-bandwidth are available in the market, a certain amount of error is unavoidable in the distribution ratios.

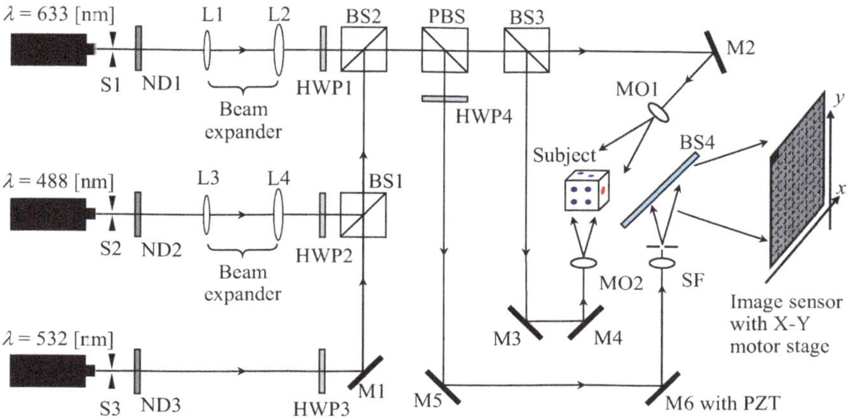

Fig. 14.15 An example of actual experimental setup used to capture object fields by lensless-Fourier synthetic aperture DH at three wavelengths [80]. M, mirror; S, shutter; ND, neutral density filter; HWP, half-wavelength plate; BS, beam splitter; PBS, polarizing beam splitter; MO, microscope objective; SF, spatial filter

The illumination arm is also subdivided into two beams and illuminates the subject through objective lenses. The reference beam is converted into a spherical field using the spatial filter (SF) and then irradiates the image sensor through a plate beam splitter (BS4). Here, beam expanders are installed into the paths of red and blue beams to spread the spherical wave sufficiently, because the output beams of these lasers were too narrow. Mirror M6 is installed on a PZT to vary the phase of the reference field. The image sensor is installed on a mechanical X-Y translation stage to apply the technique of the synthetic aperture.

To use the phase-shifting technique, at least three fringe images must be captured almost simultaneously. Therefore, the shutter of a specific laser is first opened and records three fringe images by varying the reference phase at the same wavelength. Then, the opened shutter was switched and three more fringe images were recorded at another wavelength at the same position. After completing image capture at the three wavelengths at the same position, the image sensor was moved using the X-Y mechanical stage, and the object fields were then captured again at the three wavelengths at the new position. This procedure is repeated within a specific area in which optical fringes are generated.

Examples of the amplitude images of three captured fields are shown in Fig. 14.16a. Since the number of capturing positions was 12×16, capturing process was repeated 576 ($=12 \times 16 \times 3$) times for each wavelength due to three-step phase-shifting. The amplitude images of the object fields calculated by FFT and a picture of the subject are shown in Fig. 14.16b and Fig. 14.17, respectively. According to (14.23), the sampling intervals after the Fourier transform depends on the wavelength. Therefore, the object images emerging from the object field seem to have different sizes, as in Fig. 14.16b. This is an inconvenient property of the lensless-Fourier DH.

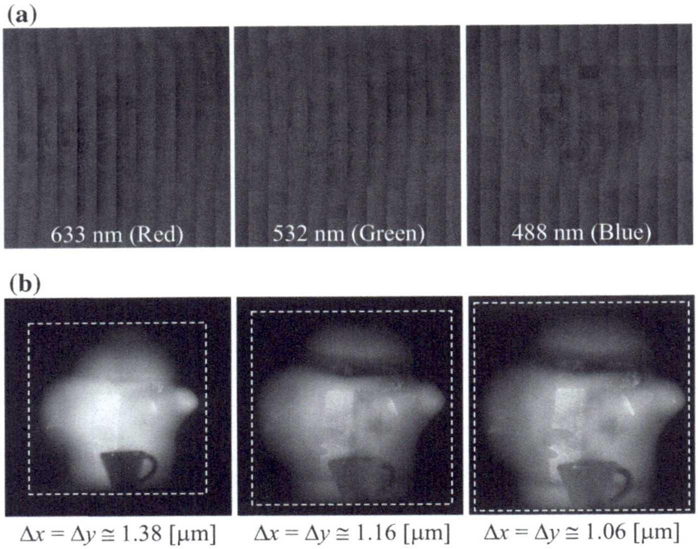

Fig. 14.16 Examples of the amplitude image of **a** captured complex fields, and **b** the object fields obtained by the Fourier transform [80]. $M_{\text{sens}} = 3{,}000$, $N_{\text{sens}} = 2{,}208$, $\delta_x = \delta_y = 3.5$ [μm], $M_0 = N_0 = 32{,}768$, and $d_R = 25.0$ [cm]

Fig. 14.17 A picture of the captured subject

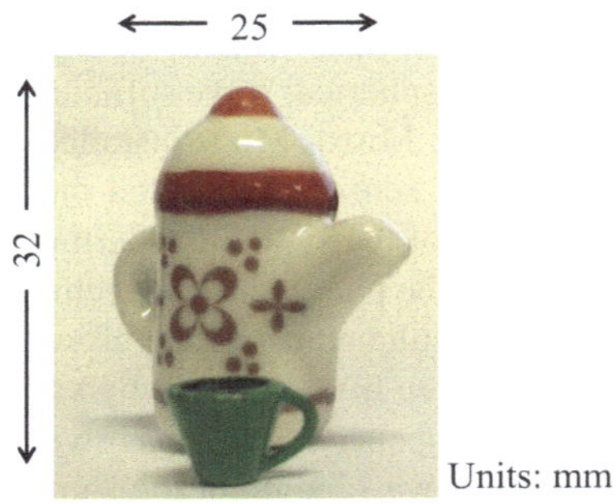

To produce a full-color CGH from these object fields, it is necessary for the three object fields to have the same sampling interval. Therefore, we have to extract the object fields within the dashed rectangles shown in Fig. 14.16b and resample them using an interpolation method, such as the cubic interpolation method described in Sect. 9.5.4, to ensure that all object fields have the same sampling interval. Figure 14.18 shows examples of full-color simulated and optical reconstructions of a CGH created from the object fields captured at three wavelengths. Here, the simulation assumes full-color reconstruction using the technique of dichroic mirrors (see Sects. 15.5 and 13.5.2.2), while the optical reconstruction uses RGB color filters (see Sect. 15.6). A multi-chip white LED is used for the illumination light source in

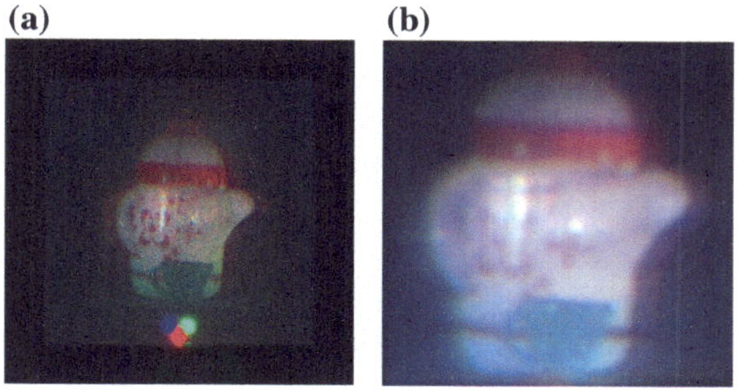

Fig. 14.18 Examples of **a** simulated and **b** optical reconstructions of a CGH created from object fields captured at three-wavelength. $M = N = 32{,}768$ and $\Delta x = \Delta y \cong 1$ [μm]. The parameters used for capturing is the same as those in Fig. 14.16

both cases. Note that the object fields used for the reconstructions are not the same as those in Fig. 14.16.[6]

14.4 Occlusion Processing Using the Silhouette Method

Digitized holography makes it possible to freely create and edit the 3D scene composed of physical objects and CG-modeled virtual objects. However, occlusion processing is necessary to arrange the real objects, more correctly, to arrange the object fields of the real objects in the 3D scene. The real objects may be reconstructed translucently unless the background light is shielded. Shielding the background field is not so difficult when using the silhouette method, described in Chap. 11.

14.4.1 The Silhouette Method Including Captured Object Fields

The silhouette method proposed for light-shielding of CG-modeled objects can also be applied to real objects. The principle of light-shielding for captured fields is shown in Fig. 14.19. The object field obtained by (14.20) is here rewritten as

$$g_{\text{obj},n}(\hat{x}, \hat{y}) \equiv g_{\text{obj}}(\hat{x}, \hat{y}; -d_{\text{R}}), \tag{14.37}$$

where $(\hat{x}, \hat{y})$ is the parallel local coordinate system for the captured object field indexed by n. This coordinate system is the same as that in the polygon-based method (see Fig. 10.6). Although $\hat{x}$ and $\hat{y}$ should have a suffix of n, it is omitted by the same reason as the polygon-based method.

The background field behind the captured object should be shielded over the cross section of the object. The background field is multiplied by binary mask $M_n(\hat{x}, \hat{y})$ that

[6]Perceptive readers may notice the handle of the cup turning to the opposite direction!

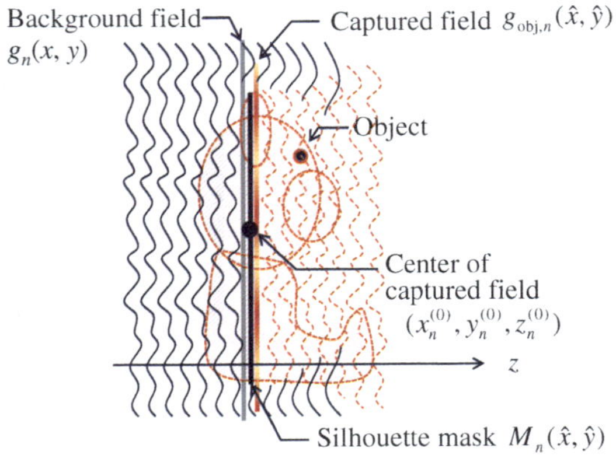

Fig. 14.19 Light shielding by the silhouette method for captured fields

corresponds to the silhouette of the captured object. The captured field $g_{obj,n}(\hat{x}, \hat{y})$ is then added to the masked background field. Therefore, the wavefield of the 3D scene is expressed by a recurrence formula as follows:

$$g_{n+1}(x, y) = \mathscr{P}_{n+1,n}\left\{M_n(x - x_n^{(0)}, y - y_n^{(0)})g_n(x, y) + g_{obj,n}(x - x_n^{(0)}, y - y_n^{(0)})\right\}, \tag{14.38}$$

where $g_n(x, y)(\equiv g_n(x, y; z_n^{(0)}))$ is a wavefield placed at $z = z_n^{(0)}$ in the global coordinate system. The object field $g_{obj,n}(\hat{x}, \hat{y})$ is embedded in the 3D scene so that the center is arranged at $(x_n^{(0)}, y_n^{(0)}, z_n^{(0)})$ of the global coordinates.

The above process is essentially the same as the O-O light shielding described in Sect. 11.2.2. Object field $g_{obj,n}(\hat{x}, \hat{y})$ can be provided by a CG model as well as a captured field. In this case, $g_{obj,n}(\hat{x}, \hat{y})$ can also be calculated by P-P light-shielding using the switch-back technique. Here, note that the object field of a CG model calculated by the switch-back technique should not be multiplied by mask function $M_n(\hat{x}, \hat{y})$. Light-shielding of the background field is properly processed by the switch-back technique in this case.

14.4.2 Making Silhouette Masks

Silhouette mask $M_n(\hat{x}, \hat{y})$ is required to shield the background field of the captured object. If the object is given by a CG model, this is very easy because we have the geometrical information. However, we usually do not have numerical data for the shape of a real object. It may be possible to produce the silhouette mask from a photograph of the subject. However, in the case, it is difficult to fit the position and size of the mask into the object field accurately. In addition, the picture may have distortion caused by a lens.

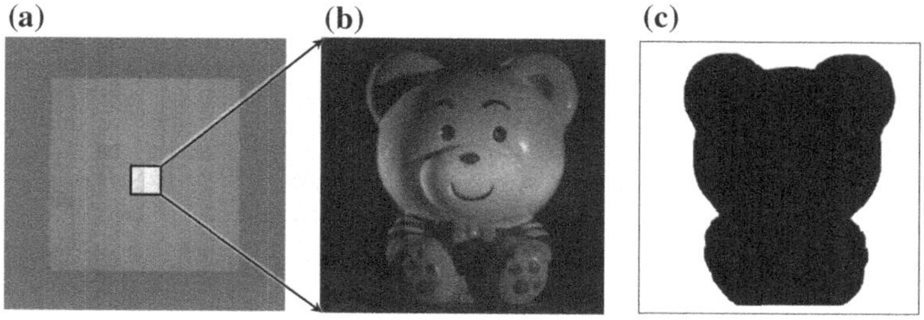

Fig. 14.20 The amplitude images of **a** a small complex field clipped from the whole field, and **b** the object field obtained by Fourier-transform of the clipped field. **c** The silhouette mask produced by painting regions black and white separately. All parameters are the same as those in Fig. 14.14

To avoid the problems of fitting and distortion, instead of the photograph, we can produce the silhouette mask from the amplitude image of the captured field by filling the inside and outside of the object with 0 and 1, respectively. However, as shown in Figs. 14.13b and 14.16b, the amplitude image of large-scale wavefields are commonly blurred because of the large ratio of the field size to object distance: W_h/d_{obj}, as mentioned in Sect. 13.2.1. Therefore, it is difficult to detect the exact outline of the object in the amplitude image.

The solution of the problem is very simple; we can control the DOF using an aperture, as described in Sect. 13.2.2. In practice, this can be simply achieved by clipping a small part of the captured complex field, as shown in Fig. 14.20a. Figure (b) shows the amplitude image of the object field obtained by the Fourier transform of the clipped small field in (a). Though the original complex field is the same as that in Fig. 14.14a, the amplitude image after the Fourier transform is much clearer than that of the whole field in Fig. 14.14b. Figure 14.20c shows the silhouette mask produced from the image in (b) by painting the inside and outside of the object black and white respectively.

It is easy to handle the mask image because the number of pixels of the amplitude image is reduced by clipping. In contrast, according to (14.23), the size of the Fourier-transformed object field is independent of the numbers of the sample points because the field size is given by

$$W_x = N_0 \Delta x = \frac{\lambda d_R}{\delta_x} \quad \text{and} \quad W_y = M_0 \Delta y = \frac{\lambda d_R}{\delta_y}. \tag{14.39}$$

This means that the mask image, e.g. Fig. 14.20c, preserves the same physical size as the object field. Therefore, even if the mask image has a very small number of pixels, we can easily fit the silhouette mask to the object field.

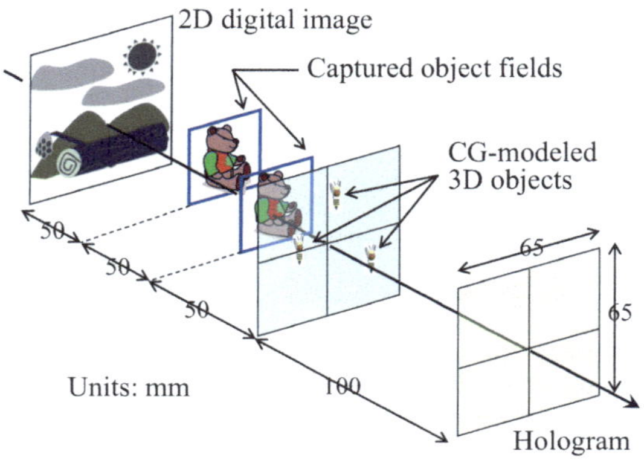

Fig. 14.21 Mixed 3D scene of HD-CGH "Bear II" [63]. The scene includes the captured object fields of the real object and the CG-modeled virtual objects

14.5 Examples of Optical Reconstruction

14.5.1 Monochrome CGH

A real object, a toy bear whose wavefield is shown in Fig. 14.14, is mixed with a virtual 3D scene. The design of the scene is shown in Fig. 14.21. Here, the bear appears twice in the scene, i.e., the same captured wavefield is used twice in the scene. Virtual objects such as the 2D wallpaper or 3D bees[7] are arranged behind or in front of the two bears. It is ensured by the silhouette method that the occluded relation is correctly reconstructed between the bears as well as between the objects behind and in front of the bears as if real objects are placed at the positions. This kind of editing of 3D scenes is almost impossible in classical holography. Only digitized holography allows us to edit the 3D scene.

The binary amplitude hologram was fabricated using a laser lithography system (see Sect. 15.3). The parameters used to create Bear II are summarized in Table 14.1. There are approximately 4 billion pixels for Bear II. Since the pixel pitches are $1\ \mu m \times 1\ \mu m$, the viewing angle is 37° both in the horizontal and vertical. Photographs and videos of the optical reconstruction of Bear II are shown in Figs. 14.22 and 14.23. Occlusion of the 3D scene is accurately reproduced in the optical reconstruction. It is verified by the movie of Fig. 14.23 that the appearance of the 3D scene varies as the point of view changes.

Black shadows that are not seen from the in-line viewpoint are visible from an off-axis view-point, as shown in Fig. 14.24.[8] This is most likely due to disagreement between the planes in which the object field is given and the object has the maximum cross section. As shown in Fig. 14.25, viewers see the silhouette mask itself in this

[7]The bees may look 2D objects rather than 3D objects because the model and its orientation are two dimensional. This is because we could not process self-occlusion of the model properly when Bear II was created. The switch-back technique was invented later.

[8]Left and right are swapped by mistake in the corresponding figure in [63].

Table 14.1 Summary of parameters used to create Bear II

Parameters	Value	Units
Number of pixels ($M \times N$)	65,536 × 65,536	
Pixel pitches ($\Delta x_h \times \Delta y_h$)	1.0 × 1.0	μm
CGH Size ($W_x \times W_y$)	65.5 × 65.5	mm
Reconstruction wavelength	633	nm
Dimension of wallpaper (W × H)	65.5 × 65.5	mm
Center position of far bear ($x_1^{(0)}, y_1^{(0)}, z_1^{(0)}$)	(15, −5, −200)	mm
Center position of near bear ($x_2^{(0)}, y_2^{(0)}, z_2^{(0)}$)	(0, −5, −150)	mm

Fig. 14.22 Photograph of the optical reconstruction of "Bear II" using reflected illumination of an ordinary red LED [63]. Video link, https://doi.org/10.1364/AO.50.00H278.m001

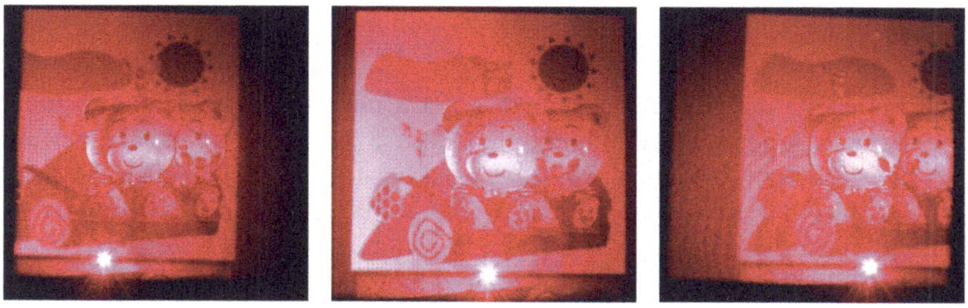

Left Center Right

Fig. 14.23 Photographs of the optical reconstruction of "Bear II" using transmitted illumination of a He-Ne laser [63]. Photographs are taken from different viewpoints. Video link, https://doi.org/10.1364/AO.50.00H278.m002

case; the background light cannot be seen, even though not hidden by the object. In this case, however, we can easily resolve the problem by numerically propagating the field at a short distance so that the field plane is exactly placed at the maximum cross section of the object.

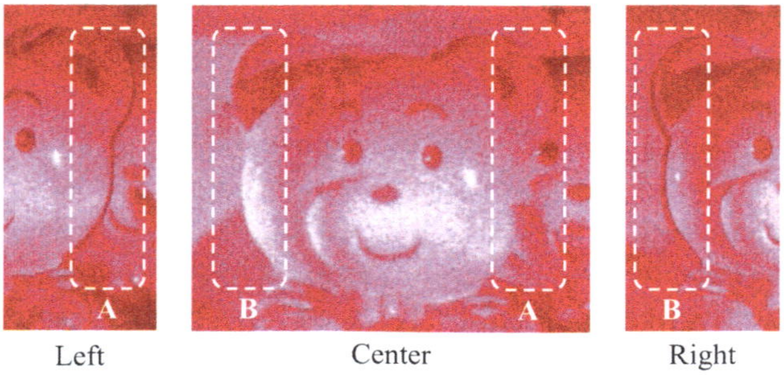

Fig. 14.24 Black shadows appearing in the reconstructed image because of occlusion errors in the case of off-axis viewpoints

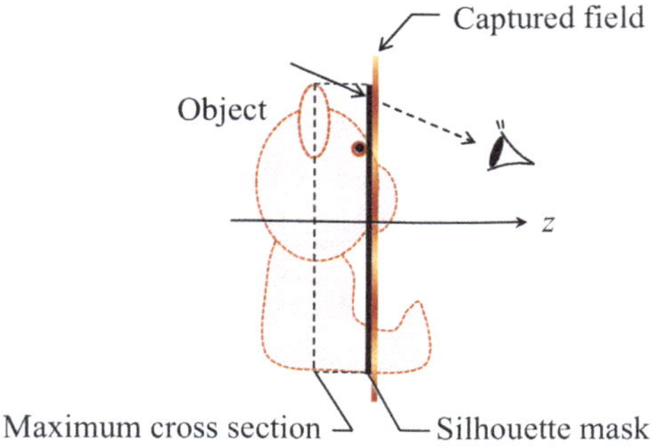

Fig. 14.25 Origin of the occlusion error in the cases where the field plane is not placed at the maximum cross section of the object

Unfortunately, the silhouette method is not a universally applicable technique for light-shielding in digitized holography. Because the silhouette method performs object-by-object light-shielding, silhouette masking does not work well in some cases where the object has severe self-occlusion or the silhouette shape of the object does not fit with the cross section.

Figure 14.26 shows the example of failure in occlusion processing. Because the captured object has a severe self-occlusion in this case, the silhouette light-shielding does not work well at off-axis viewpoints. Even in this case, we can reduce the occlusion error using multiple silhouette masks that are generated from different viewpoints. Silhouette masks for different viewpoints can be produced by changing the field clipping area correspondingly to the viewpoint. This is basically the same technique as that described in Sect. 13.2.3. Applying the silhouette masks to the field corresponding to the viewpoint and combining them, we can reduce the problem [14].

Fig. 14.26 An example of failure in occlusion processing by the silhouette light-shielding [14]

Fig. 14.27 **a** The mixed 3D scene composed of physical and nonphysical objects of a full-color HD-CGH named "Tea Time" [80], and the models of nonphysical objects: **b** polygon-meshed objects, and **c** digital image used to provide the background

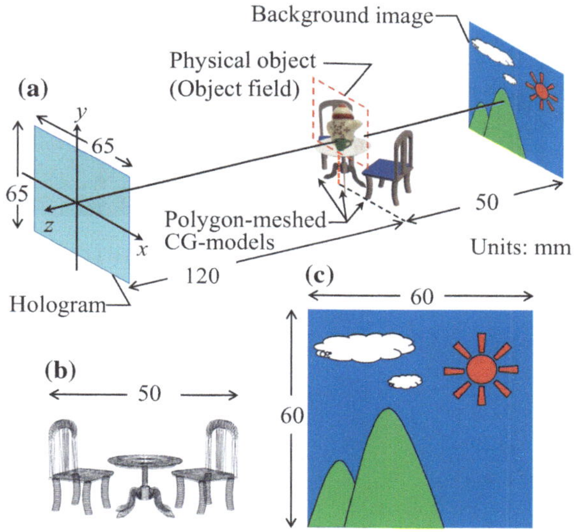

14.5.2 Full-Color CGH

A full-color CGH named "Tea Time" was created from the object fields captured at three wavelengths, shown in Fig. 14.16. Figure 14.27a shows the 3D scene. Two non-physical objects are included within the scene: a table set by polygon-meshed CG models in (b), and a digital image in (c), which is used for the background. The table set is composed of 8,328 polygons, while the background image is composed of 512×512 pixels. Here, the table set has a complex shape and thus has complex self-occlusions. The object field was therefore calculated using the switch-back technique providing P-P silhouette shielding.

The full-color HD-CGH of the mixed 3D scene was fabricated by the technique of RGB color filters described in Sect. 15.6. Here, the stripe width and the guard gap width of the RGB color filters are 80 μm and 20 μm, respectively. Figure 14.28 shows

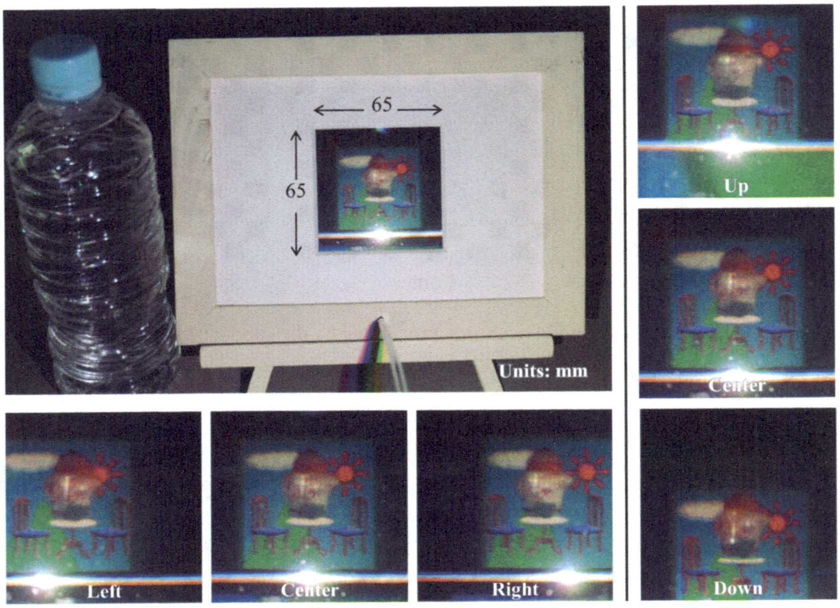

Fig. 14.28 Optical reconstruction of full-color HD-CGH "Tea Time". The pictures were taken from different angles. Video link, https://doi.org/10.6084/m9.figshare.5208961.v1

Table 14.2 Summary of parameters used to create Tea Time

Parameters	Value	Units
Number of pixels ($M \times N$)	65,536 × 65,536	
Pixel pitches ($\Delta x_h \times \Delta y_h$)	1.0 × 1.0	μm
CGH Size ($W_x \times W_y$)	65.5 × 65.5	mm
Wavelength (R, G, B)	633, 532, 488	nm
Viewing angles (R, G, B)	37, 31, 28	°
Stripe width of RGB color filters[a]	80	μm
Guard gap width[a]	20	μm

[a]See Sect. 15.6

optical reconstructions of the CGH. A multi-chip white LED is used to illuminate the CGHs. The CGH parameters are summarized in Table 14.2. The actual color of the physical object is reproduced to a certain extent. The continuous motion parallax of the physical and non-physical objects is verified in the pictures and the movie.

14.6 Resizing Object Image

Digitized holography does not only digitize the whole processes of classic holography but also allows us to digitally edit the 3D scene after recording the hologram. We can freely arrange multiple copies of a physical object in the mixed 3D scene unlike classical holography. Resizing the object image is one of digital-editings possible in digitized holography.

14.6.1 *Resizing by Change of Sampling Intervals*

The simplest technique to resize an object image is to change the sampling interval of the captured object field. Let $g_{\text{obj}}(x_m, y_n)$ be a sampled object field with sampling intervals of Δx and Δy. When the number of samplings is $M \times N$, the size of the sampling window is $W_x \times W_y = M\Delta x \times N\Delta y$ properly. Replacing Δx and Δy by $m_{\text{mag}}\Delta x$ and $m_{\text{mag}}\Delta y$ respectively, the window size is $m_{\text{mag}}W_x \times m_{\text{mag}}W_y$, and the object image is simply magnified m_{mag} times. The expanded object field is represented by $g_{\text{obj}}(x_m/m_{\text{mag}}, y_n/m_{\text{mag}})$ in this case.

Here, note that the sampling intervals of the resized field must agree with the object field of the CGH in order to superimpose it on the whole object field and perform light-shielding by the silhouette method in (14.38). This means that the resized field must be resampled using some interpolation method unless m_{mag} is the inverse of an integer. [9]

Figure 14.29a shows one-dimensional spectra $G_{\text{obj}}(u_p)(=\textbf{FFT}\left\{g_{\text{obj}}(x_m)\right\})$ of a sampled object field, where u is again the spatial frequency. When we change sampling interval Δx to $m_{\text{mag}}\Delta x$, the spatial bandwidth expands in the cases where $m_{\text{mag}} < 1$ and contracts in $m_{\text{mag}} > 1$ as in (b), according to the similarity theorem in (4.7). In both cases, when the spectrum $G_{\text{obj}}(u_m m_{\text{mag}})$ is resampled to recover the sampling interval to Δx, information of the object field is more or less lost inevitably, as shown in (c). Above all, accuracy of the scaled object fields depends on the interpolation method used for resampling.

14.6.2 *Resizing by Virtual Imaging*

The virtual imaging introduced in Chap. 13 is also a feasible technique to resize object images [15]. The simplest technique is to use image formation by a single lens. The virtual optical system in this technique are shown in Fig. 14.30a. As with real image formation by a thin lens, magnification is given by $m_{\text{mag}} = d_2/d_1$. This type of virtual imaging, however, requires a large computational effort in general. Assume that lens diameter D_{p} is large enough to transmit the whole bandwidth of

[9] In this case, resampling can be performed by just thinning out sample points.

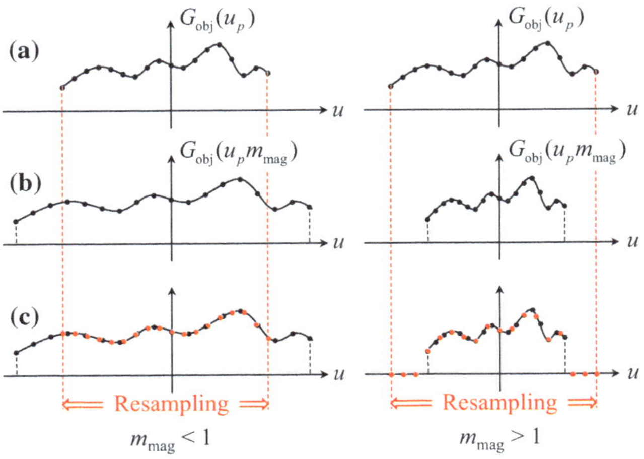

Fig. 14.29 Spectra of **a** sampled object fields, and **b** resized object fields. **c** Resampling the resized object fields

the captured object field. The viewing angle in this case is given by $2\theta_c$ indicated in Fig. 14.30a:

$$\tan\theta_c = \frac{D_p}{2d_2} = \frac{1}{2(m_{mag}+1)}\frac{D_p}{f}, \tag{14.40}$$

where the thin lens formula in (5.49) is used. According to (13.11) and (13.12), relation $D_p/f < \lambda/\Delta x$ must be satisfied to avoid aliasing errors of the lens function. As a result, the sampling interval of the lens function must satisfy

$$\Delta x < \frac{\lambda}{2(m_{mag}+1)\tan\theta_c}. \tag{14.41}$$

Thus, the lager the magnification is, the smaller the sampling interval is. If it is required that angle θ_c is equal to or larger than the maximum diffraction: $\theta_c \geq \theta_{max}$, resizing by a single virtual lens usually requires considerable computational effort.

Another technique based on image formation is use of a dual lens instead of single lens and to construct virtual afocal system, as shown in Fig. 14.30b. In this case,

$$\tan\theta_c = \frac{D_p}{2f_2} < \frac{\lambda}{\Delta x}. \tag{14.42}$$

Therefore,

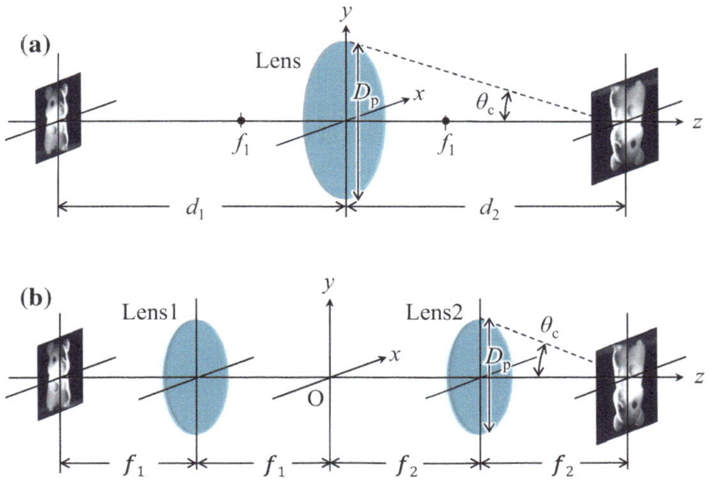

Fig. 14.30 Resizing by virtual imaging with **a** a single and **b** dual lens

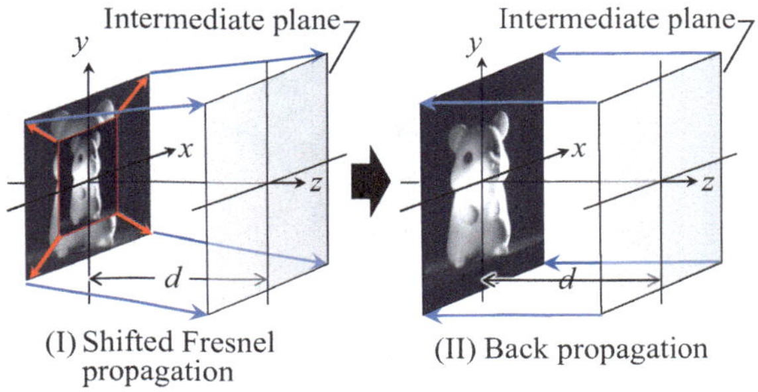

Fig. 14.31 Resizing by use of the shifted Fresnel propagation

$$\Delta x < \frac{\lambda}{2 \tan \theta_c}. \tag{14.43}$$

Thus, the sampling interval is independent of magnification. As a result, load of memory in this technique is lighter than that in the single lens configuration in many cases. However, resizing by a dual lens needs three times propagation, whereas resizing by a single lens is performed by twice propagation. Thus, resizing by a dual lens is possibly more time-consuming in some cases.

14.6.3 Resizing by Shifted Fresnel Propagation

The third technique for resizing is use of a numerical propagation method that features variable sampling intervals, i.e., the shifted Fresnel propagation described in

Sect. 12.4. In this technique, the sampling interval of the captured field is simply changed without changing the number samples. This is the same manner as that in Sect. 14.6.1. However, instead to recover the original sampling interval using an interpolation method, the scaled object field is propagated to an intermediate plane using the shifted Fresnel method, as shown in the step (I) of Fig. 14.31. Here, we set the sampling intervals in the intermediate plane to the same values as those of the original object field.

The object field in the intermediate plane is then backward propagated to the original plane. This is because we need the object field in the plane that intersects the object in order to perform light-shielding by the silhouette mask. An ordinary method that does not change the sampling interval, e.g. the BLAS method, is used for the back propagation, as shown in the step (II) of Fig. 14.31.

Distance d between the original and intermediate planes is determined so that aliasing errors do not occur in the shifted Fresnel propagation. The sampling window size $m_{\text{mag}} W_{sx} \times m_{\text{mag}} W_{sy}$ of the resized field must be equal to or smaller than the cell size of the destination field: $W_{\text{far},x} \times W_{\text{far},y}$ in (12.37) to avoid aliasing errors. Thus, the propagation distance must satisfy

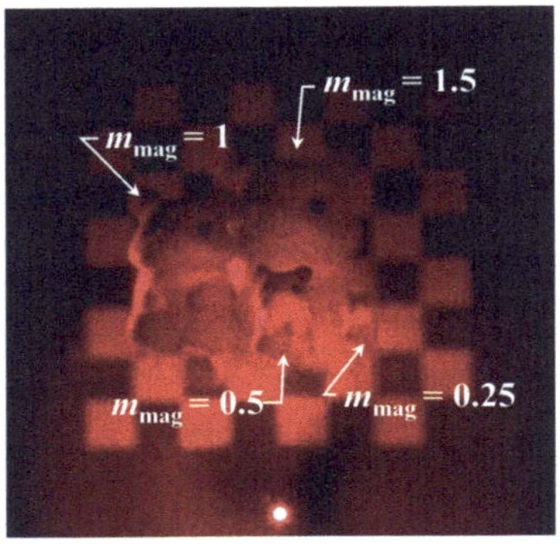

Center

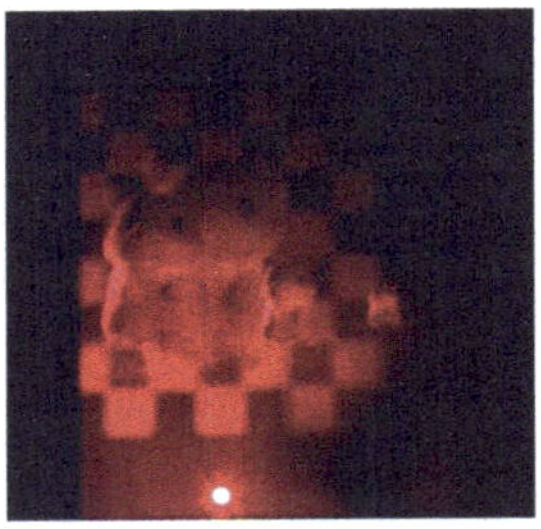

Left

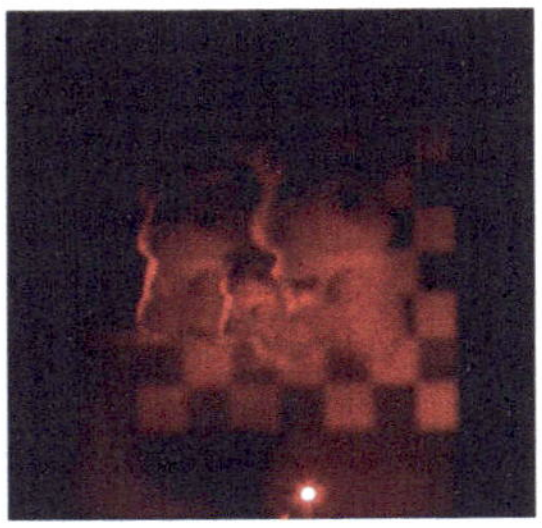

Right

Fig. 14.32 Photographs of optical reconstruction of HD-CGH “Hamsters” [15]. The pictures are taken from different viewpoints. $\lambda = 532$ [nm] (reconstruction wavelength: 633 nm), $M = N = 65{,}536$, and $\Delta x = \Delta y = 1$ [μm]. Video link, https://youtu.be/-TdA-RgMBQM

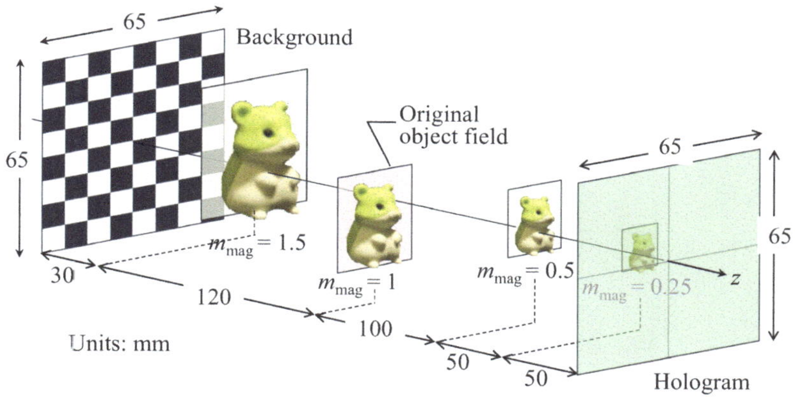

Fig. 14.33 The 3D scene of HD-CGH "Hamsters". Three resized object fields and an original captured field are arranged in the 3D scene

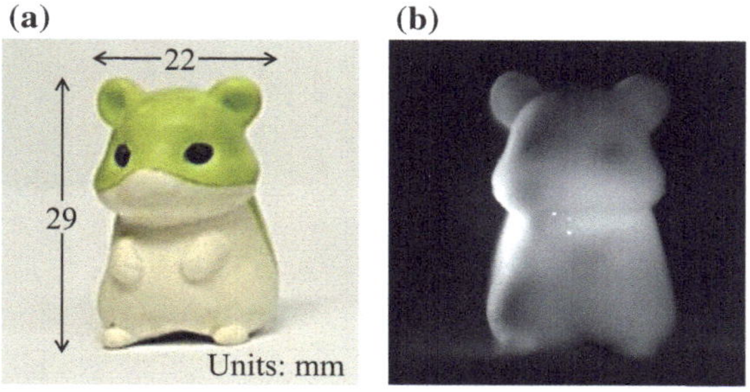

Fig. 14.34 **a** A photograph of the subject, whose object fields are embedded in the 3D scene of "Hamsters", and **b** the amplitude image of the object field captured by lensless-Fourier synthetic aperture DH [14]. $\lambda = 532$ [nm], $M_0 = N_0 = 32{,}768$, $\Delta x = \Delta y = 1$ [μm], and $d_R = 215$ [mm]

$$d \geq m_{\text{mag}} W_{sx} \frac{\Delta x}{\lambda} \quad \text{and} \quad d \geq m_{\text{mag}} W_{sy} \frac{\Delta y}{\lambda}. \tag{14.44}$$

In addition, note that the destination field in the intermediate plane needs the sampling window that at least covers the whole field propagated from the resized field having the original sampling interval.[10]

Figure 14.32 shows optical reconstruction of a HD-CGH named "Hamsters". The 3D scene of this HD-CGH is composed of resized object fields, as shown in Fig. 14.33. The original object field was captured by the technique described in the previous sections and resized using the shifted Fresnel propagation. The original captured field and subject are shown in Fig. 14.34.

[10]This means that we assume the resized field has the same sampling interval as that in the original field and obtain the maximum diffraction area at the position of the intermediate plane.

Chapter 15
Fabrication of High-Definition CGH

Abstract One of the most difficult problems in computer holography is to fabricate actual high-definition CGHs, i.e., to print the synthetic fringe pattern. In this chapter, we focus on the topics of the fabrication problem and discuss several techniques: home-made fringe printers, laser lithography, and wavefront printers. In particular, we devote pages to the technique of laser lithography because it is one of the best technique to fabricate HD-CGHs. The last three sections in this chapter are devoted to fabrication of full-color HD-CGHs. Three techniques are discussed in this chapter.

15.1 Introduction

One of the biggest problem in computer holography is, without any doubt, how to print synthetic fringe patterns. As mentioned in Sect. 8.2, the viewing angle of a CGH is determined by the pixel pitch of the printed fringe image. If using some optics such as lenses, we can enhance the viewing angle of a CGH properly; we can achieve a large viewing angle even in a low resolution fringe. However, as the CGH size increases, the optics is inevitably bulkier. As a result, the optics most likely makes it difficult to carry and display the CGH in practice. Above all, the viewers do not gain a deep impression from the CGHs installed in a rich optics in general. Only the 3D image reconstructed from a thin plate can give a sense of wonder to the viewers in holography. Accordingly, printing equipment should first of all have ability to print minute pixels.

Table 15.1 summarizes feasible printing equipment available in the market. It may be surprising that ordinary laser printers can be used for printing the synthetic fringe of CGHs [48]. In this case, the fringe pattern is commonly printed on a transparent film instead of a piece of paper. We surely see the 3D image from the printed fringe image using a transmission light source, but the viewing angle is only several degrees. We can also use an image setter to print CGH fringes. An image setter is a kind of laser printer that is used in print offices. Although this is an expensive machine, some print shops provide printing services by their image setter. We can print our fringes on a transparent film by sending the fringe data to the print shop in this case. Even in

K. Matsushima, *Introduction to Computer Holography*,
Series in Display Science and Technology,
https://doi.org/10.1007/978-3-030-38435-7_15

Table 15.1 Feasible equipments to print fringe patterns

	Resolution [dpi]	Pixel pitch [μm]	Viewing angle[a] [°]
Laser printer	$\lesssim$1,200	$\gtrsim$21	$\lesssim$1.7
Image setter	$\lesssim$4,000	$\gtrsim$6.4	$\lesssim$5.8
Laser lithography	$\gtrsim 2.5 \times 10^4$	$\lesssim$1	$\gtrsim$37
e-beam lithography	$\gtrsim 2.5 \times 10^6$	$\lesssim$0.01	~180

[a]@ 633 nm

this professional printer, however, the viewing angle is a single-digit value in units of degree.

Electron beam lithography listed at the bottom of Table 15.1 is definitely the best way to print CGH fringes from the viewpoint of resolution [21, 22, 57, 147]. The e-beam technology provides extremely high resolution to patterning. However, the equipment is very expensive especially for printing a large area, and is commonly too slow to print large-scale CGHs. In addition to these, e-beam lithography requires a vacuum chamber for printing. The laser lithography has moderate ability to print fringes, i.e., a balance between the speed and resolution, and can be handled in the air [122]. Though this is also an expensive machine, almost all HD-CGHs introduced in this book is printed using laser lithography.[1] Therefore, we devote Sect. 15.3 to describe the technique briefly.

The fringe pattern printed through both laser and e-beam lithography is made of metal films in many cases. The metallic fringes give a great advantage that the printed thin hologram can be reconstructed by reflected illuminations. We make use of this advantage to reconstruct full-color 3D images using RGB color filters (see Sect. 15.6) and to produce computer-generated volume holograms (CGVH) using contact-copy (see Sect. 15.7).

The readers may find out there is a small gap between the image setter and laser lithography[2] in Table 15.1. The gap is of the resolution as well as the price. Thus, there is a room to develop a special printer for CGHs by ourselves, as mentioned in the next section. In the end of introduction to printing fringes, we have to mention a new type of CGH printer, often called a *wavefront printer*. This type of printer seems to have the same structure as holographic printers based on direct-write digital holography [6]. However, the hologram printed by a wavefront printer reconstructs synthetic object fields unlike the holographic printers, which only print holographic stereograms. The wavefront printer is briefly explained in Sect. 15.4.

The last several sections in this chapter are devoted to fabrication of full-color CGHs: full-color reconstruction using dichroic mirrors (Sect. 15.5), RGB color filters (Sect. 15.6), and stacked CGVH (Sect. 15.7).

[1]The author beliefs that the laser lithography is one of the best technique to fabricate HD-CGHs at the current stage.

[2]An unusual fringe printer that uses a CD-R writer and may bridge the gap is also reported [106].

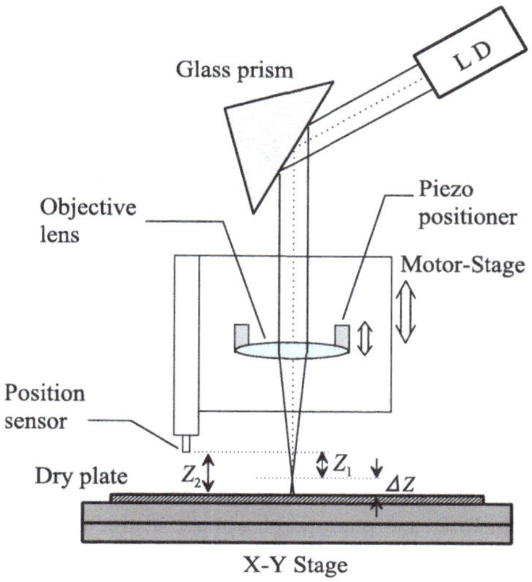

Fig. 15.1 An example of the structure of a spot-scanning fringe printer [136]

15.2 Fringe Printers

It is possible to construct a fringe printer by ourselves in order to fabricate thin transmission holograms [145]. This type of home-made fringe printers can be broadly classified into two types: a spot-scanning printer and image-tilling printer. The former printer raster-scans a laser spot to draw the fringe pattern like ordinary laser printers and laser writers used in laser lithography, while the latter printer draw the fringe pattern by projecting and tilling partial fringe images displayed on an LCD panel.

15.2.1 Spot-Scanning Fringe Printer

The author developed a spot-scanning type fringe printer to print synthetic fringes in the early first decade of 2000s [64, 136]. This is an example of the home-made fringe printer. Figure 15.1 shows the schematic structure. This fringe printer is a kind of laser printer. The output of a red laser diode (LD) is focused on a photo-sensitive material using a microscope objective whose N.A. is 0.55. The focusing spot makes a small black dot in the material after development. Because we use a holographic plate for the recording material, the sensitivity is so high that the beam is reflected by the surface of a glass prism to reduce the beam power but keep the wavefront and coherency. Here, the surface of the glass prism is polished but does not have anti-reflection coating. The holographic plate is fixed on a precision X-Y stage to raster-scan the laser spot. The objective lens is installed on another motor stage to change the vertical position roughly. A picture of the constructed fringe printer is shown in Fig. 15.2a.

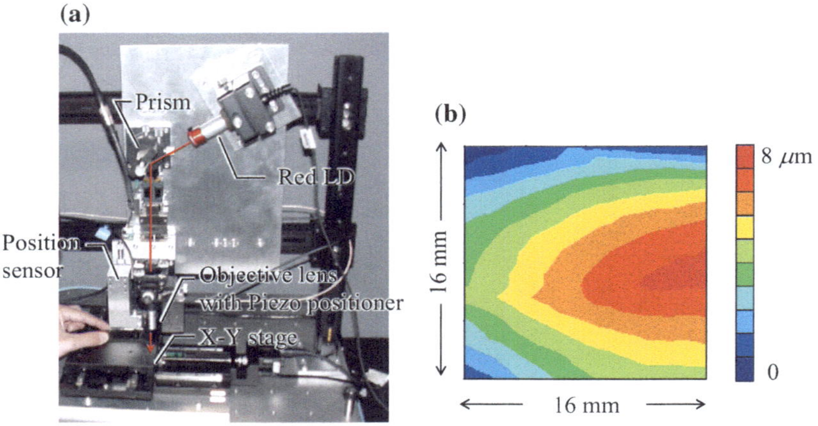

Fig. 15.2 **a** A photograph of the constructed fringe printer, and **b** an example of the contour map of a holographic plate. Holotrend PFG-03c

The material surface is not a perfect plane and sometimes tilts slightly. Therefore, this home-made fringe printer had a mechanism to keep the focusing spot on the material surface during the stage scan. The undulation and tilt of the material surface was measured using the position sensor before printing. An example of the contour map of the material surface is shown in Fig. 15.2b. The position of the objective lens was dynamically adjusted in printing using a piezoelectric actuator to compensate for the measured elevation of the material surface.

Figure 15.3b shows the photomicrograph of a point sequence printed by the fringe printer. Here, the interval of the printed dots is 2 μm. We estimated the size of a single dot at approximately 1.5 μm × 3.0 μm.[3] A microphotograph of the test pattern printed with $\Delta x = 1.5$ [μm] and $\Delta y = 3.0$ [μm] is shown in Fig. 15.3a. It is verified that the test image is printed with gray levels. This multi-level printing is realized by modulation of the LD current.

Figure 15.4 is optical reconstruction of a CGH printed by the fringe printer. The object field of the CGH is calculated by the point-based method with a linear density of 5,000 points/m. The printing speed was 200 mm/s in linear velocity. Here, note that the fringe pattern is printed with binary levels, though the printer had ability to print grayscale fringes as mentioned above. This was because binary fringe pattern gave better 3D images than that by grayscale fringes in this printer. This might be due to the problem of normalization of the fringe intensity mentioned in Sect. 8.6.

The above home-made printer drew the fringe pattern dot-by-dot. This type of printer has a simple structure but is commonly too slow to print large-scale fringes.

---

[3]A printed dot is not a perfect circle because the LD used for the light source does not output a circular beam. In this printer, elliptical dots are intentionally printed to reduce the number of dots along the vertical direction and decrease the printing time.

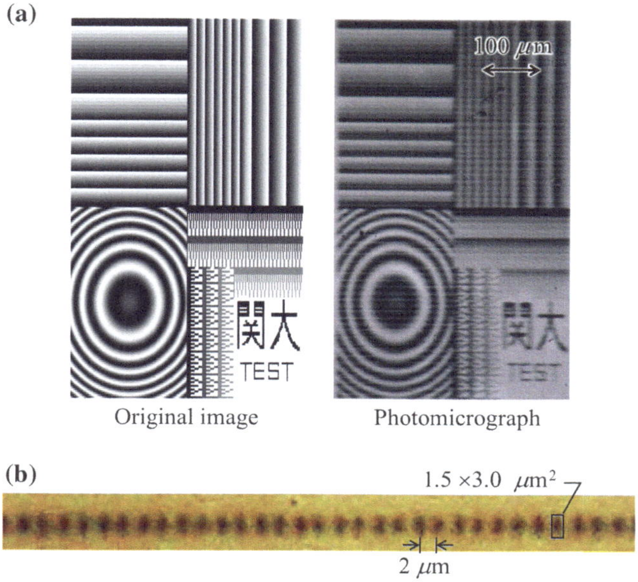

Fig. 15.3 Photomicrographs of **a** a test pattern ($M = 240$, $N = 200$, $\Delta x_h = 1.5$ [μm], and $\Delta y_h = 3.0$ [μm]), and **b** a dot sequence printed by the dot-scanning fringe printer

Fig. 15.4 Optical reconstruction of a CGH printed by the home-made fringe printer. $\lambda = 633$ [nm], $M = 8{,}129$, $N = 4{,}096$, $\Delta x_h = 1.5$ [μm], $\Delta y_h = 3.0$ [μm], $W_{obj} = 6$ [mm], and $d_{obj} = 50$ [mm]

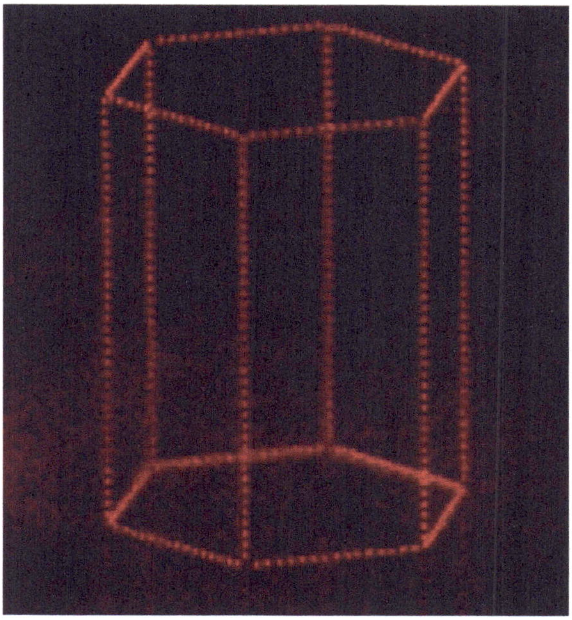

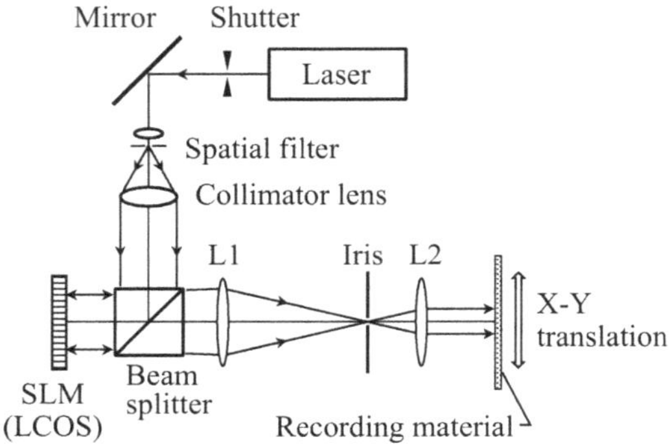

Fig. 15.5 A typical example of the structure of an image-tilling fringe printer

A fringe printer using a rotation drum instead of the X-Y stage for scanning is also proposed to speed up the spot-scanning [65].[4]

15.2.2 Image-Tilling Fringe Printer

In this type of fringe printer, the fringe pattern is printed not dot-by-dot; partial fringe images displayed on a spatial light modulator (SLM) are demagnified and projected on a photo-sensitive material. The whole fringe image is recorded by tilling the fractional fringes. Because millions pixels comprising the partial fringe image are simultaneously printed on the material, the printing time is commonly much shorter than spot-scanning fringe printers.[5]

Figure 15.5 shows a typical setup of the image-tilling type fringe printer proposed by Yoshikawa et al. [142–145].[6] A part of the fringe image is displayed on an SLM, which is the device that spatially modulates the incident beam of light. A liquid crystal display (LCD) represents a typical example of the SLM. An SLM usually modulates intensity of the incident light. It is also possible to modulate only the phase of the incident light. This type of SLM is often called a phase-only SLM. Liquid crystal on silicon (LCOS) is commonly used for the SLM in optical experiments because the pixel size of an LCOS is very small in general; thus the incident light can be modulated at considerably high spatial frequencies.

---

[4]Using a rotation drum, spot-scanning can be speeded up. However, because a film-base material must be used instead of the glass-base photographic plate in this case, it is difficult to exactly adhere the film material to the drum.

[5]Note that "image-tilling fringe printer" is a word coined by the author.

[6]Note that the setup in Fig. 15.5 is slightly changed from the original setup Yoshikawa et al. proposed, because they use a combination of polarization beam splitter and polarizer-less LCOS in order to reduce power loss caused in the beam splitter to the utmost limit. However, unfortunately, polarizer-less LCOS is not an ordinary device.

Fig. 15.6 A photograph of optical reconstruction of the rainbow hologram printed by an image-tiling fringe printer [144, 145] (From IEEE Transactions on Industrial Informatics 12(4), 1584(2016). By courtesy of Prof. Yoshikawa and Prof. Yamaguchi, Nippon University.)

Lenses L1 and L2 compose a telecentric optical system in the example of Fig. 15.5. The fringe image displayed on the SLM is demagnified by the telecentric lenses. By use of this technique, the pixel pitches of the printed fringe pattern is reduced without reducing the field of depth (DOF) so much. As a result, any active focus-control is not necessary in this type of fringe printer. Yoshikawa et al. report the minimum pixel pitch of 0.44 μm and the maximum printing speed of 3×10^9 pixels/hour in their image-tilling fringe printer using the telecentric optical system [146].

Figure 15.6 is an example of optical reconstruction of a rainbow hologram printed by the image-tilling type fringe printer [145]. Here, although the printed CGH is a thin transmission hologram, the hologram reconstructs multi-color images because of rainbow holography [4].[7]

15.3 Laser Lithography

Photolithography is a process used in microfabrication such as fabrication of large-scale integrated circuit (LSI) or flat panel display (FPD). In the process, a geometric pattern is transferred from a *photomask* to a photosensitive chemical *photoresist* coated on the product substrate, as shown in Fig. 15.7a. Here, *laser lithography* is

---

[7]Rainbow holography is a technique to reconstruct thin holograms by a white-light illumination at the sacrifice of vertical parallax. Thus, the hologram is a kind of HPO (horizontal-parallax-only) CGH (see Sect. 8.4).

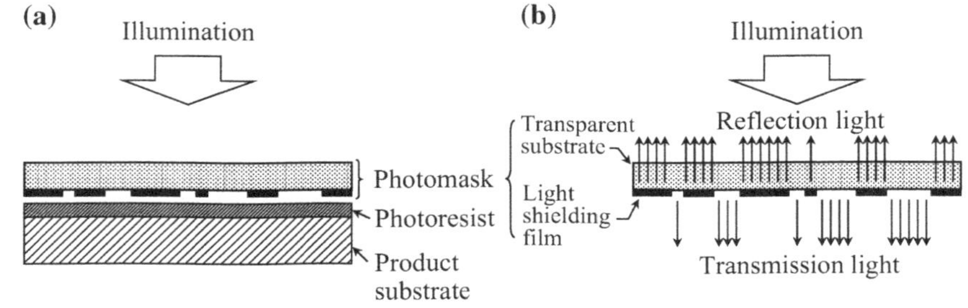

Fig. 15.7 **a** Usage of a photomask in microfabrication, and **b** the structure of a photomask

a sort of photolithography that uses a laser to make photomasks or to directly draw patterns on the photoresist.

15.3.1 Photomasks as a Binary-Amplitude CGH

The photomask in Fig. 15.7a is an optical component used in photolithography. A photomask has a role of the template of the geometric pattern transferred to the substrate surface. Drawing the fringe pattern of a CGH as the geometric pattern on a photomask, we can divert the photomask to the CGH that is encoded with binary-amplitude (see Sect. 8.6).

A photomask is basically composed of a transparent substrate and layers of light shielding films as in Fig. 15.7b. The light shielding films is made of metal thin films in ordinary photomasks to produce high-contrast transmission light. Therefore, the reflective contrast of the light shielding films is commonly as high as the transmission contrast, i.e., illumination light is highly reflected by the light shielding films.

The reflection light properly has inverse intensities to the transmission light. However, because intensity-inverted fringes reconstruct exactly the same 3D image as non-inverted fringes,[8] the CGH produced on a photomask can be reconstructed by reflection illumination as well as transmission illumination, even though the CGH is a thin hologram that is also called transmission hologram, as mentioned in Sect. 7.2. This property is one of the greatest advantage of the CGH made of a photomask.

In actual exhibition of holograms, reflection illumination drastically simplifies the setup of the illuminating light source. In addition, contact-copy is possible in reflection holograms. The technique of contact-copy is actually used to transfer a CGH to the volume hologram, as mentioned in Sect. 15.7.

[8]Inversion of a fringe pattern corresponds to taking '$-$' sign in (7.10). The major conclusions drawn in Chap. 7 do not change even in this case.

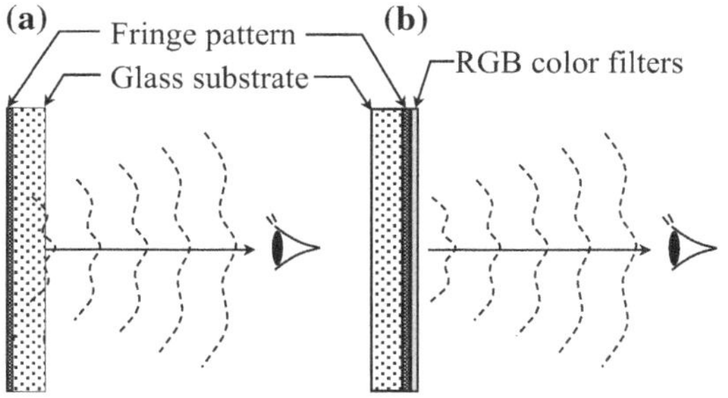

Fig. 15.8 Fringe patterns made of chromium films in **a** monochrome CGHs, and **b** full-color CGHs using RGB color filters

15.3.2 Structure of Photomasks

Both glass and film substrates are used for the transparent substrate correspondingly to the requirement of pattern resolution in manufacturing industries. Since the resolution achieved with a film substrate is roughly more than 50 μm, we have to use a glass substrate to fabricate HD-CGHs whose pixel pitches are at least less than 1 μm. There are synthetic quartz and soda-lime glass used for the transparent substrate in the industry. The former features a low thermal coefficient and is thus suited for high-resolution printing, but is more expensive than the latter. Because we do not need as high accuracy as microfabrication in the electronics industry needs, lime-soda glass is sufficient to print HD-CGHs in many cases.

Although both emulsion and metal are used for the light-shielding film in the industry, we need a metallic film because of its high reflectivity and resolution, as mentioned above. In general, chromium is used for the metal film because of its high opaqueness and chemical resistance. Optical transmittance of a chromium thin film is approximately less than 0.1% in a thickness of 100 nm. In many cases, a CrO layer is added between the chromium film and glass substrate to improve adhesion. A thick CrO layer, however, reduces the reflectivity of the chromium film.

In some cases, a CrO layer is made between the chromium film and the air in order to protect the chromium surface and reduce unnecessary reflection in the surface. In monochrome HD-CGHs, the 3D image is seen from the glass substrate side using reflection illumination as shown in Fig. 15.8a. This is mainly due to protection of the chromium film. In this case, the backside CrO layer does not cause any problem. However, the backside CrO film should not be made on full-color HD-CGHs using RGB color filters because this type of full-color CGH uses the backside reflection as in (b) in order to improve adhesion between the fringe pattern and color filters, as described in Sect. 15.6.

Unprocessed photomasks are called *mask blanks* and available in the market. Photoresist-coated mask blanks are also supplied from makers.[9] Examples of the specification of mask blanks that the author actually ordered to print HD-CGHs are

[9]The price of a mask blank that the author could get in Japan was roughly US$30 for a 5 inches photomask in 2019.

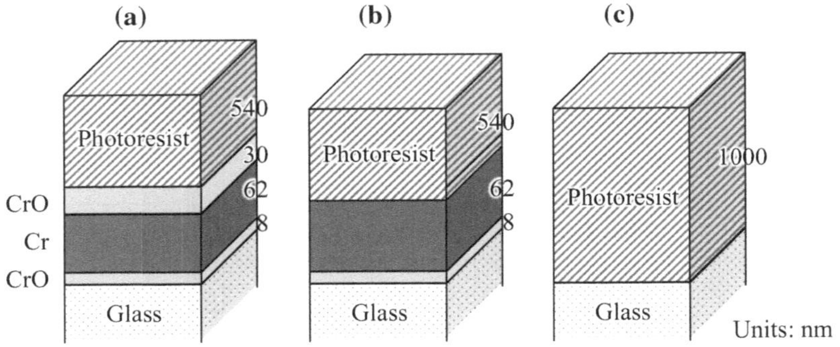

Fig. 15.9 Examples of actual mask blanks for **a** monochrome reflection/transmission binary-amplitude CGHs, **b** full-color CGHs using RGB color filters and reflection phase CGHs, and **c** transmission phase CGHs. Figures indicate a thickness of the layer in units of nano meters. The thickness of the glass substrates is several millimeters

shown in Fig. 15.9. Mask blanks in (a) are used to fabricate monochrome binary-amplitude CGHs, which are reconstructed with both reflection and transmission illuminations, as mentioned above. In this case, the metallic fringe pattern is arranged in the backside of the glass substrate, as shown in Fig. 15.8a. According to the technical data sheet, the reflectivity and optical density[10] of the Cr film whose thickness is 62 nm is approximately 45% and 3.0, respectively.

In contrast to monochrome CGHs, mask blanks for full-color CGHs using RGB color filters omit the CrO layer between the photoresist and Cr film to enhance reflectivity of the Cr-made fringe, as in Fig. 15.9b. This is because fringe patterns are arranged in front of the glass substrate in this type of CGH, as mentioned above. According to the technical data sheet, the reflectivity of the Cr film is increased from 10 to 55% by removing the 30 nm CrO layer.

As mentioned in the following section, phase CGHs can be produced by forming a surface relief made of the photoresist on the substrate. In this case, the mask blanks in Fig. 15.9b and c are used to print reflection and transmission CGHs, respectively.

15.3.3 Process to Fabricate Photomasks

The process for making photomasks is very similar to microfabrication using the photomask itself. Figure 15.10 illustrates the process. First, a pattern is drawn on the photoresist using electron beams in e-beam lithography or laser beams in laser lithography. Unnecessary portion of the photoresist is then removed by a special solution called *developer*. When exposed portion is soluble in the developer, the

[10]The optical density is given by $\log(I_0/I)$, where I and I_0 are optical intensities of light passing through a portion with and without the light-shielding film, respectively.

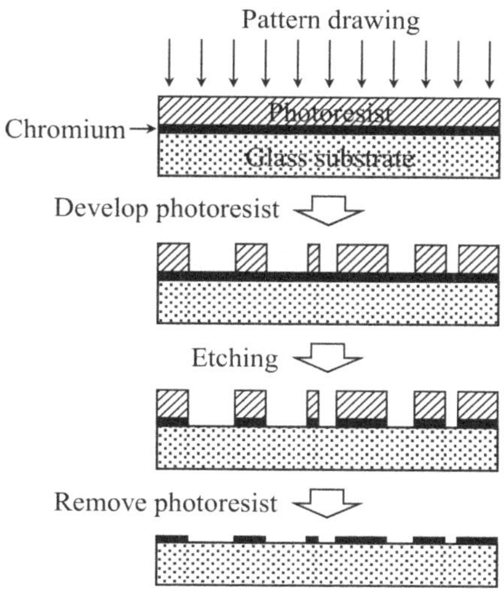

Fig. 15.10 The process for making photomasks whose light-shielding film and substrate are made of metal and glass, respectively

photoresist is called *positive* or *positive-tone* photoresist. Conversely, if unexposed portion is soluble in the developer the photoresist called *negative* or *negative-tone* photoresist.

Print of CGHs has very different properties from fabrication of photomasks used for microfabrication in the electronics industry. In photomasks for microfabrication, any defect is not allowed in general because a small defect directly leads to defective products. However, since information of an object field is recorded in a wide area of the fringe pattern in holography, a small defect of the fringe pattern commonly does not affect the 3D image. Thus, we do not need as high precision and cleanliness factor as microfabrication in the manufacturing industry needs. For example, we can develop the photoresist using a similar simple manner to that in photography, as explained in the following section. Also, although post baking is performed after development in many cases in the industry, we can commonly skip the process without severe problems.

To fabricate amplitude CGHs, a part of chromium film must be removed by etching process. Techniques of dry etching such as reactive ion etching (RIE) are usually used for the removal in the industry. However, wet etching whose apparatus is considerably cheaper than that of dry etching is generally sufficient to the removal in making CGHs. As mentioned in the following section, there is no etching process to fabricate phase CGHs.

The remaining photoresist is generally removed after etching in the industry. However, because the reflected field is not effected by the photoresist remaining in the backside of the CGH as in Fig. 15.8a, the process can be omitted in monochrome reflection CGHs in general. In contrast, when fabricating full-color CGHs using RGB color filters, the remaining photoresist should be removed because it reduces reflectivity of the chromium film and adhesiveness of color filters.

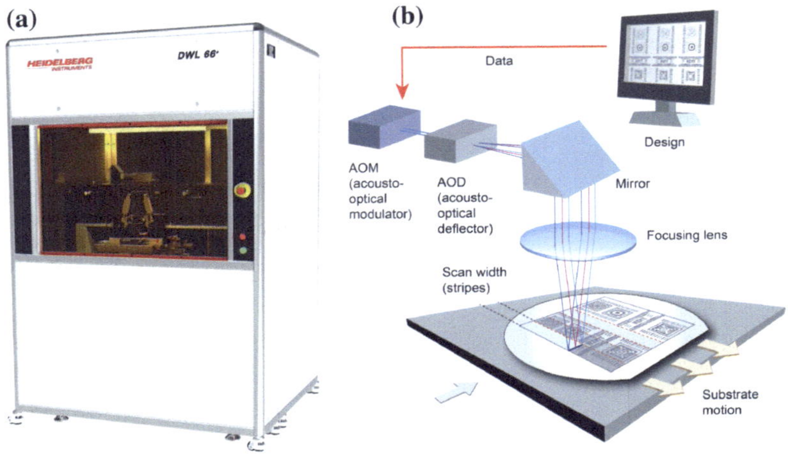

Fig. 15.11 Heidelberg Instruments DWL 66$^+$ laser writer: **a** an appearance, and **b** a schematic illustration of the optical system [26]. Illustrations courtesy of Heidelberg Instruments

15.3.4 Pattern Drawing by Laser Writer

Fringe patterns can be drawn on the photoresist using both electron and laser beams, as mentioned in the preceding section. However, the e-beam lithography needs a vacuum chamber for processing, and the resolution is somewhat over specification for printing the fringe pattern of HD-CGHs. From perspectives of the drawing speed and extent, laser writers used for fabricating FPD are suitable for printing HD-CGHs rather than high-resolution e-beam writers for fabricating LSI.

Figure 15.11 shows an example of the laser writer: DWL 66$^+$ made by Heidelberg Instruments. The basic mechanism is surprisingly similar to the spot-scanning type fringe printer shown in Fig. 15.1. The output of a violet laser[11] whose wavelength is 405 nm is focused on the photoresist of the mask blank using a write lens, and makes a small spot. The mask blank is fixed to a precision X-Y stage to raster-scan the focusing spot.[12] The stroke length and resolution of the X-Y stage, in which a laser interferometer is installed, are 200 mm × 200 mm and 10 nm, respectively.

The laser spot is not only scanned by the X-Y stage but also deflected using an acousto-optical deflector (AOD), as shown in Fig. 15.11b. Using the AOD, the scan line in this laser writer is not a narrow single-spot line but a multi-spot stripe with a given thickness. This mechanism contributes to speedup of the pattern drawing. The focus of the spot is automatically controlled by several methods depending on the write mode. The write speed and minimum structure size depends on the write mode, i.e., the write head used for drawing. Table 15.2 summarizes a few important specifications of DWL 66$^+$ for major write modes.

[11] UV laser sources can be selected optionally.

[12] Vector scan is also possible in DWL 66$^+$ but not used for printing CGHs.

Table 15.2 Summary of representative specifications of DWL 66$^+$ for several major write modes [26]

Write mode[a]	HiRes	I	II	III
Minimum structure size [μm]	0.3	0.6	0.8	1
Address grid [nm]	5	10	25	50
Maximum write speed (@405 nm) [mm^2/min]	3	13	40	150
Wavelength of light source	405 nm or 375 nm			
Maximum exposure area	200 × 200 mm^2			

[a] DWL 66$^+$ has additional write modes IV and V omitted in this table

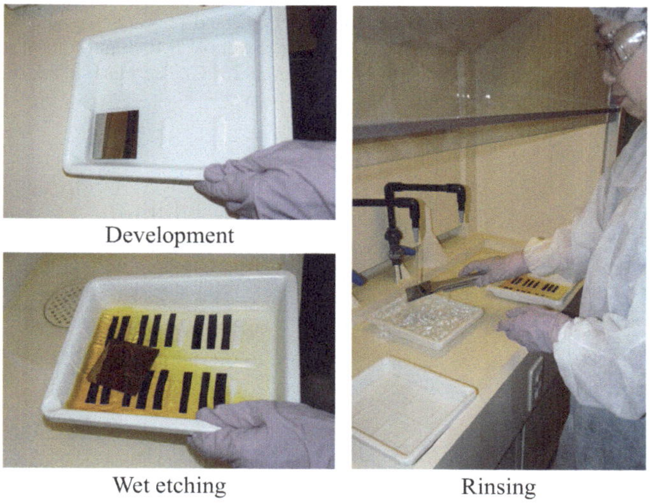

Fig. 15.12 Photographs of actual processes for fabricating a CGH after pattern drawing

15.3.5 Actual Processes of Development and Etching

The laser writer is an expensive apparatus. However, as for the processes of development and etching, we do not need so expensive equipments and rigorous handling as LSI/FPD manufacturing needs because of the nature of holography, as mentioned previously. These processes in printing CGHs resemble those in photography rather than microfabrication, as shown in Fig. 15.12.

After pattern drawing by the laser writer, a HD-CGH is finished with processes: Development → Rinsing → Wet etching → Rinsing → Drying. We first soak the mask blank having a latent image in a developing solution: developer designated by the mask blanks provider. The mask blank is then rinsed with purified water. To remove unnecessary chromium films, we soak the whole mask blank in chromium etching solution and then rinse it again. The CGH is completed by drying the mask blank with air flow.

(a) Binary fringe (b) Write power: 40 % (c) Write power: 70 %

Fig. 15.13 **a** An original binary-amplitude fringe pattern, and photomicrographs of the fringes printed by DWL 66^+ (Write mode II) with relative laser powers of **b** 40% and **c** 70%. All processing times: development, etching and rinsing are just 1 min at 23 °C. The mask blank in Fig. 15.9a is used with photoresist of AZP1350. $\Delta x_h = \Delta y_h = 0.8$ [μm]

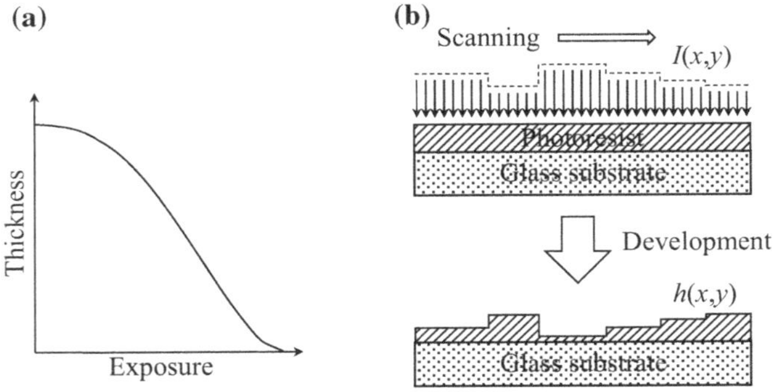

Fig. 15.14 **a** Schematic property of a positive-tone photoresist, and **b** fabrication of a surface relief using laser lithography

Optimal processing time for each process depends on various parameters such as the liquid temperature, the strength of the solution, and so on. Above all, optical intensity of the laser spot strongly affects the processing time. There are so many parameters related to the processing time that we should fix some and optimize others. In the case of the author's lab, every processing time is fixed at just one minute at 23 °C, and only the write laser power is adjusted for drawing the sharpest fringe pattern. Photomicrographs of fringes printed by DWL 66^+ with different write powers are shown in Fig. 15.13b, c. The fringe printed with a relative laser power of 40% is definitely better than that with 70% in this example.

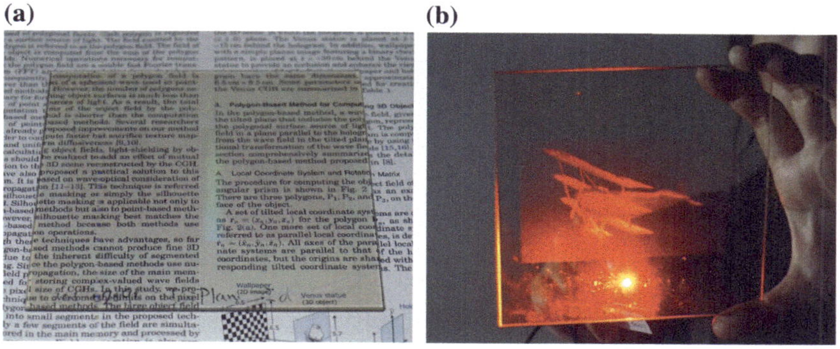

Fig. 15.15 Photographs of **a** a binary phase CGH "Triplane I", and **b** its optical reconstruction by transmission illumination using a red LED. $M = 131{,}072$, $N = 65{,}536$, and $\Delta x_{\rm h} = \Delta y_{\rm h} = 0.8\,\mu$m (see Table A.3 for the detailed parameters)

15.3.6 Creation of Phase CGHs

Development rate of a photoresist is almost linear to the exposure dose of the laser writer within a given range. Thus, the thickness of the photoresist is also nearly linear to the intensity of the write beam within the range, as shown in Fig. 15.14a. Using this fact, modulation of the write laser power produces a surface relief, as shown in (b).

Because ordinary laser writers apply on-off modulation to the laser in order to produce binary amplitude photomasks, binary phase modulation is always possible in all laser writers. Figure 15.15a is a photograph of the binary phase CGH fabricated from a mask blank without chromium thin films like Fig. 15.9c. The fabricated CGH is almost transparent because of no light-shielding film but reconstructs the 3D image by transmission illumination, as shown in Fig. 15.15b. Here, note that the reconstructed CGH, "Triplane I", is not a phase-encoded CGH but a binary-amplitude CGH in practice. The fringe pattern is generated by the technique described in Sect. 8.6.3. This is because binary-amplitude encoding is equivalent to binary-phase encoding.

Unfortunately, since a binary phase encoded CGH has almost the same property as amplitude modulation, there is no other advantage than its high efficiency, as mentioned in Sect. 8.7.3. However, some laser writers can change the beam power at multi-level in scanning the laser spot,[13] and make it possible to produce multi-level surface reliefs made of the developed photoresist. Examples of the actual multi-level surface reliefs fabricated by laser lithography are shown in Figure 15.16.

Multi-level phase modulation is realized by the multi-level surface relief. In this case, we can remove the conjugate image from optical reconstruction of the HD-CGH, as mentioned in Sect. 8.7.2. Phase-encoded CGHs can be fabricated as reflec-

[13]DWL 66$^+$ has the capability of multi-level modulation. The maker: Heidelberg Instruments call the function *Grayscale lithography*.

Fig. 15.16 Examples of surface reliefs fabricated using DWL 66+ [26]. Photos courtesy of Heidelberg Instruments

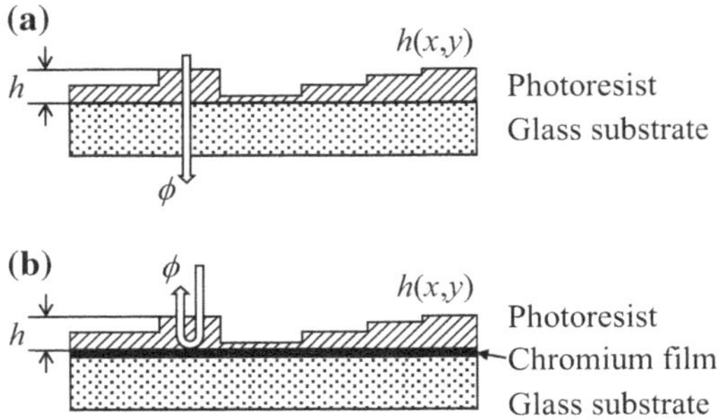

Fig. 15.17 Phase modulation produced by the surface relief in **a** transmission and **b** reflection CGHs

tion CGHs as well as transmission CGHs. Mask blanks similar to that in Fig. 15.9c is used for transmission CGHs, while type (b) having a chromium reflection layer are used for reflection CGHs.

Figure 15.17 illustrates the phase modulation produced by a surface relief. Here, the surface relief is made of the developed photoresist: resin such as novolac resins (AXP1350 photoresist). Suppose that n_{p} is a refractive index of the resin. A plane wave traveling in the z direction in the resin is written by $\exp[in_{\mathrm{p}}kz]$, while a plane wave in the air is $\exp[ikz]$, where k is a wave number in vacuum and also in the air. Thus the phase difference between the plane waves, which occurs in the travel at distance d, is given by

$$\phi = (n_{\mathrm{p}} - 1)kd. \tag{15.1}$$

As a result, relations between the spatial phase modulation and the surface elevation are written as

$$\phi(x, y) = \begin{cases} \dfrac{2\pi(n_{\mathrm{p}} - 1)}{\lambda} h(x, y) & \text{(transmission CGH)} \\ \dfrac{4\pi(n_{\mathrm{p}} - 1)}{\lambda} h(x, y) & \text{(reflection CGH)} \end{cases}. \tag{15.2}$$

The maximum depth of the surface relief is obtained by $\phi(x, y) = 2\pi$. However, it is very difficult to fabricate an exact surface relief that realizes accurate modulation depth of 2π. This is because there are many parameters that affect the modulation depth; furthermore expensive instruments are necessary to measure the surface profile accurately.

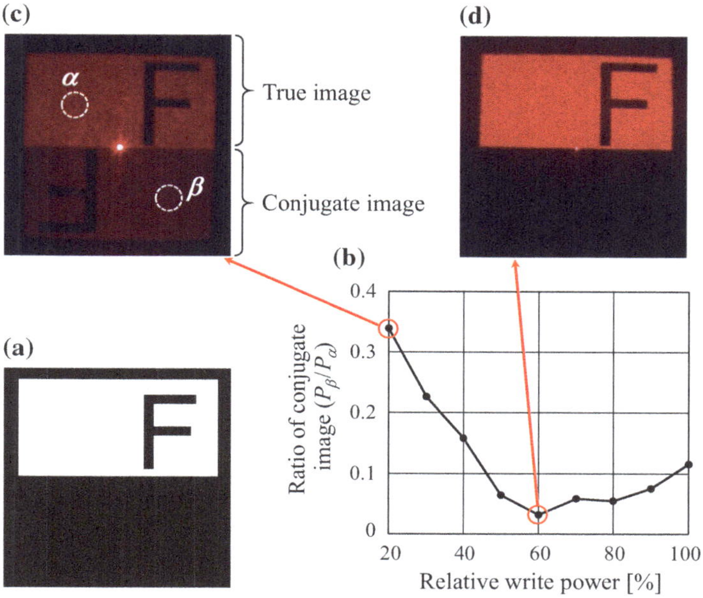

Fig. 15.18 An experiment of phase-encoded transmission Fourier transform CGHs fabricated by laser lithography: **a** the planar object, **b** power ratios of the conjugate image to the true image, and optical reconstructions of Fourier CGHs at a relative write power of **c** 20% and **d** 60%. CGH: $\lambda = 640$ [μm], $f = 200$ [mm], $M = N = 16{,}384$, and $\Delta x_h = \Delta y_h = 0.8$ [μm]

Figure 15.18 summarizes a simple experiment of phase-encoded Fourier transform CGHs. As described in Sect. 7.7.3, Fourier transform amplitude holograms generally reconstruct a twin image. In contrast, phase holograms with exact phase modulation do not reconstruct any conjugate image. Thus, phase-encoded Fourier CGH of the planar object, shown in Figure 15.18a, was calculated and fabricated with a various laser power of DWL 66^+. A CGH whose optical reconstruction is shown in (c) was fabricated with a relative write power of 20%. This CGH reconstructs not only the true image but also the conjugate image a little. We actually measured the power of the reconstructed fields inside the region α and β indicated in (c). The ratio of the conjugate image P_β/P_α was approximately 0.35 in this case.

Figure 15.18b shows ratios of the conjugate image of CGHs fabricated with different laser powers. As a result, we found the ratio of the conjugate image takes the minimum at a relative write power of 60%. The minimum ratio was approximately 0.03. The conjugate image almost disappears in this case.

In the end of this section, it should be emphasized that phase-only HD-CGHs using the surface relief is highly suitable for mass production of the HD-CGH. We can produce a metal mold by use of electroforming from the surface relief fabricated by laser lithography, and transfer the surface relief to a plastic plate by using an

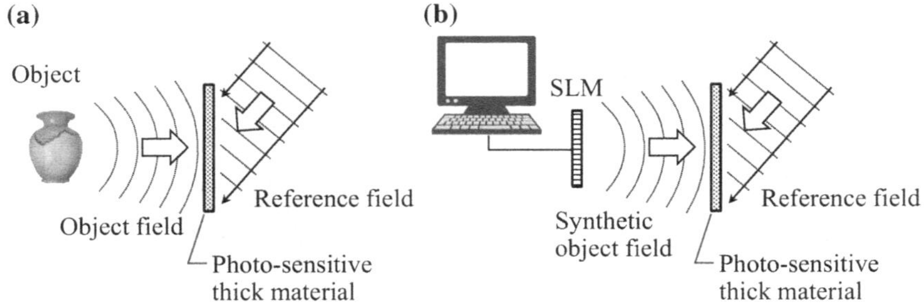

Fig. 15.19 Volume holograms **a** recorded by ordinary optical holography, and **b** printed by a wavefront printer

inprint technology. Using the technique, phase-only HD-CGHs can be produced in large quantities at a low price.

15.4 Wavefront Printer

The techniques explained in this chapter so far are methods to print encoded 2D fringe patterns of CGHs exactly like ordinary printers. The fabricated CGH is properly a thin hologram, and is only reconstructed using a monochromatic illumination. In contrast, the *wavefront printer* introduced in this section is literally a printer that prints the wavefront of a synthetic object field on a volume hologram [28, 42, 97, 126, 135]. Therefore, the printed hologram can be reconstructed using a white-light illumination.

15.4.1 Principle and Difference from Holographic Printer

Figure 15.19b illustrates the principle of a wavefront printer when compared with recording of a volume hologram in (a). Volume holograms are recorded by optical interference of the object field with a reference field. Here, the reference field illuminates the recording material from the opposite direction to the object field, as described in Sect. 7.2. Because the fringe vector is nearly perpendicular to the hologram surface, i.e., the fringes are almost parallel to the hologram surface, the resultant 3D fringes cannot be printed by any 2D printer described in the previous sections.

The basic idea of a wavefront printer is to generate the object field using an SLM instead of physical subjects, and record it in a volume hologram exactly like optical holography. The created hologram is a kind of volume hologram, but the object field is calculated from CG models or captured by digital holography. Therefore, the holo-

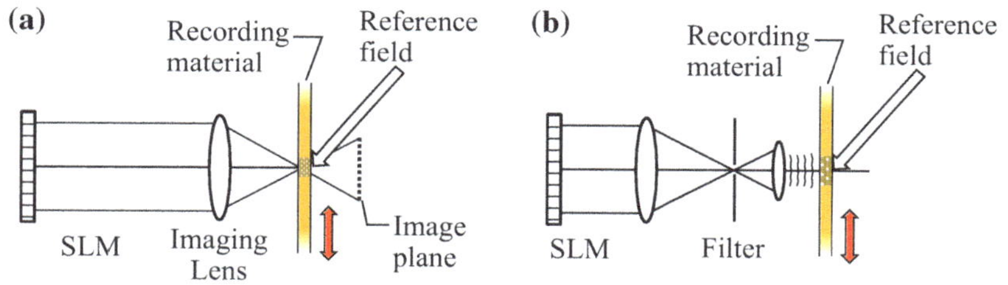

Fig. 15.20 Comparison of the structures between **a** a holographic 3D printer and **b** wavefront printer

gram printed by wavefront printers should be called a *computer-generated volume hologram* (CGVH). Here, note that the space-bandwidth product of the synthetic object field generated by an SLM is very small. A tiling technique like image-tilling fringe printers mentioned in Sect. 15.2.2 is necessary to create a given large CGVH.

Because a CGVH inherits advantages such as wavelength selectivity from optical volume holograms, the 3D image can be reconstructed using a white illumination. When a CGVH is printed with wavelength multiplexing, i.e., a CGVH is printed at three wavelength corresponding to RGB primary colors, the CGVH can reconstruct full-color images using a white illumination [42].

Here, it should be noted that there is another highly-developed holographic printing technology, called *direct-write digital holography* (DWDH) [6, 107].[14] The apparatus is sometimes called simply a *holographic 3D printer* [134, 139]. In this technique, object fields generated by an SLM is recorded on the material as a volume hologram. From this point of view, there is no difference between the holographic 3D printer and wavefront printer. However, the holographic printer differs in the object field recorded on the material. A wavefront printer records the synthetic object field of 3D/2D objects, whereas holographic 3D printers always record the field emitted from 2D planar objects. As a result, the resultant hologram is a holographic stereogram in holographic 3D printers.

Holographic 3D printers commonly use an intensity-modulation SLM and imaging lenses to form SLM images, as shown in Fig. 15.20a. A wavefront printer also uses lenses but the purpose is to enhance the viewing angle by demagnifying the field size and to filter unnecessary fields emitted from the phase-only or intensity SLM. Wavefront printers do not have any imaging lenses because the printer does not record a 2D image but the wavefield of a 3D object.

The reason that holographic 3D printers print holographic stereograms is to avoid difficulty in calculating the object field with large-scale SBP. However, modern holographic 3D printer can produce surprising full-color stereograms. The reader interested in the holographic 3D printer should refer to [6]. The disadvantage of

[14] Word "digital holography" used here is much different from that recently used for digital handling and processing of fringes and wavefields. DWDH is considered as an analog technology from the aspect of recent digital holography.

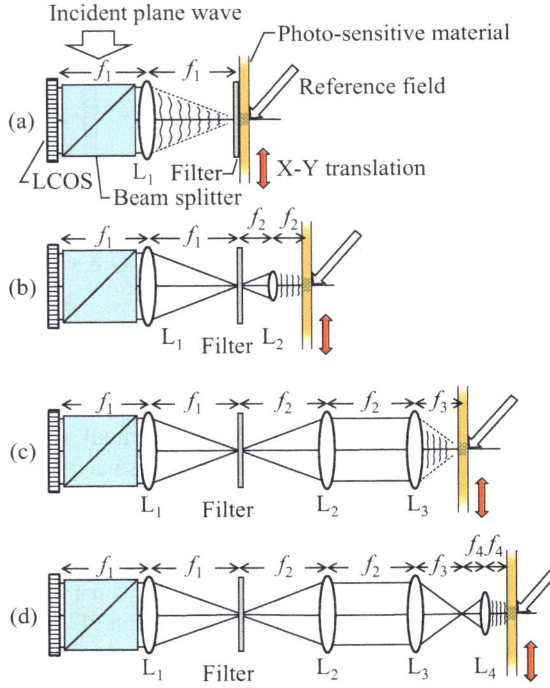

Fig. 15.21 Several optical systems used for generating the write field in wavefront printers

the holographic 3D printer is, however, that the printed hologram is a stereogram. Because the vergence-accommodation conflict is not solved in stereograms, it is difficult to reconstruct deep 3D scenes unlike HD-CGHs.

15.4.2 Optical Systems for Generating Object Fields

There are variations of the optical system for generating the object field, here called a write field, in wavefront printers. Figure 15.21 illustrates typical optical systems used for generating the write field in wavefront printers.

15.4.2.1 Comparison of Optical Systems

The optical system in Fig. 15.21a is a minimal optics using a single lens and filter [135]. In this setup, the incident plane wave illuminates the SLM through the cubic beam splitter. The field modulated by the SLM is Fourier-transformed by lens L1 and then recorded on the photo-sensitive thick material. The setup (a) is very simple but has a disadvantage that the viewing angle of the printed CGVH is mainly determined by focal length f_1 of lens L1.

The sampling intervals of the write field in setup (a) are basically given by Δx and Δy in (8.38) of Fourier-transform CGHs:

$$\Delta x = \frac{\lambda f_1}{M \Delta x_h} \quad \text{and} \quad \Delta y = \frac{\lambda f_1}{N \Delta y_h}, \tag{15.3}$$

where, Δx_h, Δy_h, M, and N are the pixel pitches and numbers of pixels of the SLM. Thus, the viewing angles are given by substituting (15.3) into (8.2):

$$\theta_{\text{view},x} = 2\sin^{-1}\left(\frac{M \Delta x_h}{2 f_1}\right) \quad \text{and} \quad \theta_{\text{view},y} = 2\sin^{-1}\left(\frac{N \Delta y_h}{2 f_1}\right). \tag{15.4}$$

Accordingly, focal length f_1 must be short for the printed CGVH to have large viewing angles. However, because an LCOS is commonly used for the SLM, a beam splitter is inevitably installed between the LCOS and lens L1. As a result, we cannot use a short focus Fourier lens for L1. In addition, in setup (a), the filter is arranged so as to contact or nearly contact with the material. This most likely causes some problem or at least makes it difficult to design and construct the printer.

Setup (c) in Fig. 15.21 is also a Fourier transform optical system. When $f_1 = f_2$, the filter and the design method are the same as those in setup (a), but the viewing angle is determined by f_3 instead of f_1 in this case. Therefore, the design freedom is extended very much. In both setup (a) and (c), when the SLM is intensity-modulation type, the size of a tile is one half of that in phase-only SLMs because the single-sideband filter (see Sect. 8.10) is used to remove the conjugate image. Figure 15.22 is a photograph of optical reconstruction of a CGVH printed by a wavefront printer employing setup (c) [145].

Unlike setups (a) and (c), setups (b) and (d) are not Fourier transform optics. The lenses simply demagnify the field. In setup (b), the sampling interval of the write field is given by (8.49), i.e., by the ratio of the focal lengths:

$$\Delta x = \frac{f_2}{f_1} \Delta x_h \quad \text{and} \quad \Delta y = \frac{f_2}{f_1} \Delta y_h. \tag{15.5}$$

Thus, the viewing angles of the printed CGVH are

$$\theta_{\text{view},x} = 2\sin^{-1}\left(\frac{\lambda}{2\Delta x_h}\frac{f_1}{f_2}\right) \quad \text{and} \quad \theta_{\text{view},y} = 2\sin^{-1}\left(\frac{\lambda}{2\Delta y_h}\frac{f_1}{f_2}\right). \tag{15.6}$$

In setup (d), the viewing angles are given by replacing f_1 and f_2 by f_3 and f_4, respectively. Even if an intensity SLM and the single-sideband method are used in these setup, each tile has the same size as that in cases where phase-only SLMs are used. However, note that the vertical viewing angle is reduced to one half that in the case of phase-only SLMs without the single-sideband technique.

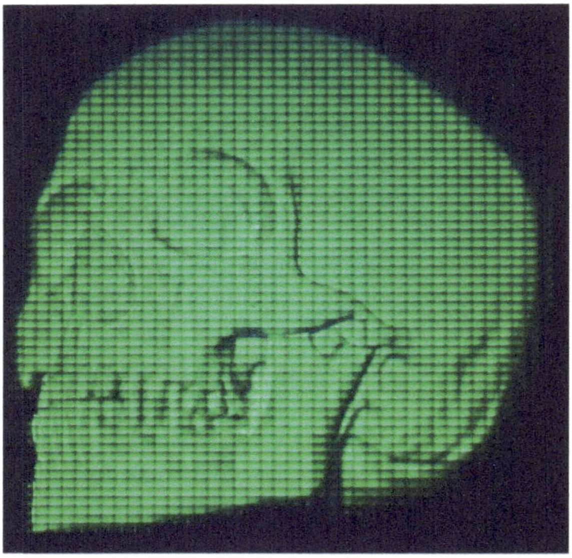

Fig. 15.22 A photograph of the CGVH printed by a wavefront printer [145] (From IEEE Transactions on Industrial Informatics 12(4), 1584(2016). By courtesy of Prof. Yoshikawa and Prof. Yamaguchi, Nippon University.)

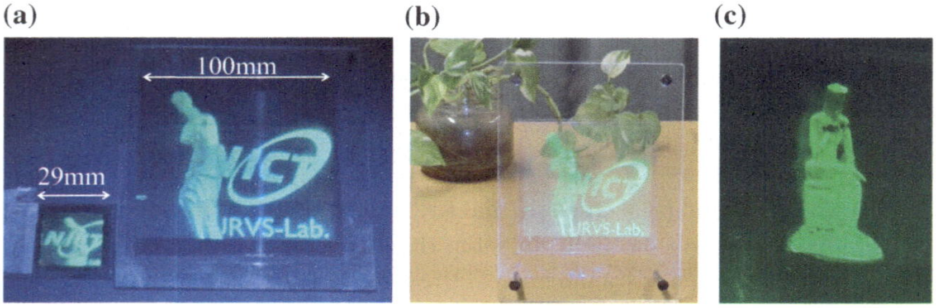

Fig. 15.23 Photographs of optical reconstruction of CGVHs printed by a wavefront printer [126]. (From SPIE Proc. 9867, 137(2016). By courtesy of Dr. Wakunami and Dr. Yamamoto, NICT)

Figure 15.23 show examples of optical reconstruction of CGVHs printed by a wavefront printer employing setup (d) [126].

15.4.2.2 Filter Structure

The minimal setup (a) in Fig. 15.21 provides a basis for the filter structure. When using an intensity SLM, the fringe displayed on the SLM is a kind of Fourier-transform amplitude CGH, described in Sect. 8.9. Therefore, the field in the Fourier plane

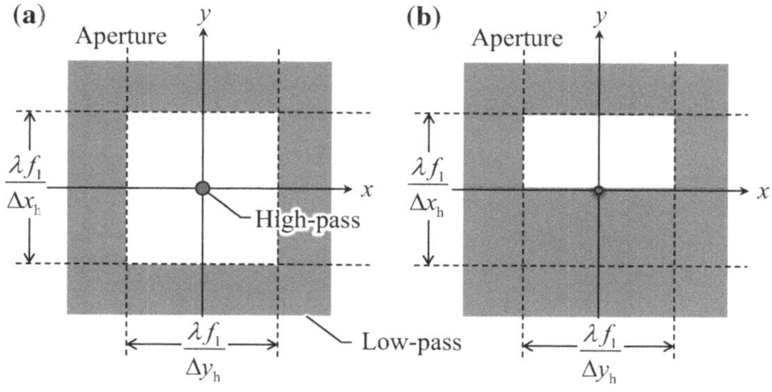

Fig. 15.24 The structure of filters in the Fourier plane of wavefront printers: **a** The bandpass filter for phase-only SLMs, and **b** the single-sideband filter for intensity SLMs

includes unnecessary light such as non-diffraction light (0th order light), higher-order diffraction images and the conjugate image in addition to the 1st order true image. Even if a phase-only SLM is used for spatial modulation, the field include at least the higher-order images and non-diffraction light caused by the pixel structure of the SLM device, such as pixel electrode and data buslines. Therefore, an aperture is generally placed in the Fourier plane to shield the unnecessary fields.

In phase-only SLMs, because the conjugate image is not generated in principle, the filter can pass the whole of the 1st order image whose size is given by (8.42). Therefore, in all cases of Fig. 15.21, the size of the low-pass filter is

$$W_x = \frac{\lambda f_1}{\Delta x_h} \quad \text{and} \quad W_y = \frac{\lambda f_1}{\Delta y_h}. \tag{15.7}$$

In contrast, the size of the high-pass filter for shielding 0th order non-modulated light depends on the structure of the SLM. The size is determined experimentally in many cases. The structure of the bandpass filter used in phase-only SLMs is shown in Fig. 15.24a.

When using an intensity SLM, the SLM generates not only 0th and higher-order images but also the conjugate image. Thus, the single-sideband method described in Sect. 8.10 is commonly used to remove the conjugate image. The structure of the single-sideband filter is shown in Fig. 8.28. The single-sideband filter in the case of Fig. 15.21 is shown in Fig. 15.24b for reader's convenience. These filter structures are in common to all optical systems in (a) to (d) of Fig. 15.21.

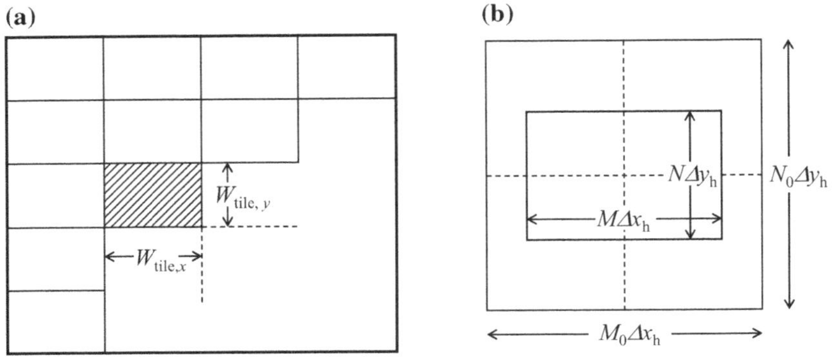

Fig. 15.25 **a** Tilling write fields, and **b** clipping a Fourier-transformed object field in Fourier-type setups

15.4.3 Calculation of Object Fields and Encoding of Fringes

A large CGVH is printed by tiling in wavefront printers. Let $W_{\text{tile},x} \times W_{\text{tile},y}$ represent the size of a tile as shown in Fig. 15.25a. To generate the fringe pattern displayed on the SLM, we have to calculate the object field for the corresponding tile in the beginning.[15] When Fourier-transform optical systems are not used, the tile size is directly obtained from the SLM size.

In setup (b) and (d) of Fig. 15.21, the number of sample points of a tile is exactly the same as that of the SLM: $M \times N$. The sampling intervals are given by (15.5), i.e., simply given by multiplying f_2/f_1 and f_4/f_3 by those of the SLM in setup (b) and (d),[16] respectively. In phase-only SLMs, the phase-encoded fringe pattern of object field $g_{\text{obj}}[m, n]$ in the $M \times N$ array is simply displayed on the SLM. Here, note that the coordinates is inverted by a double Fourier transform. When using the single-sideband method, follow the procedure described in Sect. 8.10.2.

When a Fourier-transform optics is used, calculation of the write field is a little complicated. In setup (a) of Fig. 15.21, the tile size definitely agrees with the low-pass and single-sideband filter, i.e., $W_{\text{tile},x} \times W_{\text{tile},y} = W_x \times W_y$ in phase-only SLMs and $W_x \times W_y/2$ in intensity SLMs with the single-sideband method, respectively. The sampling interval of a write field is basically given by (15.3), as mentioned above. M and N are again the numbers of pixels of the SLM. However, in general, M and N are not a power of 2 in ordinary SLMs. Thus, we first choose integers M_0 and N_0 that are a power of 2 and more than M and N, respectively. Then, we calculate the object field $g_{\text{obj}}[p, q]$ inside the tile at sampling intervals of

$$\Delta x' = \frac{\lambda f_1}{M_0 \Delta x_h} \quad \text{and} \quad \Delta y' = \frac{\lambda f_1}{N_0 \Delta y_h}. \tag{15.8}$$

[15] In practice, we first calculate the whole object field and then extract the window of $W_{\text{tile},x} \times W_{\text{tile},y}$.
[16] Here, we assume $f_1 = f_2$ in setup (d) of Fig. 15.21.

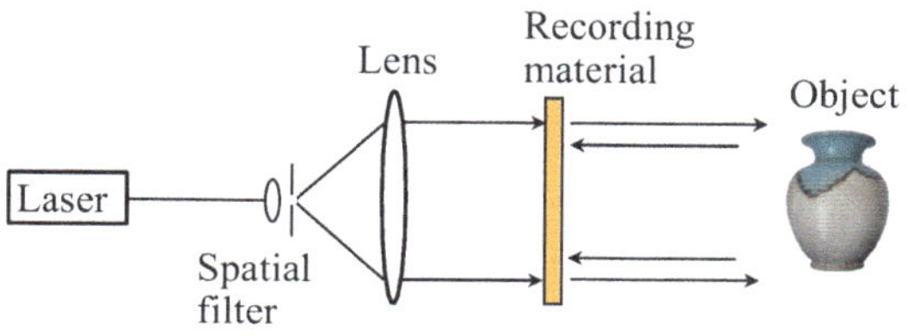

Fig. 15.26 Schematics of recording Denisyuk holograms

After inverse FFT in (8.37), the center region of $M \times N$ points is extracted from the inverse transformed field of the $M_0 \times N_0$ array, as shown in Fig. 15.25b, and is displayed on the SLM. In setup (c), f_1 is replaced by f_3 properly.

Here, the viewing angle of the printed CGH seems to be determined by the sampling interval of (15.8) in this case. However, the SLM cannot reconstruct the whole bandwidth of the calculated object filed, which is given by $\Delta x'$ and $\Delta y'$. The write field generated by the actual SLM only has a bandwidth given by Δx and Δy of (15.3). Therefore, it should be noted that even if we increase M_0 and N_0 the viewing angle does not change and keeps the value given by (15.4), and thus it is better to choose the minimum M_0 and N_0 that are equal to or more than M and N.

When the SLM is an intensity-modulation type, the fringe pattern can be encoded by the technique described in Sect. 8.9.3. In the case, the object field is obtained inside the $W_x \times W_y/2$ window and combine it to produce the Hermitian symmetry, as shown in (8.46) and Fig. 8.26.

15.4.4 Denysyuk-Type Wavefront Printer

As an example of the wavefront printer employing setup (b) in Fig. 15.21, let us take up a Denysyuk-type wavefront printer [105]. The optical system of a wavefront printer is sometimes very complicated because we have to introduce the reference field in the opposite direction to the recording material and use filters to generate the write field. As a result, the optical system is sometimes unstable and easily affected by vibration. In addition, intensity of the write field is sometimes so weak due to splitting the laser output at least twice that the volume hologram cannot be recorded for a short time. These lead to low quality of the printed CGVH.

Optical holography usually suffers from similar problems; the optics for recording is unstable and vulnerable to vibration due to the complicated optical path. The optical intensity is sometimes impractically weak because of splitting the light. Thus, the technique of Denisyuk holograms is often used in optical holography in order to reduce the problems and record volume holograms more easily [108]. Figure 15.26 is a schematic illustration of recording Denisyuk holograms. The output of a laser is not split into arms in this technique, and illuminates a subject by passing through the recording material. Since the object wave enters the opposite side of the material to the reference wave, a volume hologram is recorded in this technique. Because beam

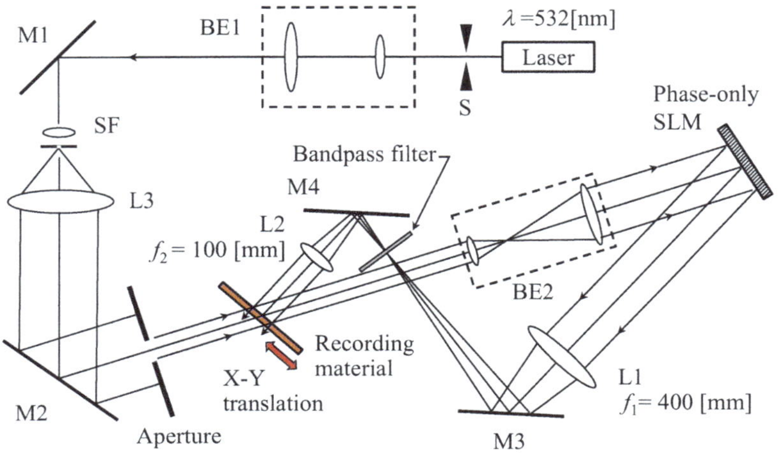

Fig. 15.27 Setup of the wavefront printer employing a Denisyuk-type optical system. M, mirror; L, lens; SF, spatial filter; BE, beam expander

splitters are not necessary in the setup, the optics can be simplified and any beam power is not lost. In optical holography, many brilliant color holograms are recorded by employing the technique.

A setup adopting the ideas of Denisyuk holograms is shown in Fig. 15.27 [105]. The output of the laser is expanded by the spatial filter and collimator lens L3, and then, enters the SLM through the recording material. Here, an aperture and beam expanders are used for adjusting the incident beam size. The slightly off-axis incident beam illuminates the phase-only SLM whose pixel pitch is $\Delta x_h = \Delta y_h = 8\,[\mu m]$. The write field is generated from the modulated output using setup (b) in Fig. 15.21. Here, mirrors M3 and M4 are used to fold the optical path and project the write field into the recording material perpendicularly. Since $f_1 = 400$ [mm] and $f_2 = 100$ [mm], effective sampling intervals of the write field are reduced to 2 μm. As a result, the viewing angle of the printed CGVH is estimated at 15.3° in both horizontal and vertical directions. Since the size of a tile is $W_{\text{tile},x} \times W_{\text{tile},y} = 3.84 \times 2.16$ [mm^2], the aperture of the reference beam is set to the same size.

Figure 15.28 shows the optical reconstruction of a CGVH printed with 16 × 16 tiles. The 3D scene is shown in Fig. 15.29. The object field is calculated with 32,768 × 32,768 total sample points using the point-source method at a density of 80,000 points/m. The 8-bit phase-encoded fringe patterns were displayed on the SLM. The size of the printed CGVH is approximately 61.4 × 34.6 [mm^2]. The parameters of the setup and printed CGVH are summarized in Table 15.3.

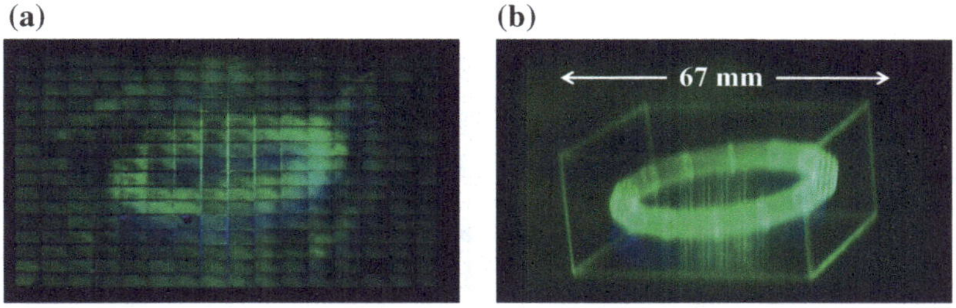

Fig. 15.28 Photographs of optical reconstruction of the printed CGVH named "Ring in Cube" [105]. The camera is focused on **a** the hologram plane, and **b** the object. A white LED is used for the illumination light source. Video link, https://youtu.be/YnjAuhjBDSA

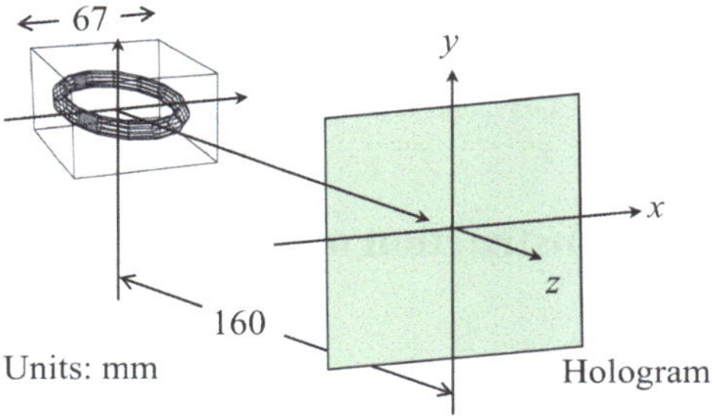

Fig. 15.29 The 3D scene of the printed CGVH: "Ring in Cube" [105]

Table 15.3 Parameters of the denisyuk-type wavefront printer and the printed CGVH: "Ring in Cube"

Parameter	Value
Number of pixels of SLM	1,920 × 1,080
Pixels pitches of SLM ($\Delta x_h \times \Delta y_h$) [μm]	8.0 × 8.0
Wavelength [nm]	532
Number of tiles	16 × 16
Number of total printed pixels	30,720 × 17,280
Sampling intervals of wright field [μm]	2.0 × 2.0

Fig. 15.30 Optical reconstruction of HD-CGH "Moai II" with **a** a monochromatic and **b** white illumination light

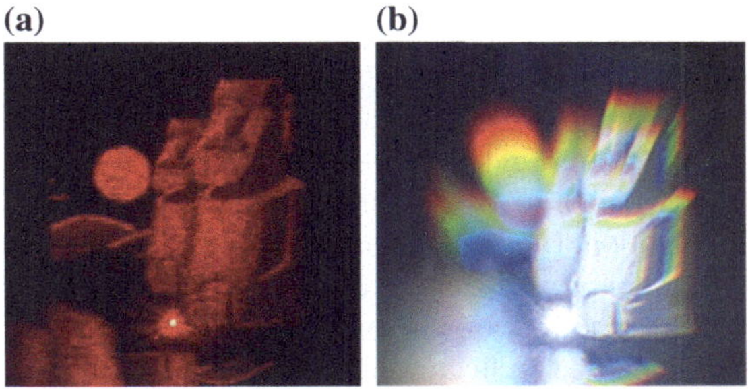

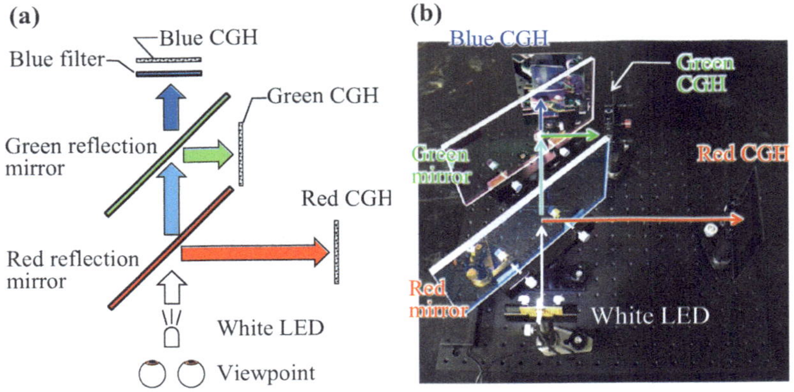

Fig. 15.31 **a** Experimental setup for an optical combiner using dichroic mirrors to reconstruct full-color images from three HD-CGHs, and **b** the photograph of the actual optical combiner [83]

15.5 Full-Color Reconstruction of HD-CGHs Using Optical Combiner

Figure 15.30 shows optical reconstruction of a HD-CGH printed by laser lithography. A monochromatic light source is used for the reflected illumination in (a), where a broadband white-light source is used in (b). If any filter is not used in reconstruction, a broadband illumination commonly generates severe color smears due to chromatic aberration. This is because every printed CGH except for CGVHs is generally a thin hologram that has no wavelength selectivity.

It is easy in principle to reconstruct HD-CGHs in full-color. Full-color reconstruction is simply realized by combining three images reconstructed at three wavelengths corresponding to primary colors. However, it is not easy in practice to fabricate full-color HD-CGHs that do not need massive optics. A practical full-color HD-CGH must be simple, light and portable enough to bring them to an exhibition in museums and galleries. There are two actual techniques that meet the requirements: full-color CGHs using RGB color filters and stacked CGVHs. These are described in the following sections. Before proceeding to these sections, let us see the basic technique to reconstruct a full-color image from monochrome HD-CGHs.

An example of the actual optical system to combine primary color images reconstructed by three HD-CGHs is shown in Fig. 15.31. In this optical system, we use two dichroic mirrors and a white LED. Red illumination light branches off from the original white-light by the red dichroic mirror, and then, green light also branches off by the green mirror. Light transmitted through two dichroic mirrors is used for blue illumination. Here, note that we insert additional blue filter after the green mirror in order to improve spectrum of the blue illumination light.

When the fringe patterns of HD-CGHs are printed using laser lithography, the CGH works as a reflection hologram, as mentioned repeatedly. Therefore, each image

Fig. 15.32 A photograph of full-color reconstruction of "Rubik Cube" using the optical combiner in Fig. 15.31. $M = N = 65{,}536$, $\Delta x = \Delta y = 0.8\,[\mu\text{m}]$, $\lambda_R = 620\,[\text{nm}]$, $\lambda_G = 540\,[\text{nm}]$, and $\lambda_B = 450\,[\text{nm}]$. Video link, https://youtu.be/0W0Wmh6SW4s

reconstructed by reflection of the CGH is backwardly traveled along the illumination path and combined by the dichroic mirrors again. The reconstructed monochromatic images finally form a full-color 3D image. The photographs of reconstruction, which are taken from different viewpoints, are already shown in Fig. 13.23. Here, Figure 15.32 is an enlarged photograph taken from the in-line viewpoint.

The full-color image reconstructed using the optical combiner is fairly clear and vivid, as shown in Fig. 15.32 and the video. However, the optical combiner is bulky and too heavy to carry it. This is mainly due to the large dichroic mirrors, whose sizes are determined so as to cover the whole viewing angle of the HD-CGHs. In the case of the combiner shown in Fig. 15.31b, the CGH size is approximately 5×5 cm^2, while the dimension of the dichroic mirrors is 22×12 cm^2.

15.6 Full-Color CGH Using RGB Color Filters

A liquid-crystal display (LCD) commonly displays full-color images using color filters attached to each pixel. Techniques employing combination of a single hologram and RGB color filters are also proposed for full-color reconstruction of CGHs. Although the SBP was very small in the early days of the technique [35, 37], the same technique can be applied to fabricating full-color HD-CGHs [80, 123].

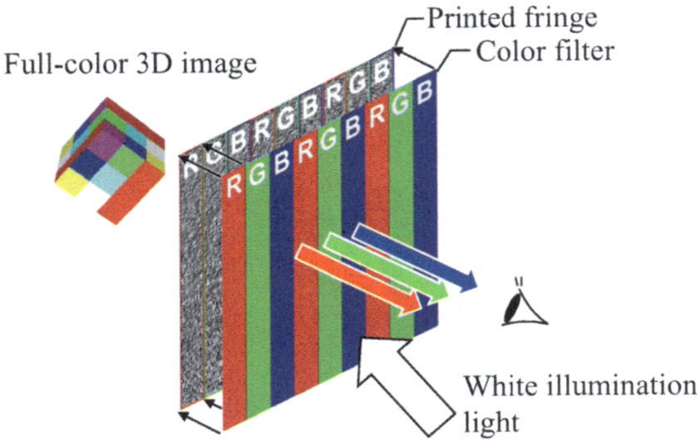

Fig. 15.33 Schematic illustration of the principle and structure of a full-color CGH using RGB color filters

15.6.1 *Principle and Structure*

Figure 15.33 schematically illustrates the principle of full-color reconstruction of a printed HD-CGH using RGB color filters. In this technique, RGB color filters are closely attached to the fringe patten printed by the laser lithography in order to filter illumination light. Each fringe pattern behind the RGB filters is calculated at a wavelength that corresponds to the RGB primary colors. Thus, the three monochromatic 3D images are independently reconstructed in the RGB colors. Viewers are able to see a full-color image by combining these RGB color images.

However, several problems had to be solved to realize this type of full-color CGH. The main challenges are:

- Low spatial resolution of the RGB color filters
- Alignment errors of the RGB color filters with the fringe pattern
- The broadband properties of the RGB color filters.

15.6.1.1 Filtering of Illumination Light on Fringe Blocks

Because the resolution of the color filter is commonly lower than that of the fringe pattern, it is difficult to attach a color filter to each individual fringe pixel as is done in image sensors and LCD panels. Therefore, the fringe pattern is divided into RGB blocks. Each block is composed of many pixels, as shown in Fig. 15.34. The fringe pattern for each block is generated at a wavelength representing either the R, G, or B color, and the corresponding color filter is attached to each fringe block.

When the CGH is irradiated with white illumination, light filtered by the attached color filter illuminates the correct fringe block. As a result, a full-color 3D image that is a combination of the RGB images reconstructed by RGB blocks emerges from the CGH. The layered structure of a full-color CGH is shown in Fig. 15.35. Unlike monochrome CGHs, the fringe pattern is formed so as to face the viewers, as already shown in Fig. 15.9.

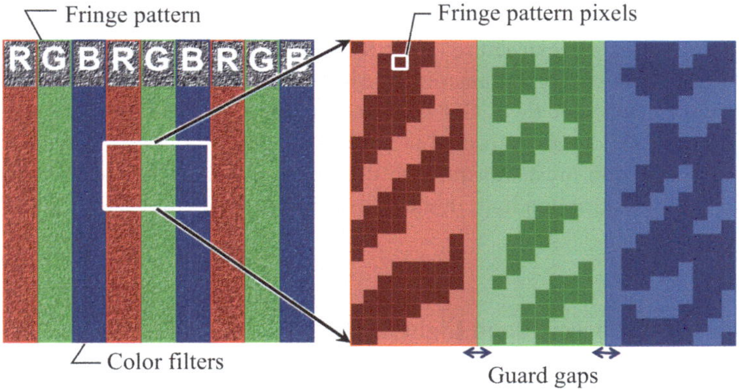

Fig. 15.34 Filtering of illumination light for each fringe block corresponding to its color

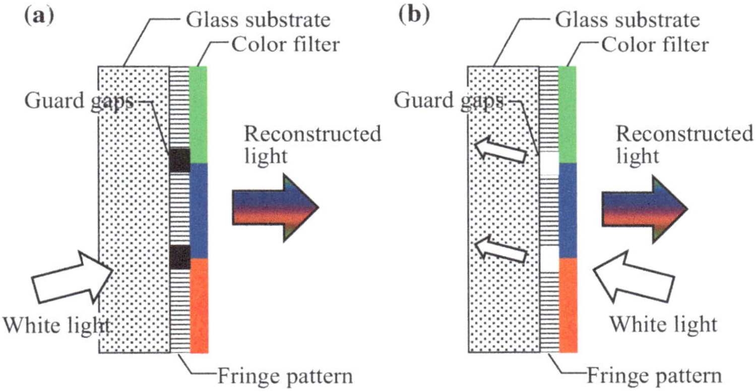

Fig. 15.35 Layered structure of a full-color HD-CGH for **a** transmission and **b** reflection illumination

15.6.1.2 Guard Gap

Because the resolution of the fringe pattern is on the scale of sub-microns, tolerance for alignment of RGB color filters is also in sub-microns. In addition, there may be color transition areas along the border between the colors of the RGB color filters. Thus, a "guard gap" should be created along the border between RGB blocks of the fringe pattern, as shown in Figs. 15.34 and 15.35. The fringe pattern is not formed in the guard-gap area.

This guard gap allows us to increase the positioning tolerance of the RGB color filters and avoid the problem of the color transition areas. In color LCD panel technology, a black matrix (BM) is usually formed on the color filter to avoid similar problems. However, in the case of CGHs, the guard gap is a simpler solution than a BM because it makes the structure of the RGB color filters very simple.

Note that the reflectivity or transmittance in the guard-gap area of the fringe pattern is dependent on whether the CGH is reconstructed by reflection or transmission. When the color CGH is reconstructed as a reflection-type hologram, the fringe pattern in the guard-gap area must transmit the illumination light, whereas the guard gap must shield the illumination light in the transmittance-type color CGHs, as shown in Fig. 15.35.

15.6.1.3 Use of Narrow Band Light Sources

One of the most difficult challenges of applying RGB color filters to the full-color reconstruction of CGHs is their broadband filter properties. Color filters are commonly used for LCD panels or image sensors. Narrow spectral transmittance is not required in these applications. However, broad spectral illumination causes severe chromatic aberration in high-definition CGHs.

Therefore, it is essential to use a narrow band light source as the illumination. RGB lasers that output three wavelengths corresponding to RGB colors are definitely the best choice from the viewpoint of the bandwidth. However, unfortunately they are usually expensive and sometimes dangerous to the viewers' eyes.

When LEDs are used for the light source, multi-chip type white LEDs should be selected instead of other types of white-light sources, such as single-chip white LEDs with phosphor or halogen lamps. Here, a multi-chip white LED is an LED in which three LED chips corresponding to the RGB primary colors are integrated into a single package. This type of LED gives a narrower spectrum for each color than single-chip LEDs and other incoherent white-light sources.

Figure 15.36a shows examples of the emission spectrum of single- and multi-chip LEDs, where the drive current of the multi-chip LED is adjusted so that the peak intensities have the same value. For reflection CGHs, effective illumination is obtained from the product of LED intensity and the square of the filter transmittance because the illumination light passes through the filter twice. Figure 15.36b shows examples of the spectra of the effective illuminations estimated by the spectra of LED outputs and filter properties in Fig. 15.36a. The multi-chip white LED produces a narrower spectra than the single-chip LED. Therefore, chromatic aberration can be reduced by adopting multi-chip type white LEDs as the illumination light source.

Here, note that the position of the light source slightly shifts depending on the RGB colors in the multi-chip white LEDs, i.e., the positions of three chips are not the same. However, this can be easily taken into account when generating the fringe patterns. We simply change reference waves for each RGB fringe block in its encoding.

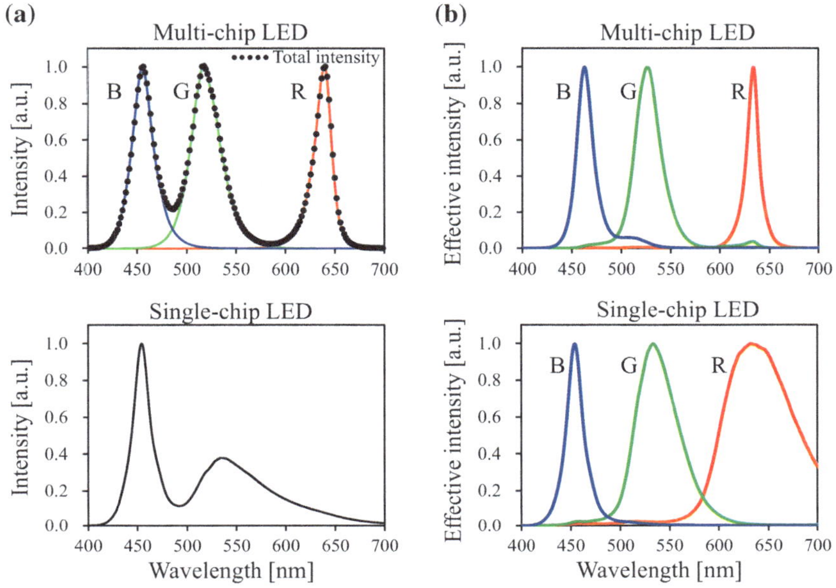

Fig. 15.36 Spectral examples of **a** measured LED outputs and **b** effective illumination estimated in single- and multi-chip LEDs [123]. Note that the peak values of the outputs of the multi-chip LED and the effective illumination are normalized to unity

15.6.2 Fringe Pattern

Figure 15.37 illustrates generation of the fringe pattern in this technique. Here, we adopt a simple stripe pattern as the shape of the RGB color filters in order to make it easier to adjust the position of the RGB color filters to the corresponding fringe block. In addition, for simplicity, we discuss only the reflection-type CGH below.

Using the techniques described in Sects. 8.6 and 8.7, fringe pattern $T(x, y; \lambda_c)$ for amplitude CGHs or $t(x, y; \lambda_c)$ for phase CGHs is first generated at three wavelengths corresponding to RGB colors from object fields that are also calculated at three wavelengths. Here, c = R, G, or B; λ_c is a design wavelength for color c. The three fringe patterns are then combined into a single fringe pattern by

$$T(x, y) = \sum_{c=R,G,B} \mathbf{Blk}_c \{T(x, y; \lambda_c)\}, \tag{15.9}$$

where $\mathbf{Blk}_c\{\}$ is a clipping operator that clips the block for color c out of the corresponding fringe pattern. Supposing that W_{str} and W_{gap} are widths of a filter stripe and gap respectively, the clipping operator for R color is, for example, defined as follows.

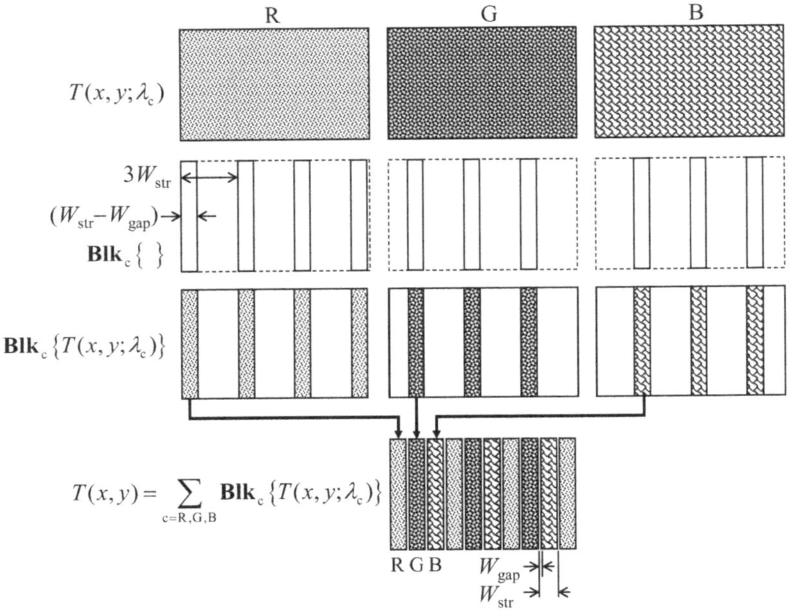

Fig. 15.37 Generation of a single fringe pattern from three fringe patterns corresponding to RGB colors

$$\mathbf{Blk}_R\{T(x, y; \lambda_R)\} \equiv \sum_n \text{rect}\left(\frac{x - 3nW_{str}}{W_{str} - W_{gap}}\right) T(x, y; \lambda_R). \tag{15.10}$$

To create phase-encoded CGHs, $T(x, y; \lambda_R)$ is replaced by $t(x, y; \lambda_R)$ properly.

15.6.3 Design Parameters of RGB Color Filters

It is important to optimize the widths of the color stripe and guard gap because the color filters may be perceived by viewers if these widths are unnecessarily wide. The reconstructed 3D images will be degraded if the filter is detected. On the contrary, narrower widths make alignment more difficult. These parameters should be determined by repeating the simulated reconstruction mentioned in Sect. 13.5.2.3.[17]

Furthermore, spectral properties of the filters must be determined by choosing the materials and thickness of the RGB color filters. The simulation should also be repeated for various materials and thicknesses of the color filters to optimize them.

[17]More accurate simulation than that described in Sect. 13.5.2.3 is desired for optimization of filter design. The reader interested in the more accurate simulation technique should refer to chapter 3 of [123].

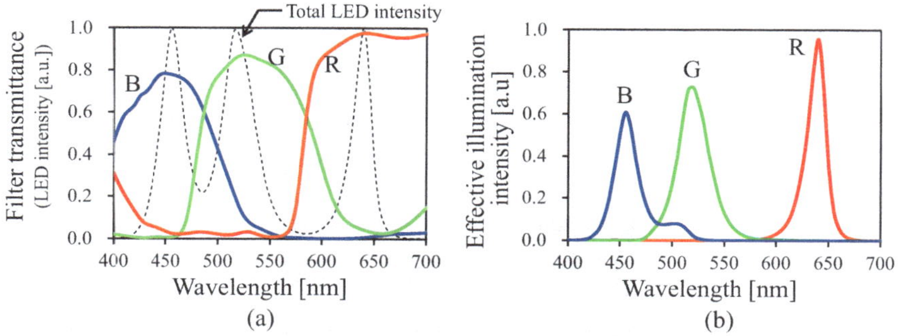

Fig. 15.38 Spectra of **a** transmittance of actual RGB color filters and **b** the effective illuminations estimated from the spectra in (**a**) [123]

The designed color filter is actually fabricated using conventional techniques used for color LCD panels, i.e., photolithography using photomasks and color photoresist. Figure 15.38a shows the spectral transmittances of the RGB color filters that the author finally adopted and fabricated. Here, the total intensity of the multi-chip LED in Fig. 15.36a is again depicted for reference. Note that the drive currents are again adjusted so that the peaks have the same intensity. The spectra of the effective illuminations estimated from the spectra of the LED intensity and filter transmittance are shown in Fig. 15.38b.

15.6.4 Examples of Optical Reconstruction

The HD-CGH named "Color Cube" is a full-color CGH using RGB color filters whose spectral property is shown in Fig. 15.38. Here, the widths of the gap and stripe are 20 μm and 80 μm, respectively. Color Cube, whose dimension is approximately $5 \times 5\ \text{cm}^2$, is designed for reconstruction of a multi-chip LED. The optical reconstruction and 3D scene are shown in Figs. 15.39 and 15.40.

"Casino Chips" is also designed for reconstruction by the multi-chip white LED, but the size is approximately $10 \times 10\ \text{cm}^2$ [123]. Figure 15.41 is a photograph of the optical reconstruction. "Color Shion II" shown in Fig. 15.42 is also a 10 cm full-color HD-CGH, but is designed for reconstruction by an RGB laser light source. Here, an ultra short focus projector in which LD light sources are installed is used for the illumination light source. In addition, the technique of fringe oversampling (see Sect. 8.8.3) is used for encoding of the CGH to realize reconstruction by the overhead illumination. The parameters of these full-color HD-CGHs are summarized in Table 15.4.

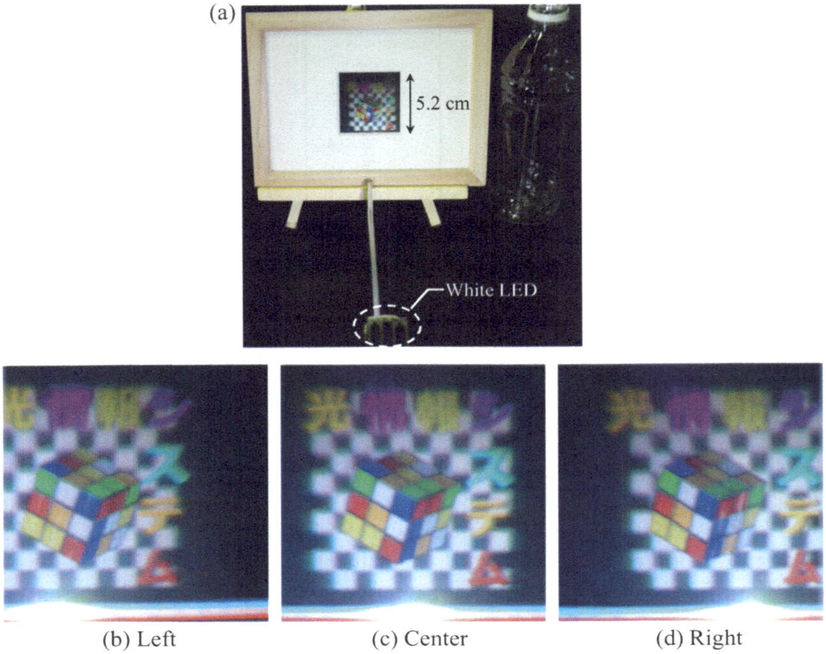

Fig. 15.39 Optical reconstruction of full-color HD-CGH "Color Cube" [123]. Photographs are taken from **a** a distant view and **b–d** close-up views from different angles. Video link, https://doi.org/10.1364/OE.25.002016.v001

Fig. 15.40 The 3D scene of full-color CGH "Color Cube" [123]

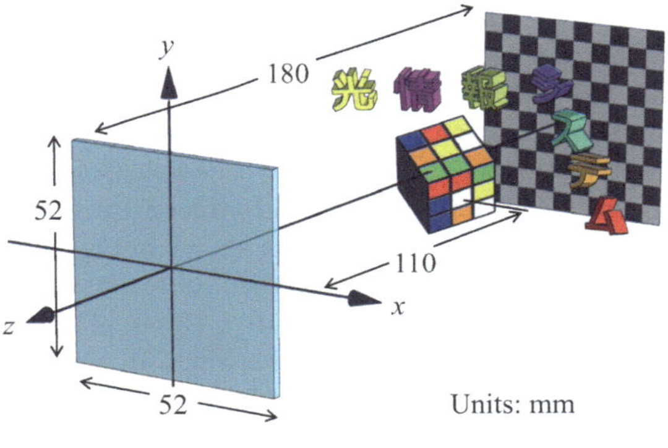

15.7 Full-Color Stacked CGVH

The CGHs produced by the technique described in the preceding section are portable and able to reconstruct full-color images easily. However, the image is not very sharp or vivid. There are mainly two reasons: (a) The spectral property of the RGB color

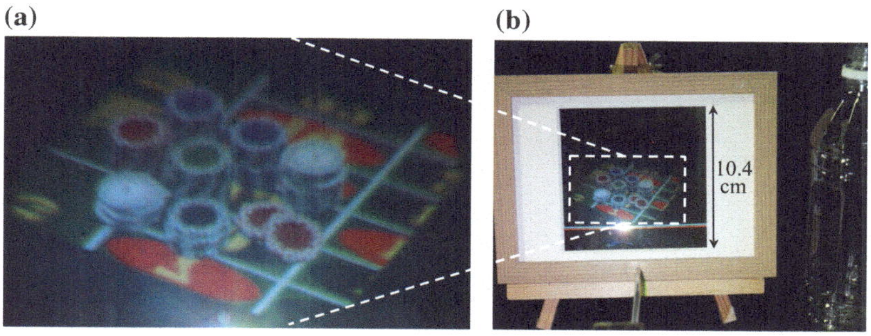

Fig. 15.41 Optical reconstruction of full-color HD-CGH "Casino Chips" at **a** close-up and **b** distant views [123]. Video link, https://doi.org/10.1364/OE.25.002016.v002

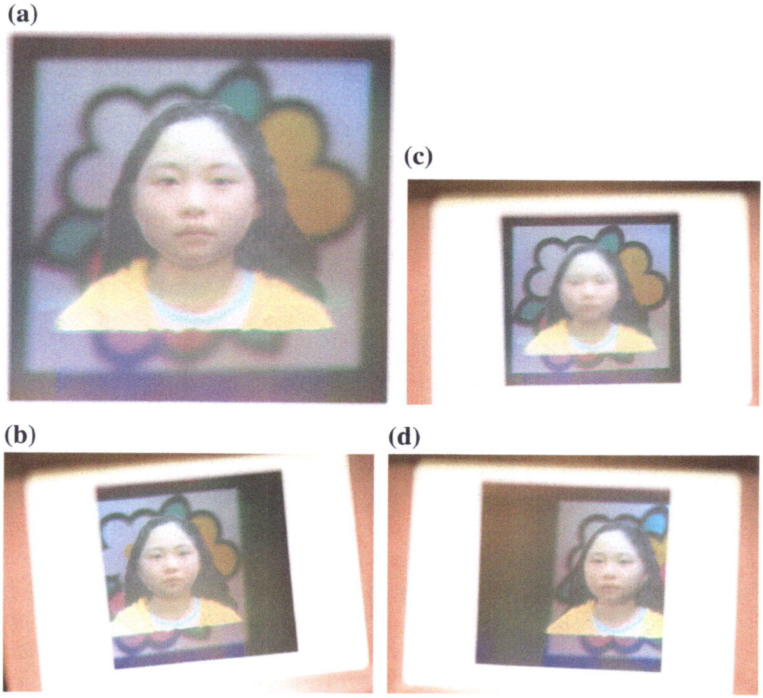

Fig. 15.42 Optical reconstruction of full-color HD-CGH "Color Shion II" at **a** a close-up view, and distant views from **b** left and **d** right viewpoints. The camera is focused on the wallpaper in **c**. Video link, https://youtu.be/DmC93CRdhFQ

Table 15.4 Parameters of full-color HD-CGHs using RGB color filters

Name	Color cube	Casino chips	Color Shion II
Width of stripe (W_{str}) [μm]	80		
Width of guard gap (W_{gap}) [μm]	20		
Illumination light source	White LED	White LED	RGB laser
Number of pixels[a]	64K × 64K	128K × 128K	128K × 256K
Pixels pitches [μm]	0.8 × 0.8	0.8 × 0.8	0.8 × 0.4
Design wavelength (R, G, B) [nm]	(630, 520, 460)	(630, 520, 460)	(635, 517, 443)

[a] 1K = 1,024

filters is commonly not narrow enough to reconstruct thin CGHs, and (b) the fringe pattern is spatially divided to fit with color filters. The former problem can be solved by use of an RGB laser light source. However, this kind of laser is expensive and sometimes not good for hologram illumination because of speckle noise as well as a risk to viewers' eyes. To use a white-light illumination, better color filtering is required. In addition, the fringe pattern should not be divided into blocks or sub-holograms. The technique described in this section meets these requirements.

15.7.1 Principle

An excellent technique that meets the above conditions has been proposed [38]. In the technique of a *stacked CGH*, fringe patterns for RGB colors are individually printed on the corresponding dichroic mirrors; more exactly, the fringe patterns are formed by the pattern of dichroic mirror films. Then, three fringe pattens are stacked on a substrate. When illuminating the stacked CGH by white light, the R layer, for example, reflects and modulates only the red component of the white light. The G and B components passing through the R layer are reflected at the corresponding layer. As a result, it is expected that a very clear full-color image is reconstructed because of the narrow bandwidth and lack of discontinuance of the dichromatic fringe pattern. However, very sophisticated film and pattern formation technology is needed to realize the stacked CGH.

15.7.1.1 Computer-Generated Volume Holograms Fabricated by Contact-Copy

Instead of dichromatic reflection fringe patterns, we can use volume holograms to provide the narrow band reconstruction [46]. Figure 15.43 shows comparison be-

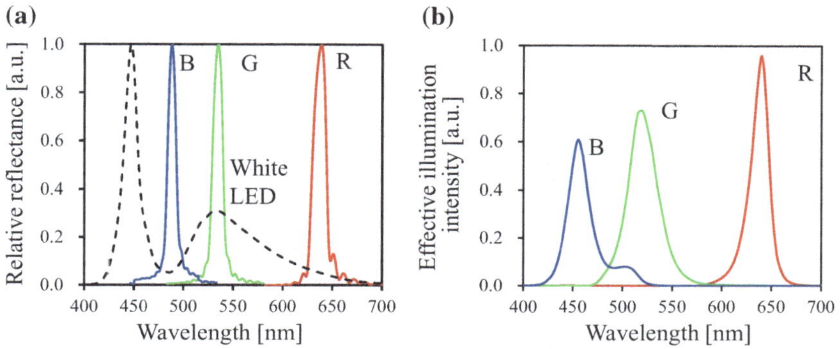

Fig. 15.43 Comparison between spectra of **a** measured reflectance of volume holograms and **b** estimated effective illuminations with use of RGB color filters (Fig. 15.38b) [46]

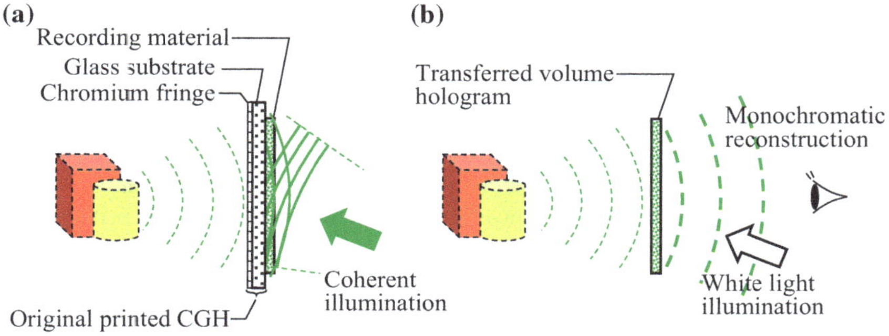

Fig. 15.44 **a** Contact-copy of an original reflection CGH printed by laser lithography and **b** its reconstruction by a white-light source

tween measured reflection spectra of volume holograms recorded by a two-beam interference exposure and effective illumination spectra of RGB color filters in the previous section. The bandwidths of the volume holograms are 15, 10, and 9 nm in order of RGB, respectively, and are quite narrower than those of the effective illumination for the CGH using RGB color filters. Thus, the volume hologram most likely provides better wavelength selectivity and less chromatic aberration to the CGH.

The HD-CGHs printed by laser lithography features reflection reconstruction, as mentioned in Sect. 15.3.1. This fact can be used to transfer the original CGH to a volume hologram. A thick photosensitive material, such as a photo-polymer, is attached to the surface of the original printed CGH to transfer the image, as shown in Fig. 15.44a. When the original CGH is reconstructed using a coherent light source, which illuminates the CGH through the recording material, the reconstructed light of the CGH is recorded on the material.

(a) (b) (c)

Fig. 15.45 Photographs of optical reconstruction of the original printed CGH using **a** monochromatic and **b** white-light illumination, and **c** the transferred CGVH using white-light illumination

Because the hologram transferred by this technique is a computer-generated volume hologram: CGVH; it has the property of wavelength selectivity. Thus, the hologram is only reconstructed at the same wavelength as that used in the transfer. Accordingly, when we illuminate the transferred CGVH with a white-light source, the 3D image is reconstructed in a single color, as shown in Fig. 15.44b. This is the same technique as *contact-copy* used in optical holography. The contact-copy technique features simplicity of the optical system and excellent vibration resistance.

Figure 15.45a shows optical reconstruction of an original HD-CGH printed using laser lithography. Here, because a green laser is used for the illumination light source, the reconstructed image is very clear and bright. However, the same CGH produces much color smear under a white-light illumination as in (b). In contrast, as shown in Fig. 15.45c, the CGVH transferred from the original CGH reconstructs almost the same monochromatic image as that of the original CGH illuminated by a laser, even if a white-light source is used for the illumination.

15.7.1.2 Stacking CGVH

The contact-copy allows us to reconstruct the CGHs by white-light illumination because of its wavelength selectivity. The fabricated CGVH is nearly transparent at other wavelengths. Thus, we can realize almost the same structure as the stacked CGH mentioned above.

Figure 15.46 schematically illustrates the principle of the stacked CGVH. In the fabrication of the stacked CGVH, three volume holograms are transferred from the three original CGHs at wavelengths corresponding to the RGB colors, and then, they are stacked as shown in Fig. 15.46. By using white-light illumination, full-color images are reconstructed by the superposition of the RGB color images. Here, note that the transferred CGHs must be aligned carefully so that the reconstructed 3D images do not cause any color shift.

It may also be possible to transfer all RGB images to a single volume hologram instead of three holograms. In this case, because the aligning work of RGB color

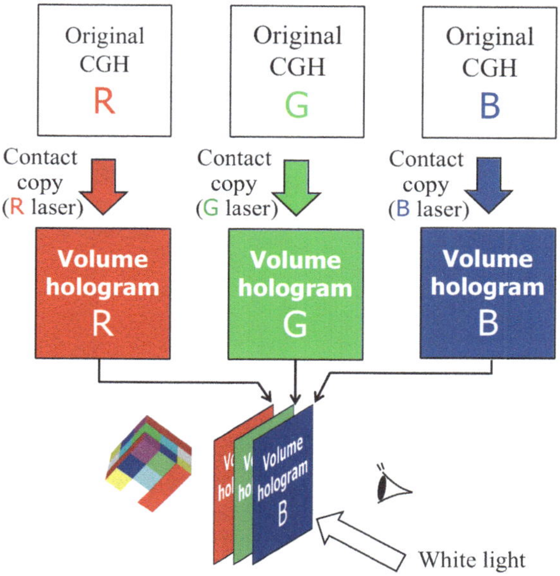

Fig. 15.46 The principle of full-color reconstruction by stacking three CGVHs

images must be done in a dark room, special aligning equipment is required. In addition, the recording material commonly has problems of limited writing-capacity. In multiple exposures, we must control the degree of exposure very carefully to produce a well-balanced and bright full-color image. These difficulties are not presented when stacking multiple holograms.

15.7.2 Compensation for Thickness and Refractive Index of Substrates

Each transferred CGVH is recorded on the photosensitive material coated on a glass substrate. Therefore, when we simply generate the three original CGHs corresponding to RGB wavelengths and stack them as shown in Fig. 15.47, the images reconstructed by the three CGVHs are not superimposed perfectly because of the position shift and aberration caused by the substrates. In the original CGHs, the fringe pattern made of chromium is also printed on a glass substrate and thus may cause a little aberration. To compensate for the thickness and refractive index of the glass substrates, the fringe patterns of the three original CGHs must be generated by the following procedure.

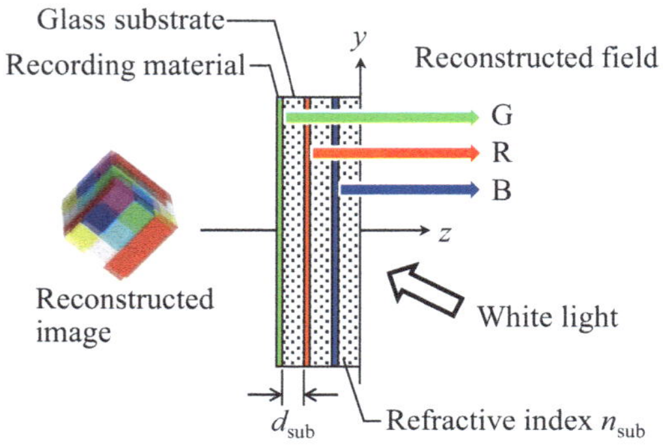

Fig. 15.47 Aberration caused by the thickness and refractive index of glass substrates of three CGVHs

15.7.2.1 Compensation for Substrates of Recording Material

Suppose that the origin of the coordinate system is placed at the forefront of a stacked CGVH. Object fields at three wavelengths are first calculated from the object model in a plane including the origin, as shown in Fig. 15.48a. Note that only the object field for green is depicted for example in Fig. 15.48. Let $g_{obj}(x, y; 0, \lambda_c)$ represent the the object field in the $(x, y, 0)$ plane. Here, λ_c is a wavelength corresponding to color c in vacuum, and c = R, G, or B again. This object field is referred to as an ideal object field in the following discussion.

In the optical reconstruction of a stacked CGVH, the field reconstructed by the volume hologram reaches the forefront after passing through the glass substrates. Therefore, we numerically propagate the ideal field backward to the position of the recording material through the glass substrates, as shown in Fig. 15.48b. Here, we assumed that the transferred holograms are stacked in the order of BRG from the forefront because the diffraction efficiency of the recording material increases in this order.[18] We also assume that the recording material is coated on the back of the substrate and thin enough to ignore the thickness.

To compensate for the aberration caused by the substrate, the object field $g_{obj}(x, y; 0, \lambda_c)$ must be propagated numerically at the wavelength inside the glass substrates. The wavelength is properly given by the refractive index of the glass:

$$\lambda'_c = \frac{\lambda_c}{n_{sub,c}}, \tag{15.11}$$

where $n_{sub,c}$ is a refractive index of the substrate for color c. The wavefield at the recording material for green, for example, is given by

$$g_{obj}(x, y; -3d_{sub}, \lambda_G) = \mathscr{P}^{(\lambda'_G)}_{-3d_{sub}} \left\{ g_{obj}(x, y; 0, \lambda_G) \right\} \tag{15.12}$$

[18]It is assumed that the recording material is a photo-polymer: Bayfol®HX200.

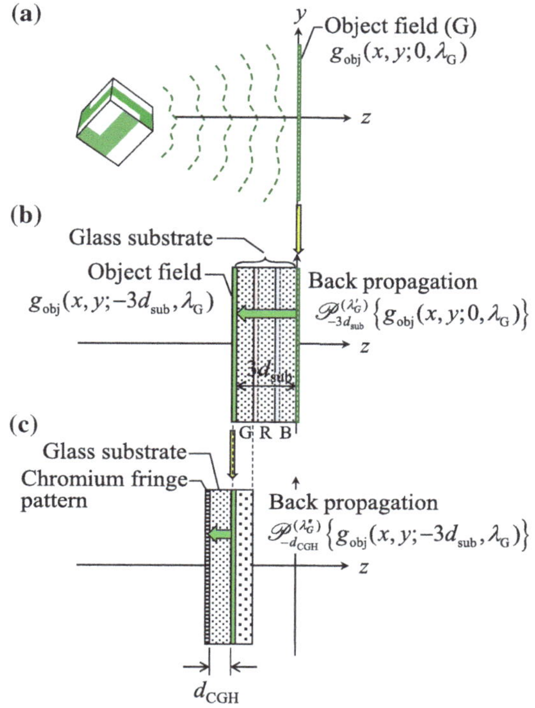

Fig. 15.48 Compensation for the thickness and refractive index of glass substrates: **a** an ideal object field, and compensation for substrates of **b** the recording material and **c** the original CGH. The object field for green is depicted for example

where propagation operator $\mathscr{P}_d^{(\lambda')}\{\}$ denotes that numerical propagation at a distance of d is performed with wavelength λ', i.e., the numerical propagation is performed with a propagation kernel in which the wavelength is λ'. The propagation distance depends on the color and is equal to $3d_{sub}$, in the case of the G field, as shown in Fig. 15.48b. Here, it is assumed that the all substrates have the same thickness.

15.7.2.2 Compensation for Substrate of Original CGH

In the original CGHs, the fringe pattern of the chromium thin films is also printed on the backside of the glass substrate, as shown in Fig. 15.44a. However, when we transfer the original CGH, the recording material is attached to the front side of the substrate of the original CGH, as in Fig. 15.48c. Thus, the field $g_{obj}(x, y; -3d_{sub}, \lambda_G)$ given at the position of the recording material should again be propagated backwardly at wavelength λ''_G in the glass substrate of the original CGH:

$$O(x, y; \lambda_G) = \mathscr{P}^{(\lambda''_G)}_{-d_{CGH}}\left\{g_{obj}(x, y; -3d_{sub}, \lambda_G)\right\} \tag{15.13}$$

where d_{CGH} is the thickness of the glass substrate of the original CGH.

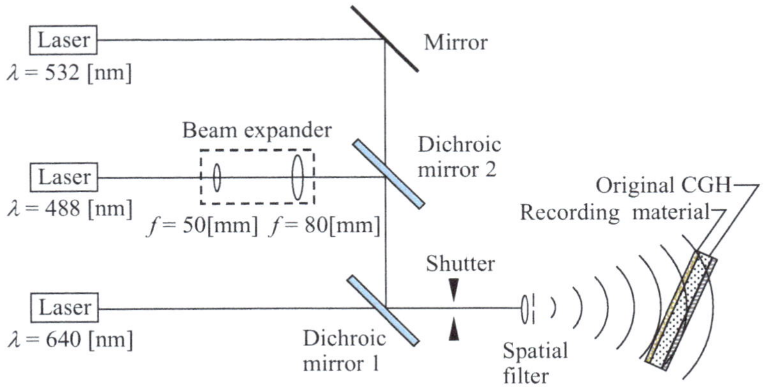

Fig. 15.49 An example of the optical system used to transfer the original CGHs to volume holograms [46]

Table 15.5 Actual parameters used for transferring the original CGHs to volume holograms

	R	G	B
Photo-polymer	Bayfol®HX200		
Intensity [mW/cm^2]	2.0	1.0	1.5
Exposure time [s]	30	30	30
Exposure [mJ/cm^2]	60	30	45

Finally, the fringe pattern of the original CGHs for each color is generated by encoding $O(x, y; \lambda_c)$ for binary amplitude CGHs (see Sect. 8.6.3).

15.7.3 Fabrication of Stacked CGVH

Figure 15.49 shows an example of the optical system to transfer the original CGHs to volume holograms. All light sources used for transfer are narrow-band diode-pumped solid-state (DPSS) lasers. The outputs of the lasers are combined into a coaxial beam using dichroic mirrors and then converted to spherical waves using a spatial filter, because the fringe pattern of the original CGHs are generated with a reference spherical wave. The beam expander, inserted into the path of the blue laser, is necessary to spread the spherical wave sufficiently; this is because the blue laser used in the experiment outputs a very narrow beam. The position of the pinhole of the spatial filter must be adjusted carefully so as to agree with the center of the spherical field used in encoding of the fringe pattern.

Here it should be noted that exposure should be optimized depending on the color sensitivity of the material. Table 15.5 shows examples of exposure parameters

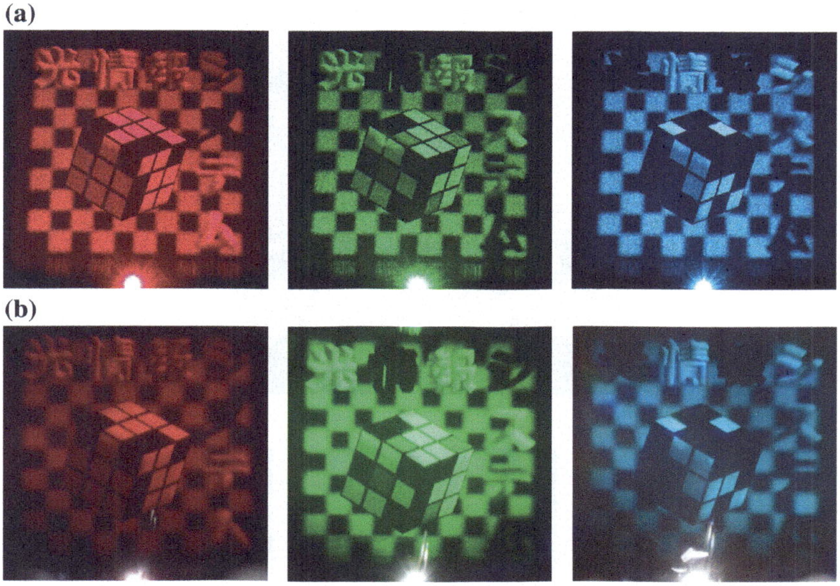

Fig. 15.50 Optical reconstructions of **a** the original CGHs illuminated by corresponding lasers and **b** individual CGVHs illuminated by a pigtail white LED [46]. Original CGHs: $m = n = 65{,}536$, $\Delta x_h = \Delta y_h = 0.8$ [μm], $\lambda_R = 640$ [nm], $\lambda_G = 532$ [nm], and $\lambda_B = 488$ [nm]. CGVH material: Bayfol®HX200

Fig. 15.51 Setup for optical reconstruction using a pigtail white LED

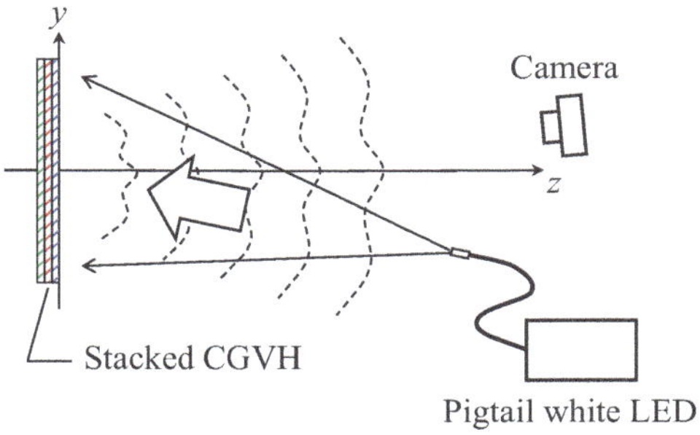

actually used for contact-copy. Optical reconstructions of the individual transferred CGH are shown in Fig. 15.50. Here, a pigtail white LED is used for the illuminating light source of CGVHs, as shown in Fig. 15.51.

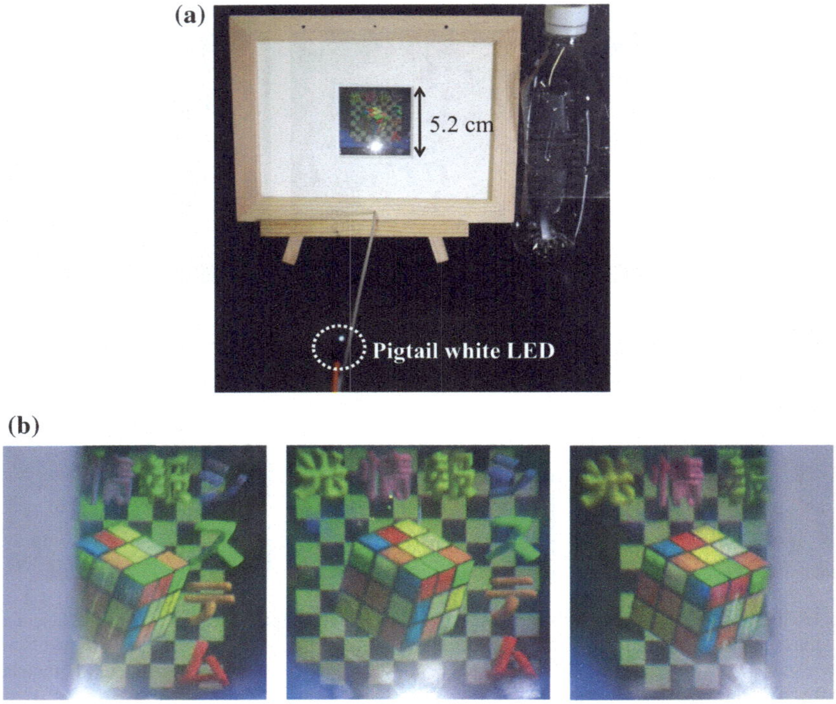

Fig. 15.52 Photographs of the 3D image reconstructed by the stacked CGVH at **a** a distant view and **b** close-up views from different viewpoints [46]. Video link, https://doi.org/10.6084/m9.figshare.8870177.v1

15.7.4 Optical Reconstruction of Stacked CGVH

When stacking the transferred CGHs to fabricate the full-color CGVH, the position of the CGVHs should be adjusted while looking the RGB image reconstructed by the white-light illumination. Optical reconstructions of the fabricated stacked CGVH are shown in Fig. 15.52, in which a pigtail white LED is again used for the illuminating light source because of its excellent spatial coherency. Close-up photographs of the 3D image reconstructed by the stacked CGVH are shown in Fig. 15.52b. The pictures are taken from different viewpoints. Little color shift is detected at any viewpoint. It is verified that the stacked CGVH reconstructs high-quality full-color 3D images in full-parallax. Continuous motion parallax, inherited from the original CGHs, is properly reconstructed, as shown in the online video.

Figure 15.53 shows a comparison of the 3D image reconstructed by the stacked CGVH with that of the full-color CGH using RGB color filters [123]. The reconstruction of the stacked CGVH definitely results in a clearer 3D image than that using RGB color filters. The 3D image by the stacked CGVH is also brighter than that by

(a) Stacked CGVH (b) RGB color filters

Fig. 15.53 Comparison between optical reconstructions of **a** the stacked CGVH and **b** full-color CGH using RGB color filters in Fig. 15.39

(a) **(b)**

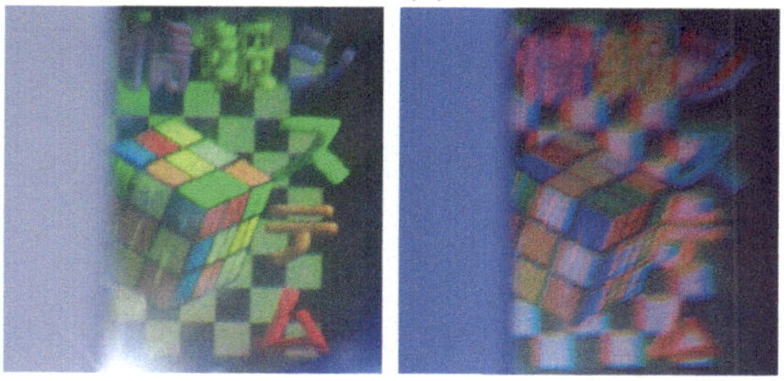

Fig. 15.54 Comparison between optical reconstructions of the stacked CGVHs **a** with and **b** without the substrate compensation [46]. Both pictures are taken from a left viewpoint

the CGH using RGB color filters. This is because the wavelength selectivity of the volume holograms is better than that by the RGB color filters, as mentioned above.

Figure 15.54 verifies the technique of compensation for the aberration caused by glass substrates. The fringe pattern in Fig. 15.54b is not generated with compensation for the aberration, whereas compensation is involved in the fringe in Fig. 15.54a. Here, the pictures are taken from a left viewpoint. Color shifts are clearly detected and the 3D image is drastically degraded in Fig. 15.54b. This means that the compensation is always necessary to decrease the aberration by the glass substrates in stacked CGVHs.

Appendix
Data of Major HD-CGHs

A.1 Parameters of Major HD-CGHs

Tables A.1 and A.2 summarize parameters of some major HD-CGHs created before 2012. The table is reprinted from [73] with reformatting. Here, 'Creator' in the table means programmer and/or designer. 'Date' usually indicates the printed date of the work. However, there is ambiguity because several versions are sometimes reprinted for improvement. Some newer HD-CGHs appearing in this book are listed in Table A.3. These CGHs were created from 2013.

K. Matsushima, *Introduction to Computer Holography*,
Series in Display Science and Technology,
https://doi.org/10.1007/978-3-030-38435-7

Table A.1 Major HD-CGHs before 2012 (From J. Phys.: Conf. Ser. 415, 012053 (2013) with reformatting)

Title	Dimensions $W \times H$ [cm]	No. of pixels $W \times H$	Pixel pitches [μm]	Creator[a]	Date	References	Notes
The Venus	6.5 × 6.5	65,536 × 65,536	1.0 × 1.0	KM	Jan. 2009	[67–69]	The first computer hologram created by our method
Moai I	6.5 × 6.5	65,536 × 65,536	1.0 × 1.0	KM	Jun. 2009	[70]	Shifted Fresnel method [87] for numerical propagation
Moai II	6.5 × 6.5	65,536 × 65,536	1.0 × 1.0	KM	Oct. 2009	[70, 74, 77]	Shifted angular spectrum method [62] for numerical propagation
Aqua 2	10.5 × 6.5	131,072 × 65,536	0.8 × 1.0	MN	Dec. 2009	[88]	Acceleration algorithm for the silhouette method in sparse 3D scenes
The Moon	10.5 × 6.5	131,072 × 65,536	0.8 × 1.0	KM	Mar. 2010	[74, 77]	Texture-mapping
Shion	10.5 × 5.2	131,072 × 65,536	0.8 × 0.8	KM, HN	Dec. 2010	[77, 92]	Texture-mapping, Gouraud shading, Shape measurement by 3D laser scanner
Bear II	6.5 × 6.5	65,536 × 65,536	1.0 × 1.0	YA, HN	Oct. 2010	[2, 63, 92]	The first major work in digitized holography
Mountain	3.3 × 3.3	32,768 × 32,768	1.0 × 1.0	KH	Nov. 2010	[92]	Big object for mountain scenery
The Metal Venus I	6.5 × 6.5	65,536 × 65,536	1.0 × 1.0	HN	Nov. 2010	[92, 95]	Specular flat shading
The Metal Venus II	6.5 × 6.5	65,536 × 65,536	1.0 × 1.0	HN	Jan. 2011	[94, 96]	Specular smooth shading

[a] *KM* Kyoji MATSUSHIMA, *MN* Masaki NAKAMURA, *HN* Hirohito NISHI, *YA* Yasuaki ARIMA, *KH* Kentaro HIGASHI, *HY* Hiroshi YAMASHITA. All holograms are fabricated by Sumio NAKAHARA

Table A.2 (Continued) Major HD-CGHs before 2012 (From J. Phys.: Conf. Ser. 415, 012053(2013) with reformatting)

Title	Dimensions $W \times H$ [cm]	No. of pixels $W \times H$	Pixel pitches [μm]	Creator[a]	Date	References	Notes
The Hands	3.3 × 3.3	32,768 × 32,768	1.0 × 1.0	MN	Jan. 2011	[76]	Prototype using the switch-back technique for accelerating silhouette method
iShion V2	5.2 × 5.2	65,536 × 65,536	0.8 × 0.8	HY	Dec. 2011	[138]	Image hologram, Reconstruction by white light
iShion V3	10.5 × 10.5	131,072 × 131,072	0.8 × 0.8	HY	May 2012		Prototype of floating hologram
Hamsters	6.5 × 6.5	65,536 × 65,536	1.0 × 1.0	DF	Apr. 2012		Resizing object in digitized holography
Parked car 1	6.5 × 6.5	65,536 × 65,536	1.0 × 1.0	YY	Jun. 2012		Prototype using multi-viewpoint images for background
Penguin	13.1 × 6.5	131,072 × 65,536	1.0 × 1.0	YA	Nov. 2011		Large hologram in digitized holography
Rose in Ring	10.5 × 10.5	131,072 × 131,072	0.8 × 0.8	KM, HN, KO	Nov. 2011		The switch-back technique Specular smooth shading
Brothers	12.6 × 10.5	196,608 × 131,072	0.64 × 0.8	KM	Apr. 2012	[71]	Texture-mapping, Gouraud shading, Two live faces measured by 3D laser scanner

[a] *KM* Kyoji MATSUSHIMA, *MN* Masaki NAKAMURA, *HN* Hirohito NISHI, *YA* Yasuaki ARIMA, *HY* Hiroshi YAMASHITA, *YY* Yutaka YOSHIZAKI, *DF* Daichi FUJITA, *KO* Kohei OGAWA. All holograms are fabricated by Sumio NAKAHARA

Table A.3 Recent HD-CGHs appearing in this book (from 2013)

Title	Dimensions $W \times H$ [cm]	No. of pixels $W \times H$	Pixel pitches [μm]	Creator[a]	Date	References	Notes
Five Rings	5.2 × 5.2	65,536 × 65,536	0.8 × 0.8	KM	2013	[75]	Switch-back technique
Rubik Cube	5.2 × 5.2	65,536 × 65,536	0.8 × 0.8	TM	2014	[83]	Full-color HD-CGH using dichroic mirrors
Color Cube	5.2 × 5.2	65,536 × 65,536	0.8 × 0.8	YT	2016	[123]	Full-color HD-CGH using RGB color filters
Casino Chips	10.5 × 10.5	131,072 × 131,072	0.8 × 0.8	KM	2016	[123]	Full-color HD-CGH using RGB color filter
Tea Time	6.5 × 6.5	65,536 × 65,536	1.0 × 1.0	NS	2016	[80]	Full-color digitized holography, RGB color filters
Triplane IIb	18 × 12	225,000 × 200,000	0.6 × 0.4	KM	2017		Fringe oversampling
Sailing Warship II	18 × 12	225,000 × 300,000	0.8 × 0.4	KM	2017		Fringe oversampling
Ring in Cube	6.1 × 3.5	30,720 × 17,280	2.0 × 2.0	TS	2017	[105]	Printed by a Denysyuk-type wavefront printer
Zero Gravity	5.2 × 5.2	65,536 × 65,536	0.8 × 0.8	SM	2017		Full-color HD-CGH using RGB color filters
Color Cube 2b	5.2 × 5.2	65,536 × 65,536	0.8 × 0.8	OK	2018	[46]	Full-color stacked CGVH, The same 3D model as Color Cube
Color Shion II	10.5 × 10.5	131,072 × 262,144	0.8 × 0.4	KM	2018		Full-color HD-CGH using RGB color filter, Fringe oversampling
Toy Train	18 × 18	225,000 × 450,000	0.8 × 0.4	KM	2018		Fringe oversampling

[a] *KM* Kyoji Matsushima, *TM* Takashi MIYAOKA, *YT* Yasuhiro TSUCHIYAMA, *NS* Noriaki SONOBE, *TS* Tomotaka SAITO, *SM* Syota MATSUZAKI, *OK* Orie KUNIEDA. Almost all holograms are fabricated by Miho II

References

1. Ahrenberg, L., Benzie, P., Magnor, M., Watson, J.: Computer generated holograms from three dimensional meshes using an analytic light transport model. Appl. Opt. **47**(10), 1567–1574 (2008). https://doi.org/10.1364/AO.47.001567. http://ao.osa.org/abstract.cfm?URI=ao-47-10-1567
2. Arima, Y., Matsushima, K., Nakahara, S.: Hybrid CGH by digitized holography: CGH for mixed 3d scene of virtual and real objects. In: Digital Holography and Three-Dimensional Imaging, DWC28. Optical Society of America (2011). https://doi.org/10.1364/DH.2011.DWC28. http://www.osapublishing.org/abstract.cfm?URI=DH-2011-DWC28
3. Bailey, D.H., Swarztrauber, P.N.: The fractional Fourier transform and applications. SIAM Rev. **33**, 389–404 (1991)
4. Benton, S.A., Michael Bove, Jr., V.: Holographic Imaging, Chap. 15. Wiley (2008)
5. Binet, R., Colineau, J., Lehureau, J.C.: Short-range synthetic aperture imaging at 633 nm by digital holography. Appl. Opt. **41**(23), 4775–4782 (2002). http://ao.osa.org/abstract.cfm?URI=ao-41-23-4775
6. Bjelkhagen, H.I., Brotherton-Ratcliffe, D.: Ultra-Realistic Imaging, Chaps. 7, 8. CRC Press (2013)
7. Brown, B.R., Lohmann, A.W.: Complex spatial filtering with binary masks. Appl. Opt. **5**(6), 967–969 (1966).https://doi.org/10.1364/AO.5.000967. http://ao.osa.org/abstract.cfm?URI=ao-5-6-967
8. Bryngdahl, O., Lohmann, A.: Single-sideband holography∗. J. Opt. Soc. Am. **58**(5), 620–624 (1968). https://doi.org/10.1364/JOSA.58.000620. http://www.osapublishing.org/abstract.cfm?URI=josa-58-5-620
9. Chen, J.S., Chu, D., Smithwick, Q.Y.: Rapid hologram generation utilizing layer-based approach and graphic rendering for realistic three-dimensional image reconstruction by angular tiling. J. Electron. Imaging **23**(2), 1–8 (2014). https://doi.org/10.1117/1.JEI.23.2.023016. https://doi.org/10.1117/1.JEI.23.2.023016
10. Chen, J.S., Chu, D.P.: Improved layer-based method for rapid hologram generation and real-time interactive holographic display applications. Opt. Express **23**(14), 18143–18155 (2015). https://doi.org/10.1364/OE.23.018143. http://www.opticsexpress.org/abstract.cfm?URI=oe-23-14-18143
11. Chen, R.H.Y., Wilkinson, T.D.: Computer generated hologram from point cloud using graphics processor. Appl. Opt. **48**(36), 6841–6850 (2009). https://doi.org/10.1364/AO.48.006841. http://ao.osa.org/abstract.cfm?URI=ao-48-36-6841

K. Matsushima, *Introduction to Computer Holography*,
Series in Display Science and Technology,
https://doi.org/10.1007/978-3-030-38435-7

12. Delen, N., Hooker, B.: Free-space beam propagation between arbitrarily oriented planes based on full diffraction theory: a fast Fourier transform approach. J. Opt. Soc. Am. A **15**, 857–867 (1998)
13. Fujii, T., Yoshikawa, H.: Improvement of hidden-surface removal for computer-generated holograms from CG. In: Adaptive Optics: Analysis and Methods/Computational Optical Sensing and Imaging/Information Photonics/Signal Recovery and Synthesis Topical Meetings on CD-ROM, DWB3. Optical Society of America (2007). https://doi.org/10.1364/DH.2007.DWB3. http://www.osapublishing.org/abstract.cfm?URI=DH-2007-DWB3
14. Fujita, D.: Digital resizing and reduction of occlusion errors in digitized holography (in Japanese). Master's thesis, Kansai University (2013)
15. Fujita, D., Matsushima, K., Nakahara, S.: Digital resizing of reconstructed object images in digitized holography. In: Digital Holography and Three-Dimensional Imaging, DW2A.7. Optical Society of America (2013). https://doi.org/10.1364/DH.2013.DW2A.7. http://www.osapublishing.org/abstract.cfm?URI=DH-2013-DW2A.7
16. Gabor, D.: A new microscopic principle. Nature **161**, 777–778 (1948). https://doi.org/10.1038/161777a0. https://www.nature.com/articles/161777a0
17. Ganci, S.: Fourier diffraction through a tilted slit. Eur. J. Phys. **2**, 158–160 (1981)
18. Gerchberg, R.W., Saxton, W.O.: A practical algorithm for determination of phase from image and diffraction plane pictures. Optik **35**(2), 237–246 (1972)
19. Goodman, J.W.: Introduction to Fourier Optics, 2nd ed., Chap. 2.2. McGraw-Hill (1996)
20. Goodman, J.W., Lawrence, R.W.: Digital image formation from electronically detected holograms. Appl. Phys. Lett. **11**(3), 77–79 (1967)
21. Hamano, T., Kitamura, M., Yoshikawa, H.: Computer-generated holograms with pulse-width modulation for multilevel 3D images. In: Benton, S.A. (ed.) Practical Holography XIII, vol. 3637, pp. 244–251. International Society for Optics and Photonics, SPIE (1999). https://doi.org/10.1117/12.343782. https://doi.org/10.1117/12.343782
22. Hamano, T., Yoshikawa, H.: Image-type CGH by means of e-beam printing. In: Benton, S.A. (ed.) Practical Holography XII, vol. 3293, pp. 2–14. International Society for Optics and Photonics, SPIE (1998). https://doi.org/10.1117/12.303638. https://doi.org/10.1117/12.303638
23. Hanák, I., Herout, A., Zemcik, P.: Acceleration of detail driven method for hologram generation. Opt. Eng. **49**(8), 1–9 (2010). https://doi.org/10.1117/1.3476331. https://doi.org/10.1117/1.3476331
24. Hanák, I., Janda, M., Skala, V.: Detail-driven digital hologram generation. Vis. Comput. **26**, 83–96 (2010). https://doi.org/10.1007/s00371-009-0378-5
25. Harburn, G., Ranniko, J.K., Williams, R.P.: An aspect of phase in Fraunhofer diffraction patterns. Optik **48**, 321–328 (1977)
26. Heidelberg Instruments: DWL66$^+$ fact sheet (2019)
27. Huang, T.S.: Digital holography. Proc. IEEE **59**(9), 1335–1346 (1971). https://doi.org/10.1109/PROC.1971.8408
28. Ichihashi, Y., Yamamoto, K., Wakunami, K., Oi, R., Okui, M., Senoh, T.: An analysis of printing conditions for wavefront overlapping printing. In: Bjelkhagen, H.I., Michael Bove, Jr., V. (eds.) Practical Holography XXXI: Materials and Applications, vol. 10127, pp. 153–158. International Society for Optics and Photonics, SPIE (2017). https://doi.org/10.1117/12.2251285. https://doi.org/10.1117/12.2251285
29. Ichikawa, T., Sakamoto, Y., Subagyo, A., Sueoka, K.: Calculation method of reflectance distributions for computer-generated holograms using the finite-difference time-domain method. Appl. Opt. **50**(34), H211–H219 (2011). https://doi.org/10.1364/AO.50.00H211. http://ao.osa.org/abstract.cfm?URI=ao-50-34-H211
30. Ichikawa, T., Yoneyama, T., Sakamoto, Y.: CGH calculation with the ray tracing method for the Fourier transform optical system. Opt. Express **21**(26), 32019–32031 (2013). https://doi.org/10.1364/OE.21.032019. http://www.opticsexpress.org/abstract.cfm?URI=oe-21-26-32019

31. Ichioka, Y., Izumi, M., Suzuki, T.: Scanning halftone plotter and computer-generated continuous-tone hologram. Appl. Opt. **10**(2), 403–411 (1971). https://doi.org/10.1364/AO.10.000403. http://ao.osa.org/abstract.cfm?URI=ao-10-2-403
32. Ito, T., Eldeib, H., Yoshida, K., Takahashi, S., Yabe, T., Kunugi, T.: Special purpose computer for holography HORN-2. Comput. Phys. Commun. **93**, 13–20 (1996)
33. Ito, T., Masuda, N., Yoshimura, K., Shiraki, A., Shimobaba, T., Sugie, T.: Special-purpose computer HORN-5 for a real-time electroholography. Opt. Express **13**, 1923–1932 (2005)
34. Ito, T., Yabe, T., Okazaki, M., Yanagi, M.: Special purpose computer holography HORN-1 for reconstruction of virtual image in three dimensions. Comput. Phys. Commun. **82**, 104–110 (1994)
35. Iwami, S., Sakamoto, Y.: A study of glasses-type color CGH using a color filter considering reduction of blurring. In: Bjelkhagen, H.I., Kostuk, R.K. (eds.) Practical Holography XXIII: Materials and Applications, vol. 7233, pp. 291–298. International Society for Optics and Photonics, SPIE (2009). https://doi.org/10.1117/12.808804. https://doi.org/10.1117/12.808804
36. Juárez-Pérez, J.L., Olivares-Pérez, A., Berriel-Valdos, L.R.: Nonredundant calculations for creating digital fresnel holograms. Appl. Opt. **36**(29), 7437–7443 (1997). https://doi.org/10.1364/AO.36.007437. http://ao.osa.org/abstract.cfm?URI=ao-36-29-7437
37. Kajiki, Y., Okamoto, M., Ando, T., Yamasaki, K., Shimizu, E.: Computer-generated color holograms using RGB color filters. In: Benton, S.A. (ed.) Practical Holography XII, vol. 3293, pp. 125–130. International Society for Optics and Photonics, SPIE (1998). https://doi.org/10.1117/12.303645. https://doi.org/10.1117/12.303645
38. Kämpfe, T., Kley, E.B., Tünnermann, A., Dannberg, P.: Design and fabrication of stacked, computer generated holograms for multicolor image generation. Appl. Opt. **46**(22), 5482–5488 (2007). https://doi.org/10.1364/AO.46.005482. http://ao.osa.org/abstract.cfm?URI=ao-46-22-5482
39. Kang, H., Yamaguchi, T., Yoshikawa, H.: Accurate phase-added stereogram to improve the coherent stereogram. Appl. Opt. **47**(19), D44–D54 (2008). https://doi.org/10.1364/AO.47.000D44. http://ao.osa.org/abstract.cfm?URI=ao-47-19-D44
40. Kim, H., Hahn, J., Lee, B.: Mathematical modeling of triangle-mesh-modeled three-dimensional surface objects for digital holography. Appl. Opt. **47**(19), D117–D127 (2008). https://doi.org/10.1364/AO.47.00D117. http://ao.osa.org/abstract.cfm?URI=ao-47-19-D117
41. Kim, S.C., Kim, E.S.: Effective generation of digital holograms of three-dimensional objects using a novel look-up table method. Appl. Opt. **47**(19), D55–D62 (2008). https://doi.org/10.1364/AO.47.000D55. http://ao.osa.org/abstract.cfm?URI=ao-47-19-D55
42. Kim, Y., Stoykova, E., Kang, H., Hong, S., Park, J., Park, J., Hong, J.: Seamless full color holographic printing method based on spatial partitioning of SLM. Opt. Express **23**(1), 172–182 (2015). https://doi.org/10.1364/OE.23.000172. http://www.opticsexpress.org/abstract.cfm?URI=oe-23-1-172
43. Kogelnik, H.: Coupled-wave theory for thick hologram grating. Bell Syst. Tech. J. **48**, 2909–2947 (1969)
44. Kollin, J.S., Benton, S.A., Jepsen, M.L.: Real-time display Of 3-d computed holograms by scanning the image of an acousto-optic modulator. In: Morris, G.M. (ed.) Holographic Optics II: Principles and Applications, vol. 1136, pp. 178–185. International Society for Optics and Photonics, SPIE (1989). https://doi.org/10.1117/12.961683. https://doi.org/10.1117/12.961683
45. Kondoh, A., Matsushima, K.: Hidden surface removal in full-parallax CGHs by silhouette approximation. Syst. Comput. Jpn. **38**, 53–61 (2007)
46. Kunieda, O., Matsushima, K.: High-quality full-parallax full-color three-dimensional image reconstructed by stacking large-scale computer-generated volume holograms. Appl. Opt. **58**(34), G104–G111 (2019). https://doi.org/10.1364/AO.58.00G104. http://ao.osa.org/abstract.cfm?URI=ao-58-34-G104

47. Lalor, É.: Conditions for the validity of the angular spectrum of plane waves. J. Opt. Soc. Am. **58**(9), 1235–1237 (1968). https://doi.org/10.1364/JOSA.58.001235. http://www.osapublishing.org/abstract.cfm?URI=josa-58-9-1235
48. Lee, A.J., Casasent, D.P.: Computer generated hologram recording using a laser printer. Appl. Opt. **26**, 136–138 (1987)
49. Lehmann, T.M., Gönner, C., Spitzer, K.: Large-sized local interpolators. In: Proceedings of IASTED International Conference on Computer Graphics and Imaging, pp. 156–161 (1999)
50. Leseberg, D.: Computer-generated three-dimensional image holograms. Appl. Opt. **31**, 223–229 (1992)
51. Leseberg, D., Frère, C.: Computer-generated holograms of 3-D objects composed of tilted planar segments. Appl. Opt. **27**, 3020–3024 (1988)
52. Liu, Y.Z., Dong, J.W., Pu, Y.Y., Chen, B.C., He, H.X., Wang, H.Z.: High-speed full analytical holographic computations for true-life scenes. Opt. Express **18**(4), 3345–3351 (2010). https://doi.org/10.1364/OE.18.003345. http://www.opticsexpress.org/abstract.cfm?URI=oe-18-4-3345
53. Lohmann, A.W.: Three-dimensional properties of wave-fields. Optik **51**, 105–117 (1978)
54. Lohmann, A.W., Dorsch, R.G., Mendlovic, D., Zalevsky, Z., Ferreira, C.: Space-bandwidth product of optical signals and systems. J. Opt. Soc. Am. A **13**(3), 470–473 (1996). https://doi.org/10.1364/JOSAA.13.000470. http://josaa.osa.org/abstract.cfm?URI=josaa-13-3-470
55. Lohmann, A.W., Paris, D.P.: Binary Fraunhofer holograms, generated by computer. Appl. Opt. **6**(10), 1739–1748 (1967). https://doi.org/10.1364/AO.6.001739. http://ao.osa.org/abstract.cfm?URI=ao-6-10-1739
56. Lucente, M.E.: Interactive computation of holograms using a look-up table. J. Electron. Imaging **2**(1), 28–34 (1993). https://doi.org/10.1117/12.133376. https://doi.org/10.1117/12.133376
57. M. Kitamura, T.H.: Computer-generated holograms for multilevel 3D images with complex amplitude modulation. In: SPIE Proceedings of the Practical Holography XVI and Holographic Materials VIII, vol. 4659, pp. 196–204 (2002)
58. Masuda, N., Ito, T., Tanaka, T., Shiraki, A., Sugie, T.: Computer generated holography using a graphics processing unit. Opt. Express **14**(2), 603–608 (2006). https://doi.org/10.1364/OPEX.14.000603. http://www.opticsexpress.org/abstract.cfm?URI=oe-14-2-603
59. Matsushima, K.: Computer-generated holograms for three-dimensional surface objects with shade and texture. Appl. Opt. **44**(22), 4607–4614 (2005). https://doi.org/10.1364/AO.44.004607. http://ao.osa.org/abstract.cfm?URI=ao-44-22-4607
60. Matsushima, K.: Exact hidden-surface removal in digitally synthetic full-parallax holograms. In: Jeong, T.H., Bjelkhagen, H.I. (eds.) Practical Holography XIX: Materials and Applications, vol. 5742, pp. 25–32. International Society for Optics and Photonics, SPIE (2005). https://doi.org/10.1117/12.592520. https://doi.org/10.1117/12.592520
61. Matsushima, K.: Formulation of the rotational transformation of wave fields and their application to digital holography. Appl. Opt. **47**(19), D110–D116 (2008). https://doi.org/10.1364/AO.47.00D110. http://ao.osa.org/abstract.cfm?URI=ao-47-19-D110
62. Matsushima, K.: Shifted angular spectrum method for off-axis numerical propagation. Opt. Express **18**(17), 18453–18463 (2010). https://doi.org/10.1364/OE.18.018453. http://www.opticsexpress.org/abstract.cfm?URI=oe-18-17-18453
63. Matsushima, K., Arima, Y., Nakahara, S.: Digitized holography: modern holography for 3D imaging of virtual and real objects. Appl. Opt. **50**(34), H278–H284 (2011). https://doi.org/10.1364/AO.50.00H278
64. Matsushima, K., Joko, A.: High-resolution printer for fabricating computer-generated display holograms (in Japanese). J. Inst. Image Inf. Telev. Eng. **56**, 1989–1994 (2002)
65. Matsushima, K., Kobayashi, S., Miyauchi, H.: A high-resolution fringe printer for studying synthetic holograms. In: Bjelkhagen, H.I., Lessard, R.A. (eds.) Practical Holography XX: Materials and Applications, vol. 6136, pp. 347–354. International Society for Optics and Photonics, SPIE (2006). https://doi.org/10.1117/12.649472. https://doi.org/10.1117/12.649472

66. Matsushima, K., Kondoh, A.: A wave-optical algorithm for hidden-surface removal in digitally synthetic full-parallax holograms for three-dimensional objects. In: Jeong, T.H., Bjelkhagen, H.I. (eds.) Practical Holography XVIII: Materials and Applications, vol. 5290, pp. 90–97. International Society for Optics and Photonics, SPIE (2004). https://doi.org/10.1117/12.526747. https://doi.org/10.1117/12.526747
67. Matsushima, K., Nakahara, S.: Extremely high-definition full-parallax computer-generated hologram created by the polygon-based method. Appl. Opt. **48**(34), H54–H63 (2009). https://doi.org/10.1364/AO.48.000H54. http://ao.osa.org/abstract.cfm?URI=ao-48-34-H54
68. Matsushima, K., Nakahara, S.: A high-definition full-parallax CGH created by the polygon-based method. In: Advances in Imaging, PDWB38. Optical Society of America (2009). https://doi.org/10.1364/DH.2009.PDWB38. http://www.osapublishing.org/abstract.cfm?URI=DH-2009-PDWB38
69. Matsushima, K., Nakahara, S.: Region segmentation and parallel processing for creating large-scale CGHs in polygon source method. In: Bjelkhagen, H.I., Kostuk, R.K. (eds.) Practical Holography XXIII: Materials and Applications, vol. 7233, pp. 104–111. International Society for Optics and Photonics, SPIE (2009). https://doi.org/10.1117/12.808633. https://doi.org/10.1117/12.808633
70. Matsushima, K., Nakahara, S.: High-definition full-parallax CGHs created by using the polygon-based method and the shifted angular spectrum method. In: Bjelkhagen, H.I., Kostuk, R.K. (eds.) Practical Holography XXIV: Materials and Applications, vol. 7619, pp. 295–302. International Society for Optics and Photonics, SPIE (2010). https://doi.org/10.1117/12.844606. https://doi.org/10.1117/12.844606
71. Matsushima, K., Nakahara, S.: Stepping closer to the perfect 3D digital image. SPIE Newsroom (2012). https://doi.org/10.1117/2.1201210.004526. https://dx.doi.org/10.1117/2.1201210.004526
72. Matsushima, K., Nakahara, S.: Frequency filtering for reduction of memory usage in computer holography. In: Digital Holography & 3-D Imaging Meeting, DM4A.6. Optical Society of America (2015). https://doi.org/10.1364/DH.2015.DM4A.6. http://www.osapublishing.org/abstract.cfm?URI=DH-2015-DM4A.6
73. Matsushima, K., Nakahara, S., Arima, Y., Nishi, H., Yamashita, H., Yoshizaki, Y., Ogawa, K.: Computer holography: 3D digital art based on high-definition CGH. J. Phys.: Conf. Ser. **415**, 012053 (2013). https://doi.org/10.1088/1742-6596/415/1/012053. https://doi.org/10.1088%2F1742-6596%2F415%2F1%2F012053
74. Matsushima, K., Nakamura, M., Nakahara, S.: Novel techniques introduced into polygon-based high-definition cghs. In: Biomedical Optics and 3-D Imaging, JMA10. Optical Society of America (2010). http://www.osapublishing.org/abstract.cfm?URI=DH-2010-JMA10
75. Matsushima, K., Nakamura, M., Nakahara, S.: Silhouette method for hidden surface removal in computer holography and its acceleration using the switch-back technique. Opt. Express **22**(20), 24450–24465 (2014). https://doi.org/10.1364/OE.22.024450. http://www.opticsexpress.org/abstract.cfm?URI=oe-22-20-24450
76. Matsushima, K., Nishi, H., Arima, Y., Nakahara, S.: Computer holography: 3D imaging of virtual and real objects. In: International Workshop on Holography and related technologies 2011 (IWH2011), pp. 1–2 (2011)
77. Matsushima, K., Nishi, H., Nakahara, S.: Simple wave-field rendering for photorealistic reconstruction in polygon-based high-definition computer holography. J. Electron. Imaging **21**(2), 1–9 (2012). https://doi.org/10.1117/1.JEI.21.2.023002. https://doi.org/10.1117/1.JEI.21.2.023002
78. Matsushima, K., Schimmel, H., Wyrowski, F.: Fast calculation method for optical diffraction on tilted planes by use of the angular spectrum of plane waves. J. Opt. Soc. Am. A **20**(9), 1755–1762 (2003). https://doi.org/10.1364/JOSAA.20.001755. http://josaa.osa.org/abstract.cfm?URI=josaa-20-9-1755
79. Matsushima, K., Shimobaba, T.: Band-limited angular spectrum method for numerical simulation of free-space propagation in far and near fields. Opt. Express **17**(22), 19662–19673 (2009).

https://doi.org/10.1364/OE.17.019662. http://www.opticsexpress.org/abstract.cfm?URI=oe-17-22-19662
80. Matsushima, K., Sonobe, N.: Full-color digitized holography for large-scale holographic 3D imaging of physical and nonphysical objects. Appl. Opt. **57**(1), A150–A156 (2018). https://doi.org/10.1364/AO.57.00A150. http://ao.osa.org/abstract.cfm?URI=ao-57-1-A150
81. Matsushima, K., Takai, M.: Recurrence formulas for fast creation of synthetic three-dimensional holograms. Appl. Opt. **39**(35), 6587–6594 (2000). https://doi.org/10.1364/AO.39.006587. http://ao.osa.org/abstract.cfm?URI=ao-39-35-6587
82. Mishina, T., Okui, M., Okano, F.: Calculation of holograms from elemental images captured by integral photography. Appl. Opt. **45**(17), 4026–4036 (2006). https://doi.org/10.1364/AO.45.004026. http://ao.osa.org/abstract.cfm?URI=ao-45-17-4026
83. Miyaoka, T., Matsushima, K., Nakahara, S.: Optimization of design-wavelength for unobtrusive chromatic aberration in high-definition color computer holography. In: Bjelkhagen, H.I., Michael Bove, Jr., V. (eds.) Practical Holography XXIX: Materials and Applications, vol. 9386, pp. 147–158. International Society for Optics and Photonics, SPIE (2015). https://doi.org/10.1117/12.2076876. https://doi.org/10.1117/12.2076876
84. Moharam, M.G., Gayloard, T.K.: Coupled-wave analysis of reflection gratings. Appl. Opt. **20**, 240–244 (1981)
85. Moharam, M.G., Gayloard, T.K.: Rigorous coupled-wave analysis of planar-grating diffraction. J. Opt. Soc. Am. **71**, 811–818 (1981)
86. Moharam, M.G., Gayloard, T.K.: Diffraction analysis of dielectric surface-relief gratings. J. Opt. Soc. Am. **72**, 1385–1392 (1982)
87. Muffoletto, R.P., Tyler, J.M., Tohline, J.E.: Shifted Fresnel diffraction for computational holography. Opt. Express **15**, 5631–5640 (2007)
88. Nakamura, M., Matsushima, K., Nakahara, S.: Hidden-surface removal based on Babinet's principle and partial wave field propagation in high-resolution computer-generated holograms (in Japanese). Inst. Image Inf. Telev. Eng. **66**(5), J136–J143 (2012). https://doi.org/10.3169/itej.66.J136
89. Nakatsuji, N., Matsushima, K.: Comparison of computation time and image quality between full-parallax 4G-pixels CGHs calculated by the point cloud and polygon-based method. In: Bjelkhagen, H.I., Michael Bove, Jr., V. (eds.) Practical Holography XXXI: Materials and Applications, vol. 10127, pp. 256–265. International Society for Optics and Photonics, SPIE (2017). https://doi.org/10.1117/12.2251766. https://doi.org/10.1117/12.2251766
90. Nakatsuji, T.: Generation of free-viewpoint images by phase-shifting synthetic aperture digital holography (in Japanese). Master's thesis, Kansai University (2007)
91. Nakatsuji, T., Matsushima, K.: Free-viewpoint images captured using phase-shifting synthetic aperture digital holography. Appl. Opt. **47**(19), D136–D143 (2008)
92. Nishi, H., Higashi, K., Arima, Y., Matsushima, K., Nakahara, S.: New techniques for wave-field rendering of polygon-based high-definition CGHs. In: Bjelkhagen, H.I. (ed.) Practical Holography XXV: Materials and Applications, vol. 7957, pp. 332–342. International Society for Optics and Photonics, SPIE (2011). https://doi.org/10.1117/12.876362. https://doi.org/10.1117/12.876362
93. Nishi, H., Matsushima, K.: Rendering of specular curved objects in polygon-based computer holography. Appl. Opt. **56**(13), F37–F44 (2017). https://doi.org/10.1364/AO.56.000F37. http://ao.osa.org/abstract.cfm?URI=ao-56-13-F37
94. Nishi, H., Matsushima, K., Nakahara, S.: A novel method for rendering specular and smooth surfaces in polygon-based high-definition CGH. In: Digital Holography and Three-Dimensional Imaging, DWC29. Optical Society of America (2011). https://doi.org/10.1364/DH.2011.DWC29. http://www.osapublishing.org/abstract.cfm?URI=DH-2011-DWC29
95. Nishi, H., Matsushima, K., Nakahara, S.: Rendering of specular surfaces in polygon-based computer-generated holograms. Appl. Opt. **50**(34), H245–H252 (2011). https://doi.org/10.1364/AO.50.00H245
96. Nishi, H., Matsushima, K., Nakahara, S.: Smooth shading of specular surfaces in polygon-based high-definition CGH. In: 3DTV-Conference 2011 (2011)

97. Nishii, W., Matsushima, K.: Stepping closer to the perfect 3D digital image. In: Bjelkhagen, H.I., Michael Bove, Jr., V. (eds.) Practical Holography XXVIII: Materials and Applications, pp. 323–330. International Society for Optics and Photonics, SPIE (2014). https://doi.org/10.1117/12.2039174. https://doi.org/10.1117/12.2039174
98. Pan, Y., Wang, Y., Liu, J., Li, X., Jia, J.: Fast polygon-based method for calculating computer-generated holograms in three-dimensional display. Appl. Opt. **52**(1), A290–A299 (2013). https://doi.org/10.1364/AO.52.00A290. http://ao.osa.org/abstract.cfm?URI=ao-52-1-A290
99. Pan, Y., Wang, Y., Liu, J., Li, X., Jia, J.: Improved full analytical polygon-based method using Fourier analysis of the three-dimensional affine transformation. Appl. Opt. **53**(7), 1354–1362 (2014). https://doi.org/10.1364/AO.53.001354. http://ao.osa.org/abstract.cfm?URI=ao-53-7-1354
100. Patorski, K.: Fraunhofer diffraction patterns of tilted planar objects. Optica Acta **30**, 673–679 (1983)
101. Petz, C., Magnor, M.: Fast hologram synthesis for 3D geometry models using graphics hardware. In: Jeong, T.H., Stevenson, S.H. (eds.) Practical Holography XVII and Holographic Materials IX, vol. 5005, pp. 266–275. International Society for Optics and Photonics, SPIE (2003). https://doi.org/10.1117/12.476879. https://doi.org/10.1117/12.476879
102. Phong, B.T.: Illumination for computer generated pictures. Commun. ACM **18**(6), 311–317 (1975). https://doi.org/10.1145/360825.360839. http://doi.acm.org/10.1145/360825.360839
103. Rabal, H.J., Bolognini, N., Sicre, E.E.: Diffraction by a tilted aperture: coherent and partially coherent cases. Optica Acta **32**, 1309–1311 (1985)
104. Ritter, A., Böttger, J., Deussen, O., König, M., Strothotte, T.: Hardware-based rendering of full-parallax synthetic holograms. Appl. Opt. **38**(8), 1364–1369 (1999). https://doi.org/10.1364/AO.38.001364. http://ao.osa.org/abstract.cfm?URI=ao-38-8-1364
105. Saito, T.: Development of wavefront printer using Denysyuk-type optical system (in Japanese). Master's thesis, Kansai University (2017)
106. Sakamoto, Y., Morishima, M., Usui, A.: Computer-generated holograms on a CD-R disk. In: SPIE Proceedings of the Practical Holography XVIII: Materials and Applications, vol. 5290, pp. 42–49 (2004)
107. Saxby, G., Zacharovas, S.: Practical Holography, 4th ed., Chap. 19. CRC Press (2016)
108. Saxby, G., Zacharovas, S.: Practical Holography, 4th ed., Chap. 7. CRC Press (2016)
109. Schnars, U., Jüptner, W.: Digital Holography. Springer (2005)
110. Schnars, U., Jüptner, W.: Direct recording of holograms by a CCD target and numerical reconstruction. Appl. Opt. **33**(2), 179–181 (1994). https://doi.org/10.1364/AO.33.000179. http://ao.osa.org/abstract.cfm?URI=ao-33-2-179
111. Shen, F., Wang, A.: Fast-fourier-transform based numerical integration method for the Rayleigh-Sommerfeld diffraction formula. Appl. Opt. **45**(6), 1102–1110 (2006). https://doi.org/10.1364/AO.45.001102. http://ao.osa.org/abstract.cfm?URI=ao-45-6-1102
112. Shimobaba, T., Ito, T.: An efficient computational method suitable for hardware of computer-generated hologram with phase computation by addition. Comput. Phys. Commun. **138**(1), 44 – 52 (2001). https://doi.org/10.1016/S0010-4655(01)00189-8. http://www.sciencedirect.com/science/article/pii/S0010465501001898
113. Shimobaba, T., Masuda, N., Ito, T.: Simple and fast calculation algorithm for computer-generated hologram with wavefront recording plane. Opt. Lett. **34**(20), 3133–3135 (2009). https://doi.org/10.1364/OL.34.003133. http://ol.osa.org/abstract.cfm?URI=ol-34-20-3133
114. Shimobaba, T., Shiraki, A., Ichihashi, Y., Masuda, N., Ito, T.: Interactive color electroholography using the FPGA technology and time division switching method. IEICE Electron. Express (ELEX) **5**, 271–277 (2008)
115. St-Hilaire, P., Benton, S.A., Lucente, M.E., Jepsen, M.L., Kollin, J., Yoshikawa, H., Underkoffler, J.S.: Electronic display system for computational holography. In: Benton, S.A. (ed.) Practical Holography IV, vol. 1212, pp. 174–182. International Society for Optics and Photonics, SPIE (1990). https://doi.org/10.1117/12.17980. https://doi.org/10.1117/12.17980

116. St-Hilaire, P., Benton, S.A., Lucente, M.E., Underkoffler, J.S., Yoshikawa, H.: Real-time holographic display: improvements using a multichannel acousto-optic modulator and holographic optical elements. In: Benton, S.A. (ed.) Practical Holography V, vol. 1461, pp. 254–261. International Society for Optics and Photonics, SPIE (1991). https://doi.org/10.1117/12.44734. https://doi.org/10.1117/12.44734
117. Stein, A.D., Wang, Z., Leigh, Jr., J.S.: Computer-generated holograms: a simplified ray-tracing approach. Comput. Phys. **6**, 389–392 (1992)
118. Takaki, Y., Kawai, H., Ohzu, H.: Hybrid holographic microscopy free of conjugate and zero-order images. Appl. Opt. **38**(23), 4990–4996 (1999). https://doi.org/10.1364/AO.38.004990. http://ao.osa.org/abstract.cfm?URI=ao-38-23-4990
119. Takaki, Y., Okada, N.: Hologram generation by horizontal scanning of a high-speed spatial light modulator. Appl. Opt. **48**(17), 3255–3260 (2009). https://doi.org/10.1364/AO.48.003255. http://ao.osa.org/abstract.cfm?URI=ao-48-17-3255
120. Tommasi, T., Bianco, B.: Computer-generated holograms of tilted planes by a spatial frequency approach. J. Opt. Soc. Am. A **10**, 299–305 (1993)
121. Tricoles, G.: Computer generated holograms: an historical review. Appl. Opt. **26**(20), 4351–4360 (1987). https://doi.org/10.1364/AO.26.004351. http://ao.osa.org/abstract.cfm?URI=ao-26-20-4351
122. Tsuchitani, T., Nakahara, S., Tottori, H., Kanazawa, K., Fujita, T.: The formation of large scale reconstructed images through the use of holograms. Opt. Rev. **15**, 44–50 (2008)
123. Tsuchiyama, Y., Matsushima, K.: Full-color large-scaled computer-generated holograms using RGB color filters. Opt. Express **25**(3), 2016–2030 (2017). https://doi.org/10.1364/OE.25.002016. http://www.opticsexpress.org/abstract.cfm?URI=oe-25-3-2016
124. Underkoffler, J.S.: Occlusion processing and smooth surface shading for fully computed synthetic holography. In: Benton, S.A., Trout, T.J. (eds.) Practical Holography XI and Holographic Materials III, vol. 3011, pp. 19–30. International Society for Optics and Photonics, SPIE (1997). https://doi.org/10.1117/12.271359. https://doi.org/10.1117/12.271359
125. Wagner, C., Seebacher, S., Osten, W., Jüptner, W.: Digital recording and numerical reconstruction of lensless fourier holograms in optical metrology. Appl. Opt. **38**(22), 4812–4820 (1999). https://doi.org/10.1364/AO.38.004812. http://ao.osa.org/abstract.cfm?URI=ao-38-22-4812
126. Wakunami, K., Oi, R., Senoh, T., Sasaki, H., Ichihashi, Y., Yamamoto, K.: Wavefront printing technique with overlapping approach toward high definition holographic image reconstruction. In: Javidi, B., Son, J.Y. (eds.) Three-Dimensional Imaging, Visualization, and Display 2016, vol. 9867, pp. 137–144. International Society for Optics and Photonics, SPIE (2016). https://doi.org/10.1117/12.2229517. https://doi.org/10.1117/12.2229517
127. Wakunami, K., Yamaguchi, M.: Calculation for computer generated hologram using ray-sampling plane. Opt. Express **19**(10), 9086–9101 (2011). https://doi.org/10.1364/OE.19.009086. http://www.opticsexpress.org/abstract.cfm?URI=oe-19-10-9086
128. Wakunami, K., Yamashita, H., Yamaguchi, M.: Occlusion culling for computer generated hologram based on ray-wavefront conversion. Opt. Express **21**(19), 21811–21822 (2013). https://doi.org/10.1364/OE.21.021811. http://www.opticsexpress.org/abstract.cfm?URI=oe-21-19-21811
129. Waters, J.P.: Holographic image synthesis utilizing theoretical methods. Appl. Phys. Lett. **9**, 405–407 (1966)
130. Yamaguchi, I., Zhang, T.: Phase-shifting digital holography. Opt. Lett. **22**(16), 1268–1270 (1997). https://doi.org/10.1364/OL.22.001268. http://ol.osa.org/abstract.cfm?URI=ol-22-16-1268
131. Yamaguchi, K., Ichikawa, T., Sakamoto, Y.: Calculation method for computer-generated holograms considering various reflectance distributions based on microfacets with various surface roughnesses. Appl. Opt. **50**(34), H195–H202 (2011). https://doi.org/10.1364/AO.50.00H195. http://ao.osa.org/abstract.cfm?URI=ao-50-34-H195
132. Yamaguchi, M.: Ray-based and wavefront-based holographic displays for high-density light-field reproduction. In: Javidi, B., Son, J.Y. (eds.) Three-Dimensional Imaging, Visualization, and Display 2011, vol. 8043, pp. 40–49. International Society for Optics and Photonics, SPIE (2011). https://doi.org/10.1117/12.884500. https://doi.org/10.1117/12.884500

133. Yamaguchi, M., Hoshino, H., Honda, T., Ohyama, N.: Phase-added stereogram: calculation of hologram using computer graphics technique. In: Benton, S.A. (ed.) Practical Holography VII: Imaging and Materials, vol. 1914, pp. 25–31. International Society for Optics and Photonics, SPIE (1993). https://doi.org/10.1117/12.155027. https://doi.org/10.1117/12.155027
134. Yamaguchi, M., Ohyama, N., Honda, T.: Holographic three-dimensional printer: new method. Appl. Opt. **31**(2), 217–222 (1992). https://doi.org/10.1364/AO.31.000217. http://ao.osa.org/abstract.cfm?URI=ao-31-2-217
135. Yamaguchi, T., Miyamoto, O., Yoshikawa, H.: Volume hologram printer to record the wavefront of three-dimensional objects. Opt. Eng. **51**(7), 1–8 (2012). https://doi.org/10.1117/1.OE.51.7.075802. https://doi.org/10.1117/1.OE.51.7.075802
136. Yamanaka, S., Matsushima, K.: A new high-resolution printer for digitally synthetic holograms (in Japanese). J. Inst. Image Inf. Telev. Eng. **58**, 1665–1668 (2004)
137. Yamashita, H.: Binary coding of high-definition CGHs reconstructed by white-light (in Japanese). Master's thesis, Kansai University (2012)
138. Yamashita, H., Matsushima, K., Nakahara, S.: Image-type high-definition CGHs encoded by optimized error diffusion. In: Bjelkhagen, H.I., Michael Bove, Jr., V. (eds.) Practical Holography XXVI: Materials and Applications, vol. 8281, pp. 226–232. International Society for Optics and Photonics, SPIE (2012). https://doi.org/10.1117/12.908353. https://doi.org/10.1117/12.908353
139. Yang, F., Murakami, Y., Yamaguchi, M.: Digital color management in full-color holographic three-dimensional printer. Appl. Opt. **51**(19), 4343–4352 (2012). https://doi.org/10.1364/AO.51.004343. http://ao.osa.org/abstract.cfm?URI=ao-51-19-4343
140. Yatagai, T.: Stereoscopic approach to 3-d display using computer-generated holograms. Appl. Opt. **15**(11), 2722–2729 (1976). https://doi.org/10.1364/AO.15.002722. http://ao.osa.org/abstract.cfm?URI=ao-15-11-2722
141. Yoshikawa, H., Iwase, S., Oneda, T.: Fast computation of Fresnel holograms employing difference. In: Stevenson, S.H., Trout, T.J., Benton, S.A., Stevenson, S.H., Trout, T.J. (eds.) Practical Holography XIV and Holographic Materials VI, vol. 3956, pp. 48–55. International Society for Optics and Photonics, SPIE (2000). https://doi.org/10.1117/12.380022. https://doi.org/10.1117/12.380022
142. Yoshikawa, H., Tachinami, M.: Development of direct fringe printer for computer-generated holograms. In: Jeong, T.H., Bjelkhagen, H.I. (eds.) Practical Holography XIX: Materials and Applications, vol. 5742, pp. 259–266. International Society for Optics and Photonics, SPIE (2005). https://doi.org/10.1117/12.589550. https://doi.org/10.1117/12.589550
143. Yoshikawa, H., Takei, K.: Development of a compact direct fringe printer for computer-generated holograms. In: Jeong, T.H., Bjelkhagen, H.I. (eds.) Practical Holography XVIII: Materials and Applications, vol. 5290, pp. 114–121. International Society for Optics and Photonics, SPIE (2004). https://doi.org/10.1117/12.526516. https://doi.org/10.1117/12.526516
144. Yoshikawa, H., Yamaguchi, T.: Computer-generated holograms for 3D display. Chin. Opt. Lett. **7**(12), 1079–1082 (2009). http://col.osa.org/abstract.cfm?URI=col-7-12-1079
145. Yoshikawa, H., Yamaguchi, T.: Review of holographic printers for computer-generated holograms. IEEE Trans. Ind. Inform. **12**(4), 1584–1589 (2016). https://doi.org/10.1109/TII.2015.2475722
146. Yoshikawa, H., Yamaguchi, T., Kajiro, S.: Direct fringe printer for computer-generated holograms: improvement of printing speed. In: Bjelkhagen, H.I., Michael Bove, Jr., V (eds.) Practical Holography XXVII: Materials and Applications, vol. 8644, pp. 208–213. International Society for Optics and Photonics, SPIE (2013). https://doi.org/10.1117/12.2002023. https://doi.org/10.1117/12.2002023
147. Yoshikawa, N., Itoh, M., Yatagai, T.: Binary computer-generated holograms for security applications from a synthetic double-exposure method by electron-beam lithography. Opt. Lett. **23**, 1483–1485 (1998)
148. Zhang, H., Cao, L., Jin, G.: Computer-generated hologram with occlusion effect using layer-based processing. Appl. Opt. **56**(13), F138–F143 (2017). https://doi.org/10.1364/AO.56.00F138. http://ao.osa.org/abstract.cfm?URI=ao-56-13-F138

149. Zhang, H., Collings, N., Chen, J., Crossland, B.A., Chu, D., Xie, J.: Full parallax three-dimensional display with occlusion effect using computer generated hologram. Opt. Eng. **50**(7), 1–6 (2011). https://doi.org/10.1117/1.3599871. https://doi.org/10.1117/1.3599871
150. Zhang, H., Zhao, Y., Cao, L., Jin, G.: Layered holographic stereogram based on inverse Fresnel diffraction. Appl. Opt. **55**(3), A154–A159 (2016). https://doi.org/10.1364/AO.55.00A154. http://ao.osa.org/abstract.cfm?URI=ao-55-3-A154
151. Zhang, Y.P., Wang, F., Poon, T.C., Fan, S., Xu, W.: Fast generation of full analytical polygon-based computer-generated holograms. Opt. Express **26**(15), 19206–19224 (2018). https://doi.org/10.1364/OE.26.019206. http://www.opticsexpress.org/abstract.cfm?URI=oe-26-15-19206

Index

K. Matsushima, *Introduction to Computer Holography*,
Series in Display Science and Technology,
https://doi.org/10.1007/978-3-030-38435-7

The manufacturer's authorised representative in the EU is Springer Nature Customer Service Centre GmbH, Europaplatz 3, 69115 Heidelberg, Germany. If you have any concerns regarding our products, please contact ProductSafety@springernature.com

Printed and bound by CPI Group (UK) Ltd, Croydon, CR0 4YY
05/10/2026
02228120-0015